Studies in Systems, Decision and Control

Volume 263

Series Editor

Janusz Kacprzyk, Systems Research Institute, Polish Academy of Sciences, Warsaw, Poland

The series "Studies in Systems, Decision and Control" (SSDC) covers both new developments and advances, as well as the state of the art, in the various areas of broadly perceived systems, decision making and control–quickly, up to date and with a high quality. The intent is to cover the theory, applications, and perspectives on the state of the art and future developments relevant to systems, decision making, control, complex processes and related areas, as embedded in the fields of engineering, computer science, physics, economics, social and life sciences, as well as the paradigms and methodologies behind them. The series contains monographs, textbooks, lecture notes and edited volumes in systems, decision making and control spanning the areas of Cyber-Physical Systems, Autonomous Systems, Sensor Networks, Control Systems, Energy Systems, Automotive Systems, Biological Systems, Vehicular Networking and Connected Vehicles, Aerospace Systems, Automation, Manufacturing, Smart Grids, Nonlinear Systems, Power Systems, Robotics, Social Systems, Economic Systems and other. Of particular value to both the contributors and the readership are the short publication timeframe and the world-wide distribution and exposure which enable both a wide and rapid dissemination of research output.

** Indexing: The books of this series are submitted to ISI, SCOPUS, DBLP, Ulrichs, MathSciNet, Current Mathematical Publications, Mathematical Reviews, Zentralblatt Math: MetaPress and Springerlink.

More information about this series at http://www.springer.com/series/13304

Gianni Bosi · María J. Campión ·
Juan C. Candeal · Esteban Indurain
Editors

Mathematical Topics on Representations of Ordered Structures and Utility Theory

Essays in Honor of Professor Ghanshyam B. Mehta

Editors
Gianni Bosi
Dipartimento di Scienze Economiche
Aziendali, Matematiche e Statistiche
Università degli Studi di Trieste
Trieste, Italy

María J. Campión
Departamento de Estadística
Informática y Matemáticas
Universidad Pública de Navarra
Pamplona, Navarra, Spain

Juan C. Candeal
Facultad de Economía y Empresa
Departamento de Análisis Económico
Universidad de Zaragoza
Zaragoza, Spain

Esteban Indurain
Departamento de Estadística
Informática y Matemáticas
Universidad Pública de Navarra
Pamplona, Spain

ISSN 2198-4182   ISSN 2198-4190   (electronic)
Studies in Systems, Decision and Control
ISBN 978-3-030-34228-9   ISBN 978-3-030-34226-5   (eBook)
https://doi.org/10.1007/978-3-030-34226-5

This Springer imprint is published by the registered company Springer Nature Switzerland AG
The registered company address is: Gewerbestrasse 11, 6330 Cham, Switzerland

Preface

We met Prof. Dr. G. B. Mehta for the first time in 1992, in a congress held in Paris, France. Since then we have fruitfully and continuously contributed with him in the publication of several papers on Utility Theory, as well as in the preparation of congresses, workshops, and seminars. He has always encouraged us to go ahead in our research on Utility Theory and, in particular, on the mathematical theory of the numerical representability of ordered structures. Before his retirement, we met him for the last time in Madrid, Spain in 2006 (August), where the International Congress of Mathematicians ICM 2006 was held.

The idea of preparing a book in his honor is old, so-to-say. However, for different reasons, we were obliged to delay it until now. Professor Mehta reached the age of 75 y.o. on July 8, 2018, and just a few months before, we started to convert the idea of editing a festschrift in something real, having also in mind the possibility of this new book to be also a tribute and recognition to a whole lifetime dedicated to research, on the occasion of his 75th anniversary.

Professor Ghanshyam B. Mehta (Bombay, now Mumbai, India 1943) belongs to a family with plenty of people devoted to Mathematics. His wife Meena, brother Vikram (1946–2014), and daughter Maithili (to cite only those of whom we know personally, probably there are even more) are mathematicians, too. He got the titles of Bachelor in Economics in Bombay University in 1963, Master of Arts in Economics in the University of California in Berkeley, USA. in 1965, and Doctor of Philosophy in Economics also in the University of California in Berkeley, USA. in 1971, with a thesis about the structure of the Keynesian Revolution. Then he worked as assistant professor at the University of New Brunswick, Canada in the period 1971–1972. Perhaps fascinated by billabongs, kangaroos, and coolibah trees, he arrived in Australia in 1974 where he settled, founded a home, and currently lives there. He got a Master of Science in Mathematics at the University of Queensland in 1978, and finally the title of Doctor in Philosophy in Mathematics, again in the University of Queensland, in 1987, with a thesis on Topological Methods in Equilibrium Analysis. He has been the advisor of several Ph.D. theses, both in Economics and Mathematics. Some of them are clearly devoted to Utility Theory (e.g., to put just one example, "Preference and utility in economic theory and the history of economic thought", by K. L. Mitchener, School of Economics, University of Queensland, 2007).

Thus, Prof. Mehta features the merit of having a wide formation and knowledge in Mathematics as well as in Economics, with doctorates in both disciplines. Obviously, this fact has allowed him to be a leader in those areas where Mathematics and Economics meet, and in particular in all that has to do with Utility Theory, understood as the mathematical problem of converting qualitative scales in which agents declare their preferences in order to compare things (usually, elements of a set of alternatives), into numerical or quantitative ones. That is, instead of just comparing things one another, the agents may alternatively and equivalently (when this is possible) compare numbers, in the sense of "the bigger the better". Mathematically, this corresponds to the theory of numerical representability of ordered structures in the Euclidean real line endowed with its usual linear order $\leq$.

Most of the classical problems arising in Utility Theory have been analyzed by Prof. G. B. Mehta in some of his valuable papers and contributions:

(i)   Numerical representations of different kinds of orderings.
(ii)   Existence and nonexistence of utility representations.
(iii)   Existence of maximal elements for suitable orderings.
(iv)   Construction of utility functions by means of scales and separable systems.
(v)   Particular constructions of utility functions on classical spaces arising in Economics.
(vi)   Infinite-dimensional utility theory.
(vii)   Mathematical Intuitionism and Constructive Utility Theory.
(viii)   Continuous representability properties of topological spaces.
(ix)   Numerical representations of orderings in codomains different from the real line.
(x)   Utility in Economics vs. Entropy in Physics: the common mathematical background.

His book (jointly written with D. S. Bridges) entitled *Representations of Preference Orderings* (Springer. Berlin. 1995) is undoubtedly a classical one in the theory of numerical representations of ordered structures, header book and the starting point of many past and ongoing researches in this framework.

Apart from this, Prof. G. B. Mehta has also paid a relevant attention to the analysis of classical works and authors in Utility Theory, showing the necessity of understanding well, and periodically reviewing, the contributions made by the pioneers (Arrow, Birkhoff, Cooper, Debreu, Fleischer, Hahn, Keynes, Milgram, Newman, Peleg, Rader, Read, Wold, etc.). He has been a specialist in showing how many modern ideas recently issued in key papers in Utility Theory sometimes already had an implicit intuition that could be noticed in some of the papers of the most classical authors. Thus, it is crucial to (re)-read the classical papers in order to get new insights and inspiration in forthcoming research on these topics related to Utility Theory.

This festschrift has been distributed in four parts. However, the distribution could perhaps be considered a bit whimsical: the reason is that they are not closed or watertight compartments, but instead, they have many connections and interactions sharing basic ideas. Anyways, we have tried to distribute the material in parts where the corresponding papers could have some key features in common. Thus, the final distribution goes as follows:

(i)   General Utility Theory.
(ii) Particular Kinds of Preferences and Representations.
(iii) Extensions of Utility.
(iv) Applications into Economics.

The first part, on General Utility Theory, consists of three contributions that can be considered as surveys on this framework. The first one, by Alan F. Beardon, is a really nice study on the mathematical background that supports the numerical representability of orderings, mainly leaning on General Topology. The second one, by Juan C. Candeal, analyzes the mathematical reasons that provoke the existence or the nonexistence of utility functions in several kinds of environments, as the algebraic one (looking for utility functions that, in addition, are also homomorphisms on the real line endowed with the corresponding algebraic structure), and the topological one as well (looking mainly for continuous utility functions defined on some ordered topological space). Finally, the third and last contribution in this part, by María J. Campión and Esteban Induráin, is unusual so-to-say, but, in our opinion, quite necessary for any potential researcher in the theory of the numerical representation of ordered structures. It is a survey in which the most classical theorems in Utility Theory are stated with the corresponding references, and then it is said that "beyond this point, everything is an open question." That is, the contribution furnishes a wide list of open problems in Utility Theory, explaining in each case what is known, and what is still unknown, and paying a special attention to the "boundary of knowledge" here, that is, to the points from which the terrain is virgin, still unknown or unexplored. This will invite and encourage the potential researcher in this theory to develop some of her/his further research in the next future on these topics.

The second part deals with particular kinds of preferences and representations. Basically, the most typical orderings that frame preferences are the total preorders (i.e., transitive and complete binary relations), the interval orders, and the semiorders. Therefore, in this part we have included those contributions that are directly related to the numerical representability of either a total preorder (through a suitable real-valued utility function), an interval order (through two intertwined utility functions), or a semiorder (by means of a real-valued utility function and a strictly positive real constant called threshold of discrimination). This part consists of seven contributions. The first one, by Y. Rebillé, as well as the second one, by A. Estevan, deal with particular features of the numerical representability of interval orders defined on topological spaces, paying particular attention to continuity. The third one, by D. Bouyssou and M. Pirlot deals with semiorders, analyzing in-depth a recently discovered condition that is necessary for the existence of a Scott-Suppes numerical representation. The fourth one, by D. Bouyssou and J.-P. Doignon, deals with biorders: this structure relates two given sets, generalizing the concept of an interval order (that appears when those two given sets coincide). This contribution studies different aspects related to numerical representations of nested families of biorders. The fifth one, by C. Chis, H.-P. A. Künzi and M. Sanchis, deals with the existence of utility representations on ordered topological groups, so relating order, algebra, and topology.The sixth contribution in this part, due to M. Cardin, studies preferences defined on convex structures, discussing in-depth the role played by convexity in different contexts of Utility Theory, so that is those settings, the economic agents exhibit an inclination for diversification and so they prefer a more balanced bundle to bundles with a more extreme composition. Finally, the seventh contribution, by C. Hervés and P. K. Monteiro, analyzes monotonicity properties of preferences, and how in remarkable situations of infinite dimensionality, so that the commodity space is rich enough, strictly monotonic preferences are neither representable by utility functions nor continuous in any linear topology.

The third part has been named "Extensions of Utility". One of these extensions is multi-utility. Roughly speaking, in order to represent a certain binary relation $\mathcal{R}$ defined on a nonempty set X we consider a family $\{f_i : i \in I\}$ of real-valued functions $f_i : X \rightarrow \mathbb{R}$ such that, given $x, y \in X$, it holds true that $x\mathcal{R}y \Leftrightarrow f_i(x) \leq f_i(y)$, for every $i \in I$. (Here I is a suitable set of indices). Depending on different features of the set of indices I (e.g., it is finite, or countable, etc.) as well as of the functions f_i involved (e.g.: if all of them are continuous as regards some given topology on X and the usual one on the real line), a new appealing and interesting panorama to represents different kinds of binary relations appears now. For instance, multi-utility representations can be used to deal with non-total (incomplete) binary relations, as, e.g., partial orders. The first four contributions in this part deal with some special kind of multi-utility: The first one, by Gianni Bosi and M. Zuanon, studies the continuous representability of non-total preorders and, in particular, shows that continuous multi-utility representations are actually very restrictive. The second one, by A. Caterino, R. Ceppittelli and L. Holá, also deals with non-total preorders, studying topological aspects related to the joint continuity of multi-utility numerical representations. The third one, by A. Kochov, also deals with incomplete preferences, extending a model of uncertainty due to Bewley (1986) on Knightian decision theory to a framework without an exogenous state space. The fourth one, due to O. Evren, analyzes the notion of preference for flexibility and extends to the continuous case an ordinal representation due to Kreps (1979), also dealing with incomplete preferences. The last two contributions in this part are of a very different nature: The fifth one, due to D. S. Bridges, uses constructive analysis. It is an unusual paper where items coming from Utility Theory are studied in the intuitionistic setting of constructive mathematics. It is a really worthwhile contribution to compare the classical Utility Theory to that really a different framework, which requires the inclusion of a variety of computationally empowered assumptions. Finally, the sixth and last one in this part, due to V. Knoblauch, extends the concept of a preference to the fuzzy approach, looking for continuous numerical representations of fuzzy preference relations defined on a universe.

The fourth and last part in this festschrift deals with applications into Economics. In fact, some of the contributions that appear here could also be viewed as extensions of utility or particular kinds of representations, so that they could also have been included in the second or third part. Nevertheless, we have created this new part because the style and context of this contributions uses a language, notation, etc., that is more typical of books and papers devoted, say, to Microeconomics or more generally to Economic Theory, rather than those that deal with Mathematics (even if it is Mathematical Economics). The first contribution here, due to C. Vázquez, studies the concept of money-metric functions as a cardinal utility to define a similarity binary relation between alternatives. That relation is taken into account for ranking opportunity sets. The second one, due to J. C. R. Alcantud and A. Giarlotta studies, the interplay between intergenerational justice and mathematical utility theory, drawing a bridge between both theories and, in particular, showing that they deal with the existence,or nonexistence of preferences, and social welfare functions, on a Cartesian product of certain subsets of the real line (whose members are usually known as utility streams). Finally, the third and last one, due to J. Quiggin and R. G. Chambers, deals with the notion of risk-aversion for state-dependent preferences that agents define over uncertain outcomes. This is, perhaps, the contribution to this festschrift that fits better in the section of Economics rather than Mathematics. Its first co-author (Quiggin) did his Ph.D. doctorate in Economics under the advisement of Prof. Mehta. As editors, we wanted to include here at least one contribution made by persons that had been direct disciples of Prof. Mehta in their formation as economists or mathematicians.

Trieste, Italy
Pamplona, Spain
Zaragoza, Spain
Pamplona, Spain
June 2019

Gianni Bosi
María J. Campión
Juan C. Candeal
Esteban Indurain

Acknowledgements

Apart from the authors of the contributions included in this festschrift, many other persons have given us the courage to go ahead with our idea of editing this festschrift. Not only we contacted people that have been co-authors of some paper with Prof. Mehta, but also colleagues of him at Queensland University in Brisbane (Australia), some of his doctorate students, and of course part of his direct family, with a special mention to his daughter Maithili and wife Meena who have kept the secret till now, so that Ghanshyam is, as far as we know, unaware of our initiative, that as far as possible, is intended to be a pleasant and affectionate surprise for him.

Other persons could certainly have contributed to this festschrift but for different reasons, mainly health, declined the possibility of sending a paper. Instead, they helped us to contact other possible contributors. Thus, we had formerly contacted with other persons that are well known, and indeed classical, in the theory of mathematical representations of ordered structures, namely F. Aleskerov, H. Bustince, A. Chateauneuf, M. Estévez, J. Gans, J. L. García-Lapresta, J. M. Gutiérrez, J. Gutiérrez García, G. Herden, B. Littleboy, F. Maccheroni, T. Marchant, A. Marley, A. McLennan, E. Minguzzi, E. Ok, S. Ovchinnikov, C. Rodríguez-Palmero, A. Tangyan, U. Schmidt, B. Subiza, K. K. Tang, A. Verdejo, P. Wakker, G. Yuan, etc. We want to express our gratitude to all of them. Incidentally, when this process had already started, we received with great sorrow the sad news of the death of Prof. G. Herden, a usual co-author of Prof. Mehta (and us, the editors) in the recent past years. Rest in peace.

The edition of this festschrift has been partially supported by the research project ECO2015-65031-R, MTM2015-63608-P, TIN2016-77356-P and TIN2017-87600-P (MINECO/ AEI-FEDER, UE)

Trieste, Italy   Gianni Bosi
Pamplona, Spain   María J. Campión
Zaragoza, Spain   Juan C. Candeal
Pamplona, Spain   Esteban Indurain
June 2019

GBM, My Father

Ghanshyam Bhagvandas Mehta was born in 1943, to Nandini and Bhagvandas Chunilal Mehta. He was the middle of three children, with an older sister, Devyani and a younger brother Vikram. Ghanshyam grew up in Mumbai, in a palatial home. He was educated in the best English-medium schools of their day. Ghanshyam and his siblings were brought up to deeply respect all academic learning. Ghanshyam and Vikram, in particular, had a deep interest and love for mathematics. Given their passion, ironically both boys performed quite poorly in school examinations!

When it came time to go to university, the brothers both decided to study in the US and made their way to Berkeley University in California. Ghanshyam chose to pursue economics, with a mathematical focus, while Vikram studied mathematics. Ghanshyam's Ph.D. supervisor was Gerard Debreu; an eminent economist who went on to win the 1983 Nobel Memorial Prize in Economic Sciences. Ghanshyam's Ph.D. was about the structure of the Keynesian Revolution.

Ghanshyam married Meena Munim in 1972. Meena also had grown up with a keen interest in mathematics and science, but as a young woman growing up in India, she had little opportunity to pursue further studies. Ghanshyam always talked about and enthused about his academic interests and soon infected Meena with a thirst for mathematical knowledge. In 1974, Ghanshyam and Meena welcomed a new addition to their little family—me!

As I grew up, my father and my uncle taught me how to read by getting me to read the titles of their maths textbooks. Where most little kids were learning to read the words "the cat sat on the mat" I was learning to say the words "calculus", "geometry", and "algebraic topology". Most little girls of that era grew up wanting to be a ballet dancer, I proclaimed, loud and proud that I wanted to be a "poffessoor of albega and topology" (it's very hard to say "professor of algebra and topology" when you're only 3 years old). My father taught me elementary probability when I was 6, by the time I was 10 I could recite several winners of the Field's medal. I grew up steeped in maths—it was everywhere, my father talked about it and taught mathematical economics, my mother had gone back to study university maths courses and my uncle was a maths professor in Mumbai. To top it all off, my father decided that one Ph.D. wasn't quite enough for him, and enrolled in a mathematics Ph.D. at the University of Queensland with a focus on Algebraic topology.

I guess I considered other career options as I grew up, but I certainly never took any of them very seriously. Thanks to my father, maths was part of the fabric of my being. Inspired by his staggering intellect, I continued my maths studies and eventually completed my own Ph.D. in mathematics in 2003.

In addition to his mathematical acumen, he has also been a great lover of the English language—he owned the complete Oxford English dictionary, a tome so hefty and detailed that it came with its very own magnifying glass—and he often delighted in using words that others had never heard of, let alone understood.

When I was awarded my degree, my father sent me a most amusing e-mail with the following line in it: "you are not a wastrel of a glorious patrimony," and indeed I wasn't! I have continued my love for mathematics, and have pursued a career in software engineering with a specialization in mathematical optimization. I now have 2 kids of my own and they're both keen little mathematicians themselves. And so a great family tradition continues on…

Brisbane, Australia   Maithili Mehta
August 2019

GBM, My Advisor

It took me 9 years to complete my Ph.D.

That seems an odd place to start an essay on my years studying with Dr. Ghanshyam Mehta at the University of Queensland, but it perfectly captures my relationship with my endlessly patient doctoral supervisor.

Despite my many years of study, I did the best academic work of my career during my Honours year at UQ, albeit almost exclusively during the final month before the dreaded thesis was due. Mathematics had been my strongest subject throughout my undergraduate economics degree, taking several subjects taught by GBM, and so I tentatively approached him to supervise my thesis. It is likely that he didn't actually recognize me as a student—I had missed more lectures than I attended—but GBM was certainly too polite to show it.

In our preliminary discussion, I told him I had no thoughts for a topic for my thesis, no idea I wanted to explore, no burning question that I sought to answer. An inauspicious start. Unconcerned, GBM agreed to a weekly meeting, and suggested a small selection of books and articles to help me explore my interests and find my "question". Economics, philosophy, mathematics, history, sociology, politics; Ghanshyam exhibited a vast array of interests and a prodigious thirst for knowledge, eager to exchange opinions, and always keen to debate dissenting ideas.

I read everything he suggested, and anything related that I could find, and was eager to demonstrate my enthusiasm when I returned for our first official meeting. GBM was taken aback. I had already read them all? Clearly, I had not given the texts their due consideration. Had I recognized the inconsistencies in this paper? Had I distinguished the gray arguments from the black-and-white in that article? I quickly learned that GBM was not impressed by students turning in papers citing hundreds of references in the bibliography. How could they possibly have understood what the writers were trying to say, simply by skim-reading the abstract, and quoting a few randomly chosen lines?

Through weeks of careful reading and thoughtful discussions with GBM, my interest in microeconomics was piqued. I was keen to maintain my mathematical focus, and so from the literature we chose an optimization problem for me to work on. I went away, uncertain how to proceed, and too proud to ask for guidance.

So began 6 months of a sort of one-sided game of hide-and-seek, in which GBM was a baffled and unwilling participant. I studiously avoided GBM in the department. I had begun tutoring undergraduate microeconomics and econometrics, and was fulfilling my New Year's resolution to attend all lectures and tutorials (discovering with delight that it was infinitely easier to do the assignments when I had actually been to class). So avoiding GBM was quite the task.

But I was up to the challenge. I sat in the back of his lectures, avoiding eye contact (with tiny class sizes, it was hardly a covert operation), and dashed from the room a few minutes before the end. I spent as little time as possible in my office in the economics department, taking my Available for Students hours in one of the tutorial rooms, and having my office-mates cover for me whenever GBM stopped by. He must have thought I was inexcusably rude, quite mad, or both. But the truth was, I was embarrassed. Since primary school, mathematics has been my strength, and yet I stared for weeks on end at the optimization problem, with no idea where to begin. I plowed through my coursework, achieving better grades than in undergrad, at last beginning to find my feet. But the optimization problem eluded me.

And then, sometime around the beginning of October, the maths started to make sense. Over the course of a weekend, I pieced together my proof, keeping track of my work by color-coding the equations with bright marking pens on large sheets of butchers' paper, the work pouring out of me now. On Monday morning, I was waiting outside GBM's office when he arrived in the department. By this point, it is likely that he thought I had dropped out of university. Or quite possibly was dead. But he received the rolled-up sausage of colorful sheets with good grace, and promised to unravel them by the end of the week. Around lunchtime that same day, he came to my office–I was no longer afraid to be there—to tell me I had produced an elegant solution to the problem. I was giddy with disbelief.

The final month was a blur of studying for finals and writing up the complete thesis, finding it miraculous that the work I had done over the course of the year was now fitting together in a logical scheme in my head and on the page. The literature review magically formed out of the reading I had done in those initial weeks. GBM was supportive throughout. There's plenty of time, don't panic. Drafts were refined, with GBM gently providing thoughtful guidance. The thunderstorm and resulting power cut on the night before the thesis was due is still the stuff of legend, with increasingly perilous stories told at student reunions of frantic trips to find an open printshop at midnight, and handwritten bibliographies.

I found the progression to Ph.D. a less daunting matter, in all probability because I completely underestimated the task at hand. I was delighted at the prospect of continuing my work with GBM, and again we began with long discussions on a variety of topics, and I took home a (short) reading list every night.

I had worked two jobs during my Honours year, alongside tutoring in the economics department. At my retail job, I was assigned shifts that worked around my tutorials, and my colleagues made lasagne for me during swotvac to ensure I didn't starve. For a period, five of us shared a two-bed flat near the university, a recipe neither for stellar academic achievement nor a happy landlord. This was clearly an unsustainable model for my Ph.D., and so when I was offered a full-time job in banking toward the end of my first year (courtesy of a recommendation by another supportive UQ economics professor), I jumped at the chance to earn a regular wage.

So began nearly a decade of 70-hour working weeks, business travel, one wedding, one interstate move, and two international moves. My books and papers traveled with us, packed neatly into removalists' boxes, sometimes never being opened for months on end.

As a remote student, I needed to find a local university that would give me a "home" and, importantly, research rights in their library. In Sydney, it was UNSW, in the Netherlands, it was University van Amsterdam, and in the UK it was Oxford. The choice of Oxford over Cambridge vexed GBM somewhat; having been visiting professor at Cambridge several times over the years, he held an instinctively dim view of "the other place". However, Oxford was an easier commute on my very few days off, and I persisted with my choice.

Even now as I write this on an unusually hot summer day in London, I recall with a something not even remotely akin to nostalgia the need to wear gloves and heavy sweaters in the library at Oxford. On applying for my Bodleian Reader's Card I had to promise, among other things, "not to bring into the Library or kindle therein any fire or flame." It had seemed an odd undertaking, an easy promise to make, until the first week in Autumn when I seriously considered setting fire to my own clothes in the Bod. It was only September, and yet I could see my own breath. The Rad Cam was no better. GBM chuckled with faux sympathy, noting the apparently cozy environments of the Marshall Library at Cambridge.

The life of a remote Ph.D. student is a disjointed one. It takes extra work and dedication on everybody's part, but in particular on the supervisor. GBM was supportive of my banking career from the beginning, eager that I should be putting my skills to work, and making an impact. He was one of my first calls when I scored that first "real" job in Sydney, and each time I was promoted in London. He was always delighted.

GBM loved language, and enjoyed conversing with Australian students in particular as he enjoyed our accents. Over the years he teased me, with a tinge of disappointment, as my Australian accent morphed into something much more British.

He was keen to discuss my work, the markets, the political situation in whichever country I was living at the time, how it all tied together and what it all could mean. Timezones added to the challenge over the years, but the telephone and email kept us in regular contact. I traveled to Brisbane as frequently as my budget allowed, and we would ensure we met up when GBM traveled to the UK; we met in Cambridge, not Oxford.

I was working on existence proofs of utility functions, grinding away slowly, but sporadic bursts of work interspersed with weeks or months without opening a book is not a recommended formula for academic success. I had been working for years, but had nothing to show for it. I was in despair. I told GBM I thought it was time I accepted defeat. He sent me back. Pull out the papers I had sourced in my literature review, find the common themes, think about the origins of the concepts, there must be an untold story there. Slowly, a history of the development of the joint concepts of preference and utility emerged. Those early years of reading across a broad range of topics revealed the usages of preference and utility across other disciplines, adding depth to my research. I enjoyed our long discussions as GBM helped me rationalize competing arguments and unearth linkages between the concepts. Again, we spent countless hours discussing language. GBM taught me to write simply. I learned that to be able to take complex concepts and arguments, and express them in simple language, is a discipline to be valued.

Thanks to GBM's endless patience and encouragement, I earned my Ph.D., 13 years after I first sat in his lecture theatre.

On the occasion of his 75th birthday, I would like to thank GBM for his contributions to economics, for his enduring friendship, and for never giving up on an oft-despairing student.

London, UK   Kerrie Mitchener-Nissen, Ph.D.
July 2019   Executive Director at J. P. Morgan

Contents

Contributors

José Carlos R. Alcantud BORDA Research Unit and Multidisciplinary Institute of Enterprise (IME), University of Salamanca, Salamanca, Spain

A. F. Beardon Centre for Mathematical Sciences, University of Cambridge, Cambridge, England

Gianni Bosi DEAMS, University of Trieste, Trieste, Italy

Denis Bouyssou LAMSADE, UMR 7243, CNRS, Université Paris-Dauphine, PSL Research University, Paris, France

Douglas S. Bridges School of Mathematics & Statistics, University of Canterbury, Christchurch, New Zealand

María J. Campión Departamento de Estadística, Informática y Matemáticas, Inarbe (Institute for Advanced Research in Business and Economics), Universidad Pública de Navarra, Pamplona, Spain

Juan C. Candeal Departamento de Análisis Económico, Facultad de Economía y Empresa, Universidad de Zaragoza, Zaragoza, Spain

Marta Cardin Department of Economics, Ca' Foscari University of Venice, Venezia, VE, Italy

Alessandro Caterino Perugia, Italy

Rita Ceppitelli Department of Mathematics and Computer Sciences, University of Perugia, Perugia, Italy

Robert G. Chambers University of Maryland, College Park, USA

C. Chis Departament de Matemàtiques, Institut Universitari de Matemàtiques i Aplicacions de Castelló (IMAC), Universitat Jaume I, Castellón de la Plana, Spain

Jean-Paul Doignon Département de Mathématique, Université Libre de Bruxelles, Bruxelles, Belgique

Asier Estevan Dpto. Estadística, Informática y Matemáticas, Instituto INAMAT, Universidad Pública de Navarra, Iruña-Pamplona, Navarra, Spain

Özgür Evren New Economic School, Moscow, Russia

Alfio Giarlotta Department of Economics and Business, University of Catania, Catania, Italy

Carlos Hervés-Beloso RGEAF-ECOBAS, Universidad de Vigo, Vigo, Spain

Lubica Holá Institute of Mathematics, Academy of Science, Bratislava, Slovakia

Esteban Indurain Departamento de Estadística, Informática y Matemáticas, InaMat (Institute for Advanced Materials), Universidad Pública de Navarra, Pamplona, Spain

Vicki Knoblauch Department of Economics, University of Connecticut, Storrs, CT, USA

Asen Kochov Department of Economics, University of Rochester, Rochester, NY, USA

H.-P. A. Künzi Department of Mathematics and Applied Mathematics, University of Cape Town, Rondebosch, Cape Town, South Africa

Paulo K. Monteiro FGV EPGE, Rio de Janeiro, Brazil

Marc Pirlot Université de Mons, rue de Houdain 9, Mons, Belgium

John Quiggin University of Queensland, Brisbane, Australia

Yann Rébillé LEMNA, IAE of Nantes, University of Nantes, Nantes Cedex 3, France

M. Sanchis Departament de Matemàtiques, Universitat Jaume I, Castellón de la Plana, Spain

Carmen Vázquez Facultade de Ciencias Económicas e Empresariais, Departamento de Matemáticas, Universidade de Vigo, Vigo, Spain

Magalì Zuanon DEM, University of Brescia, Brescia, Italy

General Utility Theory

Utility theory; Qualitative and quantitative scales; Preferences; Numerical representations; Existence of utility functions; Special kinds of utility

Topology and Preference Relations

A. F. Beardon

Abstract In this essay we assume familiarity with the basic ideas in mathematical economics, and we discuss the influence of topological ideas on the theory of preference relations and utility functions.

1 Introduction

To those, like the author, who are not an expert in the theory of utility functions, there seem to be a bewildering collection of different results available on the representation of a preference relation by a continuous utility function. In this expository essay we consider how much of the theory can be developed from a purely topological perspective. We focus on those ideas which provide a link between utility theory and topology, and we leave the economic interpretations to others. Briefly, we give priority to results that seem to be topologically important, so we pay more attention to the quotient space of indifference classes than is usual, and more attention to the order topology than other topologies. Although much of this material is already known, we hope that an exposition from a different perspective than usual may be of interest. In any case, the author wishes to thank Ghanshyam Mehta for introducing him to the subject, and for the many memorable hours spent discussing these issues together.

A. F. Beardon (✉)
Centre for Mathematical Sciences, University of Cambridge, Cambridge CB3 0WB, England
e-mail: afb@dpmms.cam.ac.uk

© Springer Nature Switzerland AG 2020
G. Bosi et al. (eds.), *Mathematical Topics on Representations of Ordered Structures and Utility Theory*, Studies in Systems, Decision and Control 263,
https://doi.org/10.1007/978-3-030-34226-5_1

2 Binary Relations

There are many accounts of relations in the literature but unfortunately, and especially in the older literature, the terminology is not used consistently. We note that Halmos writes that "the only thing to remember is that the primary motivation comes from the familar properties of *less than or equal to*, and not *less than*" [10, p. 54], but this recommendation is certainly not observed in some of the older literature. For this reason we define the terms that we shall use in this article, and these are taken from [7, 10]. A *relation R* on a non-empty set Ω is a non-empty subset of the Cartesian product space $\Omega \times \Omega$, and we prefer to write xRy and $x\not\!R\,y$ instead of $(x, y) \in R$ and $(x, y) \notin R$, respectively. The reader will know what it means for a relation R to be *reflexive*, *symmetric*, and *transitive*. In addition to these, we say that R is *anti-symmetric* if xRy and yRx implies $x = y$, and that R is *total* if, for all x and y in Ω (including $x = y$), either xRy or yRx. Then we say that a relation is

- a *preorder* if it is reflexive and transitive;
- a *total preorder* if it is reflexive, transitive, and total;
- a *partial order* if it is reflexive, transitive, and anti-symmetric;
- a *linear order* if it is reflexive, transitive, total, and anti-symmetric;
- an *equivalence relation* if it is reflexive, transitive, and symmetric.

It may be helpful to keep the following scheme (which we leave the reader to interpret) in mind (Fig. 1).

In economics, a total preorder is known as a *preference relation*, and we shall adopt the common use of the symbol $\preccurlyeq$ for this. Following the words of Halmos given above, we regard this as the primary relation in the theory, although in this article we shall only consider *total preorders* (so it will always be possible to compare any two points in Ω). Let $\preccurlyeq$ be a total preorder on Ω; then $x \not\preccurlyeq y$ means that $x \preccurlyeq y$ is false. It is well known that $\preccurlyeq$ induces two more relations on Ω, namely an equivalence relation $\sim$, called *indifference*, and a relation $\prec$, called a *strict preference*. Explicitly, $x \sim y$ if and only if $x \preccurlyeq y$ and $y \preccurlyeq x$, while $x \prec y$ if and only if $x \preccurlyeq y$ and $y \not\preccurlyeq x$ (or, equivalently, $x \preccurlyeq y$ and $y \not\succ x$). We will use elementary properties of $\preccurlyeq$, $\prec$ and $\sim$ as and when we need them (usually without proof); in particular, the strict preference relation $\prec$ satisfies the *trichotomy law*: for all x and y in Ω, *exactly one of the following holds*: $x \prec y, x \sim y, y \prec x$.

Fig. 1 The relationships between orders

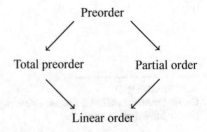

We will use $x \succcurlyeq y$ and $x \succ y$ to mean $y \preccurlyeq x$ and $y \prec x$, respectively, and we shall also use the following notation, where $a, b \in \Omega$:

$$(a, b) = \{x \in \Omega : a \prec x \prec b\}, \qquad [a, b] = \{x \in \Omega : a \preccurlyeq x \preccurlyeq b\},$$
$$(a, b] = \{x \in \Omega : a \prec x \preccurlyeq b\}, \qquad [a, b) = \{x \in \Omega : a \preccurlyeq x \prec b\},$$
$$(-\infty, a) = \{x \in \Omega : x \prec a\}, \qquad (a, +\infty) = \{x \in \Omega : a \prec x\},$$
$$(-\infty, a] = \{x \in \Omega : x \preccurlyeq a\}, \qquad [a, +\infty) = \{x \in \Omega : a \preccurlyeq x\}.$$

We shall refer to these (and only these) subsets as *intervals*, and we shall use the descriptions *bounded*, *unbounded*, *open*, and *closed*, exactly as we do for intervals on the real line (although their properties in this general context may differ from those in the real line). We shall also say that the intervals $(a, b]$ and $[a, b)$, where neither a nor b is $-\infty$ or $+\infty$, are *half-open* intervals.

3 Utility Functions

A total preorder $\preccurlyeq$ on a set Ω is *represented* by a function $u : \Omega \to \mathbb{R}$, and u is a *utility function* for $\preccurlyeq$, if u has the property that, for all x and y in Ω, $u(x) \leqslant u(y)$ if and only if $x \preccurlyeq y$. If this is so, then $u(x) = u(y)$ if and only if $x \sim y$; thus an indifference class (that is, an equivalence class) in Ω is one of the *level sets* of the function u, and conversely. More generally, *any* function $f : X \to Y$, between any pair of sets, defines an equivalence relation $\sim$ on X whose equivalence classes are the level sets of f.

Historically, a utility function was regarded as a way of assigning a numerical value to a level of preference or satisfaction. However, Debreu's famous example of the lexicographic ordering on $\mathbb{R}^2$ showed that not every preference relation can be represented by a utility function, and this prompted a shift in emphasis away from the classical view that satisfaction could be quantified numerically, and towards a more topological approach. However, even without Debreu's example, there is no good *mathematical* reason to require a utility function to be real-valued since the definition of a utility function only requires that its codomain be ordered in some way. Moreover, the fact that a particular total preorder $\preccurlyeq$ cannot be represented by a utility function does not mean that we should look no further; instead, we should examine whether or not there might exist a suitable (and preferably canonical) image space into which $\preccurlyeq$ can be represented. This view is entirely analogous to Riemann's profound view that to fully understand a many-valued analytic function f of a complex variable we have to build a Riemann surface for f (that is, we have to find a new and suitable domain for f). Indeed, in both cases the problem lies in a poor initial specification of the domain, or the codomain, of the function; in short, the optimal domain and codomain of a function should always be constructed from the function itself and *not prescribed in advance*. Another trivial, but compelling, observation is that if the collection of indifference classes has a larger cardinality than $\mathbb{R}$ then there cannot possibly be a function $u : \Omega \to \mathbb{R}$ whose level sets coincide with the indifference classes. For all these reasons it is clear that, from a theoretical perspective, the classical requirement that a utility function should be real-valued is unnecessarily restrictive. In recent years this view has attracted more interest, and among many papers on the topic we mention [4, 5] and especially [12]. Thus, inevitably, we are led to the following definition in which $(\mathbb{R}, \leqslant)$ is replaced by a more general set Σ equipped with a linear order $\leqslant$.

Definition 3.1 Suppose that $\preccurlyeq$ is a total preorder on Ω, and that $\leqslant$ is a linear order on Σ. Then $\preccurlyeq$ is *represented* by a function $u : \Omega \to \Sigma$ if and only if, for all x and y in Ω, $u(x) \leqslant u(y)$ if and only if $x \preccurlyeq y$. Such a function u is a *utility function* for $\preccurlyeq$.

Of course, such a utility function u need not be surjective. However, the total order $\leqslant$ on Σ restricts to a total order on the subset $u(\Omega)$ of Σ, so we may, if we wish, replace the space $(\Sigma, \leqslant)$ by the smaller space $\big(u(\Omega), \leqslant \big)$, and if we do this then u is surjective. We shall see later that this apparently trivial change has important theoretical consequences.

4   The Order Topology

We now consider a preference relation $\preccurlyeq$ on a set Ω, together with its induced relations $\sim$ (of indifference) and $\prec$ (of strict preference). The *order topology* $\mathcal{O}$ on $(\Omega, \preccurlyeq)$ is the smallest topology on Ω that contains all open intervals in Ω. This means that a subset of Ω is open in the order topology if and only if it is a union, say $\cup_{j \in J} U_j$, in which each U_j is the intersection of a finite number of open intervals. Now the intersection of a finite number of open intervals is itself an open interval; thus a subset of Ω is open in the order topology if and only if it is a union of open intervals. This implies that each open interval is an open set, and each closed interval is a closed set, in the order topology. Also, *each indifference class* $[x]$ *is a closed set*, for it is the complement of the open set $(-\infty, x) \cup (x, +\infty)$, and *the indifference classes partition Ω into a collection of mutually disjoint, closed sets*. This is a particular case of a general topic that has been studied by topologists under the name of *decomposition spaces*. Briefly, this is the study of topological spaces that can be expressed as the union of mutually disjoint, closed sets. If $f : X \to Y$ is any continuous function, with Y Hausdorff and y in Y, then $\{y\}$ is closed in Y so its pre-image in X is a closed (level) set of f, and this gives rise to a partition of X by the collection of mutually disjoint, closed sets $f^{-1}(y)$, where $y \in Y$. Now not every such partition of X is the collection of level sets of some continuous function $u : X \to Y$, and the problem of deciding when a given partition arises in this way is discussed in many places in the literature. We shall not pursue this any further except to draw the obvious comparison with the theory of utility functions.

For more information on the order topology, the reader can consult the pioneering work [19] by Nachbin. We end this section with two important properties of the order topology.

Theorem 4.1 *Each open set, and each closed set, in the order topology $\mathcal{O}$ is a union of equivalence classes. If U is an open, or a closed, subset of Ω, then $U = \bigcup_{x \in U} [x]$.*

Proof It is easy to see that if $y \sim x$ and $x \prec a$ then $y \prec a$; thus if $x \in (-\infty, a)$ then $[x] \subset (-\infty, a)$. Similarly, if $x \in (a, +\infty)$ then $[x] \subset (a, +\infty)$. Clearly these imply that if U is an open set, and $x \in U$, then $U = \bigcup_{x \in U} [x]$. Now suppose that V is a closed subset of Ω. Then the open complement U of V is a union of equivalence classes, and as Ω is a union of equivalence classes, so too is V. $\qquad\square$

The continuity of a utility function (indeed, of any function) depends entirely on the topologies given to its domain and codomain. For *surjective* utility functions (but not, as we show below, for non-surjective utility functions), continuity follows automatically if we use the order topologies in both cases.

Theorem 4.2 *Let $\preccurlyeq$ be a total preorder on Ω, and $\leqslant$ a linear order on Σ, and let $\mathcal{O}_\Omega$ and $\mathcal{O}_\Sigma$ be their respective order topologies. If $u : \Omega \to \Sigma$ is a surjective utility function that represents $\preccurlyeq$, then the function $u : (\Omega, \mathcal{O}_\Omega) \to (\Sigma, \mathcal{O}_\Sigma)$ is continuous.*

Proof As u is surjective, and as $x \prec y$ if and only if $u(x) < u(y)$, we see, for example, that if $y \in \Sigma$, then the pre-image under u of the subset $(-\infty, y)$ of Σ is the subset $\left(-\infty, u^{-1}(y)\right)$ of Ω. The same holds for other open intervals, so $u : (\Omega, \mathcal{O}_\Omega) \to (\Sigma, \mathcal{O}_\Sigma)$ is continuous. $\qquad\square$

The following example shows that we cannot omit the requirement of surjectivity in Theorem 4.2. This should not be surprising; indeed, the codomain of a function must be specified *before* we can decide whether or not it is surjective, and a similar situation applies to continuity.

Example 4.1 Suppose that $\Omega = [0, 1]$ (the interval in $\mathbb{R}$), and $\Sigma = \mathbb{R}$, both with the usual linear order $\leqslant$ and their corresponding order topologies. Then the function $u : [0, 1] \to \mathbb{R}$, defined by

$$u(x) = \begin{cases} x & \text{if } 0 \leqslant x < 1; \\ 3 & \text{if } x = 1, \end{cases} \tag{4.1}$$

is a utility function that represents $\leqslant$ on Ω. Now $u : \Omega \to \Sigma$ is *not* continuous because the pre-image of the open interval $(2, +\infty)$ in $\mathbb{R}$ is $\{1\}$, and this is not open in Ω. Nevertheless, by Theorem 4.2, the function $u : [0, 1] \to [0, 1) \cup \{3\}$ *is* continuous.

This example, together with Theorem 4.2, lies at the heart of Debreu's well-known gap lemma because, for example, $[1, 3)$ is a component of the complement of $u(\Omega)$, and $(2, +\infty) \cap u(\Omega)$ is not open in $u(\Omega)$ *with its order topology*.

5 The Quotient Space

Let Ω be any topological space, and let $\sim$ be any equivalence relation on Ω. We denote the equivalence class which contains x by $[x]$, and the quotient space (that is, the space of all equivalence classes) by $\Omega^{\sim}$; explicitly, $\Omega^{\sim} = \{[x]: x \in \Omega\}$. If $\sim$ is the equivalence relation derived from a preference relation $\preccurlyeq$ then, in economic terms, $\Omega^{\sim}$ is the space of all indifference classes. Throughout, the composition of maps is denoted by juxtaposition: thus $fg(x) = f(g(x))$. Now let $q: \Omega \to \Omega^{\sim}$ be the quotient map $x \mapsto [x]$. The *quotient topology* $\mathcal{Q}$ on $\Omega^{\sim}$ is the largest topology on $\Omega^{\sim}$ such that $q: \Omega \to \Omega^{\sim}$ is continuous; equivalently, a subset W of $\Omega^{\sim}$ is open (in the quotient topology $\mathcal{Q}$) if and only if $q^{-1}(W)$ is an open subset of Ω. The importance of the quotient topology is derived from the next result [15, p. 95].

Theorem 5.1 *Let Ω and $\Omega^{\sim}$ be as above and let Σ be any topological space. Then a function $g: \Omega^{\sim} \to \Sigma$ is continuous if and only if the composition $gq: \Omega \to \Sigma$ is continuous.*

We shall now discuss the quotient map before we apply these ideas to preference relations.

Theorem 5.2 *Let $\preccurlyeq$ be a total preorder on Ω, with order topology $\mathcal{O}$, and let q be the quotient map for the induced relation $\sim$. Then $q: (\Omega, \mathcal{O}) \to (\Omega^{\sim}, \mathcal{Q})$ is a continuous, open, and closed, map.*

Proof By definition, q is continuous. To show that q is an open map we have to show that if U is open in Ω then $q(U)$ is open in $\Omega^{\sim}$, and similarly for a closed map. Theorem 4.1 shows that if U is open in Ω then U is a union of equivalence classes, and this implies that $q^{-1}(q(U)) = U$. It follows that $q(U)$ is open in $\Omega^{\sim}$, so that q is an open map. Now suppose that V is closed in Ω and let U be its complement. As the open set U is a union of equivalence classes, so too is V, so that the sets $q(U)$ and $q(V)$ partition $\Omega^{\sim}$. As $q(U)$ is open in $\Omega^{\sim}$, so $q(V)$ is closed in $\Omega^{\sim}$. □

Theorem 5.3 *Let $\preccurlyeq$ be a total preorder on Ω. Then the quotient space $(\Omega^{\sim}, \mathcal{Q})$ is Hausdorff.*

Proof Take two equivalence classes, say $[x]$ and $[y]$ with $[x] \neq [y]$; then $[x]$ and $[y]$ are disjoint. Without loss of generality, we may assume that $x \prec y$. If $(x, y) = \emptyset$ then $(-\infty, y)$ and $(x, +\infty)$ are disjoint open sets that contain x and y, respectively. If $(x, y) \neq \emptyset$ there is some z with $x \prec z \prec y$, and then the disjoint open sets $(-\infty, z)$ and $(z, +\infty)$ separate x and y. In each case there are disjoint open sets $\mathcal{X}$ and $\mathcal{Y}$ in Ω such that $x \in \mathcal{X}$ and $y \in \mathcal{Y}$. As q is an open map, the sets $q(\mathcal{X})$ and $q(\mathcal{Y})$ are disjoint open sets in $\Omega^{\sim}$ that separate $[x]$ and $[y]$, so $(\Omega^{\sim}, \mathcal{Q})$ is Hausdorff. □

The same argument gives more than this. Let $\preccurlyeq$ be a total preorder on Ω, with $\mathcal{O}$ its order topology; then the relation $\preccurlyeq$ is a subset of the product space $\Omega \times \Omega$ equipped with the corresponding product topology. Now it is known [15, p. 98] that any equivalence relation R on Ω is a closed subset of $\Omega \times \Omega$ if and only if, whenever x and y are in Ω but not in the same equivalence class, there are open neighbourhoods U and V, of x and y, respectively, such that no point of U is equivalent to a point of V. In the case of an indifference relation $\sim$ formed from a preference relation $\preccurlyeq$, this is exactly what we have proved above. It is also shown in [15, p. 98] that if the quotient map $q : \Omega \to \Omega/R$ is an open map, and if R is closed in $\Omega \times \Omega$, then Ω/R is a Hausdorff space. Thus we have the following extension of Theorem 5.3 (which is now seen as a special case of a much more general topological result).

Theorem 5.4 *Let $\preccurlyeq$ be a total preorder on Ω, and let $\Omega \times \Omega$ have the product topology derived from the order topology on Ω. Then the relation $\preccurlyeq$ is a closed subset of $\Omega \times \Omega$, and the quotient space $\Omega^\sim$ is Hausdorff.*

Finally, it is well known that a total preorder $\preccurlyeq$ on Ω induces a linear order $\lesssim$ on $\Omega^\sim$, where $[x] \lesssim [y]$ if and only if for some (and hence all) a in $[x]$, and b in $[y]$, we have $a \preccurlyeq b$. It is important to note that whereas $\preccurlyeq$ is a *total preorder* on Ω, the induced relation $\lesssim$ is a *linear order* on $\Omega^\sim$ (that is, $[x] \lesssim [y]$ and $[y] \lesssim [x]$ implies $[x] = [y]$). In effect, in passing from Ω to $\Omega^\sim$ we are combining all the points in a single indifference class into a single point without destroying their 'levels of satisfaction' and this suggests (at least to the author) that the quotient space $\Omega^\sim$ should perhaps be given much more prominence than it usually is.

Starting with a total preorder $\preccurlyeq$ on Ω, we have created (each in a natural way) *two* topologies on the quotient space $\Omega^\sim$, namely the *quotient topology* $\mathcal{Q}$, and the *order topology* constructed from the induced order $\lesssim$ on $\Omega^\sim$. Fortunately, these two topologies are the same, and although the proof is easy, the result is necessary for the smooth development of the theory.

Theorem 5.5 *Let $\preccurlyeq$ be a total preorder on Ω. Then the quotient topology $\mathcal{Q}$ on the quotient space $\Omega^\sim$ coincides with the order topology $\mathcal{O}^\sim$ constructed from the linear order $\lesssim$ on $\Omega^\sim$.*

Proof First, we observe that $(-\infty, [a]) = \{[x] \in \Omega^\sim : x \prec a\} = q((-\infty, a))$. Similarly, $q((a, +\infty)) = ([a], +\infty)$. The proof of Theorem 5.5 follows immediately.
$\square$

6 Utility Functions and the Quotient Space

Let $\preccurlyeq$ be a preference relation on a set Ω, and let q be the quotient map. Then, given any function $v : \Omega^\sim \to \Sigma$, we can always define a function $u : \Omega \to \Sigma$ by $u = vq$ as illustrated below (Fig. 2).

Conversely, given any function $u : \Omega \to \Sigma$ which is constant on each equivalence class in Ω (that is, a utility function), we can define v unambiguously by the rule that, for each equivalence class ζ in $\Omega^\sim$, if $u = k$ on ζ, then $v(\zeta) = k$. This construction produces a bijection $\Theta : v \mapsto u = vq$ from the class of all functions $v : \Omega^\sim \to \Sigma$

Fig. 2 Identifying the
quotient space

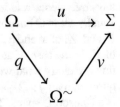

onto the class of all functions $u \colon \Omega \to \Sigma$ that are constant on each equivalence class. Theorem 5.1 implies that if either u or v is continuous, then so is the other, so we have the next (general) result.

Theorem 6.1 *Let $\sim$ be an equivalence relation on Ω, and q the quotient map, and give $\Omega^\sim$ the quotient topology. Then, given any function $u \colon \Omega \to \Sigma$ that is constant on each equivalence class, there is a unique function $u^\sim \colon \Omega^\sim \to \Sigma$ such that $u = u^\sim q$. The map $u \mapsto u^\sim$ is a bijection between these classes of functions, and, for any topology on Σ, u is continuous if and only if $u^\sim$ is continuous.*

In the context of mathematical economics, if u is a utility function for $\precsim$, then v is injective, and if u is surjective, then v is also surjective. Thus the problem of finding a surjective utility function that represents $\precsim$ in $(\Sigma, \leqslant)$ is equivalent to the problem of finding a bijection of $\Omega^\sim$ onto $(\Sigma, \leqslant)$. Moreover, a similar statement holds if we require both u and v to be continuous. While this is a trivial observation, it does imply that, if we wish, we may always work with the apparently simpler situation of a *linear order* on $\Omega^\sim$ rather than a *total preorder* on Ω. It is not immediately clear whether or not this provides a real advantage, but it is worth further consideration as it may possibly simplify some of the existing arguments. Of course, we can always take q as a utility function, and Σ to be the quotient space $\Omega^\sim$, and while this achieves nothing in a technical sense, it does emphasize that the problem of finding a utility function is precisely the same problem that is frequently faced in topology, namely that of *identifying the quotient space $\Omega^\sim$*.

7 Natural Topologies

In many applications the commodity space Ω (that is, a space Ω that supports a preference relation) is equipped with a topology $\mathcal{T}$ which reflects the quantities of goods rather than the preferences between them; for example, the elements of Ω may be vectors in $\mathbb{R}^n$, and $\mathcal{T}$ the Euclidean topology. In some sense, this creates a conflict because although $\mathcal{T}$ may be important in applications, from a theoretical perspective the order topology $\mathcal{O}$ is more important than $\mathcal{T}$ because it is the only topology which properly reflects the preference relation $\precsim$. In fact, in many (or perhaps most) applications it is assumed that, for each x, the sets $(-\infty, x]$ and $[x, +\infty)$ are closed in $\mathcal{T}$, and then $\mathcal{T}$ is said to be a *natural topology*. Here, we prefer to emphasize open sets instead of closed sets, and express the basic content of this assumption in the more decisive and inclusive form $\mathcal{O} \subset \mathcal{T}$. Of course, it is *precisely this condition* which implies that if $f: \Omega \to \mathbb{R}$ is continuous with respect to the order topology on Ω, then it is continuous with respect to any natural topology on Ω.

In [21, p. 41], Whyburn considers a connected topological space X, say with topology $\mathcal{T}$, and defines a point p in X to be a *cut-point* of X if its complement $X \backslash \{p\}$ is disconnected. He also assumes (implicitly) that each singleton subset $\{x\}$ of X is closed in X. He then considers two distinct points, say a and b, in X, and defines $E(a, b)$ to be the set of points x in X that separate a and b (in the sense that a and b lie in different components of $X \backslash \{x\}$). He then creates a binary relation which, to avoid confusion, we denote by $\ll$ on $E(a, b)$ that is *irreflexive* (it is false that $x \ll x$), *assymmetric* (if $x \ll y$ then $y \ll x$ is false), and *transitive* (see [7, p. 3]). Whyburn calls this an 'order', but it is not the same as a linear order that we have defined in Sect. 1. Next, Whyburn then says that his order is a *natural order* if, for all x in $E(a, b)$, we have

$$\overline{(-\infty, x)} \cap (x, +\infty) = \emptyset = (-\infty, x] \cap \overline{(x, +\infty)}, \tag{7.1}$$

where $\overline{A}$ refers to the closure of A in $\mathcal{T}$ (not $\mathcal{O}$). Now it is easy to see that as X is connected, (7.1) is equivalent to the statement that $\mathcal{O} \subset \mathcal{T}$ (that is, $\mathcal{T}$ is a natural topology). Thus it seems that the idea of a natural topology occurred, in a topological context, before it appeared in mathematical economics. Further, we shall see later that the idea of a cut point in topology is central to Debreu's gap lemma in mathematical economics and, in an entirely similar way, to a discussion of the existence of Jordan arcs in a topological space.

8 Monotone and Light Mappings

To motivate a discussion of monotone functions (which are *not* the same as monotonic functions in real analysis), we begin with the *Cantor function* $\Phi: [0, 1] \to [0, 1]$. Briefly, Φ is constant on each component of the complement of the Cantor middle–third set $\mathcal{C}$, and is constructed as follows. Let

$$\Phi = \tfrac{1}{2} \text{ on } [\tfrac{1}{3}, \tfrac{2}{3}], \quad \Phi = \tfrac{1}{4} \text{ on } [\tfrac{1}{9}, \tfrac{2}{9}], \quad \Phi = \tfrac{3}{4} \text{ on } [\tfrac{7}{9}, \tfrac{8}{9}],$$

and so on. This process defines Φ on the complement of $\mathcal{C}$, and it is easy to extend Φ so that its extension, which we continue to denote by Φ, is a continuous, increasing map of $[0, 1]$ onto itself. This is illustrated below, but see [13, pp. 129–131] for more details (Fig. 3).

Now let X and Y be topological spaces with Y Hausdorff, and let $f: X \to Y$ be any function (not necessarily continuous). Here we are interested in the inverse image $\{x \in X: f(x) = y\}$ of each y in Y, and we denote this by $f^{-1}(y)$. As Y is

Fig. 3 The Cantor function
$\Phi: [0, 1] \to [0, 1]$

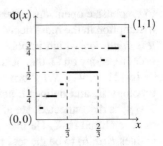

Hausdorff, each set $\{y\}$ is closed, so that if f is continuous then $f^{-1}(y)$ is also closed. As we remarked earlier, this discussion falls under a general discussion of *decomposition spaces*. Here, we shall need the following concepts which, perhaps, are not as widely known as they might be.

Definition 8.1 The function $f: X \to Y$ is

(i) *monotone* if, for each y in Y, $f^{-1}(y)$ is connected;
(ii) *light* if, for each y in Y, $f^{-1}(y)$ is totally disconnected (that is, each component is a single point).

If $f: X \to Y$ is injective then it is both monotone and light. The Cantor function Φ is monotone but not light; each non-constant polynomial of a complex variable is light but not monotone. Our interest here is in monotone functions and, in particular, with the following result.

Theorem 8.1 *If $f: [0, 1] \to Y$, where Y is a Hausdorff topological space, is a non-constant, continuous, monotone map of $[0, 1]$ onto Y, then Y is homeomorphic to $[0, 1]$.*

Although we shall not need it, we also mention the following remarkable, and basic, result on light and monotone maps; for more details of both of these theorems, see, for example, [13, pp. 136–137] and [22, pp. 25–26].

Theorem 8.2 *Suppose that X and Y are compact, Hausdorff spaces and $f: X \to Y$ is continuous. Then f can be expressed as a composition $f = \ell m$ of continuous maps ℓ and m, where $m: X \to m(X)$ is monotone, and $\ell: m(X) \to Y$ is light. Further, the maps ℓ and m are unique up to homeomorphisms.*

Our objective is to show how monotone functions are related to the theory of preference relations. As a simple example of this, it is sometimes assumed in economics that the indifference classes arising from a preference relation $\preccurlyeq$ on Ω are connected sets. In this case, any real-valued utility function u that represents $\preccurlyeq$ is such that, for each t in $u(\Omega)$, the set $u^{-1}(t)$ is a connected set, so that the utility function u is a monotone function. However, before we pursue this link between monotone functions and utility theory any further, we can develop a better understanding of monotone functions by looking at real-valued functions of a real variable. Specifically, we shall compare the definition of a *monotonic function* (in the usual sense) with the definition of a *monotone function*.

Consider a function $f : [a, b] \to \mathbb{R}$, where $[a, b]$ is a compact subinterval of $\mathbb{R}$. As usual, f is *increasing* if $x \geqslant y$ implies $f(x) \geqslant f(y)$, and *strictly increasing* if $x > y$ is equivalent to $f(x) > f(y)$. There are similar definitions for decreasing, and strictly decreasing, functions, and f is *monotonic* if and only if it is increasing or decreasing. The definition of 'monotone functions' perhaps originated from the following result.

Theorem 8.3 *Let $f : [a, b] \to \mathbb{R}$ be any function. If f is monotonic then it is monotone, but the converse is false. If f is continuous then it is monotonic if and only if it is monotone.*

Proof Suppose that f is monotonic, and take x and y in $[a, b]$ with $x < y$ and $f(x) = f(y) = k$, say. As f is monotonic, $f = k$ on the entire interval $[x, y]$. It follows that $f^{-1}(k)$ is a subinterval of $\mathbb{R}$; thus f is monotone. The converse is false because any injective map $f : [a, b] \to \mathbb{R}$ is monotone (because a singleton set is connected) but clearly not necessarily monotonic.

Now suppose that $f : [a, b] \to \mathbb{R}$ is continuous. We only have to show that if f is monotone then it is monotonic, so suppose that f is monotone. By replacing f by $-f$ (if necessary) we may assume that $f(a) \leqslant f(b)$, and we shall now show that f is increasing. Of course, if we did actually replace f by $-f$, then the original function f must be decreasing.

We now start from the premise that $f(a) \leqslant f(b)$. The Intermediate Value theorem shows that the image $f([a, b])$ contains the interval $[f(a), f(b)]$. Now suppose that for some x in $[a, b]$, $f(x) < f(a) \leqslant f(b)$. Then the Intermediate Value theorem applied to $[x, b]$ shows that there is some y in $[x, b]$ such that $f(y) = f(a)$. However, this gives $a < x < y$ and $f(x) < f(a) = f(y)$, which contradicts the assumption that f is monotone. We conclude that $f(x) \geqslant f(a)$ on $[a, b]$ and, similarly, $f(x) \leqslant f(b)$ on $[a, b]$. Thus f maps $[a, b]$ *onto* $[f(a), f(b)]$. Now consider any t in $[a, b]$. Then, by the argument above (with $b = t$), we see that f maps $[a, t]$ onto $[f(a), f(t)]$, and this easily implies that f is increasing.   □

9 Continuity of the Inverse Function

An important feature of Theorem 8.3 is that the domain $[a, b]$ of f is *connected*; indeed, any constant function whose domain is disconnected is certainly not monotone. Another consequence of the connectedness of $[a, b]$ is that *if $f : [a, b] \to \mathbb{R}$ is strictly increasing and continuous, then f maps $[a, b]$ onto $[f(a), f(b)]$, and f^{-1} is continuous on $[f(a), f(b)]$.* This result is proved in almost every basic text on real analysis but, sadly, authors omit to mention that f^{-1} is continuous *precisely because $[a, b]$ is connected*, and that the continuity of f is *entirely irrelevant* [2].

This observation suggests that we should now explore how the connectedness of the domain X of a general injective function $f: X \to \mathbb{R}$ affects the continuity of its inverse $f^{-1}: f(X) \to X$. Thus we shall now consider a non-empty subset X of $\mathbb{R}$, and a strictly increasing (but otherwise general) function $f: X \to \mathbb{R}$ together with its inverse f^{-1}. Readers familiar with Debreu's theory of utility functions will not fail to recognize the relevance of the following discussion.

Theorem 9.1 *Let $\mathbb{R}$ be given the Euclidean topology, and let X be a non-empty subset of $\mathbb{R}$ with the subspace topology. Then the following are equivalent:*

(i) *the complement $\mathbb{R} \backslash X$ of X has no components of the form $[p, q)$ or $(p, q]$, where $p < q$;*
(ii) *every strictly increasing function $f: X \to \mathbb{R}$ has a continuous inverse $f^{-1}: f(X) \to \mathbb{R}$;*
(iii) *the order topology on X coincides with its subspace topology.*

As we have already observed, when discussing the continuity of functions it is essential to give both the codomain (as a set) *and also its topology*. Thus in (ii), continuity is interpreted in the usual sense of real analysis, and not in the sense of the order topology. Of course, the Euclidean, and order, topologies on $\mathbb{R}$ are the same, but the subspace topology on X may not be the same as its order topology; indeed, Theorem 9.1 gives conditions under which these two topologies *are* the same. Notice also that Theorem 9.1 contains the statement that every strictly increasing map $f: [a, b] \to \mathbb{R}$ has a continuous inverse (whether f is continuous or not).

Before we give the proof it is perhaps worth remarking that, in the usual statements about a real-valued continuous utility function u on Ω, it is almost always (or, perhaps, always) assumed that the topology on $u(\Omega)$ is the *subspace topology* induced by the Euclidean topology on $\mathbb{R}$. This is reasonable enough in any situation in which one wishes to suppress the topological aspects, but it is not in any situation in which topology is being used. We repeat, the set $u(\Omega)$ carries **two** 'obvious' topologies, namely the subspace topology and its own 'intrinsic' order topology, and it must be made clear which of these is being used.

The proof of Theorem 9.1. Let $f: X \to \mathbb{R}$ be any strictly increasing function, and select any x_0 in X. Let $E = \{x \in X: x > x_0\} = X \cap (x_0, +\infty)$ and, when E is not empty, put $\lambda = \inf E$ so that $\lambda \geq x_0$. Now consider the following four possibilities:

(I) $E = \emptyset$;
(II) $E \neq \emptyset$ and $\lambda = x_0$;
(III) $E \neq \emptyset$, $\lambda > x_0$ and $\lambda \in X$;
(IV) $E \neq \emptyset$, $\lambda > x_0$ and $\lambda \notin X$.

Clearly, one, and only one, of these possibilities must hold. It is elementary to show that in cases (I), (II) and (III), f^{-1} is continuous from above at $f(x_0)$, while in case (IV), $(x_0, \lambda]$ is a component of $\mathbb{R} \backslash X$. We conclude that f^{-1} is continuous from above at every point of $f(X)$ if and only if $\mathbb{R} \backslash X$ has no components of the form $(x, b]$ where $x \in X$ and $b \notin X$. A similar argument holds for continuity from below, and intervals of the form $[a, x)$, so we have proved that (i) holds if and only if (ii) holds.

Next, we assume that (i) holds and prove (iii). To verify (iii) it is enough to show that, for all a in $\mathbb{R}$, the sets $(-\infty, a) \cap X$ and $(a, +\infty) \cap X$ are in $\mathcal{O}_X$, for then $\mathcal{E}_X \subset \mathcal{O}_X$. As we always have $\mathcal{O}_X \subset \mathcal{E}_X$, this will show that $\mathcal{O}_X = \mathcal{E}_X$ which is (iii). We need only consider $(-\infty, a) \cap X$ as the other case is similar. Take a in $\mathbb{R}$ and let $U = (-\infty, a) \cap X$. If $a \in X$, then U is in $\mathcal{O}_X$. Also, if U is empty then U is in $\mathcal{O}_X$. Thus we may now assume that $a \notin X$ and $U \neq \emptyset$. Let J be the component of $\mathbb{R} \backslash X$ that contains a; then J is bounded below. There are now three remaining cases to consider, and if $u = \inf J$, these are

(I) $u \in X$ and $J = (u, +\infty)$;
(II) $u \in X$ and $J = (u, v)$ for some v;
(III) $u \notin X$.

In case (I), $U = X$ so $U \in \mathcal{O}_X$. In case (II), $U = X \cap (-\infty, v)$ which is in $\mathcal{O}_X$. In case (III),

$$U = \bigcup_{t \in X, \, t < a} X \cap (-\infty, t),$$

which is in $\mathcal{O}_X$.

Finally, we show that (iii) implies (ii). To prove this let $\mathcal{O}_X$ and $\mathcal{E}_X$ be the order topology, and the Euclidean (subspace) topology, on X, respectively, and let $\mathcal{O}_{f(X)}$ and $\mathcal{E}_{f(X)}$ be the corresponding topologies on $f(X)$. Now consider the three maps

$$f^{-1} \colon \left(f(X), \mathcal{O}_{f(X)} \right) \to \left(X, \mathcal{O}_X \right),$$
$$f^{-1} \colon \left(f(X), \mathcal{O}_{f(X)} \right) \to \left(X, \mathcal{E}_X \right),$$
$$f^{-1} \colon \left(f(X), \mathcal{E}_{f(X)} \right) \to \left(X, \mathcal{E}_X \right).$$

By Theorem 4.2, the first map is continuous. If (ii) holds then $\mathcal{O}_X = \mathcal{E}_X$, so the second map is continuous. Finally, as $\mathcal{O}_{f(X)} \subset \mathcal{E}_{f(X)}$, the third map is continuous, and this conclusion is (ii). $\qquad \square$

The reader is invited to reconsider Example 4.1 in the light of Theorem 9.1.

10 The Existence of Utility Functions

In this section we briefly discuss how the topological ideas discussed in this paper might influence the theory of preference relations, and we make some tentative suggestions for further investigations.

The first proof of the existence of a continuous utility function seems to be in a fundamental paper by Wold [23] in 1943. Until the 1950s (and perhaps because of Wold's result) the general expectation seemed to be that the level of satisfaction could be quantified numerically; that is, every preference relation could be represented by a utility function. However, in 1954 Debreu published his example of the lexicographic order on $\mathbb{R}^2$ which cannot be represented by a utility function [8], and since then the theory of utility functions has been a very active area of research. Wold assumed that the commodity space Ω was a subset of the 'positive cone' $\mathcal{C}$ (that is, the set of points $(x_1, \ldots, x_m)$, where each x_j is positive) in Euclidean space $\mathbb{R}^m$, and he only sketched a proof of his construction. His argument is perhaps more subtle than he was given credit for, and for an expanded account of his proof, see [6]. In essence, Wold argued as follows. Consider a preference relation on $\mathcal{C}$ *which allows the possibility of 'thick' indifference sets* (that is, in which an indifference class may have a non-empty interior), and let D be the diagonal (given by $x_1 = \cdots = x_m$) in $\mathcal{C}$. Then, under reasonable economic assumptions, he showed that any point x in $\mathcal{C}$ is indifferent to a point x_D on D. Of course, as indifference sets may be thick, the intersection of D with an indifference class may be an interval of positive length on D. In effect, he mapped his commodity space $\mathcal{C}$, monotonically with respect to $\preccurlyeq$, into a new commodity space D, namely a line. In essence, he then proved the following theorem (and a similar result holds with [0, 1] replaced by [0, $+\infty$)).

Theorem 10.1 *Let $\succcurlyeq$ be a preference relation on [0, 1] with the property that each indifference class is a closed interval. Then there exists a continuous, increasing function $h: [0, 1] \to [0, 1]$ with the property that $h(x) = h(y)$ if and only if $x \sim y$.*

If we now apply the map $x \mapsto x_D$ from $\mathcal{C}$ to D (which Wold showed is continuous), and then follow it with the map h obtained from Theorem 10.1, we obtain a continuous utility function on $\mathcal{C}$. Note that the function h in Theorem 10.1 is a monotone function; thus, in retrospect, it is clear that Wold had seen a connection between preference relations and *monotone functions* (which were available earlier) although, of course, he did not explicitly use the term 'monotone function'. In fact, what Wold actually did was first to 'collapse' the commodity space onto a linear commodity space D (while preserving the comparison of preferences), and then create a monotone map from D into $\mathbb{R}$ (see [1] for another account of Wold's work). Indeed, if we rewrite Wold's result (Theorem 10.1) by replacing the phrase '*with the property that $h(x) = h(y)$ if and only if $x \sim y$*' with the equivalent phrase '*that is constant on each indifference class*', we see that to prove his result, Wold actually had to construct a monotone function. In fact, and as we shall see in the next section, this is the same process that is used to replace the motion of a particle, which may be at rest during certain time intervals, by a motion that is never at rest: we simply have to construct a monotone function that collapses each such time interval into a single point!

We turn now to Debreu's work. In the 1950s Debreu addressed the problem of the existence of a utility function for given preference relation, and proved that that if X is a second countable topological space then any preference relation $\succcurlyeq$ on X can be represented by a utility function. Debreu then turned his attention to the existence of a *continuous* utility function, and proved that if some utility function exists, then a utility function that is continuous with respect to the order topology also exists. It is very well known that this result was based on what is now universally known as Debreu's *Gap Lemma*: *if $E \subset \mathbb{R}$ then there exists a strictly increasing map $g: E \to \mathbb{R}$ such that no component of the complement of $g(E)$ is of the form $(a, b]$ or $[a, b)$, where a and b are finite (i.e. real numbers).* The idea behind this is that if there were any such 'gaps' we should be able to 'close' all such 'gaps' by 'sliding' then sideways, and that this would then lead to the desired result.

Unfortunately (but perhaps fortunately in the light of what followed) there is an error in this reasoning, and the error arises because after such a compression we may be left with a residual set of measure zero which is not, therefore, an interval. In fact if we consider the Cantor function described earlier, and collapse each interval on which Φ is constant to a single point, then we are left with the disjoint union of the Cantor set $\mathcal{C}$ and a countable set, and this union has measure zero. This is essentially the same as saying that the Cantor function Φ does all of its 'climbing' on the set $\mathcal{C}$ of measure zero. This aspect also features in probability theory where it shows that there do exist probability distributions which are neither discrete (with only point masses) nor continuous (with a continuous probability density function). In any event, the error in Debreu's original proof spawned a large number of ideas and papers which, in various ways, corrected and extended his argument and result. Before leaving this matter we should record the fact that there are now many proofs of Debreu's gap lemma available, but this does not mean that each of the proofs exposes the real reason why it is true. While it is true that the error is best explained (as above) in terms of measure theory, it does not follow that a correct proof should use measure theory. Indeed, it seems clear (to this author) that this result is essentially topological in nature, and if so, then the most revealing proof should surely also be topological in nature.

Beyond the work of Wold and Debreu, there are many other interesting results available on the possible existence of a (possibly continuous) utility function, and we refer the reader to [7, Chap. 4] for more information on these. Among these results, perhaps three stand out as being likely candidates for further reflection in the context of the discussion in this paper, namely those by Eilenberg, Jaffray and Monteiro [9, 14, 18] (there is no need to quote these results explicitly for they can all be found in [7]). It is important to understand here that in reflecting on these (and other) results we should *not* be asking for the most 'elementary' proofs of them (for, in the author's view at least, such proofs often fail to reveal some of the most important aspects of the result); instead we should be seeking a general framework into which these and other results appear together in a natural, coherent, and efficient way.

It is interesting to consider how some of these ideas might be generalised to arbitrary commodity spaces. For example, let $(\Omega, \preccurlyeq)$ be a commodity space, and let X be a non-empty subset of Ω. Then X carries *two* 'obvious' topologies, namely the subspace topology $\mathcal{S}_X$ (induced by the order topology on Ω), and the order topology $\mathcal{O}_X$ obtained by restricting the preference relation $\preccurlyeq$ to X and then taking $\mathcal{O}_X$ to be the order topology constructed from this restricted relation. Now these two topologies need not be the same, but we can still ask whether or not it is true that $\mathcal{O}_X = \mathcal{S}_X$ if and only if there are no non-empty sets of the form $\{x \in \Omega \backslash X : a \prec x \preccurlyeq b\}$, where $a \in X$ and $b \notin X$, or $\{x \in \Omega \backslash X : a \preccurlyeq x \prec b\}$, where $a \notin X$ and $b \in X$. In effect, this is asking whether or not it would be worthwhile to redefine the idea of a half-open half-closed gap of X *in a more general setting* by the purely topological requirement that $\mathcal{O}_X = \mathcal{S}_X$.

Finally, it is perhaps worth recalling that *convexity* and *connectedness* are, at least superficially, defined in a different way. The two concepts coincide on the real line, but in a general setting they do not. Of course, in a general setting all intervals are (trivially) convex, but not necessarily connected. Thus we might consider which of these (if any) is the really important property, or are they both essential to the discussion? Further, is the possible difference between the order topology and any other topology relevant here? Indeed, do these two concepts coincide on the real line precisely because there the order topology coincides with the metric topology? We leave the interested reader to consider these matters further.

11 Jordan Arcs in a Topological Space

We end with a discussion of Jordan arcs in a topological space. At first sight this topic may appear to have little to do with economics but, in fact, it does share some common features with economics. Indeed, both topics benefit from the ideas of cut points and monotone functions, and we have already remarked on the similarity between Wold's construction, and the motion of a particle along a curve.

Definition 11.1 A *curve*, or *path*, in a topological space X is a continuous map $\alpha: [a, b] \to X$, where $[a, b]$ is a compact subinterval of $\mathbb{R}$. The *carrier* of a curve α is the set $\{\alpha(t): t \in [a, b]\}$ which we denote by $[\alpha]$. A *Jordan arc* in X is a curve $\beta: [a, b] \to X$, where β is a homeomorphism of $[a, b]$ onto $[\beta]$.

Topologists sometimes distinguish between a *pathwise connected* space (a space in which any two points can be joined by a curve in the space), and an *arcwise connected space* (a space in which any two points can be joined by a Jordan arc in the space), and we too shall use this terminology. If X is Hausdorff, and a curve $\beta: [a, b] \to X$ is injective then (see below) it is a Jordan arc.

We now ask the following question: *let p and q be points in a topological space X, and suppose that a particle moves from p to q along a curve α in X. Can this particle also move from p to q along a Jordan arc in $[\alpha]$?* Informally, this means that the particle should only visit points that have been visited before, and in the same order as before, but it must never be at rest, and never visit any point twice. We can imagine that a positive answer might be useful when discussing analytic or geometric problems (where it might be helpful to replace a curve by a Jordan arc), and this is indeed the case. For example, Hayman, in his proof of Iversen's theorem for subharmonic functions in $\mathbb{R}^m$ [11, pp. 191–192], cites Kerékjártó [16, p. 103], and gives a positive answer in the case when X is $\mathbb{R}^m$. Milnor, in his discussion of complex dynamics on the Riemann sphere [17, p. 185], gives a positive answer when X is a Hausdorff topological space. In particular, we have a positive answer when X is a metric space.

We shall now describe the role of monotone functions in answering this question. The Cantor function Φ defines a curve, but not a Jordan arc, in $\mathbb{R}$, in which the particle is at the position $\Phi(t)$ at time t. Intuitively, the particle moves from 0 to 1 along the interval $[0, 1]$ in a time interval of 1 second in such a way that, for example, it stays stationary at the point $\frac{1}{2}$ for exactly $\frac{1}{3}$ of a second. The question asked above is whether or not the particle can move along the same path (that is, along $[0, 1]$) without ever being stationary. In this case it obviously can, but the point is that the problem can also be viewed as follows. If we consider $[0, 1]$ as a commodity space with the preorder $\preccurlyeq$ defined by $x \preccurlyeq y$ if and only if $\Phi(x) \leqslant \Phi(y)$, then the indifference classes are the subintervals of $[0, 1]$ on which Φ is constant, and the problem becomes that of finding a utility function for $\preccurlyeq$.

Briefly, the argument given in [11, 16, 17] cited above is as follows. We begin with a continuous map $\alpha : I \rightarrow X$, where X is a Hausdorff space, $I = [0, 1]$, $\alpha(0) = p$ and $\alpha(1) = q$. If α is injective then it is an arc and there is nothing further to prove. Otherwise, there are points u and v with $0 \leqslant u < v \leqslant 1$ and $\alpha(u) = \alpha(v)$. Choose any such u and v that maximises the length $v - u$, and put $u = u_1, v = v_1, I_1 = (u_1, v_1)$, and $w_1 = \alpha(u_1) = \alpha(v_1)$. Then $\alpha \neq w_1$ on the intervals $[0, u_1)$ and $(v_1, 1]$ (else $v_1 - u_1$ would not be maximal). If α is injective on $I \backslash I_1$ the process terminates; otherwise we repeat this process on the set $I \backslash I_1$ to obtain an open interval I_2, say (u_2, v_2), in $I \backslash I_1$, with $\alpha(u_2) = \alpha(v_2) = w_2$, say. Continuing in this way, we obtain a (possibly finite) sequence $I_1, I_2, \ldots$ of mutually disjoint open intervals I_n, say $I_n = (u_n, v_n)$, such that α is injective on $I \backslash \cup_n I_n$, and, for each n, $w_n = \alpha(u_n) = \alpha(v_n)$. The set $I \backslash \cup_n I_n$ does not contain any isolated points, or any intervals, and we now define $\beta : [0, 1] \rightarrow [0, 1]$ by $\beta = w_n$ on I_n, $n = 1, 2, \ldots$, and $\beta = \alpha$ on $I \backslash \cup_n I_n$. It can be shown that β is continuous; thus β is a monotone function on $[0, 1]$. Geometrically speaking, β is a curve that joins p to q in $[\alpha]$, and as t increases from 0 to 1, $\beta(t)$ moves through $[\alpha]$, with $\beta(t)$ remaining stationary while t moves through each I_n, but otherwise never returning to a point it has been to before. More precisely, if ζ is a point on $[\beta]$, then the set of times t such that $\beta(t) = \zeta$ *is an interval*, so that the point $\beta(t)$ never returns to a point of $[\alpha]$ once it has left it; thus, informally, we have replaced the curve α by a monotonic (but not a strictly monotonic) curve β. The desired result now follows from Theorem 8.1.

Not surprisingly, we have assumed that X is a Hausdorff space (as otherwise, for example, limits need not be unique). Moreover, if X is Hausdorff, then any injective curve $\alpha : I \rightarrow X$ is a continuous bijection from a compact space to the Hausdorff space $[\alpha]$; thus $\alpha : I \rightarrow [\alpha]$ is a homeomorphism, and α is a Jordan arc. If X is not Hausdorff then a continuous injective curve need not be a homeomorphism. However, the apparent generality of allowing X to be any Hausdorff space is actually an illusion because it is known that *the continuous image of a compact metric space in a Hausdorff space is metrisable*; see [13, p. 126]. Thus if α is a curve in a Hausdorff space X, then $[\alpha]$ is metrisable so we are back in the situation when X (that is, $[\alpha]$) is a metric space.

Next, we mention a well known topological characterisation of when a given set is a Jordan arc. We have already said that a point x_0 in a topological space X is a *cut point* of X if its complement $X \backslash \{x_0\}$ is disconnected. Then *a compact, connected, metric space is a Jordan arc if and only if it has exactly two points that are not cut-points*. Proofs of this can be found in [20, p. 93, 96], and in [13, pp. 14, 48, 54], and it is well worth noting that these proofs are based on *constructing an order relation, and its associated order topology*, that is very much apart of the discussions of preference relations in mathematical economics. In fact, these ideas are central to the author's purely topological proof of Debreu's gap lemma; see [3, 7]. In fact, much more is known about curves and Jordan arcs. For example, a space X is *locally connected* if and only if each component of an open set is open [13, pp. 105–106]. With this, we have the Hahn–Mazurkiewicz theorem: *a metric space X is the image of some curve $\alpha : I \to X$ if and only if it is compact, connected, and locally connected.*

Finally, we remark that there is a very general theorem available which certainly suggests that there should be a positive answer to our question on Jordan arcs. First a *continuum* is a compact, connected, topological space. Now consider a continuous map $f : W \to X$, where W is a continuum, and X is a Hausdorff topological space. Without loss of generality we may replace X by $f(W)$; thus we may assume that f is surjective, and that X is a Hausdorff continuum. Then it is known that given any subset E of X, there is a (not necessarily unique) continuum K in X that contains E, and that is minimal in the sense that no strictly smaller subcontinuum of K contains E (see [13, Theorem 2-11, p. 44] and [22, p. 27]). Now suppose that $\alpha : I \to X$ is any curve in X, where $I = [0, 1]$. Let $W = I = [0, 1]$, $X = [\alpha] = \alpha(I)$ and $E = \{\alpha(0), \alpha(1)\}$. Then, by the theorem stated above, there is a subcontinuum K of $[\alpha]$ that joins the two endpoints of α, and that is minimal in the given sense. Of course, K is a set (not a function) but it is hard to imagine that K can be anything but the carrier $[\beta]$ of some Jordan arc that joins $\alpha(0)$ to $\alpha(1)$ in $[\alpha]$.

12   Concluding Remarks

Our first objective in this essay has been to explore those topological ideas that seem to be relevant to mathematical economics, and we have also tried to place the ideas from economics in the wider topological context. Of course, we have paid no attention to the purely economic aspects, and even less to the applications, of these ideas. However, the real reason for writing this essay is to raise the general question of whether or not an approach that is guided primarily by (or even determined entirely by) topological considerations could possibly lead to a more coherent development of the theory of preference relations. Of course, we have only just begun to look at the economics from this perspective, and we leave others to consider, and judge, how far we might take this idea for themselves.

References

1. Banerjee, K., Mitra, T.: On Wold's approach to Representation of Preferences. L. Math. Econ. (2018)
2. Beardon, A.F.: On the continuity of monotonic functions. Am. Math. Mon. **74**, 314–315 (1967)

3. Beardon, A.F.: Debreu's Gap theorem. Econ. Theory **2**, 150–152 (1992)
4. Beardon, A.F., Candeal, J.C., Herden, G., Induráin, E., Mehta, G.B.: The non-existence of a utility function and the structure of non-representable preference relations. J. Math. Econ. **37**, 17–38 (2002)
5. Beardon, A.F., Candeal, J.C., Herden, G., Induráin, E., Mehta, G.B.: Lexicographic decomposition of chains and the concept of a planar chain. J. Math. Econ. **37**, 95–104 (2002)
6. Beardon, A.F., Mehta, G.: The utility theorems of Wold, Debreu and Arrow-Hahn. Econometrica **62**, 181–186 (1994)
7. Bridges, D.S., Mehta, G.B.: Representations of Preference Orderings. Springer, Berlin (1995)
8. Debreu, G.: Representation of a preference ordering by a numerical function. In: Thrall, Coombs, Davis (eds.) Decision Processes. Wiley, New York (1954)
9. Eilenberg, S.: Ordered topological spaces. Am. J. Math. **63**, 39–45 (1941)
10. Halmos, P.R.: Naive Set Theory. Van Nostrand, New York (1960)
11. Hayman, W.K.: Subharmonic Functions, vol. 1. Academic Press, London (1976)
12. Herden, G., Mehta, G.B.: The Debreu Gap Lamma and some generalizations. J. Math. Econ. **40**, 747–769 (2004)
13. Hocking, J.G., Young, G.S.: Topology. Addison-Wesley, Reading (1961)
14. Jaffray, J.-Y.: Existence of a continuous utility function: an elementary proof. Econometrica **43**, 981–983 (1975)
15. Kelley, J.L.: General Topology. Van Nostrand, Princeton (1955)
16. Kerékjártó, B.v.: Vorlesungen über Topologie I. Springer, Berlin (1923)
17. Milnor, J.: Dynamics in One Complex Variable. Annals of Mathematics Studies, vol. 160, 3rd edn. Princeton University Press, Princeton (2006)
18. Monteiro, P.K.: Some results on the existence of utility functions on path connected spaces. J. Math. Econ. **16**, 147–156 (1987)
19. Nachbin, L.: Topology and Order. Van Nostrand, New York (1965)
20. Newman, M.H.A.: Elements of the Topology of Plane Sets of Points. Cambridge University Press, Cambridge (1964)
21. Whyburn, G.T.: Analytic Topology. American Mathematical Society Colloquium Publications XXVIII. American Mathematical Society, New York (1942)
22. Whyburn, G.T.: Topological Analysis, Princeton Mathematical Studies 23. Princeton University Press, New Jersey (1964)
23. Wold, H.: A synthesis of pure demand analysis I, II and III. Scandinavisk Aktuarietidskrift 26, 85–118, 220–263, 69–120, 1943–1944

The Existence and the Non-existence of Utility Functions in Order-Theoretic, Algebraic and Topological Environments

Juan C. Candeal

Abstract This paper reviews both the representation problem and the non-representation problem in utility theory. Beginning with the classical ordinal approach, I then move to also consider certain algebraic and topological environments. Some applications to social choice and measurement theory of the results presented are shown too.

Keywords Utility · Topological and algebraic ordered structures · Social choice · Measurement theory

Mathematics Subject Classification (2010) 54F05 · 06A05 · 91B16 · 91B14

1 Introduction

This article aims to review certain developments in mathematical utility theory related to the so-called representation and non-representation problems of totally preordered structures in a variety of mathematical environments.[1]

In addition to excellent books and papers which encompass the basics on utility theory (e.g., Luce and Raiffa [65] and Fishburn [44]), it is an undeniable fact that the contribution of Bridges and Mehta [15] marks a before and after concerning the theory of numerical representation of ordered structures.[2]

---

[1]Needeless to say that the material presented here reflects a personal view about the problems considered.

[2]Of course, there exists other relevant literature on this topic such as the paper by Mehta [72] and the books of Aleskerov at al. [3] and Ok [76] just to name a few. It is my modest intention with this paper to somehow contribute to complete the information provided in all these references.

J. C. Candeal (✉)
Departamento de Análisis Económico, Facultad de Economía y Empresa, Universidad de Zaragoza, Gran Vía, 2, 50005 Zaragoza, Spain
e-mail: candeal@unizar.es

© Springer Nature Switzerland AG 2020
G. Bosi et al. (eds.), *Mathematical Topics on Representations of Ordered Structures and Utility Theory*, Studies in Systems, Decision and Control 263,
https://doi.org/10.1007/978-3-030-34226-5_2

23

Two remarks about the material included below need to be pointed out. Firstly, and due to the fact that utility theory is a multi-disciplinary subject, I must report that the emphasis here is mainly on the mathematical aspects of this discipline that could be of interest for decision-making theorists, and economists. Thus, the mathematical fundamentals of utility theory, together with some theoretical applications, forms the main body of this article.[3] Secondly, the survey is just focused on a particular kind of ordered structures; to wit, total preorders. This means that no information at all is provided concerning other orderings such as quasi-orders, interval-orders or semi-orders. Relevant material about the representation problem in the latter ordered structures can be found in the books by Bridges and Mehta [15] and Aleskerov et al. [3] (see, also, Campión and Induráin [23], this volume).

Here is a brief outline of the contents of the paper. Section 2 includes some basic definitions about ordered structures that will be used later. Both the representation problem and the non-representation problem for total preorders are presented in Sect. 3. The material collected here includes a very well-known result about the existence of a utility function due to Birkhoff [8] and Debreu [37] as well as a more recent one which is based upon a familiar property of the order topology; namely, the second countability axiom. In relation with the non-representation problem, I present a slightly tighter version of the so-called structure theorem for non-representable chains as shown by Beardon et al. [6]. Related literature, in particular the one linking the material presented here with lexicographic representations, is also discussed.

The representation problem and the non-representation problem in an algebraic environment are introduced in Sect. 4. Although there are different algebraic ordered structures that could be considered, because of the potencial applications in economics and social sciences, I have focused on totally preordered real vector spaces. This requires a certain compatibility between the order and algebra. In particular, the total preorder has to be translation-invariant and homothetic. Then, the well-known result that characterizes the existence of a linear utility function, due to Birkhoff [8], is established in terms of the fulfilment of a very popular mathematical property; to wit, Archimedeaness. The non-representation problem in this algebraic context addresses the consequences of the non-existence of a linear utility function for a translation-invariant and homothetic total preorder defined on a real vector space. Quite surprisingly, this fact entails the existence of subsets of the given vector space that behave as the lexicographic plane. In particular, for a translation-invariant and homothetic total preorder the existence of a utility function implies (actually, is equivalent to) the existence of a linear utility function.

The topological approach is considered in Sect. 5. Whenever a topological space is given it seems natural to study which total preorders defined on it admit a continuous utility function. This is exactly the continuous representation problem. Because of the applications in optimization theory and economics, the resolution of this problem has been a cornerstone in utility theory. A first result was given by Debreu [38] who

---

[3] In addition to also covering theoretical aspects of utility theory, the book of Ok [76] offers interesting applications in economics and decision theory.

made use of his well-known open-gap lemma. Basically, Debreu's result tells that a total preorder defined on a topological space admits a continuous utility function if and only if it is continuous and representable. Continuity turns out to be a minimal (necessary) condition for the continuous representation problem meets. Thus, for continuous total preorders, the continuous representation problem amounts to the representation problem. The fact that continuity of a total preorder suffices for the continuous representation problem to meet, depends upon the accomplishment of nice topological properties of the space at hand. This is confirmed by the two most important results of the topological approach to utility theory; namely, Eilenberg's theorem and Debreu's theorem. The abstraction of this property gave rise to the concept of a useful topology due to Herden [55]. The main achievements related to this topic, called the continuous representation property, are also presented, including certain generalization to the semi-continuous case. Then an important result due to Estévez and Hervés [43] is used for a double purpose. On the one hand, to gain certain insights in relation to the fulfilment of the continuous representation property and, on the other hand, to discuss the non-representation problem for continuous total preorders. As to the latter problem it is shown that, for path-connected topological spaces, the only possibility for a non-representable continuous total preorder is to contain a copy of the first uncountable ordinal number.

The combination of the three approaches; namely, ordinal, algebraic and topological is discussed in Sect. 6 in the context of totally preordered topological real vector spaces. The representation problem here seeks for conditions that guarantee the existence of a continuous linear utility function. Given the results of the preceding sections, the intuition tells that, under suitable mild hypotheses, the non-representation problem in this setting is void. More precisely, any continuous and translation-invariant total preorder defined on a topological real vector space can be represented by a continuous linear utility function; i.e., by a function belonging to its topological dual space. Because of the interest in decision-making under risk and uncertainty, some results in cones are presented too.

A miscellany of theoretical applications of the previous results, shown in Sect. 7, ends the paper. In particular, two concerning the role played by utility theory in social choice and one which emphasizes the importance of order-theoretic principles in measurement theory. The ground set in the three cases considered is the Euclidean space. The first has to do with the representation of a social welfare functional by means of a numerical function. The second includes certain results about both lexicographic-type and dictatorial orderings that appear in the literature on utility theory in social choice. The third shows the resolution of a functional inequality involving a (psycho-physical) law which transforms interval scales into ordinal scales.

2 Preliminaries

Let X be a nonempty set. Denote by $\mathcal{B}$ the set which consists of all binary relations defined on X. For a given $R \in \mathcal{B}$ there are associated three binary relations called the *asymmetric part*, the *symmetric part*, and the *transpose* of R. They are denoted by R^a, R^s, and R^t, respectively, and are defined in the following way: $x\ R^a\ y$ if and only if $x\ R\ y$ and $\neg(y\ R\ x)$; $x\ R^s\ y$ if and only if $x\ R\ y$ and $y\ R\ x$; $x\ R^t\ y$ if and only if $y\ R\ x$. A binary relation $R \in \mathcal{B}$ is said to be *non-trivial* provided that $R^a \neq \emptyset$. Otherwise, it is said to be *trivial*. A transitive and complete (hence, reflexive) binary relation $R \in \mathcal{B}$ is said to be a *total preorder*. A *total order* is an antisymmetric total preorder.

Let there be given two nonempty sets X, Y equipped with binary relations R and S, respectively.

(i) If R and S are total orders, then the *lexicographic product* of X and Y is meant to be the Cartesian product $X \times Y$ endowed with the following total order: $(x_1, y_1)\ R S_l$ $(x_2, y_2) \iff x_1\ R^a\ x_2$ or $x_1 = x_2$ and $y_1\ S\ y_2$. This concept can be easily translated to include a larger (possibly infinite) number of factors, say $(X_i, R_i)_{i \in I}$, provided that the set of indexes I is well-ordered.

(ii) A map $u : X \longrightarrow Y$ is called an *isotony* provided that $x_1\ R\ x_2 \iff u(x_1)\ S$ $u(x_2)$, for all $x_1, x_2 \in X$. An *order-isomorphism* is a surjective isotony.

Let $R \in \mathcal{B}$. Then the family of all sets of the form $L(x) = \{y \in X : y\ R^a\ x\}$ and $G(x) = \{y \in X : x\ R^a\ y\}$, where x runs over X, is a subbasis for a topology τ_R on X, called the *order topology*.

3 The Ordinal Approach

I begin by presenting *the representation problem* in utility theory.[4] Let X be a nonempty set. The representation problem is as follows: Given $R \in \mathcal{B}$, when there is $u : X \longrightarrow \mathbb{R}$ such that $x\ R\ y \iff u(x) \leq u(y)$? If such a function u does exist, then R is said to be *representable* and u is a *utility function* for R. Alternatively, the terms *order-preserving function* or, *numerical representation* of R, will be also used for u.

In simple words, the representation problem consists in characterizing the binary relations defined on X that can be represented by a utility function. In order to study this problem the following notations will be helpful. Let $\mathcal{P}$ stand for the class of total preorders defined on X, and let $\mathcal{P}_r$ be the class of representable binary relations on X. It is an obvious fact that $\mathcal{P}_r \subseteq \mathcal{P}$. A pair (X, R), where $R \in \mathcal{B}$ is a total preorder (respectively, order) on X, is said to be a *totally preordered* (respectively, *ordered*) *set*. A totally preordered set (X, R) is representable whenever so is R. The next definition provides the key to get at the desired characterization.

---

[4]Certain results included in this section are also reviewed in the papers by Campión and Induráin [23], and Alcantud and Giarlotta [2], both appearing in this volume.

Definition 3.1 A binary relation $R \in \mathcal{B}$ is said to be *perfectly separable* whenever there is a countable subset $D \subseteq X$ in such a way that for any pair of elements $x, y \in X$, with $x\ R^a\ y$, there is $d \in D$ such that $x\ R\ d\ R\ y$.

It should be noted that the fact of being a perfectly separable binary relation on X turns out to be a purely ordinal condition. The following result offers an answer to the representation problem. Actually, it encompasses the class of binary relations whose *order-types* are subsets of the real line.

Theorem 3.2 ([8, 37]) $R \in \mathcal{P}_r$ *if and only if it is a perfectly separable total preorder.*

Remark 3.3 (i) Other order-countability conditions, which are equivalent to perfect separability, that characterize the representability of a binary relation are given in Proposition 1.4.4 of Bridges and Mehta [15]. By the way, in Bridges and Mehta [15] a perfectly separable total preorder on X is termed *order separable in the sense of Debreu*.

(ii) The following remarkable fact, due to Cantor [31, 32], is a direct consequence of Theorem 3.2: Any total preorder defined on a countable set is representable.

An alternative characterization of the representation problem is shown below. It is based on a familiar topological axiom referred to the order topology τ_R associated with R.

Theorem 3.4 Let $R \in \mathcal{P}$. Then $R \in \mathcal{P}_r$ if and only if τ_R is second countable.

I now move towards the so-called *non-representation problem*. This problem, that was formulated in Beardon et al. [6], is motivated by the following remark. First, note that if a binary relation R on X is representable, then, from an ordinal point of view, X can be identified with a subset of the real line. The question is: What about if it fails to be representable? In order the non-representation problem to be meaningful it will be assumed that $R \in \mathcal{P}$. The two canonical examples of non-representable total preorders are the lexicographic plane $(\mathbb{R}^2, \leq_l)$, and the first uncountable ordinal $(\omega_1, \leq_1)$. Actually, $\leq_l$ and $\leq_1$ are total orders on $\mathbb{R}^2$ and ω_1, respectively.[5] Now, it turns out that it is extremely difficult to find other non-representable totally preordered sets "essentially different" from the two just-mentioned. In order to obtain a structure theorem for non-representable totally preordered sets, Beardon et al. [6] introduced the concepts that appear in the upcoming definition. Let there be given two totally ordered sets (X, R) and (Y, S). The notation $Y \subseteq X$ means that X contains a copy of Y; that is, there is an isotony $\phi : Y \to X$. In addition, X^t will be used for X endowed with the transpose binary relation R^t.

---

[5] The lexicographic plane $(\mathbb{R}^2, \leq_l)$ is the lexicographic product of $(\mathbb{R}, \leq)$ with itself. That is, $(x, y) \leq_l (u, v) \iff x < u$ or $x = u$ and $y \leq v$, $(x, y, u, v \in \mathbb{R})$. An *ordinal* is a well-ordered set (X, R) such that each $x \in X$ can be identified with its initial segment $\{y \in X : y\ R^a\ x\}$. The finite ordinals are the natural numbers and the first infinite ordinal, usually denoted by ω, or ω_0, is the set of all natural numbers endowed with the usual order. The first uncountable ordinal, denoted by ω_1, is the set of all countable ordinals endowed with the natural order.

Definition 3.5 An uncountable totally ordered set (X, R) is said to be:

(i) *long* if $\omega_1 \subseteq X$, or $\omega_1^t \subseteq X$,
(ii) *planar* whenever $S \subseteq X$, where $S \subseteq \mathbb{R}^2$ is a non-representable subset of the lexicographic plane,
(iii) *Aronszajn* provided that neither is long nor contains an uncountable representable subset.

Remark 3.6 (i) Note that, although the previous definition is stated for totally ordered sets, it extends in a straightforward manner to totally preordered sets. Actually, the classification of the non-representable totally preordered sets amounts to the classification of the non-representable totally ordered sets. Indeed, given a totally preordered set (X, R) it suffices to consider the quotient set $[X] := X/R^s$ induced by the symmetric relation R^s which, in turn, is an equivalence relation on X. Then $[X]$ is a totally ordered set under the binary relation, also denoted by R, defined as follows: $[x] \, R \, [y] \iff x \, R \, y$, for all $[x], [y] \in X/R^s$. In addition, and by abuse of language, it will be said that R is long or planar or Aronszajn whenever so is $([X], R)$.

(ii) Note that none of the three types of totally ordered sets described in Definition 3.5 contains a copy of the other two. In particular, they are not order-isomorphic.

The following result somehow simplifies the so-called structure theorem for non-representable chains as appears in Beardon et al. [6].

Theorem 3.7 *Let (X, R) be a non-representable totally preordered set. Then R is long or planar or Aronszajn.*

Remark 3.8 (i) Unlike the statement of Theorem 6.7 in Beardon et al. [6], the previous result does not make any reference to a *Souslin* totally ordered set. This is because a Souslin totally ordered set contains a copy of an Aronszajn one.[6]

(ii) Alternative descriptions of non-representable totally ordered sets have been provided by Giarlotta [50] by means of the concept of the representability number of a chain (see also Giarlotta [49]). Other interesting results about non-representable total orders involving nonstandard analysis and mathematical logic can be found in Rizza [79].

(iii) Whereas planar totally ordered sets are usually encountered in economics and decision sciences when ranking alternatives according to hierarchical principles, the other two are unusual in applications. An exception could be the class of long totally ordered sets that appear in information economics (see Dubra and Echenique [39] and Hervés and Monteiro [60]).

---

[6]A totally ordered set that satisfies the countable chain condition (that is, every pairwise disjoint family of open intervals is countable) and contains a non-representable subset that has, at most, countably many jumps is called a Souslin totally ordered set. In Beardon et al. [6, Proposition 6.4] is proved that the non-existence of a Souslin totally ordered set amounts to the Souslin Hypothesis (SH). It is well-known that SH is independent of the axioms of ZFC (Zermelo-Fraenkel axiomatic set theory, plus the Axiom of Choice). Yet, Aronszajn totally ordered sets can be constructed in ZFC (for details, see Beardon et al. [6, Theorem 5.3]).

Lexicographic representations of totally ordered sets have shown up in the mathematical literature since long. For instance, the following classical result holds.

Theorem 3.9 ([33–35, 73, 80]) *Let (X, R) be a totally ordered set and let α be the first ordinal whose cardinality equals the cardinality of X. Then $X \subseteq \{0, 1\}^\alpha$, where $\{0, 1\}^\alpha$ is endowed with the lexicographic order.*

It is an interesting question to know if a result similar to Theorem 3.9 can be shown involving a smaller ordinal. To this respect the following result holds.

Theorem 3.10 ([82]) *Let (X, R) be a totally ordered set such that τ_R is a separable topology. Then $X \subseteq \mathbb{R}^2$, where $\mathbb{R}^2$ is equipped with the lexicographic ordering.*

The following corollary is an immediate consequence of Theorem 3.10.

Corollary 3.11 *Let (X, R) be a non-representable totally preordered set such that τ_R is a separable topology. Then R is planar and, moreover, is neither long nor Aronszajn.*

Other results similar to the ones above can be found in Ostaszewski [77], Fleisher [46] and Beardon et al. [6, 7]. A seminal article about lexicographic orders is Fishburn [45]. For additional information about this topic and its significance in economics and decision theory see Martínez-Legaz [67] and Knoblauch [62].

4 The Algebraic Approach: Totally Preordered Vector Spaces

In this section an algebraic ingredient is added to the representation problem. Now, the kind of utility functions sought should preserve certain algebraic properties. Suppose that X is endowed with a binary operation "$+$". Then *the algebraic representation problem* consists in characterizing the binary relations defined on X that admit an additive utility function. As far as I know the problem so-stated remains an open problem. Yet, if some structure properties on X, together with certain qualifications related to the compatibility between the order and algebra, are imposed, then a lot of interesting results have been provided in the literature. Because of potencial applications in economics and social sciences I will focus on the case of *real vector spaces*. Nevertheless, there exists an extensive literature on this topic covering other algebraic systems such as semigroup, groups, rings, cones,...etc. Two basic references are Birkhoff [8] and Fuchs [47].

I briefly present the algebraic representation problem in this framework. Let $(X, +, \cdot_\mathbb{R})$ be a real vector space with zero-vector $\mathbf{0}$. A binary relation $R \in \mathcal{B}$ on X is said to be *translation-invariant* provided that $x \mathrel{R} y$ entails $x + z \mathrel{R} y + z$ and $z + x \mathrel{R} z + y$, for every $x, y, z \in X$. It is said to be *homothetic* whenever $x \mathrel{R} y$ entails $\lambda x \mathrel{R} \lambda y$, for every $x, y \in X, 0 \leq \lambda$.

As in the previous section, $\mathcal{P}$ will denote the subset of $\mathcal{B}$ which consists of all total preorders defined on X. Denote also by $\mathcal{T}$ and $\mathcal{H}$ the subsets of $\mathcal{B}$ that consist of translation-invariant and homothetic binary relations on X, respectively. The linear representation problem consists in characterizing the binary relations defined on X that can be represented by a linear utility function. In other words, given $R \in \mathcal{B}$, when there is $u : X \longrightarrow \mathbb{R}$ such that:

(i) $u(x + y) = u(x) + u(y)$, for all $x, y \in X$,
(ii) $u(\lambda x) = \lambda u(x)$, for all $x \in X$, $\lambda \in \mathbb{R}$,
(iii) $x \, R \, y \iff u(x) \leq u(y)$?

If such a function u does exist, then R is said to be *linearly representable* and u is a linear representation of R. A 4-tuple $(X, +, \cdot_{\mathbb{R}}, R)$, $R \in \mathcal{THP} := \mathcal{T} \cap \mathcal{H} \cap \mathcal{P}$, is said to be a *totally preordered real vector space*. A totally preordered real vector space $(X, +, \cdot_{\mathbb{R}}, R)$ is said to be linearly representable whenever so is R.

The notation $\mathcal{P}_{lr}$ will stand for the class of binary relations on X that are linearly representable. Note that if a binary relation on X admits a linear utility function then it belongs to $\mathcal{THP}$. The following definition provides the key to establish such a characterization.

Definition 4.1 Let $(X, +, \cdot_{\mathbb{R}})$ be a real vector space. A binary relation R on X is said to be *Archimedean* if for every $x, y \in X$ with $\mathbf{0} \, R^a \, x \, R^a \, y$, there is $\lambda > 0$ such that $y \, R^a \, \lambda x$.

Denote by $\mathcal{A}$ the set of binary relations defined on X which are Archimedean and by $\mathcal{ATHP} := \mathcal{A} \cap \mathcal{THP}$. A characterization of the linear representation problem is shown next.

Theorem 4.2 ([8]) *$R \in \mathcal{P}_{lr}$ if and only if $R \in \mathcal{ATHP}$.*

Following the spirit of Theorem 3.4 an alternative characterization of the class $\mathcal{P}_{lr}$, which involves properties of the order topology τ_R, is offered below. Recall that a *topological real vector space* is a vector space equipped with a Hausdorff topology for which the two binary operations "$+$" and "$\cdot_{\mathbb{R}}$" are continuous.

Theorem 4.3 *Let $R \in \mathcal{THP}$. Then $R \in \mathcal{P}_{lr}$ if and only if τ_R makes $(X, +, \cdot_{\mathbb{R}})$ to be a topological real vector space.*

Remark 4.4 Theorem 4.3 was proven in Candeal et al. [30]. It has the following interesting consequence: The only topological totally ordered real vector spaces, up to order-preserving algebraic isomorphism, are $\mathbb{R}$, endowed with the usual order $\leq$, and $\mathbb{R}^t$, i.e., the reals with the transpose order.

The non-representation problem in this setting concerns the structure implications on a totally preordered real vector space $(X, +, \cdot_{\mathbb{R}}, R)$ whenever a linear utility function fails to exist. The next result tells that this fact is linked with the existence of planar subsets. In particular, it will be shown that $\mathbb{R}^2 \subseteq [X]$, where $\mathbb{R}^2$ is endowed with the lexicographic order $\leq_l$. Recall that $[X] := X / R^s$, and it is simple to see that

$[X]$ agrees with the algebraic quotient $X/R^s(\mathbf{0})$, where $R^s(\mathbf{0}) := \{x \in X : x \, R^s \, \mathbf{0}\}$ which is a vector subspace of X. Actually, $([X], +, \cdot_{\mathbb{R}}, R)$ turns out to be a *totally ordered real vector space*. Thus, in this algebraic context, the content of Theorem 3.7 makes narrower.

Theorem 4.5 *Let $(X, +, \cdot_{\mathbb{R}}, R)$ be a non-representable totally preordered real vector space. Then R is planar.*

Sketch of proof. Assume that there does not exist a linear utility function that represents X. According to Theorem 4.2 it amounts to say that R fails to be Archimedean. Thus, there are $x, y \in X$, with $\mathbf{0} \, R^a \, x \, R^a \, y$, in such a way that $\lambda x \, R^a \, y$, for all $\lambda > 0$. Consider the subspace of X generated by x and y; i.e., $S(x, y) := \{\alpha x + \beta y : \alpha, \beta \in \mathbb{R}\}$. Note that the algebraic dimension of $S(x, y)$ is two since x and y are linearly independent. It remains to prove that $(S(x, y), R)$ is order-isomorphic to $(\mathbb{R}^2, \leq_l)$. By using translation-invariance and homotheticity of R, it can be shown that it suffices to prove that, if $\delta > 0$, then $\mathbf{0} \, R^a \, \delta y + \gamma x$, for all $\gamma \in \mathbb{R}$. Suppose, by way of contradiction, that there is $\gamma_0 \in \mathbb{R}$ such that $\delta y + \gamma_0 x \, R \, \mathbf{0}$. Then, clearly, $\gamma_0 < 0$. Thus, by translation-invariance and homotheticity again, it holds that $y \, R \, (-\frac{\gamma_0}{\delta})x$. However, $-\frac{\gamma_0}{\delta} > 0$, which contradicts the fact that $\lambda x \, R^a \, y$, for all $\lambda > 0$. Actually, the function $\phi : S(x, y) \to \mathbb{R}^2$ given by $\phi(\alpha x + \beta y) = (\beta, \alpha)$, provides the desired order-isomorphism. $\qquad\square$

Two immediate consequences of Theorem 4.5 are now in order. The first one can be viewed as a version of the well-known Debreu's open-gap lemma in this algebraic context. A more detailed discussion involving this important result is shown in Sect. 5.

Corollary 4.6 *Let $(X, +, \cdot_{\mathbb{R}}, R)$ be a totally preordered real vector space. Then R is representable if and only if it is linearly representable.*

The second one is a refinement of Theorem 4.5.

Corollary 4.7 *Let $(X, +, \cdot_{\mathbb{R}}, R)$ be a finite-dimensional non-representable totally preordered real vector space. Then R is planar and, moreover, is neither long nor Aronszajn.*

Remark 4.8 (i) In addition to be planar, an infinite-dimensional non-representable totally preordered real vector space $(X, +, \cdot_{\mathbb{R}}, R)$ can also be long or Aronszajn or both. In order to illustrate these three situations the notation $\leq_l$ will be used to denote the (natural) lexicographic total order on the corresponding ground set. To argue the first case, consider the set $X = \mathbb{R}^{\omega_1}$ endowed with usual operations $+$ and $\cdot_{\mathbb{R}}$ defined coordinatewise, and the natural lexicographic ordering $\leq_l$. Then, it is clear that $\leq_l$ is planar and long.

For the second, let (Z, R) be an Aronszajn totally ordered set and consider the space $X = c_{00}(Z; \mathbb{R})$ which consists of all the real-valued functions defined on Z with finite support; i.e., $c_{00}(Z; \mathbb{R}) := \{f : Z \longrightarrow \mathbb{R} : f|_{Z \setminus I_f} \equiv 0$, where I_f is a finite subset of $Z\}$. Equipped with the usual operations $+$ and $\cdot_{\mathbb{R}}$ defined for real-valued functions, X is a real vector space. Define the following extension of the lexicographic total order on the Euclidean spaces to this scenario: For any $f, g \in X$, $f \leq_l g$ if and only if $f|_{I_f \cup I_g} \leq_l g|_{I_f \cup I_g}$, where the latter symbol $\leq_l$ stands for the lexicographic order on $\mathbb{R}^{I_f \cup I_g}$. Note that it makes sense because $I_f \cup I_g$ is a totally ordered finite set. Then, it can be easily proven that $\leq_l$ so-defined is a translation-invariant and homothetic total order on X. Clearly, it is planar and Aronszajn.

Finally, for the third, it suffices to consider the space $X = \mathbb{R}^{\omega_1} \times c_{00}(Z; \mathbb{R})$ endowed with the natural lexicographic order $\leq_l$ derived from the two cases just shown. Then, $\leq_l$ is a translation-invariant and homothetic total order that is planar, long and Aronszajn.

(ii) Interesting classical embedding results from totally ordered real vector spaces into a suitable lexicographic product of totally ordered groups can be found in the literature. For example, it is worth mentioning here the well-known Hahn's embedding theorem [51] (see also Hausner and Wendel [52] and Erdős [41]).

5   The Topological Approach: Totally Preordered Topological Spaces

Suppose now that a nonempty set X is equipped with a topology τ; that is, (X, τ) is a *topological space*. In this situation it seems natural to seek for utility representations that preserve certain topological properties, the most natural being continuity. Here, continuity refers to the topology τ on X and the Euclidean topology on the reals. This kind of utility representations impose certain natural restrictions between the order and topology. A binary relation $R \in \mathcal{B}$ on X is said to be *continuous* provided that, for each $x \in X$, the sets $L(x) := \{y \in X : y \ R^a \ x\}$ and $G(x) := \{y \in X : x \ R^a \ y\}$ are open of τ.

Thus, *the continuous representation problem* consists in characterizing the binary relations defined on X that can be represented by a continuous utility function. That is, let there be given a topological space (X, τ) and a binary relation R on X. The question now is: When there is a continuous order-preserving function $u : X \to \mathbb{R}$? If such a function u exists, then R is said to be *continuously representable* and u is a continuous representation of R.

Following similar notations to the ones used in the previous sections, let $\mathcal{C}$ denote the class of continuous binary relations defined on X. As above, $\mathcal{P}$ stands for the class of total preorders defined on X. A triple (X, τ, R), $R \in \mathcal{CP} := \mathcal{C} \cap \mathcal{P}$, is said to be a *totally preordered topological space*. A totally preordered topological space (X, τ, R) is said to be continuously representable whenever so is R.

The notation $\mathcal{P}_{cr}$ will stand for the class of binary relations on X that are continuously representable. Note that if a binary relation on X admits a continuous utility function then it belongs to $\mathcal{CP}$. The continuous representation problem is a fundamental question in utility theory. Its resolution is, in a sense, surprising and has to do with what is called in the literature as Debreu's open-gap lemma. There are different proofs of this result and other versions of it, none of them being easy (see Debreu [38], Bowen [14] and Beardon [4]). Maybe, the simplest approach to this result can be found in the recent contribution by Ouwehand [78]. A thorough discussion of Debreu's open-gap lemma is shown in Bridges and Mehta [15] and Mehta [72]. Further generalizations of this result are those by Herden and Mehta [58] and, more recently, Estevan [42].

Theorem 5.1 ([38]) $\mathcal{P}_{cr} = \mathcal{C} \bigcap \mathcal{P}_r$.

Remark 5.2 (i) In words, Theorem 5.1 tells that a binary relation on X is continuously representable if and only if it is continuous and representable. In particular, it is a total preorder on X. Apparently, the sentence looks like a play on words. However, it is not. Whereas the content "$\subseteq$" in the statement is obvious, the other is tricky. Therefore, it is a consequence of the previous result the fact that, for continuous total preorders, the continuous representation problem amounts to the representation problem.

(ii) For the particular case $\tau = \tau_R$, the previous result, in combination with Theorem 3.4, tells that a continuous total preorder R on X is continuously representable if and only if τ_R is second countable.

(iii) Because of the applications in economics and social sciences, a weaker form of the continuous representation problem has also been considered in the literature. This form is called *the semi-continuous representation problem*. Basically, it studies when a semi-continuous, lower or upper, utility function does exist. Recall that, if (X, τ) is a topological space, then a binary relation $R \in \mathcal{B}$ on X is said to be *lower semi-continuous* (respectively, *upper semi-continuous*) provided that, for each $x \in X$, the sets $L(x) := \{y \in X : y \ R^a \ x\}$ (respectively, $G(x) := \{y \in X : x \ R^a \ y\}$) are τ-open.

(iv) There is a large amount of literature about topology and order being Nachbin [75] a basic reference. Articles much more oriented with the approach followed here can be found in Mehta [68–71], Herden [53, 54], Herden and Mehta [56, 57] and Bosi and Mehta [12].

(v) In the context of Euclidean spaces and normed linear spaces specific geometric methods to construct continuous utility functions for continuous total preorders that satisfy additional nice properties such as monotonicity, convexity and so on, have been developed. For example, the Wold, Arrow-Hahn and money-metric approaches just to mention a few (see Beardon and Mehta [5], Mehta [72] and Alcantud and Mehta [1]).

A different, but closely related to the representation problem, is the so-called *continuous representation property*. This property was first introduced by Herden [53, 54] under the name of a *useful topology* and has focused the attention of some researchers during the last years. In order to explain it, I first recall the two most fundamental results of the topological approach to utility theory. These are Eilenberg's theorem and Debreu's theorem.

Theorem 5.3 ([40]) *Let (X, τ) be a connected and separable topological space. Then $\mathcal{CP} \subseteq \mathcal{P}_{cr}$.*

Theorem 5.4 ([38]) *Let* (X, τ) *be a second countable topological space. Then* $\mathcal{CP} \subseteq \mathcal{P}_{cr}$.

A way of viewing the contents of the two results above is as follows: Both statements establish sufficient topological conditions on (X, τ) that guarantee the existence of a continuous utility function for any continuous total preorder defined on X. Herden [55] himself realized this fact and defined a topology τ on a set X to be useful whenever any continuous total preorder defined on X can be represented by a continuous utility function. Thus, for a useful topology, the continuous representation problem fulfils for any continuous total preorder. In this way, both Theorems 5.3 and 5.4 can be re-phrased by saying that connected and separable, and second countable, respectively, topologies have the continuous representation property. Other interesting results which ensure that the latter property holds whenever the topology meets certain familiar topological properties can be seen in Campión et al. [22].

The continuous representation property has been recently characterized. In Campión et al. [20] a characterization is provided in terms of the fulfilment of the second countability axiom for every preorderable subtopology of the given topology. More recently, in Bosi and Herden [11] this property is characterized in terms of the fulfilment of the second countability axiom for certain topologies generated by complete separable systems. Complete separable systems can be considered as specializations of Nachbin-Urysohn type families of open subsets which are related to separation axioms in ordered topological spaces (see also Burgess and Fitzpatrick [16] and Herden and Pallack [59]).

Remark 5.5 (i) The continuous representation property has also been studied in algebraic contexts (see, e.g., Campión et al. [22] and Candeal et al. [30]).

(ii) Likewise the semi-continuous representation problem, there is an analogue for the continuous representation property in the semi-continuous case. It is usually referred to in the literature as *the semi-continuous representation property*. For details, see Bosi and Herden [10] and Campión et al. [19].

What about the non-representation problem in the topological context? In other words, what can be said if a continuous complete preorder defined on a topological space fails to be representable and which are the implications of Theorem 3.7 in this case. In this respect, the following important contribution was made by Estévez and Hervés [43].

Theorem 5.6 ([43]) *Let* (X, τ) *be a non-separable metric space. Then there is* $R \in \mathcal{CP} \setminus \mathcal{P}_{cr}$.

The proof provided in Estévez and Hervés paper is based upon the following nice idea. Under the conditions of the theorem they are able to construct a continuous function from X into the long line L, equipped with its order topology, whose range fills ω_1.[7] Then, the binary relation defined on X is the one given by the corresponding

---

[7]The long line L is the lexicographic product of $(\omega_1, \leq_1)$ and $([0, 1), \leq)$. Alternatively, L can be defined in the following way: Between each ordinal α and its successor $\alpha + 1$ put one copy of the real interval $(0, 1)$. Then L is the space obtained in this way endowed with its natural order.

images of the elements in L. Clearly, the binary relation so-defined turns out to be a (non-representable) long continuous total preorder on X.

Estévez and Hervés theorem has interesting applications. For example, in combination with Theorem 5.4, it allows for characterizing the continuous representation property in metric spaces in the following neat way: A metric space satisfies the continuous representation property if and only if it is second countable (see Candeal et al. [27]). Of course, this covers the cases of normed linear spaces and Banach spaces that so often appear in practise. Yet, there are other important (non-metrizable) topologies in the context of topological vector spaces for which the latter result does not apply. That is the case, for instance, for the weak topology of a Banach space. Nevertheless, by using specific techniques of functional analysis and topology, Campión et al. [17, 18] proved that the weak topology of a Banach space satisfies the continuous representation property. By abstracting the idea underlying in Estévez and Hervés proof, the following interesting result holds: A locally convex topological vector space satisfies the continuous representation property if and only if it satisfies *the countable chain condition* (see Candeal et al. [30]). Recall that a topological space satisfies the countable chain condition if every family of pairwise disjoint open subsets is countable. It is of some interest to know whether or not the idea given above can be further used to construct non-representable continuous total preorders that are planar or Aronszajn.

As discussed in Remark 4.8(i) in the algebraic context, also in the topological setting a non-representable continuous total preorder can be planar, long and Aronszajn. Yet, if the space is path-connected, then the following result holds:

Theorem 5.7 *Let (X, τ) be a path-connected topological space and $R \in \mathcal{CP} \setminus \mathcal{P}_{cr}$. Then R is long and, moreover, is neither planar nor Aronszajn.*

Sketch of proof. Suppose first that R has a least element. Then, because (X, τ) is a path-connected topological space and $R \in \mathcal{CP} \setminus \mathcal{P}_{cr}$, it follows, by a result of Campión et al. [17], that there is a continuous isotony $u : X \longrightarrow L$. Now, Monteiro [74] proved that in a path-connected topological space a continuous total preorder is representable if and only if it is countably bounded.[8] Therefore, R is not countably bounded and so it is long. Clearly, R is neither planar nor Aronszajn. The case in which R has not a least element is similar to that just shown but now taking into account that the codomain of the function u is $L + L^t$; i.e., the ordinal sum of L and L^t. $\qquad\square$

Remark 5.8 (i) In simple words, Theorem 5.7 ensures that in a path-connected topological space the only non-representable continuous total preorders are long. In particular, in the context of topological vector spaces, no non-representable continuous total preorders can be found that are planar or Aronszajn.

---

[8]Countably bounded means that there is a countable subset $D \subseteq X$ that bounds, from above and below, X according to R. That is, for any $x \in X$ there are $d_1, d_2 \in D$ such that $d_1 \, R \, x \, R \, d_2$.

(ii) Quite surprisingly, the situation described in the result above does not hold in the semi-continuous case. Indeed, in Campión et al. [21] it is shown that in any non-separable full[9] metric space there is a planar semi-continuous, lower or upper, total preorder.

6   When Order, Algebra and Topology Meet: Totally Preordered Topological Vector Spaces

In this section a suitable mixture of the approaches seen in the previous sections is considered. According to preceding results, the intuition is that if order, algebra and topology are encountered, then the existence of a utility function satisfying suitable algebraic and topological properties is guaranteed. In particular, the non-representation problem in this scenario is void. I will focus here in the context of totally preordered topological vector spaces including a reference to certain cones contained in these spaces.

Assume $(X, +, \cdot_{\mathbb{R}}, \tau)$ is a topological real vector space and let $R \in \mathcal{B}$ a binary relation on X. Then *the continuous linear representation problem* is established in the following terms: When R can be represented by a continuous linear utility function? If such a function does exist, then R is said to be *continuously linearly representable* and the corresponding function is a continuous linear representation of R. A 5-tuple $(X, +, \cdot_{\mathbb{R}}, \tau, R)$, where $(X, +, \cdot_{\mathbb{R}}, \tau)$ is a topological real vector space and R is a continuous, translation-invariant and homothetic total preorder on X, is said to be a *totally preordered topological real vector space*. A totally preordered topological real vector space $(X, +, \cdot_{\mathbb{R}}, \tau, R)$ is said to be continuously linearly representable whenever so is R.

The notation $\mathcal{P}_{clr}$ will stand for the class of binary relations on X that are continuously linearly representable. Note that if a binary relation on X admits a continuous linear utility function then it belongs to $\mathcal{CTHP} := \mathcal{C} \bigcap \mathcal{THP}$.

The next result gives a precise account of the continuous linear representation problem in the context of topological real vector spaces. Recall that $\mathcal{CTP} := \mathcal{CP} \bigcap \mathcal{T}$.

Theorem 6.1 *Let $(X, +, \cdot_{\mathbb{R}}, \tau)$ be a topological real vector space and let $R \in \mathcal{B}$. Then the following assertions are equivalent:*

(i) $R \in \mathcal{P}_{clr}$,
(ii) $R \in \mathcal{CTP}$,
(iii) $(X, +, \cdot_{\mathbb{R}}, \tau, R)$ is a totally preordered topological real vector space.

Sketch of proof. I only will prove (ii) implies (i). If R is trivial, then $u \equiv 0$ provides the desired utility function. Assume that R is a non-trivial continuous and

---

[9]A metric space (X, d) is full if for every $x, y \in X$ and $\delta > 0$ with $0 < \delta < d(x, y)$, there is $z \in X$ such that $d(x, z) = \delta$.

translation-invariant total preorder and suppose, by way of contradiction, that it fails to be linearly representable. Then, by Theorem 4.5, R is planar. In particular, there is a 2-dimensional subspace $S \subseteq X$ such that $(S, R |_S)$ is order-isomorphic to $(\mathbb{R}^2, \leq_l)$ (cf. proof of Theorem 4.5). Now, since $(X, +, \cdot_{\mathbb{R}}, \tau)$ is a topological real vector space, it is well-known that $\tau|_S$ can be identified with the Euclidean topology. In other words, S and $\mathbb{R}^2$ are order-isomorphic, topologically homeomorphic and algebraically isomorphic. In particular, since R is continuous, this would mean that the lexicographic total order on $\mathbb{R}^2$ would be continuous with respect to the Euclidean topology which is clearly false. Thus, a contradiction is reached and, therefore, R is linearly representable. It remains to see that a linear representation, say u, of R is also continuous. To that end, let $\alpha \in \mathbb{R}$ and consider the set $u^{-1}(\infty, \alpha)$. Note that, because R is non-trivial and u is linear, there is $x_0 \in X$ such that $u(x_0) = 1$. Then, $u^{-1}(\infty, \alpha) = \{x \in X : x \, R^a \, \alpha x_0\}$, which is an open subset of X since R is continuous. In a similar manner, it can be proven that $u^{-1}(\alpha, \infty)$ is open in X. Therefore, u is a linear and continuous utility function that represents R. $\qquad\square$

Remark 6.2 (i) The proof above shows that, in particular, a continuous and translation-invariant total preorder defined on a topological real vector space turns out to be homothetic and Archimedean.

(ii) The fact that, in the context of topological real vector spaces, $R \in \mathcal{CTP}$ entails $R \in \mathcal{P}_{clr}$ was proven in Candeal and Induráin [28] by using an entirely different approach.

(iii) Theorem 6.1 provides necessary and sufficient conditions for a binary relation defined on a topological real vector space X to admit a utility function that belongs to the topological dual X'.

I conclude this section by presenting a generalization of Theorem 6.1 whenever the underlying ground set is a cone of a topological real vector space. This is important because of the potential applications of these particular sets in contexts such as decision-making under risk and uncertainty. Recall that C is a *cone* of a real vector space $(X, +, \cdot_{\mathbb{R}})$ provided that $\alpha x \in C$, for all $\alpha \geq 0$, $x \in C$. In particular, $\mathbf{0} \in C$. If, in addition, C is a convex set, then C is said to be a *convex cone*. It is easy to see that $C \subseteq X$ is a convex cone whenever $\alpha x + \beta y \in C$, for all $\alpha, \beta \geq 0$, $x, y \in C$. Note that, for a binary relation R defined on a cone C, the concept of homotheticity (and that of translation-invariance whenever it is convex) remains the same because the operation "$\cdot_{\mathbb{R}}$" (and the operation "$+$" if it is convex) is algebraically closed in C.

Let C be a cone. A function $u : C \longrightarrow \mathbb{R}$ is said to be *homogeneous of degree one* provided that $u(\alpha x) = \alpha u(x)$, for all $\alpha \geq 0$, $x \in C$. If C is a convex cone, then u is said to be *semi-linear* whenever $u(\alpha x + \beta y) = \alpha u(x) + \beta u(y)$, for all $\alpha, \beta \geq 0$, $x, y \in C$. Recall that $\mathcal{CHP} := \mathcal{CP} \bigcap \mathcal{H}$.

The following result appears in Bosi et al. [9].

Theorem 6.3 *Let C be a cone of a topological real vector space and let R be a binary relation defined on C. Then R admits a continuous homogeneous of degree one utility function if and only if $R \in \mathcal{CHP}$.*

In the case of a convex cone, and by using specific tools of topological ordered semigroups, the conclusion of Theorem 6.3 strengthens. The next theorem, which generalizes Theorem 6.1, is a direct consequence of a general result established in Candeal et al. [26].

Theorem 6.4 *Let C be a convex cone of a topological real vector space and let R be a binary relation defined on C. Then R admits a continuous semi-linear utility function if and only if $R \in \mathcal{CTP}$.*

7   A Miscellany of Applications

In this section certain applications of the former results, which emphasize the importance of the representation problem and the non-representation problem in practise, are shown. These applications fall down in the framework of social choice theory and measurement theory. The first result has to do with the representation of social choice rules by means of binary relations and numerical functions. Before some concepts and notations need to be introduced.

Let there be given a nonempty set X and a natural number $n > 1$.[10] A typical (utility) function, from X to $\mathbb{R}$, for the agent j, will be denoted by u_j. A utility function reflects the preference of the corresponding individual when comparing elements of the choice set X. The set of all real-valued functions from X to $\mathbb{R}$ will be denoted by $\mathcal{U}$. A *profile* of (utility) functions will be denoted by $U = (u_j)$ and $\mathcal{U}^n$ will stand for the set of all possible profiles. For a given $x \in X$ and $U = (u_j) \in \mathcal{U}^n$, $U(x)$ will denote the following vector of individual utilities, $U(x) = (u_j(x)) \in \mathbb{R}^n$ $(j = 1, \ldots, n)$.

A *social welfare functional*[11] is a rule $F : \mathcal{U}^n \to \mathcal{P}$ that assigns a social preference $F(U) \in \mathcal{P}$ to any profile U in the domain $\mathcal{U}^n$.

Definition 7.1 A social welfare functional F satisfies:

(i) *Pareto indifference* if $U(x) = U(y)$, then $x F(U)^s y$, for every $x, y \in X$, $U \in \mathcal{U}^n$,

(ii) *binary independence* if for every $U, V \in \mathcal{U}^n, x, y \in X$ such that $U(x) = V(x)$ and $U(y) = V(y)$ it holds that $x F(U) y$ if and only if $x F(V) y$, and $y F(U) x$ if and only if $y F(V) x$.

A *social evaluation functional* is a map $G : \mathcal{U}^n \to \mathcal{U}$ that assigns a (social) utility function $F(U) \in \mathcal{U}$ to any profile U in the domain $\mathcal{U}^n$. That is, a social evaluation functional is a function mapping profiles of individual utility functions into social

---

[10] X is usually called the set of alternatives in social choice theory and the choice set, or the commodity space, in economics. In addition, in the economics literature a binary relation $R \in \mathcal{P}$ on X is usually referred to as a *preference* and as a *social welfare ordering* in social choice theory. The number n refers to the number of agents in the society.

[11] This concept was introduced by Sen [81] as a primitive for exploring the social choice problem. See, in addition, the seminal book by Luce and Raiffa [65].

utility functions. A social evaluation functional G can be viewed as a social welfare functional such that, for every $U \in \mathcal{U}^n$, $G(U)$ is a social utility function, hence a social ordering, attached to U. So, for a social evaluation functional G, it holds true that if $U \in \mathcal{U}^n$ and $x, y \in X$, then x is less preferred than or indifferent to y, according to $G(U)$, if and only if $G(U)(x) \leq G(U)(y)$.

A social evaluation functional $G : \mathcal{U}^n \to \mathcal{U}$ is said to be *separable* if, for each pair $U, V \in \mathcal{U}^n$, $x, y \in X$ such that $U(x) = V(y)$, it holds that $G(U)(x) = G(V)(y)$.

The next definition introduces the representation problem for a social welfare functional.

Definition 7.2 Let F be a social welfare functional.

(i) A binary relation R on $\mathbb{R}^n$ is said to be a *representation* of F provided that, for all $U \in \mathcal{U}^n$, $x, y \in X$, $x F(U) y$ if and only if $U(x) \, R \, U(y)$,

(ii) A real-valued function $u : \mathbb{R}^n \longrightarrow \mathbb{R}$ is said to be a *numerical representation* of F whenever $x F(U)(y)$ if and only if $u(U(x)) \leq u(U(y))$, for every $U \in \mathcal{U}^n$, $x, y \in X$.

Remark 7.3 If R represents F, then $R \in \mathcal{P}$ (i.e., R is a total preorder on $\mathbb{R}^n$). From now on, I will use the notation $\precsim$ for a total preorder defined on (a subset of) the Euclidean space $\mathbb{R}^n$. The notations $\prec$ and $\sim$, for the asymmetric part and the symmetric part of $\precsim$, respectively, will be used accordingly. This is a typical notation used in economics and social sciences and are referred to as the strict preference and the indifference relations, respectively, associated with $\precsim$.

The next definition shows that it is possible to attach a binary relation with a social welfare functional.

Definition 7.4 Let F be a social welfare functional. Then the binary relation on $\mathbb{R}^n$ induced by F, that will be denoted by R_F, is defined in the following way: Given $a, b \in \mathbb{R}^n$, $a \, R_F \, b$ whenever there are $U \in \mathcal{U}^n$, $x, y \in X$ such that $U(x) = a$, $U(y) = b$ and $x F(U) y$.

The binary relation R_F plays an important role in this context because, as can be easily seen, it is the only binary relation on $\mathbb{R}^n$ that can represent F. A characterization of the existence of a numerical representation of a social welfare functional appears in the next result. It uses the so-called *welfarism theorem* together with Theorem 3.2. The welfarism theorem is the core result of the utilitarian approach to social choice theory. In particular, it allows for reducing the complexity of the social aggregation problem to the study of a single preference relation defined on $\mathbb{R}^n$ (for details, see Bossert and Weymark [13]).

Theorem 7.5 *For a social welfare functional F the following conditions are equivalent:*

(i) *There is a numerical representation of F,*

(ii) *There is a separable social evaluation functional, say G, such that $G(U)$ is a utility function for $F(U)$, for all $U \in \mathcal{U}^n$,*

(iii) F satisfies Pareto indifference, binary independence and R_F is perfectly separable.

Remark 7.6 More general representation results of social choice rules by means of numerical functions can be seen in Candeal and Induráin [29].

The second application I am going to present has to do with the characterization of dictatorial preference relations on $\mathbb{R}^n$. In particular, two results about lexicographic-type preference relations and strongly dictatorial preferences on $\mathbb{R}^n$ are shown. Before some definitions and notations are needed. In addition to the usual binary operations $+$ and $\cdot_\mathbb{R}$ in $\mathbb{R}^n$, the vector multiplication, also defined coordinatewise and denoted by "$*$", will be used. The topology on $\mathbb{R}^n$ that I will refer to is the Euclidean one. The zero-vector of $\mathbb{R}^n$ will be symbolized by $\mathbf{0}$. The set that consists of the n first natural numbers will be denoted by N. Given $a = (a_1, \ldots, a_n)$, $b = (b_1, \ldots, b_n) \in \mathbb{R}^n$, I write $a \le b$ whenever $a_j \le b_j$ for all $j \in N$, $a \ll b$ whenever $a_j < b_j$ for all $j \in N$ and $a < b$ whenever $a \le b$ and, for some $j \in N$, $a_j < b_j$.

Definition 7.7 A preference relation $\precsim$ defined on $\mathbb{R}^n$ is said to be:

(i) *scale-independent* if $a \precsim b, \mathbf{0} \ll c$ implies $c * a \precsim c * b$, for all $a, b, c \in \mathbb{R}^n$,

(ii) *Pareto* if $a \le b$ then $a \precsim b$, for all $a, b \in \mathbb{R}^n$,

(iii) *weak Pareto* if $a \ll b$ then $a \prec b$, for all $a, b \in \mathbb{R}^n$,

(iv) *strong Pareto* if $a < b$ then $a \prec b$, for all $a, b \in \mathbb{R}^n$,

(v) *strongly dictatorial* if there is $i \in N$ such that $a \precsim b$ if and only if $a_i \le b_i$, for all $a = (a_j), b = (b_j) \in \mathbb{R}^n$,

(vi) *lexicographic-type* if there is a permutation π of N such that for any $a = (a_j), b = (b_j) \in \mathbb{R}^n, a \precsim b \iff a = b$, or else there is $i \in N$ such that $a_{\pi(j)} = b_{\pi(j)}$ for all $j < i$ and $a_{\pi(i)} < b_{\pi(i)}$.

Two results are now presented. The first one corresponds to a well-known theorem about lexical individual dictatorship obtained by Gevers [48] and d'Aspremont [36]. The second one characterizes strong dictatorships. Note that, unlike lexicographic-type preferences, in the case of a strongly dictatorial preference on $\mathbb{R}^n$ also "ties" among alternatives are determined by the individual acting as a dictator. Only a proof of the second result is outlined.

Theorem 7.8 *A preference relation on $\mathbb{R}^n$ is lexicographic-type if and only if it is translation-invariant, scale-independent and strong Pareto.*

Remark 7.9 (i) Scale-independence states that the preference between any two utility vectors should not depend upon re-parametrizations of utilities that consist of multiplications by strictly positive real numbers. Translation-invariance is usually termed in social choice theory as *zero-independence*. The interpretation is that, in order to rank two utility vectors, all that matters is the difference between them.

(ii) Remarkably, Theorem 7.8 can be widely generalized. Indeed, it can be proven that any translation-invariant and scale-independent preference relation on $\mathbb{R}^n$ is either trivial or *two-serial*. Intuitively, a preference on $\mathbb{R}^n$ is two-serial whenever exogenously given there are $m \in N$, a subset $M \subseteq N$ with cardinality m, a partition $\{A, B\}$ of M and a permutation π of M in such a way that the preference between any two utility vectors of $\mathbb{R}^n$ is only determined by the coordinates belonging to $\pi(A) \cup \pi(B) \subseteq N$. Then, the final ranking between the two utility vectors is given by a lexicographic-type ordering over the coordinates of $\pi(A)$ and an anti-lexicographic-type ordering over the coordinates of $\pi(B)$. For technical details of this result and their applications to social choice theory see Krause [63] and Candeal [24].

For the case of strongly dictatorial preferences the following result is in order.

Theorem 7.10 *A preference relation on $\mathbb{R}^n$ is strongly dictatorial if and only if it is translation-invariant, scale-independent, Archimedean and weak Pareto.*

Sketch of proof. The "only if" part is straightforward. Thus, suppose that $\precsim$ is a translation-invariant, scale-independent, Archimedean and weak Pareto preference relation on $\mathbb{R}^n$. Note that $\precsim$ is homothetic since it is scale-independent. Then, because $\precsim$ is translation-invariant and Archimedean, by Theorem 4.2 there is a linear utility function, say $\psi : \mathbb{R}^n \longrightarrow \mathbb{R}$, which represents $\precsim$. Thus, ψ is of the form $\psi(a) = \sum_{j=1}^{n} \alpha_j a_j$, for all $a = (a_j)$, where $\alpha_j \geq 0$ for all $j \in N$, with at least one of them being strictly positive because $\precsim$ is weak Pareto. It remains to show that all, but at most one, α_j are zero. Suppose, by way of contradiction, that there exist $r, s \in N$, $r \neq s$, such that $\alpha_r \alpha_s \neq 0$. Consider the vector $b = (b_1, \ldots, b_n) \in \mathbb{R}^n$ defined as follows: $b_r = -\alpha_s$, $b_s = \alpha_r$ and $b_j = 0$ otherwise. Note that $\psi(b) = 0$. Now, because $\precsim$ is scale-independent, it holds that $\psi(c * b) = 0$, for every $\mathbf{0} \ll c$. That is, $-\alpha_s c_r + \alpha_r c_s = 0$, for all $c_r, c_s > 0$. Now, by taking $c_r = \alpha_r$ and $c_s = \alpha_s + \dfrac{1}{\alpha_r}$ the latter equality entails $1 = 0$, a contradiction. Thus, there is a single $i \in N$ such that $\alpha_i > 0$. Therefore, $a \precsim b \Longleftrightarrow \alpha_i a_i \leq \alpha_i b_i \Longleftrightarrow a_i \leq b_i$, for all $a = (a_j), b = (b_j) \in \mathbb{R}^n$, which ends the proof. $\qquad \square$

Remark 7.11 Note that, by Theorem 6.1, Archimedeaness can be replaced with continuity in the statement of Theorem 7.10.

The last application has to do with the resolution of a functional inequality that appears in the context of measurement theory. In particular, an alternative proof, which is based upon order-theoretic principles, of a result by Kim [61] is shown. In the framework of measurement theory it is important to characterize real-valued functions of several variables mapping interval scales into ordinal scales. To this respect, the following condition was introduced in Kim [61]: Let $u : \mathbb{R}^n \longrightarrow \mathbb{R}$. Then u satisfies *Condition 1* whenever $u(a) < u(b)$ if and only if $u(c * a + d) < u(c * b + d)$, for all $a, b, c, d \in \mathbb{R}^n$, with $\mathbf{0} \ll c$.

The following result was proven by Kim by using arguments based on functional equations theory.

Theorem 7.12 ([61]) *Let $u : \mathbb{R}^n \longrightarrow \mathbb{R}$ be a strictly increasing continuous function. Then u satisfies Condition 1 if and only if there are $i \in N$ and a continuous and strictly increasing function $g : \mathbb{R} \longrightarrow \mathbb{R}$ such that $u(a) = g(a_i)$, for all $a = (a_j) \in \mathbb{R}^n$.*

Sketch of proof. The "if part" is obvious. For the other implication, consider the binary relation on $\mathbb{R}^n$, induced by u, defined as follows: $a \precsim_u b \Longleftrightarrow u(a) \leq u(b)$, for all $a, b \in \mathbb{R}^n$. It is clear that $\precsim_u$ turns out to be a weak Pareto continuous total preorder defined on $\mathbb{R}^n$. Moreover, $\precsim_u$ is translation-invariant and scale-independent because u satisfies Kim's Condition 1. Thus, by Theorem 7.10, $\precsim_u$ is strongly dictatorial. Therefore, there is $i \in N$ such that, for any $a = (a_j), b = (b_j) \in \mathbb{R}^n$, it holds that $a \precsim_u b \Longleftrightarrow a_i \leq b_i$. In other words, the projection over the i-component turns out to be a utility function for $\precsim_u$. Note that, obviously, u is also a representation of $\precsim_u$. Now, it is well-known that a utility representation of a preference relation is unique upon a strictly increasing transformation. Thus, there is a strictly increasing function, say $g : \mathbb{R} \longrightarrow \mathbb{R}$, such that $u(a) = g(a_i)$, for all $a = (a_j) \in \mathbb{R}^n$. Finally, note that as u is continuous so is g. $\square$

Remark 7.13 (i) In fact, in Kim [61] it is proven that any non-constant continuous function $u : \mathbb{R}^n \longrightarrow \mathbb{R}$ turns out to be the composition of a projection with a strictly monotone function. Surprisingly enough, this result can be generalized by proving that continuity is an entirely redundant assumption. For details, see Marichal and Mesiar [66] and Candeal [25].

(ii) Mappings, as those described in Theorem 7.12, are linked with the concept of a *comparison meaningful statement* in measurement theory. This concept is used to consider the different kind of transformations among the input and output variables that appear in a lot of scientific and psycho-physical laws (see, e.g., [64]). It is also important in social choice theory where is often interpreted in terms of measurability and the possibility of making inter/intra personal comparisons of well-being among individuals (see Bossert and Weymark [13]).

Acknowledgements Thanks are given to two anonymous referees for their detailed comments and remarks which led to an improvement of the paper. Also, the work of the co-editors of this volume honoring Prof. Ghanshyam Mehta is widely acknowledged.

References

1. Alcantud, J.C.R., Mehta, G.B.: Constructive utility functions on Banach spaces. J. Math. Anal. Appl. **350**, 590–600 (2009)
2. Alcantud, J.C.R., Giarlotta, A.: The interplay between intergenerational justice and mathematical utility theory. In: Bosi, G., Campión, M.J., Candeal, J.C., Induráin, E. (eds) Mathematical Topics on Representations of Ordered Structures and Utility Theory: Essays in Honor of Professor Ghanshyam B. Mehta. Studies in Systems, Decision and Control. Springer, Switzerland (2020)
3. Aleskerov, F., Bouyssou, D., Monjardet, B.: Utility Maximization, Choice and Preference, 2nd edn. Springer, Berlin (2007)
4. Beardon, A.F.: Debreu's gap theorem. Econ. Theory **2**, 150–152 (1992)

5. Beardon, A.F., Mehta, G.B.: The utility theorems of Wold, Debreu and Arrow-Hahn. Econometrica **62**, 181–186 (1994)
6. Beardon, A.F., Candeal, J.C., Herden, G., Induráin, E., Mehta, G.B.: The non-existence of a utility function and the structure of non-representable preference relations. J. Math. Econ. **37**, 17–38 (2002)
7. Beardon, A.F., Candeal, J.C., Herden, G., Induráin, E., Mehta, G.B.: Lexicographic decomposition of chains and the concept of a planar chain. J. Math. Econ. **37**, 95–104 (2002)
8. Birkhoff, G.: Lattice Theory. American Mathematical Society, Providence (1948). (1967 third edition)
9. Bosi, G., Candeal, J.C., Induráin, E.: Continuous representability of homothetic preferences by means of homogeneous utility functions. J. Math. Econ. **33**(3), 291–298 (2000)
10. Bosi, G., Herden, G.: On the structure of completely useful topologies. App. Gen. Topol. **3**(2), 145–167 (2002)
11. Bosi, G., Herden, G.: The structure of useful topologies. J. Math. Econ. **82**, 69–73 (2019)
12. Bosi, G., Mehta, G.B.: Existence of a semicontinuous or continuous utility function: a unified approach and an elementary proof. J. Math. Econ. **38**, 311–328 (2002)
13. Bossert, W., Weymark, J.A.: Utility in social choice. In: Barberà, S., Hammond, P.J., Seidl, Ch. (eds.) Handbook of Utility Theory, vol. 2, pp. 1099–1177. Kluwer Academic Publishers, Amsterdam (2004)
14. Bowen, R.: A new proof of a theorem in utility theory. Int. Econ. Rev. **9**, 374 (1968)
15. Bridges, D.S., Mehta, G.B.: Representations of Preference Orderings. Springer, Berlin (1995)
16. Burgess, D.C.J., Fitzpatrick, M.: On separation axioms for certain types of ordered topological space. Math. Proc. Camb. Philos. Soc. **82**, 59–65 (1977)
17. Campión, M.J., Candeal, J.C., Granero, A.S., Induráin, E.: Ordinal representability in Banach spaces. In: Castillo, J.M.F., Johnson, W.B. (eds.) Methods in Banach Space Theory, pp. 183–196. Cambridge University Press, Cambridge (2006)
18. Campión, M.J., Candeal, J.C., Induráin, E.: The existence of utility functions for weakly continuous preferences on a Banach space. Math. Soc. Sci. **51**, 227–237 (2006)
19. Campión, M.J., Candeal, J.C., Induráin, E.: Semicontinuous order-representability of topological spaces. Bol Soc Mat Mex **3**(15), 81–89 (2009)
20. Campión, M.J., Candeal, J.C., Induráin, E.: Preorderable topologies and order-representability of topological spaces. Topol. Appl. **159**, 2971–2978 (2009)
21. Campión, M.J., Candeal, J.C., Induráin, E.: Semicontinuous planar total preorders on non-separable metric spaces. J. Korean. Math. Soc. **46**, 701–711 (2009)
22. Campión, M.J., Candeal, J.C., Induráin, E., Mehta, G.B.: Continuous order representability properties of topological spaces and algebraic structures. J. Korean. Math. Soc. **49**(3), 449–473 (2012)
23. Campión, M.J., Induráin, E.: Open questions in utility theory. In: Bosi, G., Campión, M.J., Candeal, J.C., Induráin, E. (eds) Mathematical Topics on Representations of Ordered Structures and Utility Theory: Essays in Honor of Professor Ghanshyam B. Mehta. Studies in Systems, Decision and Control. Springer, Switzerland (2020)
24. Candeal, J.C.: Invariance axioms for preferences: applications to social choice theory. Soc. Choice Welf **41**, 453–471 (2013)
25. Candeal, J.C.: Aggregation operators, comparison meaningfulness and social choice theory. J. Math. Psychol. **75**, 19–25 (2016)
26. Candeal, J.C., De Miguel, J.R., Induráin, E.: Expected utility from additive utility on semigroups. Theory Decis. **53**(1), 87–94 (2002)
27. Candeal, J.C., Hervés, C., Induráin, E.: Some results on representation and extension of preferences. J. Math. Econ. **29**, 75–81 (1998)
28. Candeal, J.C., Induráin, E.: A note on linear utility. Econ. Theory **6**(3), 519–522 (1995)
29. Candeal, J.C., Induráin, E.: Point-sensitive aggregation operators: functional equations and applications to social choice. Int. J. Uncertain. Fuzziness Knowl Based Sys. **25**(6), 973–986 (2017)

30. Candeal, J.C., Induráin, E., Sanchis, M.: Order representability in groups and vector spaces. Expo. Math. **30**, 103–123 (2012)
31. Cantor, G.: Beiträge zur begründung der transfinite mengenlehre (I). Math. Ann. **46**, 481–512 (1895)
32. Cantor, G.: Beiträge zur begründung der transfinite mengenlehre (II). Math. Ann. **49**, 207–246 (1897)
33. Chipman, J.: The foundations of utility. Econometrica **28**, 193–224 (1960)
34. Cuesta-Dutari, N.: Teoría decimal de los tipos de orden. Revista Matemática Hispano-Americana **3**(186–205), 242–268 (1943)
35. Cuesta-Dutari, N.: Notas sobre unos trabajos de Sierpiński. Revista Matemática Hispano-Americana **7**, 128–131 (1947)
36. d'Aspremont, C.: Axioms for social welfare orderings. In: Hurwicz, L., Schmeidler, D., Sonnenschein, H. (eds.) Social Goals and Social Organization: Essays in Memory of Elisha Pazner, pp. 19–76. Cambridge University Press, Cambridge (1985)
37. Debreu, G.: Representation of a preference ordering by a numerical function. In: Thrall, R., Coombs, C., Davies, R. (eds.) Decision Processes, pp. 159–165. John Wiley, New York (1954)
38. Debreu, G.: Continuity properties of Paretian utility. Int. Econ. Rev. **5**, 285–293 (1964)
39. Dubra, J., Echenique, F.: Monotone preferences over information. Topics in Theoretical Economics 1. Article 1 (2001)
40. Eilenberg, S.: Ordered topological spaces. Am. J. Math. **63**, 39–45 (1941)
41. Erdős, J.: On the structure of ordered real vector spaces. Publ. Math. Debrecen **4**, 334–343 (1956)
42. Estevan, A.: Generalized Debreu's open gap lemma and continuous representability of biorders. Order **33**, 213–229 (2016)
43. Estévez, M., Hervés, C.: On the existence of continuous preference orderings without utility representation. J. Math. Econ. **24**, 305–309 (1995)
44. Fishburn, P.C.: Utility Theory for Decision-Making. Wiley, New York (1970)
45. Fishburn, P.C.: Lexicographic orders, utilities and decision rules: a survey. Manage. Sci. **20**(11), 1442–1471 (1974)
46. Fleischer, I.: A representation for topologically separable chains. Private commun. (See also Notices Am. Math. Soc. (1989)) **6**(6), 814 (1983)
47. Fuchs, L.: Partially Ordered Algebraical Systems. Pergamon Press, Oxford (1963)
48. Gevers, L.: On interpersonal comparability and social welfare orderings. Econometrica **47**, 75–89 (1979)
49. Giarlotta, A.: Representable lexicographic products. Order **21**(1), 29–41 (2004)
50. Giarlotta, A.: The representability number of a chain. Topol. Appl. **150**(1), 157–177 (2005)
51. Hahn, H.: Uber die nichtarchimedischen grössensysteme. S-B Akad Wiss Wien IIa **116**, 601–655 (1907)
52. Hausner, M., Wendell, J.G.: Ordered vector spaces. Proc. Am. Math. Soc. **3**, 977–982 (1952)
53. Herden, G.: On the existence of utility functions. Math. Soc. Sci. **17**, 297–313 (1989)
54. Herden, G.: On the existence of utility functions II. Math. Soc. Sci. **18**, 109–117 (1989)
55. Herden, G.: Topological spaces for which every continuous total preorder can be represented by a continuous utility function. Math. Soc. Sci. **22**, 123–136 (1991)
56. Herden, G., Mehta, G.B.: The continuous analogue and generalization of the classical Birkhoff-Milgram theorem. Math. Soc. Sci. **28**, 59–66 (1994)
57. Herden, G., Mehta, G.B.: Open gaps, metrization and utility. Econ. Theory **7**, 541–546 (1996)
58. Herden, G., Mehta, G.B.: The Debreu Gap Lemma and some generalizations. J. Math. Econ. **40**(7), 747–769 (2004)
59. Herden, G., Pallack, A.: Useful topologies and separable systems. App. Gen. Topol. **1**(1), 61–82 (2000)
60. Hervés-Beloso, C., Monteiro, P.K.: Strictly monotonic preferences on continuum of goods commodity spaces. J. Math. Econ. **46**, 725–727 (2010)
61. Kim, S.R.: On the possible scientific laws. Math. Soc. Sci. **20**, 19–36 (1990)

62. Knoblauch, V.: Lexicographic orders and preference relations. J. Math. Econ. **34**, 255–267 (2000)
63. Krause, U.: Essentially lexicographic aggregation. Soc. Choice Welf. **12**, 233–244 (1995)
64. Luce, R.D.: A generalization of a theorem of dimensional analysis. J. Math. Psychol. **1**, 278–284 (1964)
65. Luce, R.D., Raiffa, H.: Games and Decisions: Introduction and Critical Survey. Wiley, New York (1957)
66. Marichal, J.L., Mesiar, R.: A complete description of comparison meaningful aggregation functions mapping ordinal scales into an ordinal scale: a state of the art. Aequationes Math. **77**(3), 207–236 (2009)
67. Martínez-Legaz, J.E.: Lexicographic utility and orderings. In: Barberà, S., Hammond, P., Seidl, C. (eds) Handbook of Utility Theory, vol. 1, pp. 345–369. Kluwer Academic Publishers, Dordrecht (1998)
68. Mehta, G.B.: Topological ordered spaces and utility functions. Int. Econ. Rev. **18**, 779–782 (1977)
69. Mehta, G.B.: Ordered topological spaces and the theorems of Debreu and Peleg. Indian J. Pure Appl. Math. **14**, 1174–1182 (1983)
70. Mehta, G.B.: Existence of an order preserving function on a normally preordered space. Bull. Aust. Math. Soc. **34**, 141–147 (1986)
71. Mehta, G.B.: Some general theorems on the existence of order preserving functions. Math. Soc. Sci. **15**, 135–143 (1988)
72. Mehta, G.B.: Preference and utility. In: Barberà, S., Hammond, P., Seidl, C. (eds) Handbook of Utility Theory, vol. 1, pp. 1–47. Kluwer Academic Publishers, Dordrecht (1998)
73. Milgram, A.N.: Partially ordered sets, separating systems and inductiveness. In: Menger, K. (ed.) Reports of a Mathematical Colloquium, second series, No. 1, pp. 18–30. University of Notre Dame, Notre Dame (1939)
74. Monteiro, P.K.: Some results on the existence of utility functions on path-connected spaces. J. Math. Econ. **16**, 147–156 (1987)
75. Nachbin, L.: Topology and Order. Van Nostrand Reinhold, New York (1970)
76. Ok, E.: Real Analysis with Economic Applications. Princeton University Press, New Jersey (2007)
77. Ostaszewski, A.J.: A characterization of compact, separable, ordered spaces. J. Lond. Math. Soc. **7**, 758–760 (1974)
78. Ouwehand, P.: A simple proof of Debreu's gap lemma. ORiON **26**(1), 17–20 (2010)
79. Rizza, D.: Nonstandard utilities for lexicographically decomposable orderings. J. Math. Econ. **60**, 105–109 (2010)
80. Sierpiński, W.: Sur une propriété des ensembles ordonnés. Fundam. Math. **36**, 56–67 (1949)
81. Sen, A.K.: Collective choice and social welfare. Holden-Day, San Francisco (1970)
82. Wakker, P.: Continuity of preference relations for separable topologies. Int Econ Rev **29**, 105–110 (1988)

Open Questions in Utility Theory

María J. Campión and Esteban Indurain

In honour of G. B. Mehta on occasion of his 75th birthday.
Also dedicated to the memory of our colleague Gerhard Herden.

Abstract Throughout this paper, our main idea is to explore different classical questions arising in Utility Theory, with a particular attention to those that lean on numerical representations of preference orderings. We intend to present a survey of open questions in that discipline, also showing the state-of-art of the corresponding literature.

Keywords Utility theory · Ordered structures · Numerical representability · Existence of utility functions · Special kinds of utility functions

1 Introduction

Many researches in the topic of ordered structures and its numerical representations would undoubtedly agree on the fact that the book [17], by D. S. Bridges and G. B. Mehta, published in 1995, is one of the best treatises on this kind of question. In fact, that book has been a cornerstone, the bedside book of many professional colleagues. It covers from different perspectives and alternative points of view, the most important

M. J. Campión
Departamento de Estadística, Informática y Matemáticas, Inarbe (Institute for Advanced Research in Business and Economics), Universidad Pública de Navarra, 31006 Pamplona, Spain
e-mail: mjesus.campion@unavarra.es

E. Indurain (✉)
Departamento de Estadística, Informática y Matemáticas, InaMat (Institute for Advanced Materials), Universidad Pública de Navarra, 31006 Pamplona, Spain
e-mail: steiner@unavarra.es

topics that are in the basis of the search for representations of preferences through utility functions.

Needless to say, not everything has been done. Classical problems, as, e.g., the existence of a numerical representation for a given semiorder, in the classical approach introduced by Scott and Suppes (see [38, 77]), have been solved in the last ten years. But it is also true that some crucial questions still resist to the researchers, and remain as open.

Therefore, it seems necessary to dispose at hand of a commented account of open problems arising in utility theory, at least in all what has to do with the numerical representations of the main different—and classical—kinds of ordered structures, already considered in [17], namely total preorders, interval orders and semiorders.

This is the idea that inspires the present manuscript.

2 Preliminaries

2.1 Basic Definitions

Henceforward X will denote a nonempty set, and $\mathcal{R}$ will stand for a binary relation defined on X, that is $\mathcal{R}$ is a subset of X^2. Given $x, y \in X$, with the standard notation $x\mathcal{R}y$ we mean that $(x, y) \in \mathcal{R} \subseteq X \times X$.

Definition 2.1 A binary relation $\mathcal{R}$ on X is called:

  (i) *Reflexive:* If $x\mathcal{R}x$ holds for every $x \in X$.
 (ii) *Transitive:* If $(x\mathcal{R}y) \wedge (y\mathcal{R}z) \Longrightarrow x\mathcal{R}z$ holds for every $x, y, z \in X$.
(iii) *Total or complete:* If $(x\mathcal{R}y) \vee (y\mathcal{R}x)$ holds for every $x, y \in X$.
 (iv) *Irreflexive*: If for any $x \in X$, it never happens that $x\mathcal{R}x$.
  (v) *Asymmetric:* If for any $x, y \in X$ such that $x\mathcal{R}y$ holds, it never happens that $y\mathcal{R}x$.
 (vi) *Antisymmetric*: If $(x\mathcal{R}y) \wedge (y\mathcal{R}x) \Longrightarrow x = y$ holds for any $x, y \in X$.
(vii) *Negatively transitive*: If $x\mathcal{R}z \Longrightarrow (x\mathcal{R}y) \vee (y\mathcal{R}z)$ holds for all $x, y, z \in X$.

Associated to a binary relation $\mathcal{R}$ on a set X, we consider its *negation* (respectively, its *transpose*) as the binary relation $\mathcal{R}^c$ (respectively, $\mathcal{R}^t$) on X given by $(x, y) \in \mathcal{R}^c \Leftrightarrow (x, y) \notin \mathcal{R}$ for every $x, y \in X$ (respectively, given by $(x, y) \in \mathcal{R}^t \Leftrightarrow (y, x) \in \mathcal{R}$, for every $x, y \in X$). We also define the *adjoint* $\mathcal{R}^a$ of the given relation $\mathcal{R}$, as $\mathcal{R}^a = (\mathcal{R}^t)^c$.

In the particular case in which some *ordering* has been implemented, the standard notation usually changes. We include it here for the sake of completeness, since we will use it throughout the present manuscript.

Definition 2.2 A *preorder* $\precsim$ on X is a reflexive and transitive binary relation defined on X. An antisymmetric preorder is called an *order*, whereas a *total preorder* $\precsim$ is a preorder such that if $x, y \in X$ then $(x \precsim y) \vee (y \precsim x)$ holds. A total order is also known as a *linear order* in this literature.

If $\precsim$ is a preorder on X, then as usual we denote the associated *asymmetric* relation by $\prec$ and the associated *equivalence* relation by $\sim$ and these are defined by $x \prec y \Leftrightarrow (x \precsim y) \wedge \neg(y \precsim x)$ and $x \sim y \Leftrightarrow (x \precsim y) \wedge (y \precsim x)$. (Here the symbol "$\neg$" stands for negation).

Remark 2.3 In the case of a total preorder $\precsim$, we may observe that $x \precsim y \Leftrightarrow (x \prec y) \wedge (x \sim y)$ holds for all $x, y \in X$. Also, in this case the adjoint of $\precsim$ is $\prec$ and vice versa.

Definition 2.4 An *interval order* $\prec$ is an asymmetric binary relation on X such that $(x \prec y) \wedge (z \prec t) \Rightarrow (x \prec t) \vee (z \prec y)$ $(x, y, z, t \in X)$. Its corresponding adjoint will be denoted $\precsim$, so that $a \precsim b \Leftrightarrow \neg(b \prec a)$. This relation $\precsim$ is called the *weak preference* associated to $\prec$. By the way, $\prec$ is also called a *strict preference* defined on X. In addition, the binary relation $\sim$ defined by $a \sim b \Leftrightarrow (a \precsim b) \wedge (b \precsim a)$ is said to be the *indifference* associated to $\prec$.

Remark 2.5 It is well known that given an interval order $\prec$ on a set X, the associated relations $\precsim$ and $\sim$ may fail to be transitive [56–58, 66, 77].

Definition 2.6 An interval order $\prec$ is said to be a *semiorder* if $(x \prec y) \wedge (y \prec z) \Rightarrow (x \prec w) \vee (w \prec z)$ $(x, y, z, w \in X)$. A semiorder $\prec$ is said to be *typical* if its associated weak preference $\precsim$ defined by $x \precsim y \Leftrightarrow \neg(y \prec x)$, $(x, y \in X)$ is *not* a total preorder on X.

2.2 Numerical Representations of Different Kinds of Orderings

Following [9] we introduce the notion of representability for different kinds of orderings.

The underlying idea corresponds to the possibility of converting qualitative scales into quantitative ones. Thus, in a way, a binary relation on a nonempty set X could give us the idea of comparison, preference, choice, better position, etc., depending on the context. For instance, given $a, b \in X$, the fact $a \precsim b$ could be interpreted as "b is at least as good as a". Obviously, it is more practice to compare directly real numbers rather than dealing with qualitative scales. This suggest us to convert, when possible, a qualitative scale into a quantitative or numerical one.

Definition 2.7 A total preorder $\precsim$ on X is called *representable* if there is a real-valued function $u \colon X \to \mathbb{R}$ that is order-preserving, so that, for every $x, y \in X$, it holds that $x \precsim y \Leftrightarrow u(x) \leq u(y)$. The map u is said to be a *utility function* for $\precsim$.[1]

---

[1] In other quite different and multidisciplinary contexts, this is sometimes called an *isotony*, an *entropy function*, a *measurement*, a *score* or an *order-preserving map*. (See e.g. [8, 10, 17, 18, 30, 31, 43, 49, 77]).

An interval order $\prec$ defined on X is said to be *representable* (as an interval order) if there exist two real valued maps $u, v \colon X \longrightarrow \mathbb{R}$ such that $x \prec y \Leftrightarrow v(x) < u(y)$ $(x, y \in X)$. The pair (u, v) is called a *utility pair* representing $\prec$.

A semiorder $\prec$ defined on X is said to be *representable in the sense of Scott and Suppes* if there exists a real-valued map $u \colon X \to \mathbb{R}$ (again called a *utility function*) such that $x \prec y \Leftrightarrow u(x) + 1 < u(y)$ $(x, y \in X)$ (see the seminal reference [77]).

In this case, the pair $(u, 1)$ is said to be a *Scott–Suppes representation* of $\prec$.

Remark 2.8 If (u, v) is a utility pair representing an interval order $\prec$ defined on a set X, it is straightforward to see that $u(x) \leq v(x)$ for every $x \in X$. The real interval $[u(x), v(x)]$, that could degenerate to a point if $u(x) = v(x)$, is said to be the *interval of discrimination or perception corresponding to the element* $x \in X$. And the non-negative real number $v(x) - u(x)$ is said to be the *discrimination threshold for the element* $x \in X$. Notice that these thresholds *depend on the elements of X*. If $x \neq y \in X$ it may happen that $v(x) - u(x) \neq v(y) - u(y)$. In the case of a semiorder that is representable in the sense of Scott and Suppes, the discrimination thresholds are all equal to 1.

There exist interval orders that fail to be representable (as interval orders). Also, there exist semiorders that are not representable in the sense of Scott and Suppes. By the way, since every semiorder is, by definition, a special case of an interval order, it may still happen that some particular semiorder can be represented as an interval order through a utility pair, but it fails to be representable as a semiorder in the sense of Scott and Suppes (see [41, 73] for suitable examples and further information).

Definition 2.9 Associated to an interval order $\prec$ defined on a nonempty set X, we shall consider three new binary relations (see [4, 48, 55, 56]).

These binary relations are said to be the *traces* of $\prec$. They are respectively denoted by $\prec^*$ (*left trace*), $\prec^{**}$ (*right trace*) and $\prec^0$ (*main trace*), and defined as follows: $x \prec^* y \Leftrightarrow x \prec z \precsim y$ for some $z \in X$, and similarly $x \prec^{**} y \Leftrightarrow x \precsim z \prec y$ for some $z \in X$ $(x, y \in X)$. In addition, $x \prec^0 y \Leftrightarrow (x \prec^* y) \vee (x \prec^{**} y)$ $(x, y \in X)$.

Remark 2.10 We denote $x \precsim^* y \Leftrightarrow \neg(y \prec^* x)$, $x \sim^* y \Leftrightarrow x \precsim^* y \precsim^* x$, $x \precsim^{**} y \Leftrightarrow \neg(y \prec^{**} x)$ and $x \sim^{**} y \Leftrightarrow x \precsim^{**} y \precsim^{**} x$, and finally $x \precsim^0 y \Leftrightarrow (x \precsim^* y) \wedge (x \precsim^{**} y)$ and $x \sim^0 y \Leftrightarrow (x \precsim^0 y) \wedge (y \precsim^0 x)$ $(x, y \in X)$. Both the binary relations $\precsim^*$ and $\precsim^{**}$ are total preorders on X. Moreover, the indifference relation $\sim$ associated to the interval order $\prec$ is transitive if and only if $\precsim^*, \precsim^{**}$ and $\precsim$ coincide. In this case $\precsim$ is actually a total preorder on X (see [12, 48, 55, 56, 70, 73]).

In addition, the binary relation $\precsim^0$ allows us to characterize semiorders among interval orders (see [55, 56, 70]).

Indeed, if $\prec$ an interval order on X, then it is a semiorder if and only if $\precsim^0$ is a total preorder on X. A semiorder $\prec$ is not typical if and only if $\precsim^*, \precsim^{**}, \precsim^0$ and $\precsim$ coincide.

Let us recall now some characterizations of the numerical representability of total preorders, interval orders and semiorders.

Definition 2.11 Let X be a nonempty set. A total preorder $\precsim$ defined on X is said to be *perfectly separable* if there exists a countable subset $D \subseteq X$ such that for every $x, y \in X$ with $x \prec y$ there exists $d \in D$ such that $x \precsim d \precsim y$.

An interval order $\prec$ defined on X is said to be *interval order separable* if there exists a countable subset $D \subseteq X$ such that for every $x, y \in X$ with $x \prec y$ there exists $d \in D$ such that $x \precsim^* d \prec y$. Equivalently, for every $x, y \in X$ with $x \prec y$ there exists $c \in D$ such that $x \prec c \precsim^{**} y$ (see [9, 12]).

A semiorder $\prec$ defined on X is said to be *regular with respect to sequences* if for any $x, y \in X$, and sequences $(x_n)_{n\in\mathbb{N}}, (y_n)_{n\in\mathbb{N}} \subseteq X$, none of the situations $x \prec \cdots \prec x_{n+1} \prec x_n \prec \cdots \prec x_1$ and $y_1 \prec \cdots \prec y_n \prec y_{n+1} \prec \cdots \prec y$ may occur.

The following facts are well-known in this literature (see [12, 17, 73]).

Theorem 2.12 *On a nonempty set X the following statements hold:*

(a) A total preorder $\precsim$ is representable if and only if it is perfectly separable.
(b) An interval order $\prec$ is representable if and only if it is interval order separable.

And the following result has recently been proved (see [35, 38]).

Theorem 2.13 *Let X be a nonempty set. Let $\prec$ be a typical semiorder defined on X. Then, $\prec$ is representable in the sense of Scott and Suppes if and only if it is both interval order separable and regular with respect to sequences.*

Remark 2.14 When X is countable, the condition of interval order separability trivially holds. Therefore, a semiorder $\prec$ on a countable set X is representable in the sense of Scott and Suppes if and only if it is regular with respect to sequences. This fact was already known [7, 68].

2.3 Order Topology

Definition 2.15 Let $\prec$ denote an asymmetric binary relation on X. Given $a \in X$ the sets $L(a) = \{t \in X : t \prec a\}$ and $U(a) = \{t \in X : a \prec t\}$ are called, respectively, the *lower and upper contours* of a relative to $\prec$. We say that $\prec$ is *τ-continuous* if for each $a \in X$ the sets $L(a)$ and $U(a)$ are τ-open. (See also [64] for further details). The minimal topology $\tau_\prec$ for which all the contours $L(a)$ and $U(a)$ are open ($a \in X$) is said to be the *topology generated for* $\prec$. When $\prec$ is the asymmetric part of a total preorder $\precsim$, the topology $\tau_\prec$ is also known as the *order topology* associated to $\precsim$. Sometimes it is also denoted by $\tau_{\precsim}$. Notice that a subbasis for this topology is $\{\emptyset\} \cup \{X\} \bigcup\{U(a) : a \in X\} \bigcup\{L(a) : a \in X\}$.

If $\precsim$ is a total preorder on X, a topology τ on X is said to be *natural* as regards $\precsim$ if $\prec$ is τ-continuous, or equivalently, if the topology $\tau_\prec$ is coarser than τ (i.e., τ is finer than $\tau_\prec$. We also say that $\tau_\prec$ is a *subtopology* of τ).

The order topology $\tau_{\precsim}$ associated to a total preorder $\precsim$ defined on a nonempty set X characterizes in topological terms the numerical representability of $\precsim$ through a utility function, as follows.

Theorem 2.16 *A total preorder $\precsim$ on a nonempty set X is representable if and only if its associated order topology $\tau_{\precsim}$ is second countable.*

Proof See Sects. 1.4 and 1.6 in [17]. ☐

2.4 Debreu's Open Gap Lemma and the Continuous Representability of Total Preorders

Debreu's open gap lemma is a powerful tool to obtain continuous representations of a total preorder $\precsim$ defined on a nonempty set X endowed with a natural topology τ. (See Chap. 3 in [17], as well as [25, 46]). To explain this lemma, we introduce the following definition.

Definition 2.17 Let S be a subset of the real line $\mathbb{R}$. A *lacuna L* corresponding to S is a non degenerate interval of $\mathbb{R}$ that has both a lower bound and an upper bound in S and that has no points in common with S. A maximal lacuna is said to be a *Debreu gap* (see e.g. pp. 38 and ff. in [17]).

Lemma 2.18 *If S is a subset of the extended real line $\bar{\mathbb{R}}$, then there exists a strictly increasing map, called a gap function, $g : S \longrightarrow \bar{\mathbb{R}}$ such that all the Debreu gaps of $g(S)$ are open.*

Proof See Sect. 3.1 in [17]. ☐

Theorem 2.19 *A total preorder $\precsim$ on a topological space (X, τ) is representable through a continuous utility function if and only if $\precsim$ is perfectly separable and the topology τ is natural for $\precsim$.*

Proof See Theorem 3.2.9 in [17]. We outline here the main ideas of the proof of this result, for the sake of completeness. Using Debreu's open gap lemma, the classical method to get a continuous real-valued strictly isotone function goes as follows: First, one constructs a strictly isotone function (which may not necessarily be continuous) f, whose existence is guaranteed by part (a) of Theorem 2.12 above (for a further account, see e.g. [8], Theorem 24 on p. 200, or else [17], Theorem 1.4.8 on p. 14). Then Debreu's open gap lemma is applied to find a strictly increasing function $g : f(X) \longrightarrow \mathbb{R}$ such that all the Debreu gaps of $g(f(X))$ are open. Consequently, the composition $F = g \circ f : X \longrightarrow \mathbb{R}$ is also a utility representation for the preorder $\precsim$ defined on X, but now F is continuous with respect to any given natural topology τ on X. ☐

From Theorems 2.19 and 2.12, next Corollary 2.20 is easily achieved. Other curious consequences of these results, concerning the existence of means on totally ordered sets, have been got in [19].

Corollary 2.20 *Let (X, τ) be a topological space. Suppose that X is endowed with a representable total preorder $\precsim$. Assume also that the topology τ is natural for $\precsim$. Then $\precsim$ is continuously representable.*

2.5 The Problem of the Continuous Representability of Interval Orders and Semiorders

Unlike the case of total preorders, no general characterization of the continuous representability of interval orders and semiorders has been achieved yet.

At this stage one may expect some result similar to Theorem 2.19, maybe substituting perfectly separable by interval order separable or perhaps by interval order separable plus regular with respect to sequences. Moreover, we would also need to introduce a concept as natural topology with respect to an interval order or to a semiorder. But, unfortunately, these attempts do not work in general, as proved in [9].

Since in Theorem 2.12 the condition of interval order separability is crucial, and that condition involves the traces, in order to analyze the continuous representability of interval orders it seems reasonable to introduce the following definition, which adapts to interval orders the notion of a natural topology, formerly introduced for total preorders.

Definition 2.21 Let X be a nonempty set endowed with a topology τ. Let $\prec$ be an interval order on X. We say that τ is *natural as regards* $\prec$, if all the upper and lower contour sets of $\prec$, $\prec^*$ and $\prec^{**}$ are τ-open.

The main result in [9], namely Theorem 2.22 below, gives a characterization of the continuous representability of an interval order provided that some additional conditions hold. That is, it does not constitute yet a general characterization of the continuous representability of an interval order. However, up-to-date it is, perhaps, the best achievement in this direction.

Theorem 2.22 *Let $\prec$ denote an interval order defined on a nonempty set X endowed with a topology τ. There exists a pair of continuous functions $f, g : X \to \mathbb{R}$, where on X we consider the topology τ, and $\mathbb{R}$ is endowed with the usual Euclidean one, such that the pair (f, g) represents $\prec$ and in addition f is a representation for the associated trace $\precsim^{**}$ (which is a total preorder) and g is a representation for the total preorder $\precsim^*$ if and only $\prec$ is interval order separable and τ is a natural topology with respect to $\prec$.*

Proof See Theorem 1 in [9].   □

Remark 2.23 Notice that we may have continuous representations (f, g) of an interval order $\prec$ such that f does not represent $\precsim^{**}$ or g fails to represent $\precsim^*$. Furthermore, unlike the situation for total preorders, if an interval order $\prec$ on a topological space (X, τ) admits a continuous representation, the topology τ could still fail to be natural as regards $\prec$. (See [9, 11] for further details).

Other (partial) characterizations of the continuous representability of interval orders provided that some topological condition, different from the fact of the topology being natural with respect to the interval order, is accomplished a priori have been achieved in [13]. Among them, it is noticeable that a full characterization of the continuous representability of an interval order has been obtained provided that the set X is finite. (See Sect. 4.2 in [11]).

Following [11], where the results that follow below in this Sect. 2.5 were already proved, we furnish now several necessary conditions for the representability of an interval order through a pair of continuous real-valued functions, in the general case.

Lemma 2.24 *Let $\prec$ denote an interval order defined on a nonempty set X. Assume that $\prec$ is representable by means of a pair (u, v) of real-valued functions. Then, the following statements hold for any $x, y \in X$:*

  (i) $x \prec^* y \Longrightarrow v(x) < v(y)$,
  (ii) $x \prec^{**} y \Longrightarrow u(x) < u(y)$,
  (iii) $v(x) = v(y) \Longrightarrow x \sim^* y$,
  (iv) $u(x) = u(y) \Longrightarrow x \sim^{**} y$.

Proposition 2.25 *Let (X, τ) be a topological space endowed with an interval order $\prec$. If the interval order is representable through a pair (u, v) of continuous real-valued functions, then the following properties hold:*

  (i) The interval order is τ-continuous.
  (ii) If a net $(x_j)_{j \in J} \subseteq X$ converges to two points $a, b \in X$, then $a \sim^0 b$.
  (iii) If a net $(x_j)_{j \in J} \subseteq X$ converges to $a \in X$, and $b, c \in X$ are such that $x_j \prec b \precsim a$ and $x_j \prec c \precsim a$ for any $j \in J$, then $b \sim^{**} c$.
  (iv) If a net $(x_j)_{j \in J} \subseteq X$ converges to $a \in X$, and $b, c \in X$ are such that $a \precsim b \prec x_j$ and $a \precsim c \prec x_j$ for any $j \in J$, then $b \sim^* c$.

The characterization of the continuous representability of interval orders on a topological space (X, τ) of finite support goes as follows.

Theorem 2.26 *Let (X, τ) be a topological space in which the set X is finite. Let $\prec$ be an interval order defined on X. The following statements are all equivalent:*

  (i) The interval order $\prec$ has a representation by means of a pair (u, v) of continuous real-valued functions.
  (ii) The interval order $\prec$ satisfies the necessary conditions (i)–(iv) introduced in Proposition 2.25.
  (iii) The interval order $\prec$ satisfies the condition (ii) introduced in Proposition 2.25, namely, if a net $(x_j)_{j \in J} \subseteq X$ converges to two points $a, b \in X$, then $a \sim^0 b$.
  (iv) The topology τ is natural with respect to the interval order $\prec$.

The situation for semiorders, looking for continuous Scott–Suppes representations is even more deceiving. Only a few partial results are known (see [28, 53]). Among them, one of the best achievements is the characterization of the continuous representability of semiorders on a topological space (X, τ) in which the set X is finite.

Theorem 2.27 *Let (X, τ) stand for a topological space in which the set X is finite. Let $\prec$ be a semiorder on X. Then $\prec$ admits a continuous Scott–Suppes representation if and only if for any net $(x_j)_{j \in J}$ in X it happens that if the net converges to two points $a, b \in X$ then $a \sim^0 b$ holds.*

Proof See Sect. 4 in [53]. □

Matching Theorems 2.26 and 2.27, the following straightforward corollary arises.

Corollary 2.28 *Let (X, τ) be a topological space in which the set X is finite. Let $\prec$ be a semiorder defined on X. The following statements are equivalent:*

(i) *The binary relation $\prec$ admits a continuous representation as an interval order, through a pair (f, g) of continuous real-valued functions $f, g : X \to \mathbb{R}$ such that $a \prec b \Leftrightarrow g(a) < f(b)$ holds for every $a, b \in X$.*

(ii) *The binary relation $\prec$ admits a continuous representation as a semiorder, by means of a continuous function $u : X \to \mathbb{R}$ such that $a \prec b \Leftrightarrow u(a) + 1 < u(b)$ holds for every $a, b \in X$.*

3 Open Questions Related to Numerical Representability

In this section we start with the task of furnishing the reader an explicit list of open questions related to utility theory, paying a particular attention to the numerical representability (without additional conditions such as, e.g., continuity), passing then to give an account of open questions concerning the continuous representability of several kinds of ordered structures on a topological space.

3.1 Open Problems About Numerical Representability of Binary Relations

In general, most of the open questions relative to the numerical representations of binary relations appear when dealing with additional structures, in particular the topological (e.g.: concerning continuity) and the algebraic ones.

However, there are still some open questions that appear in this setting, without any additional structure on a given nonempty set X, but related to the numerical representation of some binary relation $\mathcal{R}$ (that is, a qualitative scale on X), provided that it satisfies some natural restriction (e.g.: acyclicity).

Definition 3.1 A binary relation $\mathcal{R}$ on a nonempty set X is said to be *acyclic* if for any $(x_1, \ldots, x_n) \in X^n$ such that $x_i \mathcal{R} x_{i+1}$ $(i = 1, \ldots, n-1)$ it holds that $(x_n, x_1) \notin \mathcal{R}$ (in other words, $x_n \mathcal{R} x_1$ cannot hold).

Problem 3.2 It is an open question to find some kind of numerical representation that could convert and (re)-interpret an *acyclic binary relation* (as a qualitative scale) into one or more numerical or quantitative scales. Some partial results in this direction were issued in [3].

Problem 3.3 Other open questions follow just the opposite order of ideas in Problem 3.2. That is, suppose that we have one or more numerical scales of a certain type, and we want to identify exactly which type of binary relation they represent. To put just an example, bearing in mind the definition of a representable interval order, suppose that we are given three real-valued functions $f, g, h : X \longrightarrow \mathbb{R}$, where X is a nonempty set. We wonder which are the main abstract properties of the binary relation $\mathcal{R}$ defined by $x\mathcal{R}y \Leftrightarrow f(x) < g(y) < h(x)$ $(x, y \in X)$. Conversely, given a binary relation $\mathcal{R}$ on X we may ask ourselves if there exists a trio of real-valued functions $f, g, h : X \longrightarrow \mathbb{R}$ such that $x\mathcal{R}y \Leftrightarrow f(x) < g(y) < h(x)$ $(x, y \in X)$ holds.

Some results in a close direction appeared in [79], but, in general, problems of this kind remain still open.

Another interesting source of open problems comes to our mind when we analyze non-representability. To fix our ideas, suppose that we are studying the numerical representability of total preorders through a real-valued utility function. Due to the characterization stated in Theorem 2.12, one may give examples of non-representable total preorders: consider for instance the plane $\mathbb{R}^2$ endowed with the lexicographic ordering $\precsim_L$ given by $(a, b) \precsim_L (c, d) \Leftrightarrow (a < c) \vee [(a = c) \wedge (b \leq d)]$ $(a, b, c, d \in \mathbb{R})$ (see e.g. pp. 10 and ff. in [17] for further details). At this point we may wonder if there some *minimal* non-representable total preorder (up-to-isomorphisms) such that any non-representable total preorder should contain an isotonic copy of it. Those minimal non-representable totally preordered structures (if there is any) would then act as *germs* of non-representability. Each time that a total preorder contains a copy of one of those germs, we could immediately say that it fails to be representable.

For the case of total orders this problem was studied in depth in [6], and the corresponding main germs of non-representability were identified. These are the lexicographic plane, as well as, on the one hand, the so-called long line and on the other hand the Aronszajn chains (see [6] for definitions and further details).

Problem 3.4

(i) Analyze the existence of germs of non-representability for interval orders.
(ii) Analyze the existence of germs of non-representability (in the sense of Scott and Suppes) for semiorders.

3.2 Continuous Utility Functions That Represent Different Kinds of Orderings: The State-of-Art

The study of the existence of continuous representations of different kinds of orderings is a substantial part of the analysis of the interaction between Order and Topology. It is important to realize that when we have a nonempty set X endowed with a a topology τ and some ordering of a certain class (e.g.: a total preorder $\precsim$), the topological properties of τ may force the ordering to have some special features, and vice versa. A pioneer work on this interaction was due to Leopoldo Nachbin in the 1950s but published as a book much later (see [71]), and nowadays it is a classical reference in this literature. Conceptually, it is quite important to realize that having at hand a set X endowed with two structures of different kind (namely, its topology and ordering) may give rise to new situations in which it could not be enough to study those structures (topology vs. order) independently, as watertight compartments or pigeon holes.

In addition, the mere existence of a continuous representation for a binary relation $\mathcal{R}$ defined on a topological space (X, τ) may also induce on both τ and $\mathcal{R}$ some restrictive additional properties. To put an example, and as a straightforward exercise, consider that $\mathcal{R}$ has a numerical representation through a real-valued function $F : X \to \mathbb{R}$ such that $a\mathcal{R}b \Leftrightarrow F(a) \leq F(b)$ holds for every $a, b \in X$ and, additionally, F is continuous with respect to the topology τ in X and the usual Euclidean topology on the real line. Then, it easily follows that $\mathcal{R}$ is a total preorder on X that is actually τ-continuous. (In other words, the topology τ is natural as regards the total preorder $\mathcal{R}$). Needless to say, when two or more functions are involved in the numerical representation (as, e.g., when representing interval orders), the situation could be much more subtle and feature many more nuances.

Among the classical orderings that may be defined on a nonempty set X, namely total preorders, interval orders and semiorders, the continuous representability has only been characterized for the case of total preorders (Theorem 2.19 above). In the case of interval orders and semiorders, it has only been characterized for the case in which X is finite (Theorems 2.26 and 2.27 above). Also, for the continuous representability of interval orders, other partial characterizations known use some additional restrictions in their corresponding statements (see e.g. Theorem 2.22 above). They are not valid for the general case.

For other binary relations $\mathcal{R}$ on a topological space (X, τ) the general situation about the possibility of converting the qualitative scale $\mathbb{R}$ into one of more quantitative or numerical ones, by means some ad hoc kind of continuous functions constitutes an open framework of research in this literature.

Problem 3.5 Given a topological space (X, τ) and a binary relation $\mathcal{R}$ defined on it, of particular features, analyze the existence of one or more continuous real-valued functions whose domain is X such that they fully represent $\prec$ someway.

3.3 Continuous Numerical Representability of Interval Orders and Semiorders

As aforesaid in the Sect. 2.5 above, it is still an *open problem* to find a full characterization of the continuous representability of interval orders and semiorders. Only partial results have been got up-to-date.

Moreover, many collateral open questions appear as a by-product when studying those main questions. Consider the following example.

Problem 3.6 Which could be a suitable definition of a natural topology in the case of semiorders? Should we take into account the trace $\precsim^0$?

We explore these questions now, in order to furnish a further account of open problems in this setting.

Problem 3.7

(i) Given a topological space (X, τ) endowed with an interval order $\prec$, characterize the existence of a pair of real-valued functions $f, g : X \to \mathbb{R}$ that are continuous as regards the topology τ on X and the usual one on the real line, and represent $\prec$ so that $x \prec y \Leftrightarrow g(x) < f(y)$ holds for every $x, y \in X$.

(ii) Given a topological space (X, τ) endowed with a semiorder $\prec$, characterize the existence of a real-valued utility function $u : X \to \mathbb{R}$ that is continuous as regards the topology τ on X and the usual one on the real line, and represents $\prec$ in the sense of Scott and Suppes, so that $x \prec y \Leftrightarrow u(x) + 1 < u(y)$ holds for every $x, y \in X$.

(iii) Given a topological space (X, τ) endowed with a semiorder $\prec$, characterize the existence of a pair of real-valued functions $f, g : X \to \mathbb{R}$ that are continuous as regards the topology τ on X and the usual one on the real line, and represent $\prec$ so that $x \prec y \Leftrightarrow g(x) < f(y)$ holds for every $x, y \in X$. (In other words, the semiorder $\prec$, considered just as an interval order, is continuously representable).

(iv) Characterize the topological spaces (X, τ) for which a semiorder $\prec$ that is representable in the sense of Scott and Suppes actually admits a continuous Scott–Suppes representation. (In other words, study the topological spaces $(X, \prec)$ for which a generalization of Corollary 2.20 to semiorders and Scott–Suppes representations is still possible.)

(v) Given a topological space (X, τ) and a semiorder $\prec$ that is representable as an interval order through a pair of real-valued functions $f, g : X \to \mathbb{R}$ that are continuous as regards the topology τ on X and the usual one on the real line, try to find a continuous and strictly increasing function $h : f(X) \cup g(X) \subseteq \mathbb{R} \longrightarrow \mathbb{R}$ such that $h(g(x)) = h(f(x)) + 1$ holds for every $x \in X$.

Remark 3.8 Part (v) of Problem 3.7 is closely related to the study of a classical functional equation on a single real variable, known in the literature as the *Abel's functional equation* (see [1] for further details). Notice here that when the function

h exists, we get a continuous Scott–Suppes representation of the semiorder $\prec$ since, calling $u = h \circ f$, we have that $x \prec y \Leftrightarrow g(x) < f(y) \Leftrightarrow h(g(x)) < h(f(y)) \Leftrightarrow h(f(x)) + 1 < h(f(y)) \Leftrightarrow u(x) + 1 < u(y) \ (x, y \in X)$.

4 The Continuous Representability Properties in General Topology

Given a nonempty set X, in this section we study topologies through the properties that they induce on different kinds of orderings that may be defined on the given set X.

4.1 The Continuous Representability Property (CRP)

Following [20, 23, 24, 26] we introduce the following key definition. (The corresponding concept had also been analyzed in [60, 62], but under the name of a *useful topology*).

Definition 4.1 Let (X, τ) be a topological space. We say that the topology τ satisfies the *continuous representability property* (CRP) if every τ-continuous total preorder $\precsim$ defined on X admits a continuous utility representation by means of an order-preserving real-valued function $u : X \longrightarrow \mathbb{R}$ such that $x \precsim y \Leftrightarrow u(x) \leq u(y)$ holds for every $x, y \in X$, and u is continuous with respect to the topology τ on X and the usual topology on $\mathbb{R}$.

Definition 4.2 Given a topological space (X, τ) we say that the topology τ is *preorderable* (respectively, *orderable*) if it coincides with the order topology $\tau_{\prec}$ of some total preorder (respectively, total order) $\precsim$ defined on X.

Remark 4.3 The orderability of a topology was studied and characterized in [74, 80] in terms of the existence of subbases accomplishing some list of suitable properties. Completing the panorama, the preorderability of a topology was also characterized in Theorem 3.1 (ii) in [24].

Problem 4.4 As mentioned above, topological spaces (X, τ) such that the topology τ is the order topology of some total preorder on X were characterized in [24]. But nothing similar is known for topologies induced by other different kinds of binary relations (not necessarily total preorders), see e.g. [64]. Thus, given an asymmetric binary relation $\mathcal{R}$ on a nonempty set X, we can also consider the sets of the form $U(x) = \{a \in X : x\mathcal{R}a\}(x \in X)$ as well as the sets $L(x) = \{b \in X : b\mathcal{R}x\}(x \in X)$, and consequently define a topology $\tau_\mathcal{R}$ on X, a subbasis of which is given by $\{\emptyset\} \cup \{X\} \bigcup \{U(t) : t \in X\} \bigcup \{L(y) : y \in X\}$.

Here it is no longer true, in general, that the binary relation $\mathcal{R}^a$, adjoint of $\mathcal{R}$, is a total preorder. Important particular cases correspond to interval orders and semiorders. In this framework, it is an open problem to characterize topologies that coincide with the topology $\tau_\prec$ induced by an interval order or a semiorder $\prec$ defined on the given set X.

The property CRP has also been characterized in topological terms as follows.

Theorem 4.5 *Let (X, τ) be a topological space. The topology τ satisfies the continuous representability property if and only if every topology τ' on X such that τ' is coarser than τ (it is also said that τ' is a subtopology of τ), and preorderable, satisfies the second countability axiom.*

Proof See Theorem 5.1 in [24]. □

Problem 4.6 We could also say that a topological space (X, τ) satisfies the *continuous representability property for interval orders* (respectively, *continuous representability property for semiorders*) (CRP-I.O) (respectively (CRP-S)) if for every interval order $\prec$ (respectively, for every semiorder) defined on X and such that all the sets $U(x) = \{a \in X : x\mathcal{R}a\}(x \in X)$ and $L(x) = \{b \in X : b\mathcal{R}x\}(x \in X)$ are τ-open, there exists a continuous representation as an interval order (respectively, as a semiorder, in the sense of Scott and Suppes). The analysis and characterization of topological spaces that satisfy CRP-I.O or CRP-S is still an open problem.

Definition 4.7 A topological space (X, τ) is said to have the *hereditary continuous representation property (HCRP)* if every nonempty subset $Y \subseteq X$ has the continuous representation property (CRP) with respect to the topology τ_Y that τ induces on Y.

Remark 4.8 It is obvious that HCRP implies CRP. But the converse is not true, as proved in Example 5.4 in [22] or Example 4.1 in [26].

Problem 4.9 It is an open question to characterize the topological spaces (X, τ) that satisfy the hereditary continuous representability property (HCRP).

Remark 4.10 Concerning classical topological properties encountered in General Topology, and its relationship with the continuous representability property (CRP), it is well-known that connected plus separable topological spaces satisfy CRP (see [51]). In addition, in [26] is has been proved that locally connected plus separable topological spaces also satisfy the continuous representability property. However, the converses are not true. See also Theorem 4.3 in [27] for additional information.

Problem 4.11 Are there other topological conditions (perhaps more general than separability, connectedness or local connectedness) that imply the continuous representability presentability property?

4.2   The Semicontinuous Representability Property (SRP)

If we pay attention now to semicontinuity instead of continuity, some remarkable subtleties appear. Notice that given a topological space (X, τ) and a total preorder $\precsim$ on X it could still happen that $\precsim$ admits a semicontinuous utility representation and τ fails to be a natural topology for $\precsim$. Indeed, looking at Theorem 2.19, we may observe that in order to pass from a mere utility representation of $\precsim$ to a continuous one, we needed τ to be a natural topology. But this could not be the case now. So, the following problem appears.

Problem 4.12 Let (X, τ) a topological space. Let $\precsim$ be a total preorder defined on X. Assume that the topology τ is not natural with respect to $\precsim$. Is it still possible to find a semicontinuous utility representation for $\precsim$?

Remark 4.13 An important result in this direction was obtained by T. Rader in 1963 (see [69, 75]). Namely, Rader's utility representation theorem guarantees the existence of an upper semicontinuous utility function for any upper semicontinuous total preorder on a second countable topological space.

Let us introduce now some helpful definitions, related to that question.

Definition 4.14 Let (X, τ) be a topological space. A real-valued function $f : X \longrightarrow \mathbb{R}$ is said to be *lower* (respectively *upper*) *semicontinuous* with respect to the topology τ on X and the usual topology on $\mathbb{R}$, if for every $a \in \mathbb{R}$ the set $\{x \in X : a < f(x)\}$ (respectively, the set $\{x \in X : f(x) < a\}$) is τ-open.

Definition 4.15 Let (X, τ) a topological space. Let $\precsim$ be a total preorder defined on X. Then $\precsim$ is said to be τ-*lower* (respectively, τ-*upper*) semicontinuous on X if for each $x \in X$ the set $U(x) = \{y \in X : x \prec y\}$ (respectively, the set $L(x) = \{z \in X : z \prec x\}$ is τ-open. The topology τ on X is said to have the *semicontinuous representability property (SRP)* if every semicontinuous total preorder defined on X admits a representation by means of a semicontinuous order-monomorphism (of the same type of semicontinuity). Topologies satisfying SRP were studied in [15] under a different nomenclature. Then they have been analyzed in [22–24, 27].

Definition 4.16 Let X be a nonempty set. Let $\precsim$ be a total preorder on X. The topology $\tau^l_\precsim$ on X, a subbasis of which is given by $\{\emptyset\} \cup \{X\} \bigcup \{U(x) : x \in X\}$, where $U(x) = \{a \in X : x \prec a\}$ $(x \in X)$ is said to be the *lower order topology* associated to $\precsim$. In the same way, the topology $\tau^u_\precsim$ on X, a subbasis of which is given by $\{\emptyset\} \cup \{X\} \bigcup \{L(x) : x \in X\}$, where $L(x) = \{b \in X : b \prec x\}$ $(x \in X)$, is said to be the *upper order topology* associated to $\precsim$.

Definition 4.17 Let (X, τ) be a topological space. The topology τ is said to be *lower preorderable* if there exists a total preorder $\precsim$ on X such that τ coincides with the lower order topology $\tau^l_\precsim$.

Remark 4.18 As in Remark 4.3 above, the lower preorderability of a topology has already been characterized in terms of suitable bases (see Theorem 3.1 (i) in [24]).

Problem 4.19 Again, topological spaces (X, τ) such that the topology τ is the lower order topology of some total preorder on X have already been characterized in [24]. Also, nothing similar is known for topologies induced by other different kinds of binary relations. Thus, given an asymmetric binary relation $\mathcal{R}$ on a nonempty set X, we can also consider the sets of the form $L(x) = \{b \in X : b\mathcal{R}x\}(x \in X)$, and consequently define a topology $\tau_{\mathcal{R}}^l$ on X, a subbasis of which is given by $\{\emptyset\} \cup \{X\} \bigcup\{L(y) : y \in X\}$. Here it is an open problem to characterize topologies that coincide with the topology $\tau_{\prec}^l$ induced by an interval order or a semiorder $\prec$ on X.

Furthermore, the semicontinuous representability property (SRP) has been characterized as follows.

Theorem 4.20 *Let (X, τ) be a topological space. The topology τ satisfies the semicontinuous representability property if and only if every topology τ' on X such that τ' is coarser than τ and lower preorderable, satisfies the second countability axiom.*

Proof See Theorem 5.1 in [24]. □

Problem 4.21 Here again, we could say that a topological space (X, τ) satisfies the *semicontinuous representability property for interval orders* (respectively, *semicontinuous representability property for semiorders*) (SRP-I.O) (respectively (SRP-S) if for every interval order $\prec$ (respectively, for every semiorder) defined on X and such that all the sets $L(x) = \{b \in X : b\mathcal{R}x\}(x \in X)$ are τ-open, there exists a semicontinuous representation as an interval order (respectively, as a semiorder, in the sense of Scott and Suppes). The study and characterization of the topological spaces (X, τ) that accomplish SRP-I.O or SRP-S is still an open question. Some pioneering ideas in this direction were launched in [16].

Lemma 4.22 *Let X be a nonempty set. Let $\precsim$ denote a total preorder defined on X. Then the order topology $\tau_{\prec}$ is second countable if and only if the lower order topology $\tau_{\prec}^l$ is second countable.*

Proof See Theorem 4.2 in [24]. □

Theorem 4.23 *The semicontinuous representability property (SRP) implies the continuous representability property (CRP).*

Proof Let (X, τ) be a topological space. Suppose that τ satisfies SRP. Take a subtopology of τ that coincides with the order topology $\tau_{\prec}$ of some total preorder $\precsim$ defined on X. By Theorem 4.20, the lower topology $\tau_{\prec}^l$ is second countable. Therefore, by Lemma 4.22, $\tau_{\prec}$ is second countable. Hence τ satisfies the continuous representability property (CRP) by Theorem 4.5. □

Remark 4.24 The converse of Theorem 4.23 is not true in general: Following Corollary 4.2 in [23], to see that the converse may fail, consider the first uncountable ordinal ω_1. Endowed with the lower order topology ω_1 trivially satisfies the continuous representability property (CRP). But is not second countable, so that by Theorem 4.20 it does not accomplishes the semicontinuous representability property (SRP). See [15, 22] for further details.

Problem 4.25 Another appealing problem is to identify the semicontinuous representability property (SRP) among the classical ones, or its variants, in General Topology. In this direction, in Corollary 4.6 and Theorem 4.8 in [23] it has been proved that every second countable topological space (X, τ) satisfies SRP. Also, any topological space that satisfies the semicontinuous representability property SRP is hereditarily separable and hereditarily Lindelöf. The converses are not true. (See also [67] for further details). Thus, SRP can be viewed as a property that falls in the area of separability-countability. It is an open question to identify it exactly with some known property of that kind, or, alternatively, to prove that it is indeed a new intermediate variant.

Theorem 4.26 *The semicontinuous representability property is hereditary. That is, given a topological space (X, τ) that satisfies SRP, and a nonempty subset $Y \subseteq X$, the topological space (Y, τ_Y) also satisfies SRP. (Here τ_Y stands for the topology that τ induces on Y).*

Proof See Corollary 5.2 in [22].   □

Problem 4.27 Since the semicontinuous representability property (SRP) is hereditary, and it implies the continuous representability property (CRP), it immediately follows that SRP carries the hereditary continuous representability property HCRP. It is an open question to study the converse implication. Does HCRP imply SRP or not?

4.3 Yi's Extension Property in Topological Spaces

A classical characterization of normality in topological spaces is the Tietze extension theorem, so that in a normal topological space a continuous real valued map defined on a closed subset has a continuous extension to the whole space.

At this stage, given a topological space (X, τ) we may observe that any continuous real-valued map $f : X \to \mathbb{R}$ immediately defines a continuous total preorder $\precsim$ on X by declaring that $x \precsim y \iff f(x) \le f(y)$ $(x, y \in X)$.

This obvious fact suggests to analyze a generalization of the Tietze's extension property. Consider the following key definition (see [22, 82]).

Definition 4.28 A topological space (X, τ) is said to have *Yi's extension property* provided that any total preorder defined on a closed subset A, and continuous with respect to the relative topology τ_A that τ induces on the subset A, admits a τ-continuous extension to the entire set X.

Remark 4.29 This notion was introduced in an approach of utility theory, concerning the problem of extending total preorders (preferences) from a subset of a topological space to the entire space. It was launched by the Korean mathematical economist Gyoseob Yi.

Definition 4.30 Let (X, τ) be a topological space. (X, τ) is said to be *normal* if for each pair of disjoint τ-closed subsets $A, B \subseteq X$ there exists a pair of disjoint τ-open subsets $A^*, B^* \subseteq X$ such that $A \subseteq A^*, B \subseteq B^*$. (For basic topological definitions see e.g. [50, 52]).

It is well-known that this property of being normal is equivalent to an extension property for continuous real-valued functions. This is the *"Tietze's extension theorem"* (see e.g [81], 15.8).

Theorem 4.31 (Tietze's extension theorem) *Let (X, τ) be a topological space. Then (X, τ) is normal if and only if for every τ-closed subset $A \subseteq X$, each continuous map $f : A \to \mathbb{R}$ admits a continuous extension $F : X \to \mathbb{R}$. Moreover, if $f(X) \subseteq [-a, a]$ for some $a > 0$, then F can be chosen so that $F(X) \subseteq [-a, a]$. (This topological property is known as the Tietze's extension property).*

Now suppose that (X, τ) is a normal topological space. An immediate corollary of Tietze's extension theorem states that continuous and representable preferences defined on closed subsets of X can be continuously extended to the entire set X.

Corollary 4.32 *Let (X, τ) be a normal topological space. Let $S \subseteq X$ be a τ-closed subset of X. Let $\precsim_S$ be a continuous total preorder defined on S. Then if $\precsim_S$ is representable through a continuous utility function $u_S : S \to \mathbb{R}$, it can also be extended to a continuous total preorder $\precsim_X$ defined on the whole X.*

Proof Just observe that, by Tietze's theorem, the utility function u_S admits a continuous extension to a map $u_X : X \to \mathbb{R}$. Then define $\precsim_X$ on X as $x \precsim_X y \iff u_X(x) \leq u_X(y)$ $(x, y \in X)$.   □

It can be seen that Yi's extension property implies Tietze's extension property, but the converse does not hold, in general. Therefore, this new extension property of topological spaces, based on the consideration of total preorders defined on its closed subsets, is actually a more restrictive variant of the normality property. Thus, Yi's extension property was initially understood as an strenghtening of Tietze's extension property, in a direction in which we are not interested in extending utility functions, but only total preorders (usually known as preferences in applied contexts arising, e.g., in Mathematical Economics). Observe that Yi's and Tietze's extension properties are *not* equivalent in the general case. This is because preferences could fail to be representable.

Let us see that, indeed, Yi's extension property is stronger than Tietze's extension property, as claimed before. First we introduce a helpful definition.

Definition 4.33 A topological space (X, τ) is said to be *separably connected* if for every $a, b \in X$ there exists a connected and separable subset $C_{a,b} \subseteq X$ such that $a, b \in C_{a,b}$.

Theorem 4.34 *Let (X, τ) be a topological space that satisfies Yi's extension property. Then (X, τ) is normal. The converse is not true, in general.*

Proof Let $A, B \subseteq X$ be two (nonempty) τ-closed disjoint subsets of X. Let $S = A \cup B$. S is obviously τ-closed. Consider the total preorder $\precsim_S$ defined on S as $a_1 \sim a_2$ for every $a_1, a_2 \in A$, $b_1 \sim b_2$ for every $b_1, b_2 \in B$ and $a \prec b$ for every $a \in A, b \in B$. It is plain that the total preorder $\precsim_S$ is continuous on S. Applying Yi's extension property, there exists a continuous total preorder $\precsim_X$ defined on the whole set X, and extending $\precsim_S$. We distinguish two possible situations:

1. In the first case, we assume that there exists some element $c \in X \setminus S$ such that $a \prec_X c \prec_X b$ for every $a \in A, b \in B$. We observe that $B \subseteq U(c) = \{x \in X : c \prec_X x\}$ and also $A \subseteq L(c) = \{x \in X : x \prec_X c\}$. Since $\precsim_X$ is τ-continuous, the sets $L(c), U(c)$ are τ-open. In addition, they are disjoint by its own definition.
2. Suppose that there is no element $c \in X \setminus S$ such that $a \prec_X c \prec_X b$ for every $a \in A, b \in B$. In this case, if we fix an element $\alpha \in A$ and also an element $\beta \in B$, by definition of $\precsim_S$ we immediately observe that $A \subseteq L(\beta) = \{x \in X : x \prec_X \beta\}$ and in the same way, $B \subseteq U(\alpha) = \{x \in X : \alpha \prec_X x\}$. Since $\precsim_X$ is τ-continuous, the sets $L(\beta), U(\alpha)$ are τ-open. In addition, they are disjoint because $\alpha \prec_S \beta$ by hypothesis.

Thus we see that X is a normal topological space. But normality is a property that is equivalent to Tietze's extension property.

To see that the converse is not true in general, we should have at hand a counterexample. In [37] (see also [22]) it was proved that on separably connected metric spaces Yi's extension property is equivalent to separability. On the other hand, metric spaces are always normal. Thus, an example of a separably connected metric space that is not separable would fit our purposes. This is easy: consider a non-separable Banach space (e.g., $\ell_2(\mathbb{R})$) endowed with its norm topology. □

Problem 4.35 As far as we know, Yi's extension property has not been identified yet as some of the classical notions related to normality encountered in General Topology. The general problem of characterizing those topological spaces that satisfy Yi's extension property for preferences has not been solved yet.

Under several topological restrictions, Yi's extension property also has strong connections with the continuous representability property (CRP) analyzed before. This was studied in [22]. In fact, both properties may be equivalent in some particular cases, as next theorem states.

Theorem 4.36 *On separably connected metric spaces Yi's extension property is equivalent to the continuous representability property CRP.*

Proof See Theorem 3.3 in [22]. □

Remark 4.37 In general topological spaces, Yi's extension property and the continuous representability property (see Example 4.5 in [22]).

Problem 4.38 Is there some other set of topological conditions on a topological space (X, τ) such that under those conditions the Yi's extension property amounts to the continuous representability property CRP?

Finally, suppose that we analyze the analogous of Yi's extension property for the *semicontinuous* case. We would introduce accordingly the following definition.

Definition 4.39 A topological space (X, τ) is said to have the *semicontinuous Yi's extension property* whenever any total preorder defined on a closed subset A and lower semicontinuous with respect to the relative topology τ_A that τ induces on the subset A, admits a τ-lower semicontinuous extension to the entire set X.

At this stage, perhaps surprisingly, it can be proved that any topological space satisfies the semicontinuous Yi's extension property.

Theorem 4.40 *In any topological space semicontinuous extensions of total preorders always exist, even without asking the subsets considered to be closed.*

Proof See Theorem 5.1 in [22]. □

Problem 4.41 In spite of the Yi's extension property falling in the area of separation axioms in General Topology, and, in particular, around normality, it is also true that under other topological restrictions Yi's extension property may also fall in the area of separability-countability. Thus, as aforesaid, in [37] it has been proved that on separably connected metric spaces Yi's extension property amounts to separability. It could be interesting to look for other different families of topological spaces having the same property.

5   Algebraic Representability of Different Kinds of Orderings

5.1   Algebraic Utility

Given a nonempty set X endowed with an ordering of a certain kind, it is common that the underlying set X also has some additional algebraic structure. Just to put an example and fix ideas, consider the Euclidean plane $X = \mathbb{R}^2$ endowed with the lexicographic order $\precsim_L$ given by $(a, b) \precsim_L (c, d) \Leftrightarrow (a < c) \vee [(a = c) \wedge (b \leq d)]$ $(a, b, c, d \in \mathbb{R})$, and a binary operation $\bar{+}$ defined coordinatewise as $(x_1, x_2) \bar{+} (y_1, y_2) = (x_1 + y_1, x_2 + y_2)$, where $(x_1, x_2), (y_1, y_2) \in \mathbb{R}^2$, and $+$ denotes the usual addition of real numbers. Provided that the algebraic operation is compatible with the given ordering, it is then natural to look for numerical representations that not only preserve the ordering, but, in addition, they are also algebraic homomorphism. This is what we would call *algebraic utility*.

As a matter of fact, these kind of contexts were also analyzed in several papers that, having at hand algebraic structures as the main topic of study, considered some additional compatible ordering so that the existence of order-preserving homomorphism was a key question to be analyzed (see e.g. [5, 18, 34, 59, 63]).

In this framework, the algebraic structure and the ordering should feature some sort of compatibility. A suitable starting point here could be the theory of linearly ordered groups, or, more generally, that of totally ordered semigroups, already analyzed in the classical book [59]. Following, [18], we include here some ideas in this direction, for the sake of completeness.

Definition 5.1 A *semigroup* $(S, \circ)$ is a nonempty set S equipped with a binary operation $\circ$ that is associative. If, in addition, S has a *null element* e such that $s \circ e = e \circ s = s$ holds for every $s \in S$, then $(S, \circ)$ is said to be a *monoid*. Furthermore, if each element s of a monoid $(S, \circ)$ has an inverse $\bar{s}$ such that $s \circ \bar{s} = \bar{s} \circ s = e$, then $(S, \circ)$ is said to be a *group*. A *semigroup* $(S, \circ)$ endowed with a total order $\precsim$ is said to be a *totally ordered semigroup* if there is a compatibility between the total order $\precsim$ and the binary operation $\circ$ such that $s \precsim t \Leftrightarrow s \circ u \precsim t \circ u \Leftrightarrow u \circ s \precsim u \circ t$ holds for every $s, t, u \in S$.[2]

Given a totally ordered semigroup[3] $(S, \circ, \precsim)$ an element $s \in S$ is said to be *positive* (respectively, *negative*) if $s \prec s + s$ (respectively, if $s + s \prec s$). The set of positive (respectively, *negative*) elements of S is called the *positive cone* (respectively, *negative cone*), and denoted by S^+ (respectively, S^-). Notice in addition that S may contain an element e that is neither positive nor negative. If this happens, e must be, a fortiori, the null element of $(S, \circ)$ and, in particular, this structure is a monoid. A totally ordered semigroup is said to be *positive* (respectively, *negative*) if it only has positive (respectively, negative) elements.

Finally, a totally ordered semigroup $(S, \circ, \precsim)$ is called *additively representable* if there exists a real-valued function[4] $f : S \to \mathbb{R}$ accomplishing that $f(s \circ t) = f(s) + f(t)$ and all $s \precsim t \Leftrightarrow f(s) \leq f(t)$ hold for every $s, t \in S$.

In order to characterize the totally ordered semigroups that are additively representable, we still need to introduce the following definition.

Definition 5.2 A positive semigroup $(S, \circ, \precsim)$ is said to be *Archimedean* if for every $s, t \in S$ with $s \prec t$, there exists a natural number $n \in \mathbb{N}$ such that $t \prec s^n =$

---

[2]This property is also known as the *translation-invariance* of the total order $\precsim$ as regards the binary operation $\circ$. Notice that, in particular, a totally ordered semigroup is always *cancellative*, namely $s \circ u = t \circ u \Leftrightarrow s = t \Leftrightarrow u \circ s = u \circ t$ $(s, t, u \in S)$.

[3]Despite we are working with totally *ordered* semigroups, it can be proved that we could actually be working with a totally *preordered* semigroup, where $\precsim$ is a total preorder but not necessarily a linear order (i.e.: the binary relation $\precsim$ could fail to be antisymmetric). When this happens, we might pass to be working with a quotient space $S/\sim$ whose elements are the indifference classes of the elements of S with respect to $\sim$. That is, given $s \in S$, its corresponding class is the set $\{t \in S : t \sim s\}$. Provided that there is a compatibility between the total preorder $\precsim$ and the binary operation $\circ$ such that $s \precsim t \Leftrightarrow s \circ u \precsim t \circ u \Leftrightarrow u \circ s \precsim u \circ t$ holds for every $s, t, u \in S$, it is easy to see that $S/\sim$ inherits a structure of totally *ordered* semigroup by considering in a natural way that the binary operation $\circ$ as well as $\precsim$ directly act on the indifference classes that $\sim$ induces on S.

By this reason, in what follows we will be working with totally ordered semigroups, instead of just totally preordered semigroups, unless otherwise stated.

[4]In this setting, a mapping f with these properties is said to be an *additive utility function*.

$s \circ \ldots (n - times) \ldots \circ s$. Moreover, it is called *super-Archimedean* if for every $s, t \in S$ such that $s \prec t$ there exists a natural number $n \in \mathbb{N}$ such that $s^{n+1} \prec t^n$.

In the case of a totally ordered *group*, a classical result coming from Algebra, and stated by Hölder early in 1901, characterizes the additive representability in terms of Archimedeaness, as follows. (See [63] or else [8], p. 300).

Proposition 5.3 *A totally ordered group* $(G, \circ, \precsim)$ *is additively representable if and only if its positive cone is Archimedean.*

In the case of a totally ordered semigroup, the additive representability was characterized by Alimov in 1950, in terms of the super-Archimedean property, as follows (see [5, 45] or else the reference-note 21 on p. 26 in [76]).

Proposition 5.4 *A totally ordered semigroup* $(S, \circ, \precsim)$ *is additively representable if and only if its positive cone* S^+ *is super-Archimedean and its negative cone* S^- *is also super-Archimedean when endowed with the transpose order* $\precsim^t$ *given by* $s \precsim^t u \Leftrightarrow u \precsim s$ $(s, u \in S)$.

Remark 5.5 (i) The Archimedeaness property on positive totally ordered semigroups is not good enough to warrant the existence of an additive representation. An *example* is the semigroup $S = (0 + \infty) \times (0, +\infty)$ of pairs of strictly positive real numbers, endowed with the lexicographic total order and the binary operation $(s, t) \circ (u, v) = (s + u, t + v)$. It is well-known that this ordered set does not admit a numerical representation, even if we disregard additivity (see e.g. [8], pp. 200–201).

(ii) A positive super-Archimedean totally ordered semigroup $(S, \circ, \precsim)$ is, in particular, Archimedean. (For a detailed proof, see e.g., [45].) The converse is not true: indeed $S = (0 + \infty) \times (0, +\infty)$ endowed with the lexicographic ordering $\precsim_L$ fails to be super-Archimedean since $(3, 3) \prec_L (3, 4)$ but $(3k, 4k) \prec_L (3k + 3, 3k + 3)$ $(k \geq 1 \in \mathbb{N})$.

(iii) If $(G, \circ, \precsim)$ is an Archimedean totally ordered group, its positive cone is Archimedean and also its negative cone is Archimedean with respect to the transpose order. (See e.g., [45] for a proof and further details, related properties and comments.)

Here we introduce some open problems in this direction. Needless to say, on most of them they some partial results have already appeared (See e.g. [32, 39, 40, 42, 72]).

Problem 5.6 Let X be a nonempty set endowed with a total preorder $\precsim$ and some kind of algebraic structure that may depend on one or more binary operations defined on X. Indeed some operations could be external. Thus X could be, say, a semigroup, a monoid, a ring, a vector space, an algebra, etc. Depending on the structure considered, which is the most suitable definition of *compatibility* between the total preorder $\precsim$ and the algebraic structure?

Problem 5.7 Let X be a nonempty set endowed with an interval order or a semiorder $\prec$, as well as some kind of algebraic structure. What should be understood as a suitable definition of compatibility or intertwined relationship between the ordering $\prec$ and the algebraic structure we have on X ab initio?

At this respect, it is important to point out that sometimes we get a surprising result, as next theorem states (see e.g. [18]).

Theorem 5.8 *Let $(S, \circ)$ stand for a semigroup. Let $\prec$ be an interval order defined on S. Assume that $\prec$ is translation-invariant (i.e.: $s \prec t \Leftrightarrow s + u \prec t + u \Leftrightarrow u + s \prec u + t$ holds for every $s, t, u \in S$). Then if the interval order $\prec$ admits a representation through a pair of real-valued functions (u, v) such that, in addition, both u and v are additive (i.e., $s \prec t \Leftrightarrow v(s) \prec u(t)$; $u(s \circ t) = u(s) + u(t)$; $v(s \circ t) = v(s) + v(t)$ hold for every $s, t \in S$), then the associated binary relation $\precsim$ (adjoint of $\prec$) is a total preorder. In particular, both $\precsim$ and $\sim$ are reflexive binary relations on S.*

5.2   When Algebra Meets Topology: The Continuous Algebraic Representability Property and Related Items

At this stage a next step is to consider a topological space (X, τ) that is also endowed with an ordering (e.g.: a total preorder $\precsim$) as well as with some algebraic structure. Needless to say, we will also assume some kind of compatibility between the algebraic structure and the ordering, as well as some relationship between the topology with respect to the ordering (e.g., assuming that $\precsim$ is τ-continuous), and also between the algebraic structure and the topology (e.g.: assuming that any binary operation is continuous as regards the product topology $\tau \times \tau$ on X^2 and the given topology τ on X).

To fix ideas, following [18], for the sake of completeness we include now some results on numerical representability of a semigroup $(S, \circ)$ endowed with a compatible total preorder $\precsim$ and a topology τ.

Remark 5.9 Even being representable by an additive utility function, a totally ordered semigroup could fail to admit a *continuous* and additive representation. An *example* is the semigroup $S = [2, 3) \bigcup [4, +\infty) \subset \mathbb{R}$ with the usual sum and order of the reals, and also with the relative topology inherited from the usual one in $\mathbb{R}$. It can bee seen (see e.g. [32, 33]) that the crux for the non-existence of an additive and *continuous* representation is the discontinuity with respect to the order topology, of the sum in S.

Definition 5.10 A *topological semigroup* $(S, \circ, \tau)$ is a semigroup $(S, \circ)$ endowed with a topology τ that is compatible with the binary operation $\circ$ in the following sense: it makes continuous[5] the binary operation $\circ : S \times S \to S$ that maps the pair

---

[5]Here on $S \times S$ we will consider the product topology coming from τ on S.

$(s, t) \in S \times S$ to the element $s \circ t \in S$. A totally ordered semigroup $(S, \circ, \precsim)$ is said to be a *topological totally ordered semigroup* provided that the binary operation $\circ$ is continuous with respect to the order topology $\tau_{\precsim}$ on S (and its corresponding product topology on $S \times S$).

Remark 5.11 Similarly to Definition 5.10, a *topological group* is defined as a group $(G, \circ)$ endowed with a topology τ such that the binary operation $\circ$ is continuous, and also the *unary operation of taking inverses in G*, that assigns to each element $s \in G$ its opposite $\bar{s}$ is a also continuous as regards the topology τ in G.

A famous result due to Nyikos and Reichel (see [72]) proves that *any totally ordered group is topological with respect the order topology.*[6] However, the analogous result for semigroups is no longer valid (the counterexample shown in Remark 5.9 also fits well here).

The following result is perhaps the best achievement in this setting.

Theorem 5.12 *Let $(S, \circ, \precsim)$ be a topological totally ordered semigroup. Suppose in addition that S is given a topology τ that is natural as regards the the linear order $\precsim$. Asume that the positive cone S^+ is super-Archimedean with respect to $\precsim$, whereas the negative cone S^- is super-Archimedean with respect to the transpose order. Then S is representable by an additive real-valued function $u : S \to \mathbb{R}$ such that $u(s \circ t) = u(s) + u(t)$, $s \precsim t \Leftrightarrow u(s) \leq u(t)$ holds for every $s, t \in S$, and the function u is continuous with respect to the given topology τ in S and the usual Euclidean topology in the real line $\mathbb{R}$.*

Proof See Theorem 2 in [32]. □

Problem 5.13 In spite of a few studies having already appeared in this literature, a deep analysis of the numerical representability of several kinds of compatible orderings defined on more sophisticated algebraic structures on a topological space (X, τ) should be considered as a general source of open questions.

To conclude this section, we may consider some algebraic version of the continuous representability property.

Remember that in the topological setting the continuous representability property (CRP) appears whenever we look for ordinal representations of total preorders that, in addition, preserve a nice topological property: namely, the continuity. In the algebraic context, in addition to the order and continuity properties for a utility function, we will ask for a new demanding requirement: that of being an algebraic-homomorphism. Of course, this imposes some kind of compatibility among the ordering, topology, and algebraic structure involved.

To fix ideas, we will include here some results already launched in [27, 39]. In that paper, although one could have begun with some much simpler algebraic structure

---

[6]Notice that this is, so-to-say, a theorem about *"automatic continuity"*. It actually states that on a totally ordered group $(G, \circ, \precsim)$, both the operation $\circ$ and the unary operation of taking an inverse are, a fortiori, *continuous* as regards the order topology $\tau_{\precsim}$.

(e.g.: a totally preordered semigroup) on a topological space (X, τ), the algebraic structures actually considered were much more sophisticated, namely, totally preordered real algebras.

Definition 5.14 A *totally preordered real algebra* $(X, \precsim, +, \cdot_{\mathbb{R}}, *)$ is a real algebra equipped with a total preorder $\precsim$ which is compatible with the operations $+, \cdot_{\mathbb{R}}$ and $*$, i.e.:

(i) $x \precsim y$ implies $x + z \precsim y + z$, for all $z \in X$ (*translation-invariance*).
(ii) $x \precsim y, 0 \leq \lambda$ implies $\lambda \cdot x \precsim \lambda \cdot y$ (*homotheticity*).
(iii) $x \precsim y, 0 \precsim z$ imply $z * x \precsim z*y$ and $x * z \precsim y * z$ (*multiplicativeness*. $\mathbf{0}$ denotes the null element with respect to $+$).

Next we define the notion of a *topological real algebra*.

Definition 5.15 A real algebra $(X, +, \cdot_{\mathbb{R}}, *)$ equipped with a topology τ is said to be a *topological real algebra* if $(X, \tau, +, \cdot_{\mathbb{R}})$ is a topological vector space and $*$ is a τ-continuous binary operation on $X \times X$.

Given a topological space (X, τ) now introduce the concept of continuous representability of total preorders in the algebraic environment, namely the so-called continuous algebraic representability property.

Definition 5.16 Let $(X, +, \cdot_{\mathbb{R}}, *)$ be a real algebra and τ a topology on X. Then τ satisfies the *continuous algebraic representability property* (CARP for short) if every τ-continuous and non-zero[7] total preorder $\precsim$ defined on X, for which the structure $(X, \precsim, +, \cdot_{\mathbb{R}}, *)$ becomes a totally preordered real algebra, admits a continuous utility function which is an algebra-homomorphism (shortly, a continuous algebraic utility function).

The following characterization of the continuous algebraic representation property arises.

Theorem 5.17 *Let $(X, +, \cdot_{\mathbb{R}}, *)$ be a real algebra and τ a topology on X. The the following assertions are equivalent:*

(i) τ has CARP
(ii) $(X, \tau_{\precsim}, +, \cdot_{\mathbb{R}}, *)$ is a topological real algebra, for every non-zero and continuous total preorder $\precsim$ that makes $(X, \precsim, +, \cdot_{\mathbb{R}}, *)$ to be a totally preordered real algebra.

Proof See Theorem 5.12 in [27]. □

If τ makes $(X, +, \cdot_{\mathbb{R}}, *)$ to be a topological real algebra then we have the following interesting consequence.

---

[7]A total preorder $\precsim$ on $(X, +, \cdot_{\mathbb{R}}, *)$ is said to be *non-zero* provided that there are $\bar{x}, \bar{y} \in X$ such that $\mathbf{0} \prec \bar{x} * \bar{y}$.

Corollary 5.18 *Let* $(X, +, \cdot_{\mathbb{R}}, *)$ *be a real algebra. Then any topology* τ *on* X *that makes* $(X, \tau, +, \cdot_{\mathbb{R}}, *)$ *to be a topological real algebra accomplishes the continuous algebraic representability property CARP.*

Remark 5.19 In particular, Corollary 5.18 applies to the n-dimensional Euclidean space $\mathbb{R}^n$ endowed with the usual binary operations defined componentwise. Moreover, in this case it can be easily seen that the continuous algebraic utility functions are of the form $\psi(x_1, \ldots, x_j, \ldots, x_n) = x_j$, for some $j \in \{1, \ldots, n\}$. In other words, any continuous total preorder defined on $\mathbb{R}^n$ which is translation-invariant, homothetic and multiplicative is *projective*.

Problem 5.20 This new approach of continuous algebraic representability properties generates many open questions, such as the following ones:

(i) Explore and analyze continuous algebraic representability properties for total preorders on a topological space (X, τ) endowed with different kinds of compatible algebraic structures, as, e.g., semigroups, monoids, groups, rings, vector spaces or cones.

(ii) Explore and analyze continuous algebraic representability properties on a topological space (X, τ) equipped with some algebraic structure and a compatible interval order and semiorder.

(iii) Define and study concepts relative to semicontinuous algebraic representability properties on a topological space (X, τ) endowed with some algebraic structure and a compatible ordering of a certain type (e.g.: total preorder, interval order or semiorder).

6 Alternative Versions of Utility Representations

6.1 Representations in Codomains Different from the Real Line

Following [36], we may realize that the main emphasis in the utility theory literature, which deals with the study of order-preserving numerical representations of binary relations that model preferences, different kinds of orderings, etc., has been put on the use of real-valued functions. It is implicitly accepted that the almost exclusive use of the real line $\mathbb{R}$ as the codomain on which the order-preserving representations (also known as utility functions) take their values is based on the fact that, at least intuitively, utilities are (real) numbers. In addition, it is, perhaps, also implicitly accepted that real-valued utility functions are used because $\mathbb{R}$ is much simpler than other possible codomains, and consequently it is simpler to use numerical quantitative scales rather than qualitative ones to compare elements on a set, or to formalize the preferences of an individual.

Nevertheless, as clearly stated by Herden and Mehta in [61] (see also [21]), we may say that *"it is highly desirable and even imperative to begin the development of a theory of the existence and continuity of non-real-valued utility functions"*.

Notice that the real line $\mathbb{R}$, endowed with is usual linear order $\leq$ is just one of the possible linearly ordered sets that we could choose as suitable codomain to represent orderings. And, even being the most popular one, we should not restrict ourselves to use only $\mathbb{R}$. Moreover, as analyzed in [6] there exist different classes of linear orders that fail to admit a utility representation. Of course, those linear orders (e.g., the long line or the lexicographic plane) could also be considered someway as alternative codomains to represent some ordered structures. By cardinality reasons, it is clear that there is no "supreme codomain" $(X, \preceq)$ where X is a nonempty set and $\preceq$ is a linear order on X, such that given any nonempty set Y endowed with a total preorder $\precsim$, there exists a map $f : Y \longrightarrow X$ such that $a \precsim b \Leftrightarrow f(a) \preceq f(b)$ holds for every $a, b \in X$. To prove this claim notice that, for any set X, the power set 2^X, endowed with a well ordering can never be represented into $(X, \preceq)$ because its cardinality is bigger, so that there is no room in X for a surjective function from 2^X into X to exist. However, if we restrict the cardinality of the totally preordered set $(Y, \precsim)$ we can still find some powerful result in this direction, such as the classical one obtained by J. S. Chipman in [44].

Theorem 6.1 *A total preorder $\precsim$ defined on a nonempty set X can always be represented in the classical sense in a codomain of the type $Y = \{0, 1\}^\alpha$ where α denotes a large enough cardinal number, and Y is endowed with the corresponding lexicographic ordering.*

Problem 6.2 Has Theorem 6.1 an analogous result valid for representability of interval orders and semiorders in some way?

Problem 6.3 For a given cardinality α, find all the possible codomains $(X, \preceq)$ where X is a nonempty set and $\preceq$ is a linear order on X, such that given any nonempty set Y whose cardinality is not bigger than α, and assuming that the given set Y is endowed with a total preorder $\precsim$, there exists a map $f : Y \longrightarrow X$ such that $a \precsim b \Leftrightarrow f(a) \preceq f(b)$ holds for every $a, b \in Y$.

Problem 6.4 For a given cardinality α, is there a suitable codomain where any interval order or semiorder defined on a nonempty set Y whose cardinality is smaller than or equal to α can be represented in a feasible way?

We may also study here which are the most suitable codomains to represent total preorders $\precsim$ that are τ-continuous with respect to some topology defined on a nonempty set X, and satisfying some additional properties. In this direction, a good example is furnished by the following theorem, whose proof appears in [22].

Theorem 6.5 *Every τ-continuous total preorder $\precsim$ on a separably connected topological space (X, τ) admits an isotonic representation on the double long line endowed with its natural order. Moreover, the representation can be chosen to be continuous with respect to the topology τ on X and the order topology on the double long line.*

Proof See Lemma 4.1 in [22] for the definition and properties of the double long line, as well as for the proof of this result on representability. $\square$

Problem 6.6 Given a topological space (X, τ) that satisfies some fixed additional properties (e.g, separable connectedness as in Theorem 6.5 above), find some suitable linearly ordered codomain on which any τ-continuous total preorder $\precsim$ can be continuously represented.

Another complementary idea that leads to a change of codomain was analyzed in [14, 21, 36, 54]. The key is to consider representations that use only one map, instead of two (as in the case of interval orders), or one function plus a threshold (as in the case of semiorders).

Definition 6.7 Let C stand for a particular class of binary relations (e.g.: interval orders) that have been defined on a nonempty set X. Consider also an standard fixed particular procedure P to represent the binary relations that belong to the class C (e.g.: the representability of interval orders through a pair of real-valued functions). We say that a nonempty set Y endowed with a suitable binary relation R_Y is a *universal codomain* for the class C with respect to the procedure P, if the representability of an element of C in the sense of the procedure P is equivalent to the existence of a single map $F : X \to Y$ such that $x R y \Leftrightarrow F(x) R_Y F(y)$ holds for every $x, y \in X$.

The key here is to look for representations that use a single map. The main underlying problem is the following (see [14]):

Problem 6.8 Given a particular class C of binary relations and a fixed procedure P of representation, find a universal codomain (if any) to represent the elements of that class.

Partial answers to this general problem have already been introduced in this literature for the case of total preorders, interval orders and semiorders (see [14, 21, 36, 54]).

6.2 Utility Through Functional Equations

Let us go back to the classical representability of a total preorder $\precsim$, defined on a nonempty set X, through a real-valued utility function $u : X \to \mathbb{R}$ such that $x \precsim y \Leftrightarrow u(x) \leq u(y)$ holds for every $x, y \in X$. If we consider the bivariate map $F : X \times X \to \mathbb{R}$ given by $F(x, y) = u(y) - u(x)$ $(x, y \in X)$ we immediately observe that $x \precsim y \Leftrightarrow F(x, y) \geq 0$ holds. In addition, the function F satisfies the so-called *Sincov's functional equation* (see [78]), namely $F(x, y) + F(y, z) = F(x, z)$ $(x, y, z \in X)$.

Following [29], we introduce the definition of some classical functional equations in two variables.

Let X be a nonempty set. Let $F : X \times X \to \mathbb{R}$ be a real-valued bivariate map defined on X.

Definition 6.9 The bivariate map F is said to satisfy:

(i) The *Sincov functional equation* if $F(x, y) + F(y, z) = F(x, z)$ holds for every $x, y, z \in X$,

(ii) the *Sincov functional equation (second version)* if $F(x, y) + F(y, z) + F(z, x) = 0$ holds for every $x, y, z \in X$,

(iii) the *separability functional equation* if $F(x, y) + F(y, z) = F(x, z) + F(y, y)$ holds for every $x, y, z \in X$,

(iv) the *restricted separability functional equation* if $F(x, y) + F(y, z) = F(x, z) + F(t, t)$ holds for every $x, y, z, t \in X$,

(v) the *Ferrers functional equation* if $F(x, y) + F(z, t) = F(x, t) + F(z, y)$ holds for every $x, y, z, t \in X$,

(vi) the *semitransitivity functional equation* if $F(x, y) + F(y, z) = F(x, t) + F(t, z)$ holds for every $x, y, z, t \in X$.

Proposition 6.10 *Let X be a nonempty set. Let $F : X \times F \to \mathbb{R}$ be a bivariate map. The following statements hold:*

(i)) Both versions of Sincov functional equation are equivalent.

(ii) The separability equation is equivalent to the Ferrers equation.

(iii) The restricted separability equation is equivalent to the semitransitivity equation.

Proof See Proposition 1 in [29]. □

Let us see now how those functional equations may intervene in alternative representations of total preorders, interval orders and semiorders.

Theorem 6.11 *Let X be a nonempty set. Let $\precsim$ be a total preorder on X. Then the following statements are equivalent:*

(i) The total preorder $\precsim$ is representable by means of a utility function $u : X \to \mathbb{R}$ such that $x \precsim y \Leftrightarrow u(x) \le u(y)$ $(x, y \in X)$.

(ii) There exists a real-valued bivariate map $F : X \times X \to \mathbb{R}$ that satisties the Sincov functional equation and, in addition, $x \prec y \Leftrightarrow F(x, y) > 0$ holds for every $x, y \in X$.

Proof See Proposition 5 in [29]. □

Let us see now what happens as regards interval orders.

Theorem 6.12 *Let X be a nonempty set. Let $\prec$ be an interval order on X. Then the following statements are equivalent:*

(i) The interval order $\prec$ is representable by means of a pair of real-valued functions $u, v : X \to \mathbb{R}$ such that $x \prec y \Leftrightarrow v(x) < u(y)$ $(x, y \in X)$.

(ii) There exists a real-valued bivariate map $F : X \times X \to \mathbb{R}$ that satisfies the separability functional equation and, in addition, $x \prec y \Leftrightarrow F(x, y) > 0$ holds for every $x, y \in X$.

Proof See Proposition 6 in [29]. □

Finally, with respect to semiorders, we also get an analogous result, that goes as follows.

Theorem 6.13 *Let X be a nonempty set. Let $\prec$ be a semiorder on X. Then the following statements are equivalent:*

(i) *The semiorder $\prec$ is representable in the sense of Scott and Suppes by means of a real-valued function $u : X \to \mathbb{R}$ such that $x \prec y \Leftrightarrow u(x)+1 < u(y)$ $(x, y \in X)$.*

(ii) *There exists a real-valued bivariate map $F : X \times X \to \mathbb{R}$ that satisfies the semitransitivity functional equation as well as $F(x, x) = -1$ for every $x \in X$, and, in addition, $x \prec y \Leftrightarrow F(x, y) > 0$ holds for every $x, y \in X$.*

Proof See Proposition 7 in [29]. $\qquad\qquad\qquad\qquad\qquad\qquad\qquad\qquad\qquad\qquad\qquad\Box$

Problem 6.14 Given a nonempty set X and a bivariate map $F : X \times X \to \mathbb{R}$ we may associate to F a binary relation $\mathcal{R}_F$ on X given by $x\mathcal{R}_F y \Leftrightarrow F(x, y) > 0$ $(x, y > 0)$ (see [29]). Suppose now that F satisfies some functional equation. What can be said about the corresponding binary relation $\mathcal{R}_F$? Does it correspond to some classical kind of ordering (e.g.: a total preorder, an interval order or a semiorder)?

6.3 Utility Theory in the Fuzzy Setting

Following [2], we pay now attention to the possibility of defining comparisons or preferences that are fuzzy instead of crisp. Suppose that we consider different kinds of orderings on a nonempty set X. Our idea is to establish some sort of comparison or qualitative scale between its elements. Mathematically, when this approach is formalized a variety of classical binary relations (e.g. total preorders, interval orders, etc.) naturally arise. Usually, when dealing with those different relations, if we consider that an element $x \in X$ is related to another element $y \in X$, it happens that the relationship is either void ($= 0$) or total ($= 1$): That is, either the elements are not related at all, or they are fully related. No intermediate situation is allowed. This is the *crisp* setting, and it is the most usual one in classical Utility Theory.

However, a common feature, that nowadays, in more modern studies, is arising in many models is the consideration of comparisons or suitable binary relations that are *graded*. For instance, this is done in order to describe an intensity in the relationship between two given elements. Or this setting can also appear in situations of uncertainty. In these cases, two elements could be related at any level between 0 and 1. Of course, now the binary relation becomes *fuzzy*.

Definition 6.15 Given a nonempty set U, called *universe*, a *fuzzy subset* X of U is a map $\mu_X : U \to [0, 1]$.[8] Also, a *fuzzy binary relation* defined on a nonempty universe U is a function $\mathcal{F} : U \times U \to [0, 1]$. In other words $\mathcal{F}$ becomes a fuzzy subset of the Cartesian product $U \times U$.

---

[8]It is usual that the map μ_X and the corresponding fuzzy set X are used interchangeably if this does not give rise to confusion.

The classical properties of (crisp) binary relations should then be generalized to the fuzzy setting in some suitable way. To put an example, one may say that a fuzzy relation $\mathcal{F}$ on a universe U is *reflexive* if $F(t, t) = 1$ holds for every $t \in U$. However, definitions such as asymmetry, transitivity, etc., as well as certain operations such as intersections, unions, complements, etc., depend on the choice of a suitable triangular norm (see [2] for details). In fact, there are equivalent definitions of such concepts in the crisp setting (working with classical sets, that is, non-fuzzy), that, when extended to the fuzzy approach, are no longer equivalent, and give rise to many possible different theories and approaches, depending on the definitions considered in each fuzzy context.

Problem 6.16 Extend to the fuzzy setting in a suitable way concepts such as reflexivity, asymmetry, transitivity, negation, intersection and union, as well as classical orderings arising in the crisp setting, namely total preorders, interval orders and semiorders.

Remark 6.17 In [47, 65] several different kinds of fuzzy semiorders were introduced and analyzed. In those studies it is put in evidence that many non-equivalent definitions of the concept of a fuzzy semiorder are possible.

When working with fuzzy binary relations, we would also ask ourselves about what could be a suitable way to represent them numerically, that is, the main general question is the following.

Problem 6.18 Implement a suitable way to convert fuzzy binary relations of certain types into numerical (crisp) scales.

A possible way to do so, is to consider the α-cuts of a fuzzy binary relation. Any of this α-cuts is a crisp (i.e.: non-fuzzy) binary relation. So, if we are able to represent those α-cuts in some suitable way, we could use those numerical representations as a global or full representation of the fuzzy binary relation considered a priori. This has been analyzed in [2] for the case of fuzzy total preorders.

The definition of an α-cut goes as follows.

Definition 6.19 Given a fuzzy relation $\mathcal{F}$ on a universe U, and a real number $\alpha \in [0, 1]$, we define the *α-cut* of $\mathcal{F}$ as the crisp binary relation on U given by $\mathcal{F}_\alpha = \{(s, t) \in U^2 : F(s, t) \geq \alpha\}$.

Remark 6.20 Obviously, the consideration of α-cuts could also generate new extensions of the classical concepts of orderings from the crisp to the fuzzy setting. To put an example, we could adopt the following definition of a fuzzy total preorder: a fuzzy binary relation $\mathcal{F}$ is said to be a fuzzy total preorder if $\mathcal{F}_\alpha$ is a crisp total preorder for any $\alpha \in [0, 1]$.

Problem 6.21 Study and analyze the fuzzy binary relations such that all their α-cuts are total preorders, interval orders or semiorders. Find suitable kinds of numerical representations for each case.

7 Conclusion

Many advances have been achieved in Utility Theory in recent years. In particular, new approaches arose related to numerical representability of certain kinds of qualitative scales, such as total preorders, interval orders and semiorders.

Therefore, it could be helpful for any researcher in this framework to have at hand not only a clear account of these new achievements, but, also, an exhaustive list of open questions that may lead to new streams and trends in research.

This has been the main aim of the present work. Our intention has been to highlight most of the open problem still encountered in this general theory and its most classical branches.

Acknowledgements This work is partially supported by the research projects ECO2015-65031-R, MTM2015-63608-P (MINECO/ AEI-FEDER, UE), and TIN2016-77356-P (MINECO/ AEI-FEDER, UE).

We are grateful to two reviewers for their valuable suggestions and comments.

Thanks are also given to the organizers and participants in the congress SUMTOPO 2019, 34th Summer Conference on Topology and its Applications, University of the Witwatersrand, Johannesburg (South Africa) 1–4 July 2019, for their helpful discussions and comments on the contents of our contribution, a substantial part of which was presented there.

References

1. Abrísqueta, F.J., Candeal, J.C., Catalán, R.G., De Miguel, J.R., Induráin, E.: Generalized Abel functional equations and numerical representability of semiorders. Publ. Math. Debr. **78**(3–4), 557–568 (2011)
2. Agud, L., Catalán, R.G., Díaz, S., Induráin, E., Montes, S.: Numerical representability of fuzzy total preorders. Int. J. Comput. Intell. Syst. (IJCIS) **5**(6), 996–1009 (2012)
3. Alcantud, J.C.R., Campión, M.J., Candeal, J.C., Catalán, R.G., Induráin, E.: On the structure of acyclic binary relations. In: Medina, J., et al. (eds.) IPMU 2018, CCIS, vol. 855, pp. 3–15. Springer International Publishing AG (Part of Springer Nature 2018), Berlin (2018)
4. Aleskerov, F., Bouyssou, D., Monjardet, B.: Utility Maximization, Choice and Preference, 2nd edn. Springer, Berlin (2007)
5. Alimov, N.G.: On ordered semigroups (in Russian). Izv. Akad. Nauk SSSR Ser. Mat. **14**, 569–576 (1950)
6. Beardon, A.F., Candeal, J.C., Herden, G., Induráin, E., Mehta, G.B.: The non-existence of a utility function and the structure of non-representable preference relations. J. Math. Econ. **37**, 17–38 (2002)
7. Beja, A., Gilboa, I.: Numerical representations of imperfectly ordered preferences. A unified geometric exposition. J. Math. Psychol. **36**, 426–449 (1992)
8. Birkhoff, G.: Lattice Theory, 3rd edn. American Mathematical Society, Providence (1967)
9. Bosi, G., Campión, M.J., Candeal, J.C., Induráin, E.: Interval-valued representability of qualitative data: the continuous case. Int. J. Uncertain. Fuzziness Knowl. Based Syst. **15**(3), 299–319 (2007)
10. Bosi, G., Campión, M.J., Candeal, J.C., Induráin, E., Zuanon, M.E.: Isotonies on ordered cones through the concept of a decreasing scale. Math. Soc. Sci. **54**, 115–127 (2007)
11. Bosi, G., Candeal, J.C., Induráin, E.: Continuous representability of interval orders and biorders. J. Math. Psychol. **51**, 122–125 (2007)

12. Bosi, G., Candeal, J.C., Induráin, E., Olóriz, E., Zudaire, M.: Numerical representations of interval orders. Order **18**, 171–190 (2001)
13. Bosi, G., Estevan, A., Gutiérrez García, J., Induráin, E.: Continuous representability of interval orders: the topological compatibility setting. Internat. J. Uncertain. Fuzziness Knowl. Based Syst. **23**(3), 345–365 (2015)
14. Bosi, G., Gutiérrez García, J., Induráin, E.: Unified representability of total preorders and interval orders through a single function: the lattice approach. Order **26**, 255–275 (2009)
15. Bosi, G., Herden, G.: On the structure of completely useful topologies. Appl. Gen. Topol. **3**(2), 145–167 (2002)
16. Bosi, G., Zuanon, M.E.: Semicontinuous representability of interval orders on a metrizable topological space. Int. J. Contemp. Math. Sci. **2**(17–20), 853–858 (2007)
17. Bridges, D.S., Mehta, G.B.: Representations of Preference Orderings. Springer, Berlin (1995)
18. Campión, M.J., Arzamendi, G., Gandía, L.M., Induráin, E.: Entropy of chemical processes versus numerical representability of orderings. J. Math. Chem. **54**, 503–526 (2016)
19. Campión, M.J., Candeal, J.C., Catalán, R.G., Giarlotta, A., Greco, S., Induráin, E., Montero, J.: An axiomatic approach to finite means. Inf. Sci. **457–458**, 12–28 (2018)
20. Campión, M.J., Candeal, J.C., Granero, A.S., Induráin, E.: Ordinal representability in Banach spaces. In: Castillo, J.M.F., Johnson, W.B. (eds.) Methods in Banach Space Theory, pp. 183–196. Cambridge University Press, Cambridge (2006)
21. Campión, M.J., Candeal, J.C., Induráin, E.: Representability of binary relations through fuzzy numbers. Fuzzy Sets Syst. **157**, 1–19 (2006)
22. Campión, M.J., Candeal, J.C., Induráin, E.: On Yi's extension property for totally ordered topological spaces. J. Korean Math. Soc. **43**(1), 159–181 (2006)
23. Campión, M.J., Candeal, J.C., Induráin, E.: Semicontinuous order-representability of topological spaces. Bol. Soc. Mat. Mexicana **3**(15), 81–89 (2009)
24. Campión, M.J., Candeal, J.C., Induráin, E.: Preorderable topologies and order-representability of topological spaces. Topol. Appl. **159**, 2971–2978 (2009)
25. Campión, M.J., Candeal, J.C., Induráin, E., Mehta, G.B.: Order embeddings with irrational codomains: Debreu properties of real subsets. Order **23**, 343–357 (2006)
26. Campión, M.J., Candeal, J.C., Induráin, E., Mehta, G.B.: Representable topologies and locally connected spaces. Topol. Appl. **154**, 2040–2049 (2007)
27. Campión, M.J., Candeal, J.C., Induráin, E., Mehta, G.B.: Continuous order representability properties of topological spaces and algebraic structures. J. Korean Math. Soc. **49**(3), 449–473 (2012)
28. Campión, M.J., Candeal, J.C., Induráin, E., Zudaire, M.: Continuous representability of semiorders. J. Math. Psychol. **52**, 48–54 (2008)
29. Campión, M.J., De Miguel, L., Catalán, R.G., Induráin, E., Abrísqueta, F.J.: Binary relations coming from solutions of functional equations: orderings and fuzzy subsets. Internat. J. Uncertain Fuzziness Knowl.-Based Syst. **25**(Suppl. 1), 19–42 (2017)
30. Campión, M.J., Falcó, E., García-Lapresta, J.L., Induráin, E.: Assigning numerical scores to linguistic expressions. Axioms **6**(3), UNSP 19 (2017)
31. Campión, M.J., Gómez-Polo, C., Induráin, E., Raventós-Pujol, A.: A survey of the mathematical foundations of axiomatic entropy: representability and orderings. Axioms **7**, 29 (2018)
32. Candeal, J.C., De Miguel, J.R., Induráin, E.: Extensive measurement: continuous additive utility functions on semigroups. J. Math. Psychol. **40**(4), 281–286 (1996)
33. Candeal, J.C., De Miguel, J.R., Induráin, E.: Topological additively representable semigroups. J. Math. Anal. Appl. **210**, 385–389 (1997)
34. Candeal, J.C., De Miguel, J.R., Induráin, E., Mehta, G.B.: Representations of ordered semigroups and the physical concept of entropy. Appl. Gen. Topol. **5**(1), 11–23 (2004)
35. Candeal, J.C., Estevan, A., Gutiérrez-García, J., Induráin, E.: Semiorders with separability properties. J. Math. Psychol. **56**, 444–451 (2012)
36. Candeal, J.C., Gutiérrez García, J., Induráin, E.: Universal codomains to represent interval orders. Int. J. Uncertain. Fuzziness Knowl. Based Syst. **17**(2), 197–219 (2009)

37. Candeal, J.C., Hervés, C., Induráin, E.: Some results on representation and extension of preferences. J. Math. Econ. **29**, 75–81 (1998)
38. Candeal, J.C., Induráin, E.: Semiorders and thresholds of utility discrimination: Solving the Scott-Suppes representability problem. J. Math. Psychol. **54**, 485–490 (2010)
39. Candeal, J.C., Induráin, E., Molina, J.A.: Numerical representability of ordered topological spaces with compatible algebraic structure. Order **29**, 131–146 (2012)
40. Candeal, J.C., Induráin, E., Sanchis, M.: Order representability in groups and vector spaces. Expo. Math. **30**, 103–123 (2012)
41. Candeal, J.C., Induráin, E., Zudaire, M.: Numerical representability of semiorders. Math. Soc. Sci. **43**(1), 61–77 (2002)
42. Candeal-Haro, J.C., Induráin Eraso, E.: A note on linear utility. Econ. Theory **6**(3), 519–522 (1995)
43. Candeal-Haro, J.C., Induráin Eraso, E.: Utility representations from the concept of measure. Math. Soc. Sci. **26**(1), 51–62 (1993)
44. Chipman, J.S.: The foundations of utility. Econometrica **28**(2), 193–224 (1960)
45. De Miguel, J.R., Candeal, J.C., Induráin, E.: Archimedeaness and additive utility on totally ordered semigroups. Semigroup Forum **52**, 335–347 (1996)
46. Debreu, G.: Representation if a preference ordering by a numerical function. In: Thrall, R., Coombs, C., Davies, R. (eds.) Decision Processes, pp. 159–166. Wiley, New York (1954)
47. Díaz, S., Induráin, E., De Baets, B., Montes, S.: Fuzzy semi-orders: the case of t-norms without zero divisors. Fuzzy Sets Syst. **184**, 52–67 (2011)
48. Doignon, J.P., Ducamp, A., Falmagne, J.C.: On realizable biorders and the biorder dimension of a relation. J. Math. Psychol. **28**, 73–109 (1984)
49. Droste, M.: Ordinal scales in the theory of measurement. J. Math. Psychol. **31**, 60–82 (1987)
50. Dugundji, J.: Topology. Allyn and Bacon, Boston (1966)
51. Eilenberg, S.: Ordered topological spaces. Am. J. Math. **63**, 39–45 (1941)
52. Engelking, R.: General Topology, Revised and completed edition. Heldermann Verlag, Berlin (1989)
53. Estevan, A., Gutiérrez García, J., Induráin, E.: Further results on the continuous representability of semiorders, Internat. J. Uncertain. Fuzziness Knowl. Based Syst. **21**(5), 675–694 (2013)
54. Estevan, A., Gutiérrez García, J., Induráin, E.: Numerical representation of semiorders. Order **30**, 455–462 (2013)
55. Fishburn, P.C.: Utility Theory for Decision-Making. Wiley, New York (1970)
56. Fishburn, P.C.: Intransitive indifference with unequal indifference intervals. J. Math. Psychol. **7**, 144–149 (1970)
57. Fishburn, P.C.: Intransitive indifference in preference theory: a survey. Oper. Res. **18**(2), 207–228 (1970)
58. Fishburn, P.C.: Interval Orders and Interval Graphs. Wiley, New York (1985)
59. Fuchs, L.: Partially Ordered Algebraical Systems. Pergamon Press, Oxford (1963)
60. Herden, G.: Topological spaces for which every continuous total preorder can be represented by a continuous utility function. Math. Soc. Sci. **22**, 123–136 (1991)
61. Herden, G., Mehta, G.B.: The Debreu Gap Lemma and some generalizations. J. Math. Econ. **40**(7), 747–769 (2004)
62. Herden, G., Pallack, A.: Useful topologies and separable systems. Appl. Gen. Topol. **1**(1), 61–82 (2000)
63. Hölder, O.: Der Axiome der Quantität und die Lehre vom Mass. Berichte über die Verhandlungen der Königlich Sachsischen Gesellschaft der Wissenschaften zu Leipzig, Math. Phys. Kl. **53**, 1–64 (1901)
64. Induráin, E., Knoblauch, V.: On topological spaces whose topology is induced by a binary relation. Quaest. Math. **36**(1), 47–65 (2013)
65. Induráin, E., Martinetti, D., Montes, S., Díaz, S., Abrísqueta, F.J.: On the preservation of semiorders from the fuzzy to the crisp setting. Int. J. Uncertain. Fuzziness Knowl. Based Syst. **19**(6), 899–920 (2011)

66. Luce, R.D.: Semiorders and a theory of utility discrimination. Econometrica **24**, 178–191 (1956)
67. Lutzer, D.J., Bennet, H.R.: Separability, the countable chain condition and the Lindelf property on linearly ordered spaces. Proc. Am. Math. Soc. **23**(3), 664–667 (1969)
68. Manders, K.L.: On JND representations of semiorders. J. Math. Psychol. **24**, 224–248 (1981)
69. Mehta, G.B.: A remark on a utility representation theorem of Rader. Econ. Theory **9**, 367–370 (1997)
70. Monjardet, B.: Axiomatiques et propriétés des quasi-ordres. Math. Sci. Hum. **63**, 51–82 (1978)
71. Nachbin, L.: Topology and Order. Van Nostrand Reinhold, New York (1970)
72. Nyikos, P.J., Reichel, H.C.: Topologically orderable groups. Gen. Topol. Appl. **5**(3), 195–204 (1975)
73. Olóriz, E., Candeal, J.C., Induráin, E.: Representability of interval orders. J. Econ. Theory **78**(1), 219–227 (1998)
74. Purisch, S.: A history of results on orderability and suborderability. In: Handbook of the History of General Topology, vol. 2, pp. 689–702. Kluwer, Dordrecht (1998)
75. Rader, T.: The existence of a utility function to represent preferences. Rev. Econ. Stud. **30**, 229–232 (1963)
76. Roberts, F.S., Luce, R.D.: Axiomatic thermodynamics and extensive measurement. Synthese **18**, 311–326 (1968)
77. Scott, D., Suppes, P.: Foundational aspects of theories of measurement. J. Symb. Log. **23**, 113–128 (1958)
78. Sincov, D.M.: Über eine Funktionalgleichung. Arch. Math. Phys. **3**(6), 216–227 (1903)
79. The, A.N., Tsoukias, A.: Numerical representation of PQI interval orders. Discrete Appl. Math. **147**(1), 125–146 (2005)
80. Van Dalen, J., Wattel, E.: A topological characterization of ordered spaces. Gen. Topol. Appl. **3**, 347–354 (1973)
81. Willard, S.: General Topology. Reading, Massachussets (1970)
82. Yi, G.: Continuous extension of preferences. J. Math. Econ. **22**, 547–555 (1993)

Particular Kinds of Preferences and Representations

Interval orders; Semiorders; Biorders; Chain representations; Algebraic utility; Group topologies; Convex structures; Monotonic preferences; Numerical representability

Representations of Interval Orders on Connected Separable Topological Spaces

Yann Rébillé

Abstract We establish a partial continuous interval representation for interval orders preferences on connected separable topological spaces. Then we provide a generalization of interval representations of interval orders when a continuous strict weak order is given. This allows one to obtain for incomplete preferences one-side or two-sides interval representations and also one-side or two-sides weak interval representations that are partial continuous.

1 Introduction

Interval orders received great attention regarding representation theory. The first occurrence of "interval order" is [16]. Numerous applications ranging from pure mathematics (set theory, topology, graph theory), applied mathematics (statistics), social sciences (economics) involve interval orders or related (semi-orders) (see also [21]). Interestingly, interval order relations trace back to the early works of Wiener [20, 24] and were to be rediscovered much later.

Results of first importance on interval orders are given in [18] for the general case through the countability condition imposed on a suitable quotient space. The existence of an interval representation of preferences can be characterized through the interval order separability condition (Theorem 1 p. 223 in [22] and Theorem 1 p. 175 in [5]).

A continuous interval representation of preferences is obtained in [12] on connected separable topological spaces (see Corollary p. 145) and generalizes Debreu-Eilenberg's theorem on continuous utility representations ([13–15], see [23] for a simple proof). Numerous contributions aim at obtaining continuous interval representation of preferences on general topological spaces through separability conditions, separable systems and continuity conditions [4, 6, 9].

Y. Rébillé (✉)
LEMNA, IAE of Nantes, University of Nantes, chemin la Censive du Tertre, BP 52231, 44322
Nantes Cedex 3, France
e-mail: yann.rebille@univ-nantes.fr

© Springer Nature Switzerland AG 2020
G. Bosi et al. (eds.), *Mathematical Topics on Representations of Ordered Structures and Utility Theory*, Studies in Systems, Decision and Control 263,
https://doi.org/10.1007/978-3-030-34226-5_4

Our aim is to give a weaker statement that involves partial continuity of the interval representation of preferences on connected separable topological spaces.

Our statement is closely related and can be seen as a sharpening of Theorem 6.4.5 p. 104 in [3]. This kind of results with semi-continuity occurs since [2, 8, 10]. A careful look reveals that this approach can be generalized in some way. We suggest a possible way to go beyond the traditional approach of interval orders which deals with continuity of the traces (e.g. [6]). The traces are binary relations derived from the original interval order and can be related to the precedence and the succession relations associated to an interval order.[1] For this, we shall assume that a continuous strict weak order on the underlying space is given. Typically, there exist simultaneously a *subjective* binary relation and an *objective* binary relation. The *subjective* binary relation reflects psychological perceptions that are performed loosely and the *objective* binary relation reflects physical measures that are performed accurately. The objective binary relation indicates how precedence and succession should organize. Then, the subjective binary relation can behave in accordance or not with the rules of precedence and succession.

Our first contribution is to present a partial continuous interval representation of preferences on connected separable topological spaces. Semi-continuous versions can be obtained thereafter. Our second contribution provides a possible generalization of interval order representations when a continuous strict weak order is already given. This allows one to obtain for incomplete preferences one-side or two-sides interval representations and also one-side or two-sides weak interval representations that are partial continuous.

The organization of the paper is the following. Next section introduces the definitions and terminology. Then, useful properties related to interval orders are reminded. Section 4 contains weak representation theorems for interval orders and Sect. 5 contains representation theorems for interval orders. Section 6 contains generalizations. Proofs of the various properties are relegated to an appendix.

2 Definitions

Let X be a nonempty set, $x \in X$ denotes a generic element.

The set X is a topological space if it is endowed with a topology (of open sets), denoted τ. Then, X is said to be *connected* if for all $O_1, O_2 \in \tau$, if $O_1 \cup O_2 = X$ and $O_1 \cap O_2 = \emptyset$ then $O_1 = \emptyset$ or $O_2 = \emptyset$, and X is said to be *separable* if there exists a countable subset C such that $cl(C) = X$ where cl denotes the (topological) closure.

A binary relation on X is denoted by $\succ \subset X^2$. The complement (or negation) relation of $\succ$, denoted by $\not\succ$ is defined by $x \not\succ y \iff (x, y) \notin \succ$ for $x, y \in X$. The dual (or symmetric) relation $\prec$ is defined by $x \prec y \iff y \succ x$ for $x, y \in X$.

---

[1] An element x is said to *precede* y if there exists z such that $x \sim z \prec y$ and y is said to *succeed* x if there exists t such that $x \prec t \sim y$.

Similarly, the dual complement is defined by $x \not\prec y \iff y \not\succ x$ for $x, y \in X$. Let $\succeq = \not\prec$ and $\preceq = \not\succ$. The *indifference relation* is given by $\sim = \succeq \cap \preceq = \not\prec \cap \not\succ$.

$\succ$ is *irreflexive* if for all $x \in X$, $x \not\succ x$ (denoted (IR)).

$\succ$ is *asymmetric* if for all $x, y \in X$, $x \succ y \Rightarrow y \not\succ x$. That is, $\succ \subset \succeq$.

$\succ$ is *acyclic* if for all $n \in \mathbb{N}, x_0, \ldots, x_n \in X$ if $x_0 \prec x_1, \ldots, x_{n-1} \prec x_n$ then $x_n \not\prec x_0$.

$\succ$ is *transitive* if for all $x, y, z \in X$, $x \prec y$, $y \prec z \Rightarrow x \prec z$ (denoted (TR)).

Clearly, asymmetry implies irreflexivity, acyclicity implies asymmetry, transitivity and irreflexivity imply acyclicity.

A binary relation is a *strict (partial) order* if $\succ$ is irreflexive and transitive.

$\succ$ is *negatively transitive* if for all $x, y, z \in X$, $x \not\succ y$, $y \not\succ z \Rightarrow x \not\succ z$.

A binary relation is a *strict weak order* if $\succ$ is asymmetric and negatively transitive. Strict weak orders coincide with strict partial orders having a transitive indifference relation.

A strict (partial) order is an *interval order* if it satisfies *Ferrer's condition* also known as,

(PTR) Pseudo transitivity *For all* $x, y, z, w \in X$, $x \prec y \preceq z \prec w \Rightarrow x \prec w$.

Stated equivalently, pseudo transitivity coincides with Fishburn's P2 condition (see [18]),

$$x \prec y \text{ and } z \prec w \Rightarrow z \prec y \text{ or } x \prec w, \text{ for all } x, y, z, w \in X.$$

Let $u : X \longrightarrow \mathbb{R}$ be a function and $\succ \subset X^2$. The function u is called a *utility function* if u represents $\succ$, i.e., for all $x, y \in X$,

$$x \succ y \iff u(x) > u(y),$$

the function u is called a *weak utility function* if for all $x, y \in X$,

$$x \succ y \Rightarrow u(x) > u(y).$$

Necessarily, if $\succ$ admits a weak utility representation then acyclicity holds.

Let $u, v : X \longrightarrow \mathbb{R}$ be functions with $v \geq u$ and $\succ \subset X^2$. The couple of functions (u, v) is an *interval representation* of $\succ$ (a closed-interval representation in [18]), if, for all $x, y \in X$,

$$x \succ y \iff u(x) > v(y),$$

the couple of functions (u, v) is a *weak interval representation* if for all $x, y \in X$,

$$x \succ y \Rightarrow u(x) > v(y).$$

In particular, u and v are weak utility functions representing $\succ$ since $v \geq u$.

Weak utility functions are particularly important for optimization issues.

An element $x^* \in X$ is said to be *maximal* if for all $x \in X$, $x \not\succ x^*$. An element $x_* \in X$ is said to be *minimal* if for all $x \in X$, $x \not\prec x_*$. Denote the set of $\succ$-maximal elements by $\max(\succ)$ and the set of $\succ$-minimal elements by $\min(\succ)$. A maximum (minimum) for a weak utility function provides a maximal (minimal) element for the preference relation.

3 Basic Properties

Let us recall some properties relating irreflexivity, asymmetry, transitivity and pseudotransitivity. All the results in this section can be more or less found in the existing literature (see also Chap. 6 in [3]). For sake of completeness we provide an appendix that gathers the proofs.

We may consider a weaker property than (PTR), namely (WPTR), that is based on $\sim$ instead of $\succeq$. This coincides precisely with Wiener's axiom [20, 24],

(WPTR) Weak pseudo transitivity *For all* $x, y, z, w \in X, x \prec y \sim z \prec w \Rightarrow x \prec w$.

Consider the binary relations of *precedence* and *succession*, denoted by $\prec_1, \prec_2$, and defined as follows,

$$x \prec_1 y \text{ if there exists } z \in X / x \sim z \text{ and } z \prec y,$$

$$x \prec_2 y \text{ if there exists } z \in X / x \prec z \text{ and } z \sim y.$$

We will adopt Fishburn 1973s notations in the whole paper ([18], see p. 94). These relations are denoted respectively by $\prec^-, \prec^+$ in [19] (see p. 21) and $<^L, <^R$ in [17] (see p. 211).

Other relations of importance related to $\prec_1, \prec_2$ can be considered, and defined as follows,

$$x \prec_1^* y \text{ if there exists } z \in X / x \preceq z \text{ and } z \prec y,$$

$$x \prec_2^* y \text{ if there exists } z \in X / x \prec z \text{ and } z \preceq y.$$

Clearly, $\prec_1 \subset \prec_1^*$ and $\prec_2 \subset \prec_2^*$. In fact, equalities obtain instead of inclusions under (IR) and (WPTR) (see Properties below).

Other important relations related to $\prec$ are widely adopted in the literature (e.g. [1]). These binary relations are known as the *left trace* and the *right trace*, and defined as follows:

$$x \prec^* y \text{ if there exists } z \in X / x \prec z \text{ and } z \preceq y,$$

$$x \prec^{**} y \text{ if there exists } z \in X / x \preceq z \text{ and } z \prec y.$$

Clearly, $\prec^* = \prec_2^*$ and $\prec^{**} = \prec_1^*$.

In [12] (see also [3] p. 105), $\succ_1, \succ_2$ are defined in reversed order,

$$x \succ_1 y \text{ if there exists } z \in X \: / \: x \succeq z \text{ and } z \succ y \,,$$

$$x \succ_2 y \text{ if there exists } z \in X \: / \: x \succ z \text{ and } z \succeq y \,.$$

Thus, $\prec_1 = \prec_2^*$ and $\prec_2 = \prec_1^*$.

Properties *Let* $\succ \subset X^2$.

1. *If* $\succ$ *satisfies (TR) and (WPTR) then* $\succ$ *satisfies (PTR).*
2. *If* $\succ$ *satisfies (IR) and (TR) then* $\succ$ *is asymmetric.*
3. *If* $\succ$ *satisfies (IR) and (WPTR) then* $\succ$ *satisfies (TR).*
4. *If* $\succ$ *satisfies (IR) and (WPTR) then* $\succ$ *is asymmetric.*
5. *If* $\succ$ *satisfies (IR) then* $\prec \subset \prec_1$ *and* $\prec \subset \prec_2$.
6. *If* $\succ$ *is negatively transitive then* $\prec_1 \subset \prec$ *and* $\prec_2 \subset \prec$.
7. *If* $\succ$ *satisfies (WPTR) then* $\prec_1$ *and* $\prec_2$ *are asymmetric.*
8. *If* $\succ$ *satisfies (IR) and (TR) then* $\prec_1$ *and* $\prec_2$ *are negatively transitive.*
9. *Let* $\succ^a, \succ^b \subset X^2$ *with* $\succ^b$ *asymmetric. If* $\prec^a \subset \prec^b$ *and* $\sim^a \subset \sim^b$ *then* $\prec^a = \prec^b$.
   In particular, $\prec_1 = \prec_1^*$ *and* $\prec_2 = \prec_2^*$ *whenever* $\prec$ *satisfies (IR) and (WPTR).*
10. *Assume* $\succ$ *satisfies (IR) and (TR) then,*

$$\forall x, y, z \in X, x \succeq_1 y, y \succ z \Rightarrow x \succ z \,,$$

$$\forall x, y, z \in X, x \succ y, y \succeq_2 z \Rightarrow x \succ z \,,$$

where $\succeq_1 = \not\prec_1$ *and* $\succeq_2 = \not\prec_2$.

Consequently, from Properties 7–8, if $\succ$ satisfies (IR) and (WPTR), then $\prec_1$ and $\prec_2$ are strict weak orders such that $\prec \subset \prec_1$ and $\prec \subset \prec_2$. The fact that (IR) and (WPTR) imply that $\prec_1^*$ and $\prec_2^*$ are strict weak orders is well-known (see Proposition 2.1 in [1]). A statement similar to Property 10 with $\succeq_1^*$ and $\succeq_2^*$ instead of $\succeq_1$ and $\succeq_2$ is also well-known.

So, according to the Properties we can make an observation that links (WPTR) to asymmetry of $\succ_1$. Assume $\succ$ satisfies (IR) and (TR), then

$$(\text{WPTR}) \Rightarrow \succ_1 \subset \succeq_1, \text{ by Property 7}$$

and

$$\succ_1 \subset \succeq_1 \Rightarrow (\text{WPTR}), \text{ by Property 10.}$$

4   Weak Utility Representation Theorems

We shall consider now the topological setting. Assume that (X, τ) is a connected separable topological space.

Let $\mathcal{R} \subset X^2$ be a binary relation. The *strict upper* and *strict lower sections* w.r.t. $\mathcal{R}$ are denoted and given by $(x, \rightarrow)_{\mathcal{R}} = \{y \in X : y\mathcal{R}x\}$ and $(\leftarrow, x)_{\mathcal{R}} = \{y \in X : x\mathcal{R}y\}$ for all $x \in X$. If $\mathcal{R} = \succ$ we may simply omit the subscript.

(USC) Upper Semi-continuity For all $x \in X$, $(\leftarrow, x) \in \tau$.

(LSC) Lower Semi-continuity For all $x \in X$, $(x, \rightarrow) \in \tau$.

(C) Continuity For all $x \in X$, $(x, \rightarrow) \in \tau$, $(\leftarrow, x) \in \tau$.

Let us recall the Debreu-Eilenberg's theorem which is crucial for our approach,

Debreu-Eilenberg's theorem [13–15] *Let (X, τ) be a connected separable topological space and let $\succ \subset X^2$. Then, if $\succ$ is a strict weak order and satisfies (C) then there exists a continuous function $u : X \longrightarrow \mathbb{R}$ such that for all $x, y \in X$,*

$$x \succ y \iff u(x) > u(y) .$$

Conversely, if $\succ$ is represented by a continuous function, then $\succ$ is a strict weak order and satisfies (C).

Moreover, u is ordinal *i.e. unique up to an increasing transformation.*

Let us remind that the converse part of Debreu-Eilenberg's theorem always holds on any topological space.

Example 1 Let $X = \mathbb{R}$. Let $\tau = \tau_>$ stands for the Euclidean topology. Define the binary relation $\succ$ as follows: For all $x, y \in \mathbb{R}$, $x \succ y$ if and only if $\lceil x \rceil > \lceil y \rceil$, where $\lceil . \rceil$ denotes the integer part by excess, i.e. $\lceil z \rceil = \min\{n : n \in \mathbb{Z}, n \geq z\}$ for $z \in \mathbb{R}$. Then, $\succ$ is a strict weak order that is representable and satisfies (LSC) but not (USC).

Building on Debreu-Eilenberg's theorem we have the following. A related statement is Theorem 3.3 p. 152 in [1].

Theorem 1 *Let (X, τ) be a connected separable topological space and let $\succ \subset X^2$. Assume that (IR), (WPTR) hold. If $\succ_1$ satisfies (C), then there exists a continuous function $u_1 : X \longrightarrow \mathbb{R}$ representing $\succ_1$ such that for all $x, y, z \in X$,*

$$x \succ y \Rightarrow u_1(x) > u_1(y) ,$$

$$and \ x \succeq_1 y \succ z \Rightarrow u_1(x) > u_1(z) .$$

Similarly, if $\succ_2$ satisfies (C), then there exists a continuous function $u_2 : X \longrightarrow \mathbb{R}$ representing $\succ_2$ such that for all $x, y, z \in X$,

$$x \succ y \Rightarrow u_2(x) > u_2(y) ,$$

$$and \ x \succ y \succeq_2 z \Rightarrow u_2(x) > u_2(z) \ .$$

Moreover, any increasing transformation of u_1 and u_2 is convenient, *i.e., u_1 and u_2 are ordinal representations of $\succ_1$ and $\succ_2$ respectively.*

Proof According to Properties 7–8, $\succ_1, \succ_2$ are strict weak orders and by (C), both $\succ_1, \succ_2$ are continuous. So, according to Debreu-Eilenberg's theorem, there exist continuous functions $u_1, u_2 : X \longrightarrow \mathbb{R}$ representing $\succ_1, \succ_2$ respectively. By Property 5, we have $\succ \subset \succ_1, \succ_2$. Hence, u_1, u_2 are weak utility representations of $\succ$. (Moreover). Let u'_1 represent $\succ_1$. Then, there exists an increasing transformation ϕ_1 such that $u'_1 = \phi_1 \circ u_1$ and u'_1 is a weak utility function representing $\succ$. Similarly, let u'_2 represent $\succ_2$. Then, there exists an increasing transformation ϕ_2 such that $u'_2 = \phi_2 \circ u_2$ and u'_2 is a weak utility function representing $\succ$. $\qquad \square$

If $\succ$ satisfies (LSC) or (USC) then the continuity conditions on $\succ_1$ or on $\succ_2$ can be weakened. For instance, when preferences defined on a topological space admit an interval representation and $\succ$ is continuous then there exists a representation (U, V) with U lower semi-continuous and V upper semi-continuous representing respectively $\succ_1$ and $\succ_2$ (see Theorem 3 $(ii) \Rightarrow (iii)$ p. 186 in [5]).

Proposition 1 *Let $\succ \subset X^2$. Assume that (IR), (WPTR) hold. If $\succ$ satisfies (LSC) then $\succ_1$ satisfies (LSC), and if $\succ$ satisfies (USC) then $\succ_2$ satisfies (USC).*

Proof Let $x, y \in X$. Assume $x \prec_1 y$, that is $y \in (x, \rightarrow)_1$. There exists $z \in X$ such that $x \sim z \prec y$. Since $\succ$ is lower semi-continuous, there exists $O_y \in \tau$ with $y \in O_y$ such that for all $t \in O_y$, we have $z \prec t$. Hence, for all $t \in O_y$, we have $x \prec_1 t$. That is $O_y \subset (x, \rightarrow)_1$. So $(x, \rightarrow)_1 \in \tau$.

Similarly, let $x, y \in X$. Assume $y \prec_2 x$, that is $y \in (\leftarrow, x)_2$. There exists $z \in X$ such that $y \prec z \sim x$. Since $\succ$ is upper semi-continuous, there exists $O_y \in \tau$ with $y \in O_y$ such that for all $t \in O_y$, we have $t \prec z$. Hence, for all $t \in O_y$, we have $t \prec_2 x$. That is $O_y \subset (\leftarrow, x)_2$. So $(\leftarrow, x)_2 \in \tau$. $\qquad \square$

5 Interval Order Representation Theorems

Let us provide a sufficient condition to guarantee an interval representation based on the continuity of $\succ_1$ and lower semi-continuity of $\succ$.

Theorem 2 *Let (X, τ) be a connected separable topological space and let $\succ \subset X^2$. Assume that (IR), (WPTR) hold. Then, if $\succ$ satisfies (LSC) and $\succ_1$ satisfies (C) then there exist a continuous function $u_1 : X \longrightarrow \mathbb{R}$ representing $\succ_1$ and $\psi_1 : X \setminus \max(\succ) \longrightarrow \mathbb{R}$ with $\psi_1 \geq 0$, such that for all $x \in X$ and $y \in X \setminus \max(\succ)$,*

$$x \succ y \iff u_1(x) > u_1(y) + \psi_1(y) = v_1(y) \ (\star),$$

where,

$$\psi_1(y) = \inf_{x':x' \succ y} u_1(x') - u_1(y) \ and \ v_1(y) = \inf_{x':x' \succ y} u_1(x').$$

Conversely, if $\succ$ admits an interval representation given by $(\star)$ then (IR), (WPTR), (LSC) hold and $\succ_1$ satisfies (C).

In particular, $\succ$ satisfies (USC) if and only if ψ_1 is upper semi-continuous (or v_1 is upper semi-continuous). Moreover, (u_1, v_1) is unique up to a common increasing transformation, i.e., if (u_1', v_1') represents $\succ$ with u_1' representing $\succ_1$ and v_1' defined by $v_1'(x) = \inf_{x':x' \succ x} u_1'(x')$ for $x \in X \setminus \max(\succ)$ then there exists an increasing transformation ϕ such that $u_1' = \phi \circ u_1$ and $v_1' = \phi \circ v_1$.

Proof Let us prove the first $\Leftarrow$.

Let $x \in X$ and $y \in X \setminus \max(\succ)$ such that $u_1(x) - u_1(y) > \psi_1(y) \geq 0$. Then, there exists $x' \in X$ with $x' \succ y$ such that $u_1(x) - u_1(y) > u_1(x') - u_1(y)$. Thus, $x \succ_1 x' \succ y$, so by (WPTR), $x \succ y$ holds.

Let us establish $\Rightarrow$.

Let $x \in X$ and $y \in X \setminus \max(\succ)$ with $x \succ y$. Since $\succ$ satisfies (LSC), $(y, \rightarrow) \in \tau$ and $x \in (y, \rightarrow)$. Since $\succ_1$ satisfies (C) and $\succ \subset \succ_1$ we have $(\leftarrow, x)_{\succ_1} \in \tau$ and $y \in (\leftarrow, x)_{\succ_1}$.

Let us prove that $(y, \rightarrow) \cup (\leftarrow, x)_{\succ_1} = X$. Let $z \in X$ such that $z \nsucc y$. If $y \sim z$, since $y \prec x$, we have $z \prec_1 x$. If $y \succ z$, then since $x \succ y$, by transitivity $x \succ z$, and then $x \succ_1 z$ since $\succ \subset \succ_1$. So in either case $z \in (\leftarrow, x)_{\succ_1}$.

Now X is connected, thus $(y, \rightarrow) \cap (\leftarrow, x)_{\succ_1} \neq \emptyset$. So, there exists $x' \in X$ such that $y \prec x' \prec_1 x$. And then, it follows that,

$$u_1(x) - u_1(y) > u_1(x') - u_1(y) \geq \psi_1(y) \ .$$

(Converse). Assume there exist a continuous function $u_1 : X \longrightarrow \mathbb{R}$ representing $\succ_1$ and $\psi_1 : X \setminus \max(\succ) \longrightarrow \mathbb{R}$ with $\psi_1 \geq 0$ satisfying $(\star)$. Clearly, (IR) is satisfied since for all $x \in X$, $u_1(x) \leq u_1(x) + \psi_1(x)$.

Let us check (WPTR). Let $x, y, z, t \in X$ such that $x \succ y \sim z \succ t$. Then, $x \succ_1 z \succ t$. Thus, $u_1(x) > u_1(z) > u_1(t) + \psi_1(t)$, and then $x \succ t$.

Let $y \in X \setminus \max(\succ)$. Then, $(y, \rightarrow) = u_1^{-1}((u_1(y) + \psi_1(y), +\infty))$ which is open since u_1 is continuous. So, $\succ$ satisfies (LSC).

Since u_1 is continuous, $\succ_1$ satisfies (C).

(In particular). Assume v_1 is upper semi-continuous. Let $x, y \in X$ with $x \succ y$. Then, $(\leftarrow, x) = v_1^{-1}((-\infty, u_1(x)))$ which is open since v_1 is upper semi-continuous. Hence, $\succ$ satisfies (USC).

Assume $\succ$ satisfies (USC). Let $\alpha \in \mathbb{R}$ and $x_0 \in X$ such that $v_1(x_0) < \alpha$. There exists $x' \in X$ with $x' \succ x_0$ such that $v_1(x_0) \leq u_1(x') < \alpha$. Thus, for all $x \in (\leftarrow, x')$ it holds $v_1(x) \leq u_1(x') < \alpha$. Now by (USC), $(\leftarrow, x') \in \tau$ and $x_0 \in (\leftarrow, x')$, this proves that v_1 is upper semi-continuous at x_0.

(Moreover). Any common increasing transformation $(\phi \circ u_1, \phi \circ v_1)$ with ϕ increasing on $u_1(X)$ provides an interval representation (let us notice that $v_1(X) \subset u_1(X)$ since $u_1(X)$ is a bounded or unbounded interval in $\mathbb{R}$). For all $x \in X$, we have

$\phi \circ v_1(x) = \phi(\inf_{x':x'\succ x} u_1(x')) = \inf_{x':x'\succ x} \phi \circ u_1(x')$. Thus, for all $x, y \in X$, $x \succ y \iff u_1(x) > v_1(y) \iff \phi \circ u_1(x) > \phi \circ v_1(y)$.

Conversely, let (u_1', v_1') represents $\succ$ with u_1' representing $\succ_1$ and v_1' defined for $x \in X \setminus \max(\succ)$ by,

$$v_1'(x) = \inf_{x':x'\succ x} u_1'(x') .$$

Then, there exists an increasing transformation ϕ such that $u_1' = \phi \circ u_1$. Hence, we have for $x \in X \setminus \max(\succ)$,

$$v_1'(x) = \inf_{x':x'\succ x} \phi \circ u_1(x') = \phi(\inf_{x':x'\succ x} u_1(x')) = \phi(v_1(x)) .$$

This establishes that the interval representation is unique up to a common increasing transformation. □

Remark 1 An intimately related result on connected separable topological space is Theorem 6.4.5 p. 104 in [3]. Therein, $\succ$ satisfies (C) and an openness condition[2] is required. This openness condition is equivalent to $\succ_1$ satisfying (USC). Notice that by Proposition 1, $\succ_1$ satisfies (LSC) as soon as $\succ$ satisfies (LSC). Their conclusion states that there exists a representation (u, v) with u lower semi-continuous and v upper semi-continuous. Our result exhibits a continuous u and does not necessitate that $\succ$ satisfies (USC). Then, $\succ$ satisfies (USC) if and only if v is upper semi-continuous. Our Theorem 2 can be seen as a sharpening of Bridges-Mehta's theorem.

Example 2 Let $X = \mathbb{R}$. Let $\tau = \tau_>$ stands for the Euclidean topology. Define the binary relation $\succ$ as follows: For all $x, y \in \mathbb{R}$, $x \succ y$ if and only if $\lceil x \rceil > \lceil y \rceil + 1$. Then, for all $x, y \in \mathbb{R}$, $x \succ_1 y \iff \lceil x \rceil > \lceil y \rceil$. So, $\succ$ is an interval order that is representable and satisfies (LSC), but $\succ_1$ does not satisfy (USC).

A similar result to Theorem 2 obtains if one considers $\succ_2$ instead of $\succ_1$.

Theorem 3 *Let (X, τ) be a connected separable topological space and let $\succ \subset X^2$. Assume that (IR), (WPTR) hold. Then, if $\succ$ satisfies (USC) and $\succ_2$ satisfies (C) then there exist a continuous function $u_2 : X \longrightarrow \mathbb{R}$ representing $\succ_2$ and $\psi_2 : X \setminus \min(\succ) \longrightarrow \mathbb{R}$ with $\psi_2 \geq 0$, such that for all $x \in X \setminus \min(\succ)$ and $y \in X$,*

$$x \succ y \iff u_2(x) - \psi_2(x) > u_2(y) \, (\star),$$

with

$$\psi_2(x) = \inf_{x':x'\prec x} u_2(x) - u_2(x') ,$$

that is,

$$x \succ y \iff U_2(x) > U_2(y) + \psi_2(y) ,$$

---

[2]To be more explicit: "$\{z \in X : x \in C'(z)\}$ is open", where $C'(z) = \{y : \exists x' \,/\, y \succ x' \succeq z\}$. We have that $C'(z) = (z, \rightarrow)_{\succ_1}$, and thus $\{z \in X : x \in C'(z)\} = (\leftarrow, x)_{\succ_1}$.

with

$$U_2(y) = u_2(y) - \psi_2(y) = \sup_{y':y' \prec y} u_2(y') .$$

Conversely, if $\succ$ admits an interval representation given by $(\star)$ then (IR), (WPTR), (USC) hold and $\succ_2$ satisfies (C).

In particular, $\succ$ satisfies (LSC) if and only if ψ_2 is upper semi-continuous (or U_2 is lower semi-continuous).

Moreover, (U_2, u_2) is unique up to a common increasing transformation, *i.e., if (U_2', u_2') represents $\succ$ with u_2' representing $\succ_2$ and U_2' defined by $U_2'(x) = \sup_{x':x' \prec x} u_2'(x')$ for $x \in X \setminus \min(\succ)$ then there exists an increasing transformation ϕ such that $u_2' = \phi \circ u_2$ and $U_2' = \phi \circ U_2$.*

Proof The proof is quite similar to Theorem 2. Let us prove the first $\Leftarrow$.

Let $x \in X \setminus \min(\succ)$ and $y \in X$ such that $u_2(x) - u_2(y) > \psi_2(x) \geq 0$. Then, there exists $x' \in X$ with $x' \prec x$ such that $u_2(x) - u_2(y) > u_2(x) - u_2(x')$. Thus, $x \succ x' \succ_2 y$, so by (WPTR), $x \succ y$ holds.

Let us establish $\Rightarrow$.

Let $x \in X$ and $y \in X \setminus \max(\succ)$ with $x \succ y$. Since $\succ$ satisfies (USC), $(\leftarrow, x) \in \tau$ and $y \in (\leftarrow, x)$. Since $\succ_2$ satisfies (C) and $\succ \subset \succ_2$, we have that $(y, \rightarrow)_{\succ_2} \in \tau$ and $x \in (y, \rightarrow)_{\succ_2}$.

Let us prove that $(\leftarrow, x) \cup (y, \rightarrow)_{\succ_2} = X$. Let $z \in X$ such that $z \not\prec x$. If $z \sim x$, since $y \prec x$, we have $y \prec_2 z$. If $z \succ x$, then since $x \succ y$, by transitivity $z \succ y$, and then $z \succ_2 y$ since $\succ \subset \succ_2$. So in either case $z \in (y, \rightarrow)_{\succ_2}$.

Now X is connected, thus $(\leftarrow, x) \cap (y, \rightarrow)_{\succ_2} \neq \emptyset$. So, there exists $x' \in X$ such that $y \prec_2 x' \prec x$. And then, it follows that,

$$u_2(x) - \psi_2(x) \geq u_2(x) - (u_2(x) - u_2(x')) = u_2(x') > u_2(y) .$$

(Converse). Assume there exist a continuous function $u_2 : X \longrightarrow \mathbb{R}$ representing $\succ_2$ and $\psi_2 : X \setminus \min(\succ) \longrightarrow \mathbb{R}$ with $\psi_2 \geq 0$ satisfying $(\star)$. Clearly, (IR) is satisfied since for all $x \in X$, $u_2(x) - \psi_2(x) \leq u_2(x)$.

Let us check (WPTR). Let $x, y, z, t \in X$ such that $x \succ y \sim z \succ t$. Then, $x \succ y \succ_2 t$. Thus, $u_2(x) - \psi_2(x) > u_2(y) > u_2(t)$, and then $x \succ t$.

Let $y \in X \setminus \min(\succ)$. Then, $(\leftarrow, y) = u_2^{-1}((-\infty, u_2(y) + \psi_2(y)))$ which is open since u_2 is continuous. So, $\succ$ satisfies (USC).

Since u_2 is continuous, $\succ_2$ satisfies (C).

(In particular). Assume U_2 is lower semi-continuous. Let $x, y \in X$ with $x \succ y$. Then, $(y, \rightarrow) = U_2^{-1}((U_2(y) + \psi_2(y), +\infty))$ which is open since U_2 is lower semi-continuous. Hence, $\succ$ satisfies (LSC).

Assume $\succ$ satisfies (LSC). Let $\alpha \in \mathbb{R}$ and $x_0 \in X$ such that $U_2(x_0) > \alpha$. There exists $x' \in X$ with $x' \prec x_0$ such that $U_2(x_0) \geq u_2(x') > \alpha$. Thus, for all $x \in (x', \rightarrow)$ it holds $U_2(x) \geq u_2(x') > \alpha$. Now by (LSC), $(x', \rightarrow) \in \tau$ and $x_0 \in (x', \rightarrow)$, this proves that U_2 is lower semi-continuous at x_0.

(Moreover). Any common increasing transformation $(\phi \circ U_2, \phi \circ u_2)$ with ϕ increasing on $u_2(X)$ provides an interval representation (let us notice that $U_2(X) \subset u_2(X)$

since $u_2(X)$ is a bounded or unbounded interval in $\mathbb{R}$). For all $x \in X$, we have $\phi \circ U_2(x) = \phi(\sup_{x':x' \prec x} u_2(x')) = \sup_{x':x' \prec x} \phi(u_2(x'))$. Thus, for all $x, y \in X$, $x \succ y \iff U_2(x) > u_2(y) \iff \phi \circ U_2(x) > \phi \circ u_2(y)$.

Conversely, let (U_2', u_2') represents $\succ$ with u_2' representing $\succ_2$ and U_2' defined for $x \in X \setminus \min(\succ)$ by,

$$U_2'(x) = \sup_{x':x' \prec x} u_2'(x') .$$

Then, there exists an increasing transformation ϕ such that $u_2' = \phi \circ u_2$. Hence, we have for $x \in X \setminus \min(\succ)$,

$$U_2'(x) = \sup_{x':x' \prec x} \phi \circ u_2(x') = \phi(\sup_{x':x' \prec x} u_2(x')) = \phi(U_2(x)) .$$

This establishes that the interval representation is unique up to a common increasing transformation. □

Remark 2 A natural question is: How can we combine our Theorems 2 and 3 to obtain a continuous interval representation? If one requires (IR), (WPTR) to hold and also that $\succ, \succ_1, \succ_2$ satisfy (C) then, according to Chateauneuf's Corollary, $\succ$ admits an interval order representation (u, v) with u, v continuous and where u, v represent $\succ_1, \succ_2$. An indirect argument relies on Theorem 3.5 p. 219 in [11], see also Remark 3.6 (iii), (iv) p. 224. Therein, $\succ$ admits a continuous interval order representation where u, v represent $\succ_1, \succ_2$ if and only if $\succ$ admits an interval order representation with u, v both upper (or lower) semi-continuous and $\succ_1, \succ_2$ satisfying (C). So, we may apply our Theorems 2 or 3 and suppose that $\succ, \succ_1, \succ_2$ satisfy (C) in order to conclude that a continuous interval representation exists.

6 Generalizations

Let us tackle now with a general version of interval order representations. We will also provide one-side and weak interval representations of preferences. We shall assume that there exists a given continuous strict weak order $\succ^* \subset X^2$ and that $\succ$ and $\succ^*$ behave in accordance. Here $\succ^*$ may be interpreted as an objective binary relation and $\succ$ is a subjective binary relation.

(CP) Compatibility $\succ$ *is* compatible with $\succ^*$, i.e. $\succ^*$ *is an* extension *of* $\succ$, *if for all* $x, y \in X$, $x \succ y \Rightarrow x \succ^* y$.

Since $\succ^*$ is a strict weak order, if $\succ \subset \succ^*$ then necessarily $\succ$ is acyclic. Next proposition is an immediate consequence of Debreu-Eilenberg's theorem and extends Theorem 1.

Proposition 2 *Let* (X, τ) *be a connected separable topological space and let* $\succ$, $\succ^* \subset X^2$. *Assume that (IR) holds and* $\succ^*$ *is a continuous strict weak order. If* $\succ^*$ *is an extension of* $\succ$ *then there exists a continuous function* $u^* : X \longrightarrow \mathbb{R}$ *(representing* $\succ^*$*) such that for all* $x, y, z \in X$,

$$x \succ y \Rightarrow u^*(x) > u^*(y) \,,$$

$$\text{and } x \succeq^* y \succ z \Rightarrow u^*(x) > u^*(z) \,.$$

Moreover, any increasing transformation of u^* is convenient, *i.e., u^* is an ordinal representation of $\succ^*$.*

Proof According to Debreu-Eilenberg's theorem, there exists a continuous function $u^* : X \longrightarrow \mathbb{R}$ representing $\succ^*$. Since $\succ \subset \succ^*$, for all $x, y, z \in X$, if $x \succ y$ then $x \succ^* y$ thus $u^*(x) > u^*(y)$ and if $x \succeq^* y \succ z$ then $u^*(x) \geq u^*(y) > u^*(z)$. (Moreover). Let u'^* represent $\succ^*$. Then, there exists an increasing transformation ϕ such that $u'^* = \phi \circ u^*$ and u'^* is a weak utility function representing $\succ$.   □

(LH) Lower-hereditary $\succ^*$ *is* lower-hereditary *with respect to* $\succ$, *if for all $x, y, z \in X$, $x \succ^* y \succ z \Rightarrow x \succ z$.*
(LR) Lower-regularity $\succ^*$ *is* lower-regular *with respect to* $\succ$, *if for all $x, y, z \in X$, $x \sim^* y \succ z \Rightarrow x \succ z$.*

   Otherwise put, (LH) states that lower sections should be ordered (by inclusion) under $\prec^*$ and (LR) states that lower sections should coincide under $\sim^*$ i.e., for all $x, y \in X$,

$$y \prec^* x \Rightarrow (\leftarrow, y) \subset (\leftarrow, x) \,,$$

$$\text{and } y \sim^* x \Rightarrow (\leftarrow, y) = (\leftarrow, x) \,.$$

In a concise manner, (LH) and (LR) are satisfied if,

$$\forall x, y, z \in X, x \succeq^* y \succ z \Rightarrow x \succ z \,.$$

Similarly, we may introduce upper-hereditary and upper-regularity, however we shall not repeat the upcoming results with upper versions.
(UH) Upper-hereditary $\succ^*$ *is* upper-hereditary *with respect to* $\succ$, *if for all $x, y, z \in X$, $x \succ y \succ^* z \Rightarrow x \succ z$.*
(UR) Lower-regularity $\succ^*$ *is* upper-regular *with respect to* $\succ$, *if for all $x, y, z \in X$, $x \succ y \sim^* z \Rightarrow x \succ z$.*
   In a concise manner, (UH) and (UR) are satisfied if,

$$\forall x, y, z \in X, x \succ y \succeq^* z \Rightarrow x \succ z \,.$$

We may notice that if (CP) and (LH) hold, then $\succ$ necessarily satisfies (IR) and (TR). Indeed, since $\succ^*$ satisfies (IR) and $\succ \subset \succ^*$ then $\succ$ satisfies (IR). Let $x, y, z \in X$ with $x \succ y$ and $y \succ z$. Then, by (CP) $x \succ^* y$ and then, by (LH) $x \succ z$.
   For instance consider $\succ$ and $\succ_1$ with $\succ$ satisfying (IR) and (WPTR). By Properties 7–8, $\succ_1$ is a strict weak order. Since $\succ$ satisfies (IR), by Property 5 it holds $\succ \subset \succ_1$, hence $\succ$ is compatible with $\succ_1$. Since $\succ$ satisfies (IR) and (TR) (see Property 3) then by Property 10, (LH) and (LR) are satisfied. Similarly, $\succ_2$ is a strict weak order, $\succ \subset \succ_2$ and by Property 10, (UH) and (UR) are satisfied.

Theorem 2 can be seen as a special case of the following.

Theorem 4 *Let (X, τ) be a connected separable topological space and let $\succ, \succ^* \subset X^2$. Assume that (IR), (TR) hold, that $\succ$ satisfies (LSC) and $\succ^*$ is a continuous strict weak order and that $\succ^*$ is an extension of $\succ$ (CP) and that $\succ^*$ is lower-hereditary (LH) and lower-regular with respect to $\succ$ (LR). Then, there exist a continuous function $u^* : X \longrightarrow \mathbb{R}$ representing $\succ^*$ and $\psi^* : X \setminus \max(\succ) \longrightarrow \mathbb{R}$ with $\psi^* \geq 0$, such that for all $x \in X$ and $y \in X \setminus \max(\succ)$,*

$$x \succ y \iff u^*(x) > u^*(y) + \psi^*(y) = v^*(y) \ (\star),$$

with

$$\psi^*(y) = \inf_{x':x'\succ y} u^*(x') - u^*(y) \text{ and } v^*(y) = \inf_{x':x'\succ y} u^*(x').$$

Conversely, if $\succ$ admits an interval representation given by $(\star)$ then (IR), (TR), (LSC) hold and $\succ^*$ satisfies (C), $\succ^*$ is an extension of $\succ$ and $\succ^*$ is lower-hereditary and lower-regular with respect to $\succ$.

In particular, $\succ$ satisfies (USC) if and only if ψ^* is upper semi-continuous (or v^* is upper semi-continuous).

Moreover, (u^*, v^*) is unique up to a common increasing transformation, i.e., if (u'^*, v'^*) represents $\succ$ with u'^* representing $\succ^*$ and v'^* defined by $v'^*(x) = \inf_{x':x'\succ x} u'^*(x')$ for $x \in X \setminus \max(\succ)$ then there exists an increasing transformation ϕ such that $u'^* = \phi \circ u^*$ and $v'^* = \phi \circ v^*$.

Proof According to Debreu-Eilenberg's theorem there exists $u^* : X \longrightarrow \mathbb{R}$ representing $\succ^*$.

Let us prove the first $\Leftarrow$.

Let $x \in X$ and $y \in X \setminus \max(\succ)$ such that $u^*(x) - u^*(y) > \psi^*(y) \geq 0$. Then, there exists $x' \in X$ with $x' \succ y$ such that $u^*(x) - u^*(y) > u^*(x') - u^*(y)$. Thus, $x \succ^* x' \succ y$, so by lower-heredity, $x \succ y$ holds.

Let us establish $\Rightarrow$.

Let $x \in X$ and $y \in X \setminus \max(\succ)$ with $x \succ y$. We have $x \in (y, \rightarrow)$ and $y \in (\leftarrow, x)_{\succ^*}$ since $\succ \subset \succ^*$. We have that $\succ$ satisfies (LSC), thus $(y, \rightarrow) \in \tau$ and $\succ^*$ satisfies (C), thus $(\leftarrow, x)_{\succ^*} \in \tau$.

Let us prove that $(y, \rightarrow) \cup (\leftarrow, x)_{\succ^*} = X$. Assume $z \notin (\leftarrow, x)_{\succ^*}$. That is $z \not\prec^* x$. If $z \succ^* x$, since, $x \succ y$, by (LH) we get that $z \succ y$. If $z \sim^* x$, since, $x \succ y$, by (LR) we get that $z \succ y$. In either case, we have that $z \in (y, \rightarrow)$.

Now X is connected, thus $(y, \rightarrow) \cap (\leftarrow, x)_{\succ^*} \neq \emptyset$. So, there exists $x' \in X$ such that $y \prec x' \prec^* x$. And then, since $\succ$ is compatible with $\succ^*$, we get that $y \prec^* x'$, and also $u^*(x') < u^*(x)$ since $x' \prec^* x$. Hence

$$u^*(x) - u^*(y) > u^*(x') - u^*(y) \geq \psi^*(y).$$

(Converse). Assume there exist a continuous function $u^* : X \longrightarrow \mathbb{R}$ representing $\succ^*$ and $\psi^* : X \setminus \max(\succ) \longrightarrow \mathbb{R}$ with $\psi^* \geq 0$ satisfying $(\star)$. Clearly, (IR) is satisfied since for all $x \in X, u^*(x) \leq u^*(x) + \psi^*(x)$.

Let us check (TR). Similarly, let x, y, z with $x \succ y$, $y \succ z$ then $u^*(x) > u^*(y) + \psi^*(y)$ and $u^*(y) > u^*(z) + \psi^*(z)$. Thus, $u^*(x) + u^*(y) > u^*(y) + \psi^*(y) + u^*(z) + \psi^*(z)$ and a fortiori $u^*(x) > u^*(z) + \psi^*(z)$ since $\psi^*(y) \geq 0$.

Since u^* is continuous, $\succ^*$ satisfies (C).

Let $y \in X \setminus \max(\succ)$. Then, $(y, \rightarrow) = (u^*)^{-1}((u^*(y) + \psi^*(y), +\infty))$ which is open since u^* is continuous. So, $\succ$ satisfies (LSC).

Obviously, $\succ \subset \succ^*$, if $x \succ y$ then $u^*(x) > u^*(y) + \psi^*(y)$ thus $u^*(x) > u^*(y)$ since $\psi^*(y) \geq 0$, so $x \succ^* y$.

Let us check lower-heredity and lower-regularity. Let x, y, $z \in X$ such that $x \succeq^* y \succ z$. Thus, $u^*(x) \geq u^*(y) > u^*(z) + \psi^*(z)$, and then $u^*(x) > u^*(z) + \psi^*(z)$ so $x \succ z$.

(In particular). Assume v^* is upper semi-continuous. Let x, $y \in X$ with $x \succ y$. Then, $(\leftarrow, x) = (v^*)^{-1}((-\infty, u^*(x)))$ which is open since v^* is upper semi-continuous. Hence, $\succ$ satisfies (USC).

Assume $\succ$ satisfies (USC). Let $\alpha \in \mathbb{R}$ and $x_0 \in X$ such that $v^*(x_0) < \alpha$. There exists $x' \in X$ with $x' \succ x_0$ such that $v^*(x_0) \leq u^*(x') < \alpha$. Thus, for all $x \in (\leftarrow, x')$ it holds $v^*(x) \leq u^*(x') < \alpha$. Now by (USC), $(\leftarrow, x') \in \tau$ and $x_0 \in (\leftarrow, x')$, this proves that v^* is upper semi-continuous at x_0.

(Moreover). Any common increasing transformation $(\phi \circ u^*, \phi \circ v^*)$ with ϕ increasing on $u^*(X)$ provides an interval representation (let us notice that $v^*(X) \subset u^*(X)$ since $u^*(X)$ is a bounded or unbounded interval in $\mathbb{R}$). For all $x \in X$, we have $\phi \circ v^*(x) = \phi(\inf_{x':x' \succ x} u^*(x')) = \inf_{x':x' \succ x} \phi \circ u^*(x')$. Thus, for all x, $y \in X$, $x \succ y \iff u^*(x) > v^*(y) \iff \phi \circ u^*(x) > \phi \circ v^*(y)$.

Conversely, let (u'^*, v'^*) represents $\succ$ with u'^* representing $\succ^*$ and v'^* defined for $x \in X \setminus \max(\succ)$ by,

$$v'^*(x) = \inf_{x':x' \succ x} u'^*(x') .$$

Then, there exists an increasing transformation ϕ such that $u'^* = \phi \circ u^*$. Hence, we have for $x \in X \setminus \max(\succ)$,

$$v'^*(x) = \inf_{x':x' \succ x} \phi \circ u^*(x') = \phi(\inf_{x':x' \succ x} u^*(x')) = \phi(v^*(x)) .$$

This establishes that the interval representation is unique up to a common increasing transformation.   □

The following examples show that an interval representation exists even when $\succ_1$ is not continuous whether $\succ$ admits a utility or an interval representation.

Example 3 (Example 1 continued) Let $X = \mathbb{R}$. Let $\tau = \tau_{\succ}$ stands for the Euclidean topology. Define the binary relation $\succ$ as follows: For all x, $y \in \mathbb{R}$, $x \succ y$ if and only if $\lceil x \rceil > \lceil y \rceil$. Consider $\succ^* = >$ the usual order on $\mathbb{R}$. Then, $\succ$ is compatible and $\succ^*$ is lower-hereditary and lower-regular w.r.t $\succ$. Moreover, for all x, $y \in \mathbb{R}$, we have, $x \succ y \iff \lceil x \rceil > \lceil y \rceil \iff x > \lceil y \rceil \iff x > y + \psi^*(y)$, with $\psi^*(y) = \lceil y \rceil - y \in [0, 1)$.

Example 4 (Example 2 continued) Let $X = \mathbb{R}$. Let $\tau = \tau_>$ stands for the Euclidean topology. Define the binary relation $\succ$ as follows: For all $x, y \in \mathbb{R}$, $x \succ y$ if and only if $\lceil x \rceil > \lceil y \rceil + 1$. Consider $\succ^* = \, > $ the usual order on $\mathbb{R}$. Then, $\succ$ is compatible and $\succ^*$ is lower-hereditary and lower-regular w.r.t $\succ$. Moreover, for all $x, y \in \mathbb{R}$, we have, $x \succ y \iff \lceil x \rceil > \lceil y \rceil + 1 \iff x > \lceil y \rceil + 1 \iff x > y + \psi^*(y)$, with $\psi^*(y) = \lceil y \rceil - y + 1 \in [1, 2)$.

Let us state now a *weak interval representation* result under the weaker assumption of acyclicity.

Corollary 1 *Let (X, τ) be a connected separable topological space and let $\succ$, $\succ^* \subset X^2$. Assume that $\succ$ is acyclic. If $\succ$ satisfies (LSC) and $\succ^*$ is a continuous strict weak order such that $\succ$ is compatible with $\succ^*$ (CP). Assume also that $\succ^*$ is lower-hereditary (LH) and lower-regular with respect to $\succ$ (LR). Then, there exist a continuous function $u^* : X \longrightarrow \mathbb{R}$ representing $\succ^*$ and $\psi^* : X \setminus \max(\succ) \longrightarrow \mathbb{R}$ with $\psi^* \geq 0$, such that for all $x \in X$ and $y \in X \setminus \max(\succ)$,*

$$x \succ y \Rightarrow u^*(x) > u^*(y) + \psi^*(y) = v^*(y) \ (\star),$$

with

$$\psi^*(y) = \inf_{x':x'\succ^{tr} y} u^*(x') - u^*(y) \text{ and } v^*(y) = \inf_{x':x'\succ^{tr} y} u^*(x') \,,$$

where $\succ^{tr}$ denotes the transitive closure[3] *of $\succ$.*

Conversely, if $\succ$ admits a weak interval representation given by $(\star)$ then $\succ$ is acyclic and $\succ^*$ satisfies (C), $\succ^*$ is an extension of $\succ$.

Moreover, any increasing transformation of (u^*, v^*) is convenient, i.e., any common increasing transformation (u'^*, v'^*) with $u'^* = \phi \circ u^*$ and $v'^* = \phi \circ v^*$ for some increasing transformation ϕ provides a weak interval representation.

Proof Let $\succ^{tr}$ denote transitive closure. By construction, $\succ \, \subset \, \succ^{tr}$ and $\succ^{tr}$ is transitive. The binary relation $\succ^{tr}$ is irreflexive if and only if $\succ$ is acyclic. Moreover, $\max(\succ) = \max(\succ^{tr})$. If $x \notin \max(\succ^{tr})$, there exist $y \in X$ and $n \in \mathbb{N}$, $y_0, \ldots, y_{n+1} \in X$ with $y_0 = x$ and $y_{n+1} = y$ such that $y_k \prec y_{k+1}$ for $k = 0, \ldots, n$. So, $x \prec y_1$, thus $x \notin \max(\succ)$. This proves the inclusion $\subset$. Since $\succ \, \subset \, \succ^{tr}$, the inclusion $\supset$ follows.

- Let us prove that $\succ^{tr}$ satisfies (LSC). Let $x, y \in X$ with $x \prec^{tr} y$. There exist $n \in \mathbb{N}$, $x_0, \ldots, x_{n+1} \in X$ with $x_0 = x$ and $x_{n+1} = y$ such that $x_k \prec x_{k+1}$ for $k = 0, \ldots, n$. By (LSC) there exists $O_y \in \tau$ with $y \in O_y$ such that for all $t \in O_y$, $x_n \prec t$. Thus, $x \prec^{tr} t$. So, $(x, \rightarrow)_{\succ^{tr}} \in \tau$.
- Since $\succ$ satisfies (CP) and since $\succ^*$ is transitive we have $\succ^{tr} \subset \succ^*$.

---

[3]The transitive closure of $\succ$, denoted by $\succ^{tr}$, is defined as follows: For all $x, y \in X$, $x \succ^{tr} y$ if and only if there exist $n \in \mathbb{N}$, $x_0, \ldots, x_{n+1} \in X$ with $x_0 = x$ and $x_{n+1} = y$ such that $x_k \succ x_{k+1}$ for $k = 0, \ldots, n$.

- Let $x, y, z \in X$ such that $x \succ^* y \succ^{tr} z$. There exists $n \in \mathbb{N}$, $y_0, \ldots, y_{n+1} \in X$ with $y_0 = y$ and $y_{n+1} = z$ such that $y_k \succ y_{k+1}$ for $k = 0, \ldots, n$. Thus, $x \succ^* y \succ y_1$ and by (LH) it comes $x \succ y_1$. Now, since $y_1 \succ \ldots \succ z$ we have $x \succ^{tr} z$. This establishes that $\succ^{tr}$ satisfies (LH) w.r.t. $\succ^*$.

- Let $x, y, z \in X$ such that $x \sim^* y \succ^{tr} z$. There exists $n \in \mathbb{N}$, $y_0, \ldots, y_{n+1} \in X$ with $y_0 = y$ and $y_{n+1} = z$ such that $y_k \succ y_{k+1}$ for $k = 0, \ldots, n$. Thus, $x \sim^* y \succ y_1$ and by (LR) it comes $x \succ y_1$. Now, since $y_1 \succ \ldots \succ z$ we have $x \succ^{tr} z$. This establishes that $\succ^{tr}$ satisfies (LR) w.r.t. $\succ^*$.

According to Theorem 4, there exist a continuous function $u^* : X \longrightarrow \mathbb{R}$ representing $\succ^*$ and $\psi^* : X \setminus \max(\succ^{tr}) \longrightarrow \mathbb{R}$ with $\psi^* \geq 0$, such that for all $x \in X$ and $y \in X \setminus \max(\succ^{tr})$,

$$x \succ^{tr} y \iff u^*(x) > u^*(y) + \psi^*(y) = v^*(y),$$

with

$$\psi^*(y) = \inf_{x':x' \succ^{tr} y} u^*(x') - u^*(y) \text{ and } v^*(y) = \inf_{x':x' \succ^{tr} y} u^*(x').$$

A fortiori, since $\succ \subset \succ^{tr}$ and $\max(\succ) = \max(\succ^{tr})$, we get that for all $x \in X$ and $y \in X \setminus \max(\succ)$,

$$x \succ y \Rightarrow u^*(x) > u^*(y) + \psi^*(y) = v^*(y).$$

(Moreover). It is immediate to check by composition.   □

Let us state now a conditional interval representation result without the assumption of compatibility. Here, $\succ^*$ gives a direction to comparisons between $x, y \in X$ with $x \succ^* y$ and then $\succ$ may comply or not with $\succ^*$ whether $x \succ y$ or $x \prec y$. Hence, a *one-side interval representation* obtains.

Corollary 2 *Let (X, τ) be a connected separable topological space and let $\succ, \succ^* \subset X^2$. Assume that (IR), (TR) hold, that $\succ$ satisfies (LSC) and $\succ^*$ is a continuous strict weak order and also that $\succ^*$ is lower-hereditary (LH) and lower-regular with respect to $\succ$ (LR) and that there exist $x, y \in X$ such that $x \succ y$ and $x \succ^* y$. Then, there exist a continuous function $u^* : X \longrightarrow \mathbb{R}$ representing $\succ^*$ and $\psi^* : X \setminus \max(\succ \cap \succ^*) \longrightarrow \mathbb{R}$ with $\psi^* \geq 0$, such that for all $x \in X$ and $y \in X \setminus \max(\succ \cap \succ^*)$,*

$$x \succ y \text{ and } x \succ^* y \iff u^*(x) > u^*(y) + \psi^*(y) = v^*(y) \ (\star).$$

Notice that since $\psi^* \geq 0$ if $u^*(x) > u^*(y) + \psi^*(y)$ holds then $u^*(x) > u^*(y)$, thus $x \succ^* y$.

Moreover, any increasing transformation of (u^*, v^*) is convenient.

Proof Let us introduce an auxiliary relation $\succ' = \succ \cap \succ^*$. Here $x \succ' y$ stands for "$x \succ y$ and $x \succ^* y$". Since there exist $x, y \in X$ such that $x \succ y$ and $x \succ^* y$, we have $\succ' \neq \emptyset$.

- Since $\succ$ satisfies (IR), $\succ'$ satisfies (IR). Since $\succ$ and $\succ^*$ satisfy (TR), $\succ'$ satisfies (TR).
- By construction, $\succ' \subset \succ^*$. So, $\succ^*$ is an extension of $\succ'$.
- By construction, $\succ^*$ is lower-hereditary w.r.t. $\succ'$. Indeed, let $x, y, z \in X$ with $x \succ^* y \succ' z$. Since $\succ' = \succ \cap \succ^*$, we have $x \succ^* y \succ z$ and $x \succ^* y \succ^* z$. So $x \succ z$ by (LH) and also $x \succ^* z$ by transitivity of $\succ^*$. That is, $x \succ' z$.
- By construction, $\succ^*$ is lower-regular w.r.t. $\succ'$. Let $x, y, z \in X$ with $x \sim^* y \succ' z$. Firstly, we have $x \sim^* y \succ^* z$, and since $\succ^*$ is a strict weak order it comes $x \succ^* z$. Secondly, we have $x \sim^* y \succ z$. Thus, by (LR), it comes $x \succ z$. That is, $x \succ' z$.
- Assume $\succ$ satisfies (LSC) and $\succ^*$ satisfies (C). Let $x \in X$. We have, $(x, \rightarrow)_{\succ'} = (x, \rightarrow)_{\succ} \cap (x, \rightarrow)_{\succ^*} \in \tau$. So, $\succ'$ satisfies (LSC).

Now according to Theorem 4, there exist a continuous function $u^* : X \longrightarrow \mathbb{R}$ representing $\succ^*$ and $\psi^* : X \setminus \max(\succ') \longrightarrow \mathbb{R}$ with $\psi^* \geq 0$, such that for all $x \in X$ and $y \in X \setminus \max(\succ')$,

$$x \succ' y \iff u^*(x) > u^*(y) + \psi^*(y) = v^*(y) \ (\star) ,$$

with

$$\psi^*(y) = \inf_{x' : x' \succ' y} u^*(x') - u^*(y) \text{ and } v^*(y) = \inf_{x' : x' \succ' y} u^*(x') .$$

(Moreover). It is immediate to check by composition. □

Let us state now a *weak interval representation* result without the assumption of lower-heredity.

Corollary 3 *Let (X, τ) be a connected separable topological space and let $\succ, \succ^* \subset X^2$. Assume that (IR), (TR) hold, that $\succ$ satisfies (LSC) and that $\succ^*$ is a continuous strict weak order and also that $\succ$ is compatible with $\succ^*$ (CP) and that $\succ^*$ is lower-regular with respect to $\succ$ (LR). Then, there exist a continuous function $u^* : X \longrightarrow \mathbb{R}$ representing $\succ^*$ and $\psi^* : X \setminus \max(\succ) \longrightarrow \mathbb{R}$ with $\psi^* \geq 0$, such that for all $x \in X$ and $y \in X \setminus \max(\succ)$,*

$$x \succ y \Rightarrow u^*(x) > u^*(y) + \psi^*(y) = v^*(y) \ (\star) .$$

Conversely, if $\succ$ admits a weak interval representation given by $(\star)$ then acyclicity holds.

Moreover, any increasing transformation of (u^*, v^*) is convenient.

Proof Let us introduce an auxiliary relation $\succ'' = \succ^* \circ \succ \cup \succ$. Here $x \succ^* \circ \succ \cup \succ y$ stands for "there exists z such that $x \succ^* z$ and $z \succ y$ or $x \succ y$". By construction, $\succ \subset \succ''$.

- Let us show that $\succ'$ satisfies (IR) and (TR). Let $x \in X$ with $x \succ'' x$. Since $\succ$ satisfies (IR) we have $x \not\succ x$. Hence, there exists $z \in X$ such that $x \succ^* z \succ x$. Since, $\succ \subset \succ''$, it comes $x \succ^* z \succ^* x$, so $x \succ^* x$, which is a contradiction.

Let $x, y, z \in X$ with $x \succ'' y \succ'' z$.

If $x \succ y \succ z$. Then, by (TR), $x \succ z$. So, $x \succ'' z$.

If $x \succ'' y \succ z$ and $x \not\succ y$. There exists $t \in X$ such that, $x \succ^* t \succ y$ and $y \succ z$. Then, by (TR), $x \succ^* t \succ z$. So, $x \succ'' z$.

If $x \succ'' y \succ'' z$ and $x \not\succ y, y \not\succ z$. There exists $t, s \in X$ such that, $x \succ^* t \succ y$ and $y \succ^* s \succ z$. Since, $\succ \subset \succ^*$, we have $x \succ^* t \succ^* y \succ^* s$. Then, by (TR) of $\succ^*$, $x \succ^* s$. So, $x \succ^* s \succ z$, that is $x \succ'' z$.

- Let us show that $\succ^*$ is lower-hereditary with respect to $\succ''$.

  Let $x, y, z \in X$ with $x \succ^* y \succ'' z$.

  If $y \succ z$. Then, $x \succ^* y \succ z$ that is $x \succ'' z$.

  If $y \not\succ z$. Then, there exists $t \in X$ such that $y \succ^* t \succ z$. Since $x \succ^* y$, by transitivity of $\succ^*$ it comes $x \succ^* t \succ z$. That is, $x \succ'' z$.

- Let us show that $\succ^*$ is lower-regular with respect to $\succ''$.

  Let $x, y, z \in X$ with $x \sim^* y \succ'' z$.

  If $y \succ z$, then $x \sim^* y \succ z$. Thus, by (LR) it comes $x \succ z$.

  If $y \not\succ z$, then there exists $t \in X$ such that $x \sim^* y \succ^* t \succ z$. Since $\succ^*$ is a strict weak order, we get that $x \succ^* t \succ z$. Thus, $x \succ'' z$.

- Let us show that $\succ''$ is compatible with $\succ^*$.

  Let $x, y \in X$ with $x \succ'' y$.

  If $x \succ y$. Then, $x \succ^* y$ holds by compatibility of $\succ$ with $\succ^*$.

  If $x \succ^* \circ \succ y$. There exists $z \in X$ such that $x \succ^* z \succ y$. Thus, $z \succ^* y$ by compatibility. And then, by transitivity of $\succ^*$ it follows that $x \succ^* y$.

- Let us show that $\succ''$ satisfies (LSC) whenever $\succ$ satisfies (LSC). Let $x, y \in X$ with $x \succ'' y$.

  If $x \succ y$. By (LSC), there exists $O_x \in \tau$ with $x \in O_x$ such that for all $t \in O_x$, $t \succ y$, thus, $t \succ'' y$. Hence, $O_x \subset (y, \rightarrow,)_{\succ''}$.

  If $x \succ'' y$ and $x \not\succ y$, there exists $z \in X$ such that $x \succ^* z$ and $z \succ y$. Since $\succ^*$ satisfies (LSC), there exists $O_x \in \tau$ with $x \in O_x$ such that for all $t \in O_x, t \succ^* z$, thus, $t \succ'' y$. Hence, $O_x \subset (y, \rightarrow)_{\succ''}$.

  We have established that $\succ''$ satisfies (LSC).

- Let us show that $\max(\succ'') = \max(\succ)$.

  ($\subset$). Suppose there exists $x \notin \max(\succ)$. Then, there exists $y \in X$ with $y \succ x$. Thus, $y \succ'' x$. So $x \notin \max(\succ'')$.

  ($\supset$). Suppose there exists $x \notin \max(\succ'')$. Then, there exists $y \in X$ with $y \succ'' x$. Then, either $y \succ x$ or there exists $z \in X$ such that $y \succ^* z \succ x$. So, either $y \succ x$ or $z \succ x$, thus $x \notin \max(\succ)$.

Now, according to Theorem 4, there exist a continuous function $u^* : X \longrightarrow \mathbb{R}$ representing $\succ^*$ and $\psi^* : X \setminus \max(\succ'') \longrightarrow \mathbb{R}$ with $\psi^* \geq 0$, such that for all $x \in X$ and $y \in X \setminus \max(\succ'')$,

$$x \succ'' y \iff u^*(x) > u^*(y) + \psi^*(y) \ (\star),$$

with

$$\psi^*(y) = \inf_{x'' : x'' \succ'' y} u^*(x'') - u^*(y).$$

Here, ψ^* is built with $\succ''$ and not with $\succ$. In particular, since $\succ \subset \succ''$ and $\max(\succ'') = \max(\succ)$ hold, for all $x \in X$ and $y \in X \setminus \max(\succ)$,

$$x \succ y \Rightarrow u^*(x) > u^*(y) + \psi^*(y) .$$

(Moreover). It is immediate to check by composition.   □

The following example shows that we may have a compatible continuous utility representation but without (LH) and (LR).

Example 5 Let $X = \mathbb{R}^2$. Let $\tau = \tau_>$ stands for the Euclidean topology. Define the binary relation $\succ \; = \; >>$ as follows: For all $x, y \in \mathbb{R}$, $x >> y$ if and only if $x_1 > y_1$ and $x_2 > y_2$. Consider $\succ^* \; = \; >^1$ defined as follows: For all $x, y \in \mathbb{R}$, $x >^1 y$ if and only if $x_1 > y_1$. Clearly $\succ$ satisfies (LSC) and $>> \subset >^1$. However, $(1, 0) >^1 (0, 2) >> (-1, 1)$ and also $(1, 0) =^1 (1, 2) >> (-1, 1)$ but $(1, 0) \not> \not> (-1, 1)$. Hence, neither (LH) nor (LR) holds.

We may extend Corollaries 2–3 to strict partial orders without the compatibility condition (CP) nor the heredity condition (LH). Hence, a *one-side weak interval representation* obtains.

Corollary 4 *Let (X, τ) be a connected separable topological space and let $\succ, \succ^* \subset X^2$. Assume that (IR), (TR) hold and that $\succ$ satisfies (LSC). Assume also that $\succ^*$ is a continuous strict weak order and lower-regular with respect to $\succ$ (LR) and that there exist $x, y \in X$ such that $x \succ y$ and $x \succ^* y$, then there exist a continuous function $u^* : X \longrightarrow \mathbb{R}$ representing $\succ^*$ and $\psi^* : X \setminus \max(\succ \cap \succ^*) \longrightarrow \mathbb{R}$ with $\psi^* \geq 0$, such that for all $x \in X$ and $y \in X \setminus \max(\succ \cap \succ^*)$,*

$$x \succ y \text{ and } x \succ^* y \Rightarrow u^*(x) > u^*(y) + \psi^*(y) = v^*(y) \; (\star) .$$

Notice that since $\psi^* \geq 0$ if $u^*(x) > u^*(y) + \psi^*(y)$ holds then $x \succ^* y$ necessarily holds. Moreover, any increasing transformation of (u^*, v^*) is convenient.

Similarly, if there exist $x, y \in X$ such that $x \succ y$ and $x \prec^* y$, then there exist a continuous function $u^{**} : X \longrightarrow \mathbb{R}$ representing $\prec^*$ and $\psi^{**} : X \setminus \max(\succ \cap \prec^*) \longrightarrow \mathbb{R}$ with $\psi^{**} \geq 0$, such that for all $x \in X$ and $y \in X \setminus \max(\succ \cap \prec^*)$,

$$x \succ y \text{ and } x \prec^* y \Rightarrow u^{**}(x) > u^{**}(y) + \psi^{**}(y) = v^{**}(y) \; (\star) .$$

Notice that since $\psi^{**} \geq 0$ if $u^{**}(x) > u^{**}(y) + \psi^{**}(y)$ holds then $x \prec^* y$ necessarily holds. Moreover, any increasing transformation of (u^{**}, v^{**}) is convenient.

Proof Let us introduce an auxiliary relation $\succ''' \; = \; \succ \cap \succ^*$. Here $x \succ''' y$ stands for "$x \succ y$ and $x \succ^* y$".

- By construction, $\succ''' \subset \succ^*$. Since there exist $x, y \in X$ such that $x \succ y$ and $x \succ^* y$, we have $\succ''' \neq \emptyset$.

- Since $\succ$ satisfies (IR), $\succ'''$ satisfies (IR). Since $\succ$ and $\succ^*$ satisfy (TR), $\succ''$ satisfies (TR).
- Assume $\succ$ satisfies (LSC) and $\succ^*$ satisfies (C). Let $x \in X$. We have, $(x, \to)_{\succ'''} = (x, \to)_{\succ} \cap (x, \to)_{\succ^*} \in \tau$. So, $\succ'''$ satisfies (LSC).
- Let us check that $\succ'''$ satisfies (LR). Let $x, y, z \in X$ with $x \sim^* y \succ''' z$. Then, since $x \sim^* y \succ^* z$ and $\succ^*$ is a strict weak order, we have $x \succ^* z$. And, since $x \sim^* y \succ z$, we have $x \succ z$ by (LR). Finally, $x \succ''' z$.

Now according to Corollary 3, there exist a continuous function $u^* : X \longrightarrow \mathbb{R}$ representing $\succ^*$ and $\psi^* : X \setminus \max(\succ''') \longrightarrow \mathbb{R}$ with $\psi^* \geq 0$, such that for all $x \in X$ and $y \in X \setminus \max(\succ''')$,

$$x \succ''' y \Rightarrow u^*(x) > u^*(y) + \psi^*(y) \ (\star).$$

(Moreover). It is immediate to check by composition.

Let us introduce an auxiliary relation $\succ'''' = \succ \cap \prec^*$. Here $x \succ'''' y$ stands for "$x \succ y$ and $x \prec^* y$". We may take $u^{**} = -u^*$ to represent $\prec^*$. However, ψ^{**} does not coincide with $-\psi^*$.

(Moreover). It is immediate to check by composition. □

Example 6 Let $X = \mathbb{R}$. Let $\tau = \tau_>$ stands for the Euclidean topology. Define the binary relation $\succ$ as follows: For all $x, y \in \mathbb{R}$, $x \succ y$ if and only if $x^2 > y^2$. Consider $\succ^* = >$ the usual order on $\mathbb{R}$. Then (LR) is satisfied since, $x \sim^* y \iff x = y$ for all $x, y \in X$. However, $\succ$ is not compatible with $\succ^*$, for instance $(-2)^2 = 4 > 1 = 1^2$ but $-2 \not\succ 1$. And (LH) is not satisfied, for instance $0 > -2$ and $(-2)^2 = 4 > 1 = (-1)^2$ but $0^2 = 0 \not\succ 1 = (-1)^2$.

Examples 5–6 show that (CP) and (LR) are independent conditions.

Next example shows that we may not be able to apply Corollary 4, since neither (CP), (LH) nor (LR) holds, even though $\succ$ admits a continuous utility representation.

Example 7 Let $X = \mathbb{R}$. Let $\tau = \tau_>$ stands for the Euclidean topology. Consider $\succ = >$ the usual order on $\mathbb{R}$. Define the strict weak order $\succ^*$ as follows: For all $x, y \in \mathbb{R}$, $x \succ^* y$ if and only if $|x| > |y|$. Then, $\succ$ is not compatible with $\succ^*$, for instance $-1 > -2$ but $|-1| \not\succ |-2|$. And also, (LH) is not satisfied, for instance $|-2| > |1|$ and $1 > 0$ but $-2 \not\succ 1$. Finally, (LR) is not satisfied, for instance $|-1| > |1|$ and $1 > 0$ but $-1 \not\succ 0$.

7 Final Remarks

We have first considered interval orders on connected separable topological spaces under lower (respectively upper) semi-continuity and continuity of the derived relation of precedence $\prec_1$ (resp. succession $\prec_2$). Thanks to Debreu-Eilenberg's theorem, we were able to obtain partial continuous interval representations of preferences

$(u_1, u_1 + \psi_1)$ with u_1 continuous or $(u_2 - \psi_2, u_2)$ with u_2 continuous. A natural question is to know whether similar results could hold for other topological spaces, for instance second countable [13, 14].

The second part of our contribution (i.e. Sect. 6) assumes that an objective continuous strict weak order is given on a connected separable topological spaces. This assumption is equivalent to the existence of a non-constant continuous function on X. Partial continuous interval representations of preferences can be obtained under compatibility, lower-heredity and lower-regularity conditions. The question regarding general topological spaces, for instance second countable topological spaces, poses again. Another line of research would be to relate these results with multi-utility representations (see [7] and the references therein). A possible way to make a connection with multi-utility representations is to consider an interval order defined for all $x, y \in X$ by,

$$x \succ y \iff I_{\mathcal{U}}(x) = \inf_{\mathcal{U}} f(x) > \sup_{\mathcal{U}} f(y) = J_{\mathcal{U}}(x),$$

where $\mathcal{U} \neq \emptyset, \mathcal{U} \subset \mathcal{C}$ is a family of continuous real-valued functions on X. Here, $I_{\mathcal{U}}$ denotes a lower utility function and $J_{\mathcal{U}}$ denotes an upper utility function. A fortiori if $x \succ y$ then $f(x) > f(y)$ for all $f \in \mathcal{U}$. Conversely, let $\succ$ be an interval order that admits (u, v) for a representation (with $u \leq v$). We may naturally associate the family of continuous real-valued functions on X bounded by u and v, $\mathcal{U}_{u,v} = \{f : f \in \mathcal{C}, u \leq f \leq v\}$ and $u \leq I_{\mathcal{U}_{u,v}} \leq J_{\mathcal{U}_{u,v}} \leq v$. In particular, $I_{\mathcal{U}_{u,v}} = u$ whenever u is continuous and $J_{\mathcal{U}_{u,v}} = v$ whenever v is continuous, this provides a way to obtain partial continuous interval representations.

8 Proofs of Properties

Proof (Property 1) Let $x, y, z, w \in X$ such that $x \prec y \preceq z \prec w$. If $y \sim z$ then by (WPTR) it follows that $x \prec w$. If $y \nsim z$ we have that $y \prec z$ and then by (TR) it follows that $x \prec w$. $\qquad \square$

Proof (Property 2) Suppose $\succ$ is not asymmetric. There exist $x, y \in X$ such that $x \succ y$ and $y \succ x$. By (TR), $x \succ x$, so contradicting (IR). $\qquad \square$

Proof (Property 3) Let $x, y, z \in X$ with $x \prec y, y \prec z$. By (IR) we have $y \sim y$. We get that $x \prec y \sim y \prec z$ and then by (WPTR), $x \prec z$. $\qquad \square$

Property 4 follows by conjunction of Properties 2 and 3. We provide a direct proof.

Proof (Property 4) Suppose $\succ$ is not asymmetric. There exist $x, y \in X$ such that $x \succ y$ and $y \succ x$. By (IR) we have $y \sim y$. We get that $x \prec y \sim y \prec x$ and then by (WPTR), $x \prec x$, so contradicting (IR). $\qquad \square$

Proof (*Property* 5) Let $x, y \in X$ such that $x \prec y$. By (IR) we have $x \sim x$ and we get that $x \sim x \prec y$, hence $x \prec_1 y$. Similarly, by (IR) we have $y \sim y$ and we get that $x \prec y \sim y$, hence $x \prec_2 y$. $\qquad\square$

Proof (*Property* 6) Let $x, y \in X$ with $y \not\prec x$. Assume $y \prec_1 x$, then there exists $z \in X$ such that $y \sim z \prec x$. So, $z \not\prec y$, and by negative transitivity it follows that $z \not\prec x$, which is a contradiction. Hence, $\prec_1 \subset \prec$. Similarly, let $x, y \in X$ with $y \not\prec x$. Assume $y \prec_2 x$, then there exists $z \in X$ such that $y \prec z \sim x$. So, $x \not\prec z$, and by negative transitivity it follows that $y \not\prec z$, which is a contradiction. Hence, $\prec_2 \subset \prec$. $\qquad\square$

Proof (*Property* 7) Let $x, y \in X$ such that $x \prec_1 y$. There exists $z \in X$ such that $x \sim z$ and $z \prec y$. Assume $y \prec_1 x$. There exists $t \in X$ such that $y \sim t \prec x$, hence $z \prec y \sim t \prec x$. By (WPTR) we have $z \prec x$, which is a contradiction. So $y \not\prec_1 x$.

Similarly, let $x, y \in X$ such that $x \prec_2 y$. There exists $z \in X$ such that $x \prec z$ and $z \sim y$. Assume $y \prec_2 x$. There exists $t \in X$ such that $y \prec t \sim x$, hence $y \prec t \sim x \prec z$. By (WPTR) we have $y \prec z$, which is a contradiction. So $y \not\prec_2 x$. $\qquad\square$

Proof (*Property* 8) Let $x, y, z \in X$ such that $x \prec_1 z$. Let us prove that either $x \prec_1 y$ or $y \prec_1 z$ holds. There exists $s \in X$ such that $s \sim x$ and $s \prec z$. Assume $x \not\prec_1 y$. Since $s \sim x$, we have $s \not\prec y$.

If $s \not\succ y$ it follows that $z \succ s \sim y$, so $y \prec_1 z$.

If $s \succ y$, since $s \prec z$ we have $y \prec z$ by (TR). And since $\succ$ satisfies (IR), we have $y \prec_1 z$ by Property 5.

Similarly, let $x, y, z \in X$ such that $x \prec_2 z$. Let us prove that either $x \prec_2 y$ or $y \prec_2 z$ holds. There exists $s \in X$ such that $x \prec s$ and $s \sim z$. Assume $y \not\prec_2 z$. Since $s \sim z$, we have $y \not\prec s$.

If $s \not\prec y$ it follows that $x \prec s \sim y$, so $x \prec_2 y$.

If $s \prec y$, since $x \prec s$ we have $x \prec y$ by (TR). And since $\succ$ satisfies (IR), we have $x \prec_2 y$ by Property 5. $\qquad\square$

Proof (*Property* 9) Assume there are $x, y \in X$ such that $x \prec^b y$ and $x \not\prec^a y$. Then, $y \not\prec^b x$ since $\succ^b$ is asymmetric. Thus, $y \not\prec^a x$ since $\prec^a \subset \prec^b$. Hence, $x \sim^a y$, and then $x \sim^b y$. But $x \prec^b y$, which is a contradiction.

We have to prove that $\prec_1 = \prec_1^*$ and $\prec_2 = \prec_2^*$ whenever $\prec$ satisfies (IR) and (WPTR). Clearly, $\prec_1 \subset \prec_1^*$ and $\prec_2 \subset \prec_2^*$. Let us check that $\prec_1^*$ is irreflexive. Assume there exists $x \in X$ such that $x \prec_1^* x$. Then, there exists $s \in X$ such that $x \preceq s \prec x$, that is $s \not\prec x$ and $s \prec x$, which is absurd. Let us check that $\prec_1^*$ is asymmetric. Let $x, y \in X$ with $x \prec_1^* y$. Assume $y \prec_1^* x$. Then, there exist $s, t \in X$ such that $x \preceq t \prec y \preceq s \prec x$. According to Properties 1 and 3, (WPTR) implies (PTR). Thus, $x \preceq t \prec x$, which is absurd.

Let us check $\sim_1 \subset \sim_1^*$. Assume, by contradiction, that there exist $x, y \in X$ with $x \not\sim_1^* y$. If $x \prec_1^* y$, then there exists $z \in X$ such that $x \preceq z \prec y$. If $x \prec z$, then by (TR) we get that $x \prec y$. And by Property 5, it follows that $x \prec_1 y$. If $x \not\prec z$, we have that $x \sim z \prec y$, thus $x \prec_1 y$ and then $x \not\sim_1 y$.

Similarly, if $y \prec_1^* x$ then there exists $z \in X$ such that $y \preceq z \prec x$. If $y \prec z$, then by (TR) we get that $y \prec x$. And by Property 5, it follows that $y \prec_1 x$. If $y \not\prec z$, then we have that $y \sim z \prec x$, thus $y \prec_1 x$ and then $x \not\sim_1 y$.

The proof of $\prec_2 = \prec_2^*$ is similar. $\qquad\square$

Proof (*Property 10*) Since $\succ$ satisfies (IR), by Property 5 it holds $\succ \subset \succ_1$. Let $x, y, z \in X$ be such that $x \not\prec_1 y$ and $y \succ z$. Since $x \not\prec_1 y$, we have $x \not\prec y$. And since $y \succ z$, we have $x \not\prec z$, otherwise if $x \prec z$ then by (TR) it follows that $x \prec y$, which is a contradiction. Now, if $x \sim z$ and since $y \succ z$ we get that $x \prec_1 y$, a contradiction. Thus, we have $x \succ z$.

Since $\succ$ satisfies (IR), by Property 5 it holds $\succ \subset \succ_2$. Let $x, y, z \in X$ be such that $y \not\prec_2 z$ and $x \succ y$. Since $y \not\prec_2 z$, we have $y \not\prec z$. And since $x \succ y$, we have $x \not\prec z$, otherwise if $x \prec z$ then by (TR) it follows that $z \succ y$, which is a contradiction. Now, if $x \sim z$ and since $x \succ y$ we get that $z \succ_2 y$, which is a contradiction. Thus, we have $x \succ z$. $\qquad\square$

References

1. Bridges, D.S.: Representing interval orders by a single real-valued function. J. Econ. Theory **36**, 149–155 (1985)
2. Bridges, D.S.: Numerical representation of interval orders on a topological space. J. Econ. Theory **38**, 160–166 (1985)
3. Bridges, D.S., Mehta, G.B.: Representations of Preference Orderings. Springer, Berlin (1995)
4. Bosi, G., Candeal, J.C., Induráin, E.: Continuous representability of interval orders and biorders. J. Math. Psychol. **51**, 122–125 (2007)
5. Bosi, G., Candeal, J.C., Induráin, E., Oloriz, E., Zudaire, M.: Numerical representations of interval orders. Order **18**, 171–190 (2001)
6. Bosi, G., Estevan, A., Gutiérrez García, J., Induráin, E.: Continuous representability of interval orders: the topological compatibility setting. Int. J. Uncertain. Fuzziness Knowl. Based Syst. **23**, 345–365 (2015)
7. Bosi, G., Herden, G.: On continuous multi-utility representations of semi-closed and closed preorders. Math. Soc. Sci. **79**, 20–29 (2016)
8. Bosi, G., Isler, R.: Representing preferences with nontransitive indifference by a single real-valued function. J. Math. Econ. **25**, 621–631 (1995)
9. Bosi, G., Isler, R.: A full characterization of continuous representability of interval orders. Math. Pannonica **18**, 125–134 (2007)
10. Bosi, G., Zuanon, M.: Upper semicontinuous representations of interval orders. Math. Soc. Sci. **60**, 60–63 (2014)
11. Candeal, J.C., Induráin, E., Zudaire, M.: Continuous representability of interval orders. Appl. Gen. Topol. **5**, 213–230 (2004)
12. Chateauneuf, A.: Continuous representation of a preference relation on a connected topological space. J. Math. Econ. **16**, 139–146 (1987)
13. Debreu, G.: Representation of a preference ordering by a numerical function. In: Thrall, M., Davis, R.C., Coombs, C.H. (eds.) Decision Processes, pp. 159–165. Wiley, New York (1954)
14. Debreu, G.: Continuity properties of paretian utility. Int. Econ. Rev. **5**, 285–293 (1964)
15. Eilenberg, S.: Ordered topological spaces. Am. J. Math. **63**, 39–45 (1941)
16. Fishburn, P.C.: Intransitive indifference with unequal indifference intervals. J. Math. Psychol. **7**, 144–149 (1970)
17. Fishburn, P.C.: Intransitive indifference in preference theory: a survey. Oper. Res. **18**, 207–228 (1970)
18. Fishburn, P.C.: Interval representations for interval orders and semiorders. J. Math. Psychol. **10**, 91–105 (1973)

19. Fishburn, P.C.: Interval Orders and Interval Graphs. Wiley, New York (1985)
20. Fishburn, P.C., Monjardet, B.: Norbert Wiener on the theory of measurement (1914, 1915, 1921). J. Math. Psychol. **36**, 165–184 (1992)
21. Monjardet, B.: Intervals, intervals. Order **5**, 211–219 (1988)
22. Oloriz, E., Candeal, J.C., Induráin, E.: Representability of interval orders. J. Econ. Theory **78**, 219–227 (1998)
23. Rébillé, Y.: Continuous utility on connected separable topological spaces. Econ. Theory Bull. **7**, 147–153 (2019)
24. Wiener, N.: A contribution to the theory of relative position. Proc. Camb. Philos. Soc. **17**, 441–449 (1914)

Searching for a Debreu's Open Gap Lemma for Semiorders

Asier Estevan

Abstract In 1956 R. D. Luce introduced the notion of a semiorder to deal with indifference relations in the representation of a preference. During several years the problem of finding a utility function was studied until a representability characterization was found. However, there was almost no results on the continuity of the representation. A similar result to Debreu's Lemma, but for semiorders was never achieved. In the present paper we propose a characterization for the existence of a continuous representation (in the sense of Scott-Suppes) for bounded semiorders. As a matter of fact, the weaker but more manageable concept of ε-continuity is properly introduced for semiorders. As a consequence of this study, a version of the Debreu's Open Gap Lemma is presented (but now for the case of semiorders) just as a conjecture, which would allow to remove the open-closed and closed-open gaps of a subset $S \subseteq \mathbb{R}$, but now keeping the constant threshold, so that $x + 1 < y$ if and only if $g(x) + 1 < g(y)$ $(x, y \in S)$.

1 Introduction

The concepts of an interval order and a semiorder were introduced by Wiener, but under a different nomenclature [28, 35, 36]. The notion of a semiorder is usually attributed to Luce (1956), who developed this field when dealing with applications in Economics and Psychology. On the other hand, the idea of an interval order was studied in depth by Fishburn in the 1970s [23–27].

The use of those concepts was due to the need of developing mathematical models of measurements related to situations of intransitive indifference.

If X is a nonempty set endowed with an interval order $\prec$, the classical numerical representation (if any) consists of two real-valued functions $u, v \colon X \to \mathbb{R}$ such that $x \prec y \Leftrightarrow v(x) < u(y)$ holds for all $x, y \in X$.

A. Estevan (✉)
Dpto. Estadística, Informática y Matemáticas, Instituto INAMAT, Universidad Pública de
Navarra, Campus Arrosadía, 31006 Iruña-Pamplona, Navarra, Spain
e-mail: asier.mugertza@unavarra.es

© Springer Nature Switzerland AG 2020
G. Bosi et al. (eds.), *Mathematical Topics on Representations of Ordered Structures and Utility Theory*, Studies in Systems, Decision and Control 263,
https://doi.org/10.1007/978-3-030-34226-5_5

In the special case of a semiorder, the representation *in the sense of Scott-Suppes* (*SS-representation*, for short) is defined by means of a single function $u\colon X \to \mathbb{R}$ such that $x \prec y \Leftrightarrow u(x) + 1 < u(y)$, for every $x, y \in X$. Notice that this is actually a special kind of interval order representation through a pair (u, v) in which $v(x) = u(x) + 1$ for every $x \in X$ [15, 30, 33, 34].

In addition, when the set X is also endowed with a topology τ, it may be interesting to study the semicontinuity or continuity of the numerical representations (if any) [10–12, 16].

The problem of finding a characterization of the continuous representability of a total preorder $\precsim$ defined on a topological space (X, τ) was solved dy Gerard Debreu. Given any subset S of the real line $\mathbb{R}$, Debreu defined in 1964 a *lacuna* of S as a non-degenerate interval disjoint from S and having a lower bound and an upper bound in S, and a *gap* of S as a maximal lacuna of S. The famous *Debreu's Open Gap Lemma* allows us to construct a continuous representation of a representable and continuous total preorder [20]. It reads as follows:

Lemma 1.1 (Open Gap Lemma) *If $S \subseteq \mathbb{R}$, then there is a strictly increasing function $g\colon S \to \mathbb{R}$ such that all the gaps of $g(S)$ are open.*

However, the analogous problem for interval orders and semiorders remains still open. Important results have been obtained whenever the topology τ is *natural* as regards the interval order through an ordinal condition called *interval order separability* (see Sect. 2) [3]. Other results on continuous representability of interval orders were achieved in [2, 4, 7, 8, 18, 19].

There is no characterization of the continuous Scott-Suppes representability. Here, the idea of a natural topology does not fit well for the general case [3, 22]. Some results about continuous representability of semiorders may be found in [13, 22, 29].

Although a semiorder is a particular case of an interval order, the difference between them is critical. In the case of a SS-representation (u, k) for a semiorder, there is a positive threshold $k > 0$ (we may assume $k = 1$ without loss of generality) such that $x \prec y$ if and only if $u(x) + k < u(y)$, whereas in the case of interval orders we have two functions (u, v) such that $x \prec y$ if and only if $v(x) < u(y)$. Therefore, the SS-representation implies a *geometrical* structure further than the topological one. To see this, notice that, for any representation (u, v) of an interval order and for any (any!) strictly increasing function $g\colon \mathbb{R} \to \mathbb{R}$, the composition $(g \circ u, g \circ v)$ is also a representation. Therefore, we are allowed to make contractions or expansions of the functions at different points. This technique was successfully used in order to achieve continuity (assuming some necessary conditions for the existence of a continuous representation) [6, 21]. A similar situation holds with total preorders, for which increasing functions are used in order to achieve continuity, as it is well known since Debreu's work [20]. It is known that Debreu's procedure cannot be applied to the case of semiorders [22]. In fact, the procedure applied for interval orders cannot be applied neither to SS-representations, due the fact that in this kind of representations there is a single function u used for comparisons (instead of two functions u and v) [6].

Thus, it is not possible to compose (arbitrarily) a SS-representation with increasing functions for continuity purposes. However, this is not possible (in general) for SS-representations. Of course, any linear contraction or expansion (that is, by means of a function $g: \mathbb{R} \to \mathbb{R}$ such that $g(r) = a \cdot r + b$, with $a > 0$) on $u(X)$ would keep the representation, thus, achieving another SS-representation $(g \circ u, a \cdot k)$ but now with the corresponding new threshold $k' = a \cdot k$. Nevertheless, in general any other kind of transformation w of the pair (u, k) would not represent the order structure, such that $x \prec y$ if and only if $w(u(x)) + k' < w(u(y))$. The changes made in each threshold interval must be the same. Thus, the semiorders endow X with a rigid structure at a large scale: small changes in a neighborhood of a point may imply changes too far from the point. To see that, we retrieve the following example from [22]:

Example 1.2 Let $X = \mathbb{Q} \bigcap ([0, 0'5) \cup [1, +\infty))$. Endow X with the topology τ defined by means of the subbasis $\{z \in X : z < x\}_{x \in X} \bigcup \{y \in X : x < y\}_{x \in X}$, where $<$ is the usual strict order of the real line[1] $\mathbb{R}$. Let $\prec$ denote the semiorder defined on X by declaring that $a \prec b \iff a + 1 < b$ $(a, b \in X)$. Thus, the inclusion i (where $i(x) = x$, for any $x \in X$) is a SS-representation. It can be proved that i fails to be continuous at $x = 1$. To see that, notice that the sequence $(x_n)_{n \in \mathbb{N}} = (0'5 - \frac{1}{n+1})_{n \in \mathbb{N}}$ converges to 1 in (X, τ). If we try to modify the function i at the point 1 (in a neighborhood of 1) in order to achieve continuity, this modification implies strong changes in the subset $[1'5, 2]$ too.

To see that, notice that if we want to warrant continuity (of our new function u) at 1, then it should happen that the limit $\lim_{n \to +\infty} u(x_n)$ equals $u(1)$. Now, observe that $x_n \prec 1'5 \precsim 1$, and also $x_n \prec 1'6 \precsim 1$, so that $u(x_n) + 1 < u(1'5) \leq u(1) + 1$ and also $u(x_n) + 1 < u(1'6) \leq u(1) + 1$, for every $n \in \mathbb{N}$. Since u is, by hypothesis, continuous, taking limits we get $u(1) + 1 = u(1'5) = u(1'6)$. Thus, this modification implies the contraction of the interval $[1'5, 2]$ to a point, which makes impossible the construction of the desired representation (here, notice that $1'5 \prec^0 1'6$, so it must hold for any representation u that $u(1'5) < u(1'6)$).

The aforementioned limitation in the (arbitrary) use of increasing functions is a huge handicap in the search of a continuous SS-representation. Hence, some advances in the study of continuous representability of interval orders cannot be translated to semiorders.

However, as a counterpart, we are able to introduce a new concept that generalizes and approximates the idea of continuity for semiorders: the ε-*continuity*. This new concept is useless for interval orders, but it seems crucial in the study of some other representation with a *geometrical* component (e.g. a threshold) such as in the case of SS-representations.

Dealing with real functions, it is easy to measure the length of a jump-discontinuity, so that we may aspire to construct functions such that the length of each

---

[1] Observe that τ does not coincide with the induced topology $\tau_{u|X}$ inherited from the usual topology τ_u on $\mathbb{R}$. As a matter of fact $\tau \subsetneq \tau_{u|X}$.

jump-discontinuity is smaller than a desired constant. However, as we will see, working with representations of total preorders or interval orders, this is useless.

Example 1.3 Let $\prec$ be the interval order on $S = [0, 1] \cup (1'5, 3] \subseteq \mathbb{R}$ defined by $x \prec y \iff 2x < y$. Endow now the set S with the topology $\tau_\leq$ defined by the Euclidean order $\leq$ on S. Then, the pair of functions (u, v) where $u(x) = x$ and $v(x) = 2x$ is a representation of the interval order and both are continuous on the whole set with the exception of the point $x = 1$. In fact, at $x = 1$ the function u has a jump-discontinuity of length $0'5$ and v has another one of length 1.

However, for any $n \in \mathbb{N}$, it is possible to find a representation (u', v') whose biggest gap is smaller than $\frac{1}{n}$. For that, it is enough to define the pair $(u' = \frac{1}{n} \cdot u, v' = \frac{1}{n} \cdot v)$. Nevertheless, as we said before, this is quite trivial.

On the other hand, given the semiorder $x \prec y \iff x + 1 < y$ on the same space $(S, \tau_\leq)$, the pair $(u, 1)$ is a SS-representation that fails to be continuous at $x = 1$. The length of that jump-discontinuity is $0'5$ (i.e. the ratio with respect to the constant threshold $k = 1$ is $\frac{1}{2}$). But now, if we argue as before and construct a new function $u' = \frac{1}{n} \cdot u$ in order to reduce the length of the jump-discontinuity to $\frac{0'5}{n}$, then the threshold is reduced too, so that the ratio between the jump and the threshold is the same: $\frac{0'5/n}{1/n} = \frac{1}{2}$.

In the case of SS-representations there is a threshold k (we may assume that $k = 1$), it is possible to compare the length of each jump-discontinuity with the value $k = 1$. Therefore, it makes perfect sense to say that a semiorder is r-continuous (for a positive value $r \in \mathbb{R}$) if there exists a SS-representation $(u, 1)$ such that the length of each jump-discontinuity is bounded by this constant r. It order to approximate the idea of continuity, we may say that a semiorder is ε-continuous if for any $\varepsilon > 0$ there exists a SS-representation $(u, 1)$ such that the length of each junp-discontinuity is bounded by this value ε.

In the present paper we introduce the concept of ε-continuity as a tool when dealing with semiorders that fail to be continuously representable. As a matter of a fact, through this idea we propose a characterization of the bounded semiorders which are continuously representable.

The structure of the paper goes as follows: After a section of preliminaries, necessary conditions for the existence of a continuous SS-representation are introduced. Then, in Sect. 4, the image subset $u(X)$ is studied for a given SS-representation $(u, 1)$ of a semiorder that satisfies the aforementioned necessary conditions. In next Sect. 5, the new concept of ε-continuity for semiorders is defined and justified. By means of this new concept, some conjectures on the continuous SS-representability of semiorders are presented.

2 Preliminaries

From now on X will denote a nonempty set.

Definition 2.1 A *binary relation* $\mathscr{R}$ on X is a subset of the Cartesian product $X \times X$. Given two elements $x, y \in X$, the notation $x\mathscr{R}y$ expresses that the pair (x, y) belongs to $\mathscr{R}$.

Associated to a binary relation $\mathscr{R}$ on a set X, we consider its *negation* (respectively, its *transpose*) as the binary relation $\mathscr{R}^c$ (respectively, $\mathscr{R}^t$) on X given by $(x, y) \in \mathscr{R}^c \Leftrightarrow (x, y) \notin \mathscr{R}$ for every $x, y \in X$ (respectively, given by $(x, y) \in \mathscr{R}^t \Leftrightarrow (y, x) \in \mathscr{R}$, for every $x, y \in X$). We also define the *adjoint* $\mathscr{R}^a$ of the given relation $\mathscr{R}$, as $\mathscr{R}^a = (\mathscr{R}^t)^c$.

A binary relation $\mathscr{R}$ defined on a set X is said to be:

 (i) *reflexive* if $x\mathscr{R}x$ holds for every $x \in X$,
 (ii) *irreflexive* if $\neg(x\mathscr{R}x)$ holds for every $x \in X$,
 (iii) *symmetric* if $\mathscr{R}$ and $\mathscr{R}^t$ coincide,
 (iv) *antisymmetric* if $\mathscr{R} \cap \mathscr{R}^t \subseteq \Delta = \{(x, x) : x \in X\}$,
 (v) *asymmetric* if $\mathscr{R} \cap \mathscr{R}^t = \varnothing$,
 (vi) *total* if $\mathscr{R} \cup \mathscr{R}^t = X \times X$,
 (vii) *transitive* if $x\mathscr{R}y \wedge y\mathscr{R}z \Rightarrow x\mathscr{R}z$ for every $x, y, z \in X$.

Definition 2.2 A *preorder* $\precsim$ on X is a binary relation which is reflexive and transitive. An antisymmetric preorder is said to be an *order*. A *total preorder* $\precsim$ on a set X is a preorder such that if $x, y \in X$ then $(x \precsim y) \vee (y \precsim x)$ holds. If $\precsim$ is a preorder on X, then as usual we denote the associated *asymmetric* relation by $\prec$ and the associated *equivalence* relation by $\sim$ and these are defined by $x \prec y \Leftrightarrow (x \precsim y) \wedge \neg(y \precsim x)$ and $x \sim y \Leftrightarrow (x \precsim y) \wedge (y \precsim x)$.

Definition 2.3 An *interval order* $\prec$ is an asymmetric binary relation on X such that $(x \prec y) \wedge (z \prec t) \Rightarrow (x \prec t) \vee (z \prec y)$ $(x, y, z, t \in X)$. Its symmetric part is denoted by $\precsim$, so that $a \precsim b \Leftrightarrow \neg(b \prec a)$. The binary relation $\sim$ defined by $a \sim b \Leftrightarrow (a \precsim b) \wedge (b \precsim a)$ is said to be the *indifference* associated to $\prec$.

Remark 2.4 It is well known that given an interval order $\prec$ on a set X, the associated relations $\precsim$ and $\sim$ may fail to be transitive [23, 24, 27, 30, 33].

Definition 2.5 An interval order $\prec$ is said to be a *semiorder* if $(x \prec y) \wedge (y \prec z) \Rightarrow (x \prec w) \vee (w \prec z)$ $(x, y, z, w \in X)$. A semiorder $\prec$ is said to be *typical* if $\precsim$ is *not* a total preorder on X.

Through the next definition, we introduce the notion of representability for different kinds of orderings, that makes possible to convert qualitative scales into quantitative ones.

Definition 2.6 A total preorder $\precsim$ on X is called *representable* if there is a real-valued function $u : X \to \mathbb{R}$ that is order-preserving, so that, for every $x, y \in X$, it holds that $x \precsim y \Leftrightarrow u(x) \leq u(y)$. The map u is said to be a *utility function* for $\precsim$.

An interval order $\prec$ defined on X is said to be *representable* (as an interval order) if there exist two real valued maps $u, v \colon X \longrightarrow \mathbb{R}$ such that $x \prec y \Leftrightarrow v(x) < u(y)$ $(x, y \in X)$. The pair (u, v) is called a *utility pair* representing $\prec$.

A semiorder $\prec$ defined on X is said to be *representable in the sense of Scott and Suppes* if there exists a real-valued map $u \colon X \to \mathbb{R}$ (again called a *utility function*) such that $x \prec y \Leftrightarrow u(x) + 1 < u(y)$ $(x, y \in X)$ [33].

In this case, the pair $(u, 1)$ is said to be a *Scott-Suppes representation* (*SS-representation*, for short) of $\prec$.

Remark 2.7 If (u, v) represents an interval order $\prec$ defined on a set X, it is straightforward to see that $u(x) \leq v(x)$ for every $x \in X$. And the non-negative real number $v(x) - u(x)$ is said to be the *discrimination threshold for the element* $x \in X$. In the case of a semiorder that is representable in the sense of Scott and Suppes, the discrimination thresholds are all equal to 1.

There exist interval orders that fail to be representable (as interval orders). Also, there exist semiorders that are not representable in the sense of Scott and Suppes, not even as an interval order [17, 32].

Definition 2.8 Associated to an interval order $\prec$ defined on a nonempty set X, we shall consider three new binary relations [1, 23, 25].

These binary relations are said to be the *traces* of $\prec$. They are respectively denoted by $\prec^*$ (*left trace*), $\prec^{**}$ (*right trace*) and $\prec^0$ (*main trace*), and defined as follows: $x \prec^* y \Leftrightarrow x \prec z \precsim y$ for some $z \in X$, and similarly $x \prec^{**} y \Leftrightarrow x \precsim z \prec y$ for some $z \in X$ $(x, y \in X)$. In addition, $x \prec^0 y \Leftrightarrow (x \prec^* y) \vee (x \prec^{**} y)$ $(x, y \in X)$.

Remark 2.9 We denote $x \precsim^* y \Leftrightarrow \neg(y \prec^* x)$, $x \sim^* y \Leftrightarrow x \precsim^* y \precsim^* x$, $x \precsim^{**} y \Leftrightarrow \neg(y \prec^{**} x)$ and $x \sim^{**} y \Leftrightarrow x \precsim^{**} y \precsim^{**} x$, and finally $x \precsim^0 y \Leftrightarrow (x \precsim^* y) \wedge (x \precsim^{**} y)$ and $x \sim^0 y \Leftrightarrow (x \precsim^0 y) \wedge (y \precsim^0 x)$ $(x, y \in X)$. Both the binary relations $\precsim^*$ and $\precsim^{**}$ are total preorders on X. Moreover, the indifference relation $\sim$ associated to the interval order $\prec$ is transitive if and only if $\precsim^*$, $\precsim^{**}$ and $\precsim$ coincide. In this case $\precsim$ is actually a total preorder on X [5, 23, 25, 31, 32].

Furthermore, given an interval order $\prec$ on X, it holds that $\prec$ is actually a semiorder if and only if $\precsim^0$ is a total preorder on X.

Let us recall now a characterization of the numerical representability of semiorders.

Definition 2.10 Let X be a nonempty set. An interval order (e.g. also a semiorder) $\prec$ defined on X is said to be *interval order separable* if there exists a countable subset $D \subseteq X$ such that for every $x, y \in X$ with $x \prec y$ there exists $d \in D$ such that $x \precsim^* d \prec y$.

A semiorder $\precsim$ defined on X is said to be *regular with respect to sequences* if for any $x, y \in X$, and sequences $(x_n)_{n \in \mathbb{N}}, (y_n)_{n \in \mathbb{N}} \subseteq X$, none of the situations $x \prec \cdots \prec x_{n+1} \prec x_n \prec \cdots \prec x_1$ and $y_1 \prec \cdots \prec y_n \prec y_{n+1} \prec \cdots \prec y$ may occur. It is said to be *bounded* if there is no strictly increasing or decreasing infinite sequences, i.e. there is no sequence $(x_n)_{n \in \mathbb{N}} \subseteq X$ such that $\cdots \prec x_{n+1} \prec x_n \prec \cdots \prec x_1$ or $x_1 \prec \cdots \prec x_n \prec x_{n+1} \prec \cdots$.

The following result is proved [14, 15].

Theorem 2.11 *Let X be a nonempty set. Let $\prec$ be a typical semiorder defined on X. Then, $\prec$ is representable in the sense of Scott and Suppes if and only if it is both interval order separable and regular with respect to sequences.*

Remark 2.12 If $(u, 1)$ is a SS-representation of a semiorder $\prec$, then u is strictly increasing with respect to the main trace $\precsim_0$ [6].

3 Continuous SS-Representability of Semiorders

Let (X, τ) stand for a topological space (a set X with a topology τ).

Definition 3.1 Let $\prec$ denote an asymmetric binary relation on X. Given $a \in X$ the sets $L(a) = \{t \in X : t \prec a\}$ and $U(a) = \{t \in X : a \prec t\}$ are called, respectively, the *lower and upper contours* of a relative to $\prec$. We say that $\prec$ is τ-*continuous* if for each $a \in X$ the sets $L(a)$ and $U(a)$ are τ-open.

The following facts are proved [6, 22].

Theorem 3.2 *Let $\prec$ be an interval order defined on a set X. Then the indifference $\sim^0$ associated to the main trace is an equivalence relation.*

Definition 3.3 Let (X, τ) be a topological space. Let $\prec$ be an interval order on X. The topology τ is said to be *compatible with respect to the indifference of the main trace of* $\prec$ if $x \sim^0 y \Rightarrow (x \in \mathcal{O} \iff y \in \mathcal{O})$ holds true for every $x, y \in X$ and every τ-open subset $\mathcal{O} \in \tau$.

Remark 3.4 Notice that, according to the idea before, elements that are indistinguishable with respect to $\prec$ (because they play the same role on $(X, \prec)$) should also be indistinguishable from a topological point of view.

In particular, in the main case in which $x \sim_0 y \iff x = y$, i.e. when X coincides with the quotient set $X/\sim_0$, the topology is always compatible.

With respect to continuity, the following result is proved [22].

Lemma 3.5 *Let (X, τ) be a topological space endowed with a semiorder $\prec$. Assume that $\prec$ is representable in the sense of Scott and Suppes by means of a pair $(u, 1)$ with u continuous. Then the following properties hold true:*

(a) The semiorder $\prec$ is τ-continuous.

(b) If a net $(x_j)_{j \in J} \subseteq X^2$ converges to two points $a, b \in X$, then $a \sim^0 b$.

(c) If a net $(x_j)_{j \in J} \subseteq X$ converges to $a \in X$, and $b, c \in X$ are such that $x_j \prec b \precsim a$ and also $x_j \prec c \precsim a$ for every $j \in J$, then $b \sim^0 c$.

---

[2] J denotes here a directed set of indices. Since this does not lead to confusion, we will use the same notation '$<$' of the order on the real numbers than for the partial order on the set of indices J.

*(d) If a net $(x_j)_{j \in J} \subseteq X$ converges to $a \in X$, and $b, c \in X$ are such that $a \precsim b \prec x_j$
and also $a \precsim c \prec x_j$ for every $j \in J$, then $b \sim^0 c$.*

Throughout the paper, we shall refer to these conditions (a)–(d) by (NC) (*necessary conditions*).

Next result was also proved [6].

Proposition 3.6 *Let (X, τ) be a topological space endowed with a semiorder $\prec$.
Assume that τ is compatible with respect to the indifference of the main trace of $\prec$.
Suppose also that $\prec$ is representable in the sense of Scott and Suppes by means of a
pair $(u, 1)$ with u continuous. Then the total preorder $\precsim^0$ is τ-continuous.*

A partial result was achieved (namely Theorem 3.7), that guarantees the continuous representability of a semiorder but as an interval order (that is, through a pair (u, v) of continuous real-valued functions) whenever the condition of compatibility between the topology τ and the indifference $\sim^0$ is satisfied [6].

Theorem 3.7 *Let (X, τ) be a topological space. Let $\prec$ be a representable semiorder
on X. Assume that $\prec$ satisfies the aforementioned necessary conditions (NC) and,
in addition, the topology τ is compatible with respect to the indifference of the main
trace of $\prec$. If $\precsim^0$ is τ-continuous, then $\prec$ admits a representation as an interval order,
through a pair (u, v) of continuous real-valued functions.*

3.1 A New Necessary Condition for the Existence of a Continuous SS-Representation

From now on, we shall assume that the topology of the space is compatible with respect to the indifference of the main trace of the semiorder (e.g. the quotient set $X/\sim^0$ coincides with X). Hence, by Proposition 3.6, we will assume that $\precsim^0$ is τ-continuous.

It is known that the aforementioned necessary conditions (a)–(d) of Lemma 3.5 are not sufficient in order to guarantee the existence of a continuous SS-representation [22]. There is a simple reason that clarifies this insufficiency: there is, at least, another necessary condition that must be satisfied. This new condition tries to explain the *rigid* structure of the semiorder and its representation. In order to study that, first we introduce the new concept of *adjoint nets*.

Definition 3.8 Let (X, τ) be a topological space endowed with a semiorder $\prec$. Let $(x_j)_{j \in J}$ and $(y_k)_{k \in K}$ be two nets on X. We shall say that these nets are *adjoint nets*, and we denote it by $(x_j) \precdot (y_k)$, if one of the following conditions hold:

Condition 1: If none of these two nets is constant, then the following both conditions are satisfied:

$(1.i)$ for each $j_0 \in J$ there exists $k_0 \in K$ s.t. $x_{j_0} \prec y_k$ for any $k > k_0$,
$(1.ii)$ for each $k_0 \in K$ there exists $j_0 \in J$ s.t. $y_{k_0} \precsim x_j$ for any $j > j_0$.

Condition 2: If one (and only one) of the nets is constant, that is $y_k = b$ for all $k \in K$, where b is called *adjoint point*, then any of the following conditions is satisfied:

(2.*i*) $x_j \prec b$ for each $j \in J$ and the net converges to $a \in X$ such that $b \precsim a$,
(2.*ii*) $b \prec x_j$ for each $j \in J$ and the net converges to $a \in X$ such that $a \precsim b$.

Analogously, for each $n \in \mathbb{N}$ we define the *n-adjoint nets*, and we denote them by $(x_j) \precsim^n (y_k)$, if there exists a chain of length n of adjoint nets: $(x_j) \precsim (a_{i_1}) \precsim \ldots \precsim (a_{i_{n-1}}) \precsim (y_k)$. For any m such that $-m \in \mathbb{N}$, we also say that $(x_j)_{j \in J}$ and $(y_k)_{k \in K}$ are *m-adjoint nets*, and we denote them by $(x_j) \precsim^m (y_k)$ if $(y_k) \precsim^{-m} (x_j)$.

Remark 3.9 In the case in which both nets are constant, i.e. $(x_j)_{j \in J} = (a)$ and $(y_k)_{k \in K} = (b)$, then it holds that $a \prec b$ or $b \precsim a$ (or, dually, $b \prec a$ or $a \precsim b$). In any case, given a SS-representation $(u, 1)$, nothing can say about the distance on $\mathbb{R}$ between $u(a)$ and $u(b)$. This distance is what motivates Definition 3.8, as the following Lemma 3.10 shows:

Lemma 3.10 *Let $\prec$ be a semiorder defined on a topological space (X, τ) and let $(u, 1)$ be a continuous representation. If $(x_j)_{j \in J}$ and $(y_k)_{k \in K}$ are n-adjoint nets, then $\lim_{j \in J} u(x_j) + n = \lim_{k \in K} u(y_k)$.*

Proof We prove it by induction on $n \in \mathbb{N}$. For $n = 1$, if $(x_j) \precsim (y_k)$ and none of them is a constant, then for any $j_0 \in J$ there exists $k_0 \in K$ such that $x_{j_0} \prec y_k$ for each $k > k_0$, so $u(x_{j_0}) + 1 < \lim(u(y_k))$ for each $j_0 \in J$. Hence, $\lim(u(x_j)) + 1 \leq \lim(u(y_k))$.

Similarly, for any $k_0 \in K$ there exists $j_0 \in J$ such that $y_{k_0} \precsim x_j$ for each $j > j_0$, so $u(y_{k_0}) \leq \lim(u(x_j)) + 1$ for each $k_0 \in K$. Hence, $\lim(u(y_k)) \leq \lim(u(x_j)) + 1$. So we have proved that $\lim(u(x_j)) + 1 = \lim(u(y_k))$.

If one of them is a constant net (suppose $y_k = b$ for all $k \in K$), then $x_j \prec b$ for any $j \in J$ and there exists $\lim(x_j) \in X$ such that $b \precsim \lim(x_j)$. So it holds that $\lim(u(x_j)) + 1 \leq \lim u(b) \leq (u(x_j)) + 1$. Hence, $\lim(u(x_j)) + 1 = \lim(u(y_k))$. Similarly if the case (2.*ii*) of Definition 3.8 holds.

Now, suppose that the lemma is true for a fixed $n \in \mathbb{N}$. If $(x_j)_{j \in J}$ and $(y_k)_{k \in K}$ are two $n + 1$-adjoint nets, there exists another net $(z_r)_{r \in R}$ such that $(x_j) \precsim^n (z_r) \precsim (y_k)$. So, by the induction hypothesis, it holds that $\lim(u(x_j)) + n - 1 = \lim(u(z_r))$ and $\lim(u(z_r)) + 1 = \lim(u(y_k))$. Hence, it holds that $\lim(u(x_j)) + n = \lim(u(y_k))$.

We proceed analogously if one of them is a constant net. $\qquad\square$

Remark 3.11 Notice that, in the proof, we only used the continuity in case of constant nets. Thus, for any two *n-adjoint nets* $(x_j)_{j \in J}$ and $(y_k)_{k \in K}$ (connected by adjoint nets such that none of them is constant) it holds that $\lim_{j \in J} u(x_j) + n = \lim_{k \in K} u(y_k)$, even without requiring continuity for u.

Through the following theorem we introduce a new necessary condition (we will denote it by (e)) for the continuous SS-representability of semiorders. This condition tries to describe the rigid structure of semiorders.

Theorem 3.12 *Let $\prec$ be a continuously representable semiorder defined on a topo-*
logical space (X, τ). Let $(x_j)_{j \in J}$ and $(y_k)_{k \in K}$ be two nets such that they converge
to the same point a in X and $(w_r)_{r \in R}$ (respectively $(z_s)_{s \in S}$) two n-adjoint nets of
$(x_j)_{j \in J}$ (respectively, of $(y_k)_{k \in K}$) for some $n \in \mathbb{Z} - \{0\}$.

If there are two elements $b, c \in X$ such that $w_r \prec^0 b \prec^0 z_s$ as well as $w_r \prec^0 c \prec^0$
z_s (for each $r \in R, s \in S$), then $b \sim^0 c$.

Proof Let $(u, 1)$ be a continuous SS-representation. From Lemma 3.10 it follows
that $\lim(u(x_j)) + n = \lim(u(w_r))$ and $\lim(u(y_k)) + n = \lim(u(z_s))$ in $\mathbb{R}$. The nets
$(x_j)_{j \in J}$ and $(y_k)_{k \in K}$ converge to the same point a, so if u is a continuous representation it holds that $\lim(u(z_s)) = \lim(u(y_k)) + n = u(a) + n = \lim(u(x_j)) + n =$
$\lim(u(w_r))$. Hence, for any $b, c \in X$ such that $w_r \prec^0 b, c \prec^0 z_s$ (for each $r \in R, s \in$
S) it holds that $\lim(u(w_r)) \leq u(b), u(c) \leq \lim(u(z_s)) = \lim(u(w_r))$, so $u(b) = u(c)$
and then, (see Lemma 1 in [22]) $b \sim^0 c$.   □

Remark 3.13 As a matter of a fact, conditions (c) and (d) are now particular cases of
this new condition (e), in which $n = 1$. Besides, this new condition is independent
of (a) and (b) (as well as it generalizes conditions (c) and (d)). To see that, just notice
that Examples 2, 3 and 4 introduced in [22] satisfy this new condition (e), but that
is not the case of Example 5 in [22]. From now on, we refer to all these necessary
conditions as (NC).

Hence, these necessary conditions imply a rigid structure on the set, with a geometrical component due to the existence of that constant threshold $k = 1$.

4   The Implications of the Necessary Conditions on the SS-Representation

This section is devoted to show that, under the assumption of the necessary conditions,
the function u of the SS-representations has a particular structure. This holds for any
representation, continuous or not. Again, we will assume that the topology of the
space is compatible with respect to the indifference of the main trace $\precsim^0$, as well
as the τ-continuity of the main trace. Since we are imposing some conditions on
the semiorder structure $(X, \precsim)$, these conditions are reflected in the subset $u(X)$ of
$(\mathbb{R}, \leq)$ too. In the present section we shall take advantage of these features of $u(X)$
in order to try to achieve continuity.

First, we shall take into account the following results, which are well known [9,
20] (see also Corollary 1 and Proposition 1 in [21]).

Proposition 4.1 *Let $\precsim$ be a continuous and perfectly separable total preorder. Let*
f be a utility function of $\precsim$. Then, f has at most a countably infinite number of
discontinuities (which are jump-discontinuities).

Definition 4.2 Let X be a nonempty set and $u: X \to \mathbb{R}$ a real function on X. Given
a gap $(a, b]$ (respectively, $[a, b)$) of $u(X)$, we will say that a is the *end-point* of $(a, b]$

with respect to u (respectively, *b* is the *end-point of* $[a, b)$ *with respect to u*). For an open gap (a, b), *a* and *b* are the *end-points with respect to u*, whereas a closed gap $[a, b]$ has no *end-points with respect to u*.

Corollary 4.3 *Let* (X, τ) *be a topological space endowed with a* τ*-continuous total preorder* $\precsim$. *Given any representation u of the total preorder, then u is continuous at every point of X, excluding the inverse images of the end-points with respect to u of some gaps (of* $u(X)$*) that are not closed neither open, that is, excluding some* $x \in X$ *such that* $(u(x) = a, b]$ *or* $[b, u(x) = a)$ *is a gap of* $u(X) \subseteq \mathbb{R}$.

Definition 4.4 Let (X, τ) be a topological space endowed with a τ-continuous total preorder $\precsim$. Given a representation *u* of the total preorder, we will say that a gap of $u(X) \subseteq \mathbb{R}$ is a *bad gap* if the function *u* is not continuous at the inverse images of the end-point of the gap.

We will refer to the length of the bad gap as the *length of the jump-discontinuity*.

Remark 4.5 From Definition 4.4 and using Corollary 4.3, notice that given any representation *u* of a τ-continuous total preorder on (X, τ), then any closed or open gap $([a, b], (a, b) \subseteq \mathbb{R})$ of $u(X)$ is not a bad gap.

Proposition 4.6 *Let* $\precsim$ *be a* τ*-continuous total preorder and u a representation. If u is discontinuous at* $a \in X$, *then there is a gap* $(u(a), r]$ *or* $[r, u(a))$.

Now, we retrieve Proposition 2 of [21]:

Proposition 4.7 *Let* (X, τ) *be a topological space endowed with a* τ*-continuous total preorder* $\precsim$. *Let u be a representation of* $\precsim$, $x \in X$, $\varepsilon > 0$ *such that* $(u(x), u(x) + \varepsilon]$ *is a gap of* $u(X)$ *and* $u(x)$ *is not the right end-point of a gap of* $u(X)$. *If u fails to be continuous at x then there exists a net* $(x_i)_{i \in I}$ *in X convergent to x and such that* $u(x) + \varepsilon < u(x_i)$ *for all* $i \in I$.

Let us focus now on condition (*a*).

Proposition 4.8 *Let* $\prec$ *be a* τ*-continuous semiorder on* (X, τ). *Let* $(u, 1)$ *be a SS-representation. If u is discontinuous at* $a \in X$, *then one of the following situations holds:*

(i) $(u(a), r]$ *is a gap and* $u(X) \cap (u(a) + 1, r + 1] = \emptyset$,
(ii) $[r, u(a))$ *is a gap and* $u(X) \cap [r - 1, u(a) - 1) = \emptyset$.

Proof We argue by contradiction. If there was an element $u(b) \in (u(a) + 1, r + 1]$, then it would hold that $u(a) + 1 < u(b)$, thus $a \prec b$. Since $L(b)$ is open, that implies that there is no discontinuity at *a*, arriving at a contradiction.

We use a dual argument for a gap $[r, u(a))$. $\qquad\square$

Let us focus now on conditions (*c*)–(*d*).

Proposition 4.9 *Let* $\prec$ *be a semiorder on* (X, τ) *satisfying the necessary conditions* (*c*) *and* (*d*). *Let* $(u, 1)$ *be a SS-representation. If u is discontinuous at* $a \in X$, *then one of the following situations holds:*

(i) $[r, u(a))$ *is a gap and* $u(X) \cap [r + 1, u(a) + 1]$ *has at most one point* $s \in \mathbb{R}$,
(ii) $(u(a), r]$ *is a gap and* $u(X) \cap [u(a) - 1, r - 1]$ *has at most one point* $s \in \mathbb{R}$.

Proof Suppose there is a gap $[r, u(a))$ such that u fails to be continuous at a. Then, by Proposition 4.7, there is a net $(x_i)_{i \in I}$ in X, convergent to a and such that $u(x_i) < r < u(a)$ for all $i \in I$. For any two points $u(b), u(c) \in [r + 1, u(a) + 1]$, it holds that $u(x_i) + 1 < r + 1 < u(b), u(c) \leq u(a) + 1$. Thus $x_i \prec b, c$ for any $i \in I$ as well as $b, c \precsim a$. Hence, by condition (c) it holds that $b \sim^0 c$ and, since u also represent the trace $\precsim^0$, we conclude that $u(b) = u(c)$, that is $u(X) \cap [r + 1, u(a) + 1]$ has at most one point.

A similar argument holds for a gap $(u(a), r]$. $\square$

Finally, let us study now the implications of condition (e).

Proposition 4.10 *Let* $\prec$ *be a semiorder on* (X, τ) *satisfying the necessary condition* (e). *Let* $(u, 1)$ *be a SS-representation. Assume that* u *is discontinuous at* $a \in X$, *i.e. there is a gap* $[r, u(a))$ *(or* $(u(a), r]$*). Thus, there is a net* $(u(y_i))_{i \in I}$ *converging to* r *in* $\mathbb{R}$ *and there is also another net* $(u(x_j))_{j \in J}$ *(it may be constant, i.e.* $u(x_j) = u(a)$ *for any* $j \in J$*) converging to* $u(a)$ *in* $\mathbb{R}$.

If there exist two nets $(w_r)_{r \in R}$ *and* $(z_s)_{s \in S}$ *such that* $(w_r)_{r \in R}$ *is n-adjoint to* $(y_i)_{i \in I}$ *and* $(z_s)_{s \in S}$ *is n-adjoint to* $(x_j)_{j \in J}$ *(for some* $n \in \mathbb{Z} - \{0\}$*), then* $u(X) \cap [r + n, u(a) + n]$ *has at most one point* $s \in \mathbb{R}$.

Proof Suppose there is a gap $[r, u(a))$ such that u fails to be continuous at a. Then, by Proposition 4.7, there is a net $(y_i)_{i \in I}$ in X convergent to a and such that $u(y_i) < r < u(a)$ for all $i \in I$. Thus, we have two nets $(y_i)_{i \in I}$ and $(x_j)_{j \in J}$ both converging to a such that $(u(y_i))_{i \in I}$ converges to r and $(u(x_j))_{j \in J}$ converges to $u(a)$.

If there exist two nets $(w_r)_{r \in R}$ and $(z_s)_{s \in S}$ such that $(w_r)_{r \in R}$ is n-adjoint to $(y_i)_{i \in I}$ and $(z_s)_{s \in S}$ is n-adjoint to $(x_j)_{j \in J}$ (for some $n \in \mathbb{Z} - \{0\}$), then by Lemma 3.10, it holds that $\lim u(y_i) + n = r + n = \lim u(w_r)$ as well as $\lim u(x_j) + n = u(a) + n = \lim u(z_s)$.

For any two points $u(b), u(c) \in [r + n, u(a) + n]$, it holds that $w_r \prec_0 b, c \prec_0 z_s$ for any $r \in R$ and $s \in S$. Hence, by Theorem 3.12 $b \sim_0 c$ is satisfied, so $u(b) = u(c)$. Thus, we conclude that $u(X) \cap [r + n, u(a) + n]$ has at most one point $s \in \mathbb{R}$.

A similar argument holds for a gap $(u(a), r]$. $\square$

The following corollary tries to summarize the structure of any SS-representation (i.e. not necessarily continuous) of a continuously representable semiorder. That will be made using the necessary conditions. For the sake of brevity, we include a sketch of proof instead of a more detailed one.

Corollary 4.11 *Let* $\prec$ *be a continuously representable semiorder on* (X, τ), *i.e. satisfying the necessary conditions* (NC). *Let* $(u, 1)$ *be a SS-representation. Suppose that there is a discontinuity at a point* a *such that* $[r, u(a))$ *(or* $(u(a), r]$*) is a gap.*

Then, $u(X) \cap [r + 1, u(a) + 1]$ *(or* $u(X) \cap [u(a) - 1, r - 1]$, *respectively) has at most one point* $s \in \mathbb{R}$ *and one of the following situations holds:*

(a_1) *Depending on the existence of that point s, it holds that $[r + 1, u(a) + 1]$ (there is no s), $(r + 1, u(a) + 1]$ $(s = r + 1)$ or $[r + 1, u(a) + 1)$ $(s = u(a) + 1)$ is a gap, or $[r + 1, u(a) + 1]$ is the union of two consecutive gaps $[r + 1, s) \cup (s, u(a) + 1]$. In that case, $u(X) \cap [r + 2, u(a) + 2]$ (resp. $u(X) \cap [u(a) - 2, r - 2])$ has at most one point $s' \in \mathbb{R}$ and we will continue applying these cases (a_1) or (b_1), but now on $[r + 2, u(a) + 2]$ (resp. $[u(a) - 2, r - 2]$).*

(b_1) *If the case (a_1) does not hold, then there exist $\gamma_l, \gamma_r \geq 0$ such that at least one of them is strictly positive, such that $u(X) \cap [r + 1 - \gamma_l, u(a) + 1 + \gamma_r]$ (or $u(X) \cap [u(a) - 1 - \gamma_l, r - 1 + \gamma_r]$, respectively) has at most that point s. In that case, $[r + 2, u(a) + 2]$ (resp. $[u(a) - 2, r - 2]$) may contain more than one point. In fact, if there exists that point s then $(s + 1, u(a) + 2]$ may be nonempty if $\gamma_r > 0$, dually, $[r + 2, s + 1)$ may be nonempty if $\gamma_l > 0$.*

Moreover, $u(X) \cap [r - 1, u(a) - 1) = \emptyset$ (or $u(X) \cap (u(a) + 1, r + 1] = \emptyset$, resp.). Here, again, one of the following situations holds:

(a_2) *If $[r - 1, u(a) - 1]$ or $[r - 1, u(a) - 1)$ (resp. with $[u(a) + 1, r + 1]$ or $[u(a) + 1, r + 1))$ is a gap (depending on the existence of that adjoint point $u(a) - 1$ (resp. $u(a) + 1$)), then $u(X) \cap [r - 2, u(a) - 2]$ (resp. $u(X) \cap [u(a) + 2, r + 2])$ has at most the adjoint point $u(a) - 2$ (resp. $u(a) + 2$), and we will continue applying these cases (a_2) or (b_2) but now on $[r - 2, u(a) - 2]$ (resp. $[u(a) + 2, r + 2]$).*

(b_2) *If the case (a_2) before does not hold, then there exist $\gamma_l, \gamma_r \geq 0$ such that at least one of them is strictly positive, such that $u(X) \cap [r - 1 - \gamma_l, u(a) - 1 + \gamma_r]$ (or $u(X) \cap [u(a) + 1 - \gamma_l, r + 1 + \gamma_r]$, respectively) has at most that point $u(a) - 1$. In that case, $[r - 2, u(a) - 2]$ (resp. $[u(a) + 2, r + 2]$) may contain more than one point. In fact, if there exists that point $u(a) - 1$, then $[r - 2, u(a) - 2]$ may be nonempty if $\gamma_l > 0$.*

Proof First, if there is a discontinuity at a point a such that $[r, u(a))$ (or $(u(a), r]$) is a gap, then there is a net $(u(y_i))_{i \in I}$ converging to r in $\mathbb{R}$ and there is another net $(u(x_j))_{j \in J}$ (it may be constant, i.e. $u(x_j) = u(a)$ for any $j \in J$) converging to $u(a)$ in $\mathbb{R}$.

The first statement before points (a_1) and (b_1), relative to the possible existence of a unique point in $u(X) \cap [r + 1, u(a) + 1]$ (or $u(X) \cap [u(a) - 1, r - 1]$, respectively), is proved in Proposition 4.9.

If $[r + 1, u(a) + 1]$ (resp. $[u(a) - 1, r - 1]$) is as described in case (a_1), then notice that there exist adjoint nets $(z_s)_{s \in S}$ and $(w_r)_{r \in R}$ such that $(y_i) \preccurlyeq (z_s)$ and $(x_j) \preccurlyeq (w_r)$. Hence, by Proposition 4.10 we deduce that $u(X) \cap [r + 2, u(a) + 2]$ (resp. $[u(X) \cap u(a) - 2, r - 2]$) contains at most one point s'. We will continue arguing on $[r + 2, u(a) + 2]$ (resp. $[u(a) - 2, r - 2]$).

If $[r + 1, u(a) + 1]$ (resp. $[u(a) - 1, r - 1]$) is as described in case (b_1), i.e. there is an $\varepsilon > 0$ such that $u(X) \cap [r + 1 - \varepsilon, u(a) + 1]$ or $u(X) \cap [r + 1, u(a) + 1 + \varepsilon]$ (or $u(X) \cap [u(a) - 1 - \varepsilon, r - 1]$ or $u(X) \cap [u(a) - 1, r - 1 + \varepsilon]$, respectively) has at most one point, then there are not adjoint nets with respect to $(y_i)_{i \in I}$ and $(x_j)_{j \in J}$. In that case, $[r + 2, u(a) + 2]$ (resp. $[u(a) - 2, r - 2]$) may contain more than one point without violating any necessary condition. The last part of this statement (b_1) is deduced arguing in the existence of a continuous SS-representation.

A similar argument holds for the second part $(a_2) - (b_2)$ corresponding to $[r - 1, u(a) - 1]$ and $[u(a) + 1, r + 1]$. □

Proposition 4.12 *Let $\prec$ be a bounded semiorder on (X, τ). Let $(u, 1)$ be a SS-representation. Then, there is no sequence of gaps $\{g_n\}_{n \in \mathbb{N}}$ such that the length of a gap g_n is strictly smaller than the length of g_{n+1}, for any $n \in \mathbb{N}$. Hence, there always exists a maximal gap, that is, a gap which length is the biggest.*

Proof Since the semiorder is bounded, so is its representation and, therefore, the sum of the length of the gaps (denoted by $\sum_{n \in \mathbb{N}} L(g_n)$) is finite. Hence, we conclude that the sequence $\{L(g_n)\}_{n \in \mathbb{N}}$ converges to 0. In consequence, there always exists a maximal gap. □

The next corollary is directly deduced from the proposition before.

Corollary 4.13 *Let $\prec$ be a bounded semiorder on (X, τ). Let $(u, 1)$ be a SS-representation. Then, for any gap g there exists another smaller gap g' such that there is no gap which length is strictly between the length of g and that of g'.*

5   A Continuity Approach: ε-Continuity

First of all, we recall again an important point: we shall assume that the topology of the space is compatible with respect to the indifference of the main trace of the semiorder, as well as the τ-continuity of $\precsim^0$ (see Proposition 3.6). Therefore, we may assume without loss of generality that the function u of a given SS-representation also represents the total preorder $\precsim^0$ [6]. The author believes that this case is the most common one, since if there are two elements which are equivalent (as regards to the order structure), then—at first—there is no reason to distinguish them topologically. This is also the case when the topology τ is defined on the quotient set $X/\sim^0$.

Definition 5.1 Let $\prec$ be a semiorder on (X, τ). We shall say that the semiorder is *r-continuous* (for a positive value $r \in \mathbb{R}$) if there exists a SS-representation $(u, 1)$ such that the length of each jump-discontinuity is strictly smaller than this constant r.

It order to approximate to the idea of continuity, we may let r tend to 0. This motivates the following definition.

Definition 5.2 Let $\prec$ be a semiorder on X. We shall say that the semiorder is *ε-continuous* if for any $\varepsilon > 0$ there exists a SS-representation $(u_\varepsilon, 1)$ such that the length of each jump-discontinuity is strictly smaller than the value ε.

It is trivial that this new concept is weaker than the usual continuity. Actually, the so called necessary conditions (a)–(e) for the usual continuity are not needed for the existence of an ε-continuous SS-representation.

Example 5.3 Let X be the set $[0, 0'5) \cup [1, 3]$ endowed with the topology $\tau_{\leq}$ generated by the usual Euclidean order on X. We define now the semiorder $\prec$ by $x \prec y \iff x + 1 < y$.

The inclusion function $i : (X, \precsim) \to (\mathbb{R}, \tau_u)$ is a SS-representation that fails to be continuous at the point $x = 1$. To see that, notice that the sequence $(0'5 - \frac{1}{n+1})_{n\in\mathbb{N}}$ converges to 1 in $(X, \tau_{\leq})$, whereas the image sequence fails to converge to $i(1)$ in $(\mathbb{R}, \tau_u)$. In fact, this semiorder does not satisfy condition (b), therefore it is not continuously SS-representable.

However, it is ε-continuous since for any $\varepsilon > 0$ there exists a SS-representation $(u_\varepsilon, 1)$ such that the length of the jump-discontinuities is bounded by this value ε.

Let us prove that it is actually an ε-continuously representable semiorder. Let $\delta > 0$ be the length of the bad gap of the new function we shall define. We construct it stretching the pieces proportionally, keeping the order structure, but since the gap $[0'5, 1)$ must be shortened, we have to shrink too the intervals $[1'5, 2]$ and $[2'5, 3)$, again, proportionally :

$$
u_\delta(x) = \begin{cases}
x \cdot \frac{1-\delta}{1/2} & ; x \in [0, 0'5), \\
x \cdot \frac{1-\delta}{1/2} - (1 - 2\delta) & ; x \in [1, 1'5], \\
x \cdot \frac{\delta}{1/2} + (2 - 4\delta) & ; x \in [1'5, 2), \\
x \cdot \frac{1-\delta}{1/2} + (4\delta - 2) & ; x \in [2, 2'5), \\
x \cdot \frac{\delta}{1/2} + (3 - 6\delta) & ; x \in [2'5, 3],
\end{cases}
$$

Hence, we conclude that the semiorder of the present example is ε-continuously representable and, in fact, we know how to construct the corresponding function.

Unfortunately, although for any $n \in \mathbb{N}$ we are able to construct a SS-representation $(u_n, 1)$ such that the length of each jump-discontinuity is smaller than $\frac{1}{n}$, it is straightforward to see that the limit function $u = \lim_{n \to +\infty} u_n{}^3$ fails to be a SS-representation, since the intervals $[1'5, 2]$ and $[2'5, 3]$ would be reduced to a point.

We set this idea as a corollary:

Corollary 5.4 *Let $\prec$ be a semiorder on (X, τ). If it is continuously SS-representable, then it is ε-continuous. However, there exist ε-continuous semiorders that fail to be continuously SS-representable.*

Furthermore, there exist semiorders that fail to be ε_0-continuously SS-representable, for a given $\varepsilon_0 > 0$ (with $\varepsilon_0 \leq 1$). To see this we introduce the following example.

---

[3]Here, notice that the sequence $\{u_n\}_{n\in\mathbb{N}}$ is a Cauchy sequence with respect to the sup norm.

Example 5.5 Let X be the set $X = (-10, -0'5) \cup [0, 1] \cup (1'5, 10)$ endowed with the topology $\tau_\leq$ defined by the Euclidean order $\leq$ on X. Let $\prec$ be a semiorder on $X \subseteq \mathbb{R}$ defined by $x \prec y \iff x + 1 < y$, for any $x, y \in X$. It is straightforward to see that the identity function is a SS-representation of the semiorder, which has two bad gaps of length $0'5$, a first gap $[-0'5, 0)$ and the second one $(1, 1'5]$.

Notice that the sequences $(-0'5 - \frac{1}{n})_{n \in \mathbb{N}}$, $(0'5 - \frac{1}{n})_{n \in \mathbb{N}}$ and $(1'5 + \frac{1}{n})_{n \in \mathbb{N}}$ converge to $-0'5, 0'5$ and $1'5$, respectively.

Let δ be a value in $(0, 0'5]$. If there exists a δ-continuous SS-representation u, then, since $(-0'5 - \frac{1}{n})_{n \in \mathbb{N}}$ and $(0'5 - \frac{1}{n})_{n \in \mathbb{N}}$ are adjoint sequences, it holds that $\lim_{n \to \infty} u(-0'5 - \frac{1}{n}) + 1 = \lim_{n \to \infty} u(0'5 - \frac{1}{n})$. And it also holds true that $u(0'5) + 1 < u(1'5 + \frac{1}{n})$, for any $n \in \mathbb{N}$.

Thus, if $\lim_{n \to \infty} u(-0'5 - \frac{1}{n}) > u(0) - \delta$, then $u(1'5 + \frac{1}{n}) > u(0) - \delta + 2$, for any $n \in \mathbb{N}$, as well as $u(1) \leq u(0) + 1$. Therefore, $u(1'5 + \frac{1}{n}) > u(1) + 1 - \delta$, for any $n \in \mathbb{N}$. In consequence, if the length of one gap is reduced from $0'5$ to $\delta \leq 0'5$, then the other gap must increase from $0'5$ to $1 - \delta$.

Hence, we conclude that the semiorder of this example fails to be $0'5$-continuous.

Before we introduce our conjectures, we include the following concept that could allow us to present a constructive method to handle ε-continuity under some appropriate conditions.

Definition 5.6 Let (X, τ) be a topological space and $u : X \to \mathbb{R}$ a real function on X. Let $I = [a, b]$ be a bounded interval of the real line. A subset $\mathscr{C} = u(X) \cap I$ is said to be a *discontinuous Cantor set* if it satisfies the following properties:

(i) It has measure 0,
(ii) it has an infinite number of gaps,
(iii) every gap of $\mathscr{C}$ is a bad gap.

If there is a bounded interval I such that $\mathscr{C} = u(X) \cap I$ is a discontinuous Cantor set, then we will say that $u(X)$ *contains a discontinuous Cantor set*.

Remark 5.7 Notice that, given a discontinuous Cantor set $\mathscr{C} = I \cap u(X)$, then the sum of all the gaps of $\mathscr{C}$ is the length of the interval I.

We present the following example inspired in the Cantor set, which justifies the name of this new concept.

Example 5.8 We iteratively define the following Cantor set (we denote it by X) deleting the open middle thirds from a set of line segments. First, we delete the middle third $(1/3, 2/3]$ from $I = [0, 1]$, leaving two segments: $[0, 1/3] \cup (2/3, 1]$. Next, the middle third of each of these remaining segments is deleted, leaving four segments: $[0, 1/9] \cup (2/9, 1/3] \cup (2/3, 7/9] \cup (8/9, 1]$. This process is continued, where the nth set is

$$C_n = \frac{C_{n-1}}{3} \cup \left(\frac{2}{3} + \frac{C_{n-1}}{3} \right) \text{ for } n \geq 1, \text{ and } C_0 = [0, 1].$$

We define this Cantor set by $X = \bigcap_{n=1}^{\infty} C_n$.

Now we endow X with the Euclidean order topology $\tau_{\leq}$ and apply the inclusion function $i : (X, \tau_{\leq}) \rightarrow (\mathbb{R}, \tau_u)$. We denote $i(X) \cap [0, 1] = i(X)$ by $\mathscr{C}$. It is known that $\mathscr{C}$ measures 0. Furthermore, each gap is a bad gap (with respect to the usual topology of the real line and the Euclidean order topology on $\mathscr{C}$), since the inclusion function $i : (X, \tau_{\leq}) \rightarrow (\mathbb{R}, \tau_u)$ is discontinuous at every point a of each gap $(a, b]$.

Now, we are ready to present our main conjectures.

First, we introduce the weakest one.

Conjecture 5.9 (The Weak Conjecture)
Let $\prec$ be a SS-representable and bounded semiorder on a topological space (X, τ) and $(u, 1)$ a SS-representation. If it satisfies the necessary conditions (NC) and there is no discontinuous Cantor set contained in $u(X)$, then it is ε-continuously representable.

Now, we present the *Strong Conjecture*.

Conjecture 5.10 (The Strong Conjecture)
Let $\prec$ be a SS-representable and bounded semiorder on a topological space (X, τ). If it satisfies the necessary conditions (NC), then it is continuously representable.

In case the conjecture above holds true, then we could conclude the following result, that we present as a *Debreu's Open Gap Lemma for Bounded Semiorders*.

Conjecture 5.11 (Debreu's Open Gap Lemma for Bounded Semiorders)
Let S be a bounded subset of $\mathbb{R}$. Then, there exists a strictly increasing function $g : S \rightarrow \mathbb{R}$ such that all the gaps of $g(S)$ are open or closed, and satisfying that $x + 1 < y \iff g(x) + 1 < g(y)$ if and only if the following conditions hold:

(i) *There are no open-closed or closed-open gaps which length is bigger than or equal to 1.*

(ii) *For any gap $[a, b]$:*

   (a) *$[a + n, b + n] \cap S$ may content one point, for any $n \in \mathbb{N}$ with $n < m_r$, for some $m_r \in \mathbb{N}$ such that there exist $\gamma_l, \gamma_r \geq 0$ (with at least one of them different from 0) satisfying that $S \cap [a + m_r - \gamma_l, b + m_r + \gamma_r]$ contains at most one point s. If there exists that point s then $(s + n, b + n]$ may be nonempty if $\gamma_r > 0$, for any $n > m_r$ and, dually, $[a + n, s + n)$ may be nonempty if $\gamma_l > 0$ $(n > m_r)$.*

   (b) *$[a - n, b - n]$ or $[a - n, b - n)$ are gaps, for any $n \in \mathbb{N}$ with $n < m_l$, for some $m_l \in \mathbb{N}$ such that there exist $\gamma_l, \gamma_r \geq 0$ (with at least one of them different from 0) satisfying that $S \cap [a - m_l - \gamma_l, b - m_l + \gamma_r]$ contains at most the point $b - m_l + \gamma_r$.*

(ii) *For any gap $(a, b]$:*

(a) $S \cap [a - n, -b - n]$ contains at most one point, for any $n \in \mathbb{N}$ with $n < m_l$, for some $m_l \in \mathbb{N}$ such that there exist $\gamma_l, \gamma_r \geq 0$ (with at least one of them different from 0) satisfying that $S \cap [a - m_l - \gamma_l, b - m_l + \gamma_r]$ may contain at most one point s. If there exists that point s, then $(s - n, b - n]$ may be nonempty if $\gamma_r > 0$, for any $n > m_l$ and, dually, $[a - n, s - n)$ may be nonempty if $\gamma_l > 0$ $(n > m_l)$.

(b) $[a + n, b + n]$ or $[a + n, b + n)$ are gaps, for any $n \in \mathbb{N}$ with $n < m_r$, for some $m_r \in \mathbb{N}$ such that there exist $\gamma_l, \gamma_r \geq 0$ (with at least one of them different from 0) satisfying that $S \cap [a + m_r - \gamma_l, b + m_r + \gamma_r]$ contains at most the point $a + m_r - \gamma_l$.

6 Concluding Remarks

In the present study we have shown that a semiorder implies a rigid structure on the set, with a geometrical component due to the existence of the constant threshold $k = 1$. The necessary conditions for the existence of a continuous SS-representation are extremely strict and demanding. Hence, a semiorder rarely would be continuously representable.

Due to this handicap, we introduce the weaker concept of ε-continuity as a tool when dealing with semiorders that fail to be continuously representable. Furthermore, through this idea we present some conjectures on continuous SS-representability of semiorders. In case of a positive answer, these conjectures can be summarized just as a version of the *Debreu's Open Gap Lemma* but with the additional component of a threshold.

Acknowledgements The author acknowledges financial support from the Ministry of Economy and Competitiveness of Spain under grants MTM2015-63608-P and ECO2015-65031.

I am grateful for the helpful advice of the anonymous referees, to whom I am indebted for the detailed reading of previous versions of the manuscript and their very helpful suggestions and comments.

References

1. Aleskerov, F., Bouyssou, D., Monjardet, B.: Utility Maximization, Choice and Preference, 2nd edn. Springer, Berlin (2007)
2. Bosi, G.: Continuous representations of interval orders based on induced preorders. Rivista di Matematica per le Scienze Economiche e Soziali **18**(1), 75–82 (1995)
3. Bosi, G., Campión, M.J., Candeal, J.C., Induráin, E.: Interval-valued representability of qualitative data: the continuous case. Internat. J. Uncertain. Fuzziness Knowl. Based Syst. **15**(3), 299–319 (2007)
4. Bosi, G., Candeal, J.C., Induráin, E.: Continuous representability of interval orders and biorders. J. Math. Psych. **51**, 122–125 (2007)
5. Bosi, G., Candeal, J.C., Induráin, E., Olóriz, E., Zudaire, M.: Numerical representations of interval orders. Order **18**, 171–190 (2001)

6. Bosi, G., Estevan, A., Gutiérrez-García, J., Induráin, E.: Continuous representability of interval orders: the topological compatibility setting internat. J. Uncertain. Fuzziness Knowl. Based Syst. **23**(03), 345–365 (2015)
7. Bosi, G., Zuanon, M.E.: Semicontinuous representability of interval orders on a metrizable topological space. Int. J. Contemp. Math. Sci. **2**(18), 853–858 (2007)
8. Bosi, G., Zuanon, M.E.: Representations of an interval order by means of two upper semicontinuous functions. Int. Math. Forum **6**(42), 2067–2071 (2011)
9. Bridges, D.S., Mehta, G.B.: Representations of Preference Orderings. Springer, Berlin-Heidelberg-New York (1995)
10. Campión, M.J., Candeal, J.C., Induráin, E.: On Yi's extension property for totally preordered topological spaces. J. Korean Math. Soc. **43**(1), 159–181 (2006)
11. Campión, M.J., Candeal, J.C., Induráin, E.: Semicontinuous planar total preorders on non-separable metric spaces. J. Korean Math. Soc. **46**(4), 701–711 (2009)
12. Campión, M.J., Candeal, J.C., Induráin, E., Mehta, G.B.: Continuous order representability properties of topological spaces and algebraic structures. J. Korean Math. Soc. **49**(3), 449–473 (2012)
13. Campión, M.J., Candeal, J.C., Induráin, E., Zudaire, M.: Continuous representability of semiorders. J. Math. Psych. **52**, 48–54 (2008)
14. Candeal, J.C., Estevan, A., Gutiérrez-García, J., Induráin, E.: Semiorders with separability properties. J. Math. Psychol. **56**, 444–451 (2012)
15. Candeal, J.C., Induráin, E.: Semiorders and thresholds of utility discrimination: solving the Scott-Suppes representability problem. J. Math. Psych. **54**, 485–490 (2010)
16. Candeal, J.C., Induráin, E., Sanchis, M.: Order representability in groups and vector spaces. Expo. Math. **30**, 103–123 (2012)
17. Candeal, J.C., Induráin, E., Zudaire, M.: Numerical representability of semiorders. Math. Soc. Sci. **43**(1), 61–77 (2002)
18. Candeal, J.C., Induráin, E., Zudaire, M.: Continuous representability of interval orders. Appl. Gen. Topol. **5**(2), 213–230 (2004)
19. Chateauneuf, A.: Continuous representation of a preference relation on a connected topological space. J. Math. Econom. **16**, 139–146 (1987)
20. Debreu, G.: Continuity properties of paretian utility. Internat. Econom. Rev. **5**, 285–293 (1964)
21. Estevan, A.: Generalized Debreu's open gap lemma and continuous representability of biorders. Order **33**(2), 213–229 (2016)
22. Estevan, A., Gutiérrez García, J., Induráin, E.: Further results on the continuous representability of semiorders. Internat. J. Uncertain. Fuzziness Knowl. Based Syst. **21**(5), 675–694 (2013)
23. Fishburn, P.C.: Intransitive indifference with unequal indifference intervals. J. Math. Psych. **7**, 144–149 (1970)
24. Fishburn, P.C.: Intransitive indifference in preference theory: a survey. Oper. Res. **18**(2), 207–228 (1970)
25. Fishburn, P.C.: Utility Theory for Decision-Making. Wiley, New York (1970)
26. Fishburn, P.C.: Interval representations for interval orders and semiorders. J. Math. Psych. **10**, 91–105 (1973)
27. Fishburn, P.C.: Interval Orders and Interval Graphs. Wiley, New York (1985)
28. Fishburn, P.C., Monjardet, B.: Norbert Wiener on the theory of measurement (1914, 1915, 1921). J. Math. Psychol. **36**, 165–184 (1992)
29. Gensemer, S.H.: Continuous semiorder representations. J. Math. Econom. **16**, 275–289 (1987)
30. Luce, R.D.: Semiorders and a theory of utility discrimination. Econometrica **24**, 178–191 (1956)
31. Monjardet, B.: Axiomatiques et propriétés des quasi-ordres. Math. Sci. Hum. **63**, 51–82 (1978)
32. Olóriz, E., Candeal, J.C., Induráin, E.: Representability of interval orders. J. Econom. Theory **78**(1), 219–227 (1998)
33. Scott, D., Suppes, P.: Foundational aspects of theories of measurement. J. Symbolic Logic **23**, 113–128 (1958). **54**, 485–490 (2010)

34. Vincke, P.: Linear utility functions on semiordered misture spaces. Econometrica **48**(3), 771–775 (1980)
35. Wiener, N.: Contribution to the theory of relative position. Math. Proc. Cambridge Philos. Soc. **17**, 441–449 (1914)
36. Wiener, N.: A new theory of measurement. Proc. London Math. Soc. **19**, 181–205 (1919)

A Note on Candeal and Induráin's Semiorder Separability Condition

Denis Bouyssou and Marc Pirlot

Abstract We show that the semiorder separability condition used by Candeal and Induráin in their characterization of semiorders having a strict representation with positive threshold can be factorized into two conditions. The first says that the trace of the semiorder must have a numerical representation. The second asserts that the number of "noses" in the semiorder must be finite or countably infinite. We discuss the interest of such a factorization.

Keywords Utility theory · Semiorders · Separability · Noses

1 Introduction

Although the idea of introducing a threshold into preference or perception models has distant origins (see Pirlot and Vincke [34] and Fishburn and Monjardet [19], for historical accounts of the idea) the formal definition of semiorders is due to Luce [26]. Shortly after, Scott and Suppes [41] showed that a semiorder defined on a *finite* set always has a strict representation with positive threshold (see also Scott [40]).

Because the threshold used in the representation is constant and positive, it is clear that the result does not carry over to countably infinite sets (see, e.g., Fishburn [18], p. 30). This makes semiorders at variance with what happens with many other preference structures (e.g., weak orders, biorders, interval orders, suborders, see Aleskerov et al. [2] and Bridges and Mehta [8]) for which the finite and the countably infinite cases are identical.

D. Bouyssou (✉)
LAMSADE, UMR 7243, CNRS, Université Paris-Dauphine, PSL Research University, 75016 Paris, France
e-mail: bouyssou@lamsade.dauphine.fr

M. Pirlot
Université de Mons, rue de Houdain 9, 7000 Mons, Belgium
e-mail: marc.piriot@umons.ac.be

© Springer Nature Switzerland AG 2020
G. Bosi et al. (eds.), *Mathematical Topics on Representations of Ordered Structures and Utility Theory*, Studies in Systems, Decision and Control 263,
https://doi.org/10.1007/978-3-030-34226-5_6

The characterization of semiorders on countably infinite sets having a strict representation with positive threshold was achieved by Manders [27] and Beja and Gilboa [4]. Basically, what is required is to prevent the existence of infinite (ascending or descending) chains of strict preference that are bounded.

Finally, building upon these results,[1] Candeal and Induráin [9] have achieved a general characterization of semiorders having a strict positive threshold representation by the addition of a condition called *semiorder separability* (see [11, 16] for further results).

This note is part of a long-term project aiming at giving new proofs for the existence of numerical representations of semiorders that would unify the finite, countably infinite and general cases [6, 7]. Its purpose is to show that the semiorder separability condition proposed in Candeal and Induráin [9] can be factorized into two conditions. The first says that the trace of the semiorder must have a numerical representation. This condition holds for strict as well as for nonstrict representations, as defined below. The second condition asserts that the number of "noses" in the semiorder must be finite or countably infinite. It is highly specific to strict representations. We discuss the interest of such a factorization.

The paper is organized as follows. Section 2 introduces our notation and framework. Section 3 presents our results. They are discussed in Sect. 4.

2 Notation and Framework

2.1 Notation and Definitions

In the rest of this text, we say that a set is *denumerable* if it is finite or countably infinite.

Let S be a binary relation on a set X. We often write $x\ S\ y$ instead of $(x, y) \in S$. The relation S is a *semiorder* if it is complete ($x\ S\ y$ or $y\ S\ x$, for all $x, y \in X$), Ferrers ($x\ S\ y$ and $z\ S\ w$ imply $x\ S\ w$ or $z\ S\ y$, for all $x, y, z, w \in X$) and semitransitive ($x\ S\ y$ and $y\ S\ z$ imply $x\ S\ w$ or $w\ S\ z$, for all $x, y, z, w \in X$). In the sequel, we shall often write the semiorder S as a pair (P, I) of relations, where P (resp. I) denotes the asymmetric (resp. symmetric) part of S. The asymmetric part of S is the relation P, often called the "strict preference" relation. It is a partial order on X, i.e., an asymmetric and transitive relation, which is also Ferrers and semitransitive. The symmetric part of S is the relation I, often called the "indifference" relation. It is reflexive and symmetric but not necessarily transitive. Because S is complete, notice that we could have alternatively defined a semiorder, giving its asymmetric part P, while letting I be the symmetric complement of P (i.e., $x\ I\ y$ iff [$Not[x\ P\ y]$ and $Not[y\ P\ x]$]) and $S = P \cup I$. We refer to Aleskerov et al. [2], Fishburn [18],

---

[1]Earlier results in the general case include Abrisqueta et al. [1], Candeal et al. [10], Gensemer [21–23], Narens [30].

Monjardet [28], Pirlot and Vincke [34], Roubens and Vincke [38], Suppes et al. [42] for detailed studies of the various properties of semiorders.

A *complete preorder* on X is a complete and transitive relation. A *linear* order (or *total* order) on X is a complete, antisymmetric and transitive relation.

The trace $\succsim_S$ of a semiorder S on X is the relation defined as follows: for all $x, y \in X$, $x \succsim_S y$ if for all $z \in X$, $y S z$ implies $x S z$ and $z S x$ implies $z S y$. We omit the subscript when there is no ambiguity on the underlying semiorder. It is easy to check that the trace $\succsim_S$ can be equivalently defined using P, i.e., $x \succsim_S y$ if for all $z \in X$, $y P z$ implies $x P z$ and $z P x$ implies $z P y$.

It is well-known (see, e.g., [2, 28, 34]) that the trace of a semiorder is a complete preorder (moreover, a semiorder is identical to its trace iff it is a complete preorder). We will use $\sim$, $\succ$, $\precsim$, and $\prec$, as is usual. Two elements $x, y \in X$ such that $x \sim y$ are said *equivalent*, i.e., for all $z \in X$, we have $z S x$ iff $z S y$ and $x S z$ iff $y S z$.

2.2 Chains

Let R be a binary relation on the set X. An R-chain (we use here the terminology used in the field of ordered sets, see Caspard et al. [12] or Schröder [39]. Graph theorists may prefer the term "path") is a sequence x_i of elements of X indexed by a subset of consecutive integers $J \subseteq \mathbb{Z}$ and such that any two consecutive elements of the sequence belong to the relation R. Formally, the sequence $(x_i, x_i \in X, i \in J)$, where $J \subseteq \mathbb{Z}$ and $(x_i, x_{i+1}) \in R$, for all $i, i+1 \in J$ is an R-chain. We shall consider P-chains and I-chains in the sequel.

Note that an R-chain needs neither have a first nor a last element. In other terms, it can have an infinite number of elements before or after a given element, but not between two given elements. An R-chain is said to start at $x \in X$, if the set J has a minimum element and x is the element of X indexed by the minimal element in J. In this case, the chain is said to *have a first element*, which is this x. An R-chain is said to terminate at $y \in X$, if the set J has a maximum element and y is the element of X indexed by the maximal element in J. In this case, the chain is said to *have a last element*, which is this y. An R-chain starting at x and terminating at y (we also say "an R-chain from x to y") is finite by definition.

A P-chain $(x_i, i \in J)$ has an *upper* (resp. *lower*) *bound* if there exists $a \in X$ (resp. $b \in X$) such that $a P x_i$ (resp. $x_i P b$) for all $i \in J$. If the chain has both an upper bound a and a lower bound b, we say it is *bounded*.

Note that the set $\{x_i : i \in J\} \cup \{a, b\}$ is totally ordered by P, but cannot always be indexed by the elements of a subset J' of $\mathbb{Z}$. It cannot be when the P-chain $(x_i, i \in J)$ has no first or no last element. The elements of a finite subset of X which is totally ordered by P can be indexed by a set J of consecutive integers in order to form a P-chain. If a P-chain $(x_i, i \in J)$ has no last (resp. first) element, then for all $i \in J$, x_{i+k} (resp. x_{i-k}) belongs to the chain, for all $k \in \mathbb{N}$.

2.3 Representations with Positive Threshold

Let us make precise the definition of a strict representation with positive threshold.

Definition 1 A strict representation with positive threshold of a semiorder $S = (P, I)$ on the set X is a constant $k > 0$ and a function u from X to $\mathbb{R}$, such that, for all $x, y \in X$,

$$
\begin{aligned}
x \, P \, y &\iff u(x) > u(y) + k, \\
x \, I \, y &\iff -k \le u(x) - u(y) \le k.
\end{aligned}
\tag{1}
$$

A *strict unit representation* of the semiorder $S = (P, I)$ is a strict positive threshold representation with $k = 1$.

It is clear that a strict representation with positive threshold exists iff a strict unit representation exists. We focus on strict unit representations below.

Because we will be dealing with infinite sets, it is important to keep in mind that other kinds of numerical representation using a positive threshold can be envisaged. Let us simply mention here the case of a *nonstrict unit representation* in which, for all $x, y \in X$,

$$
\begin{aligned}
x \, P \, y &\iff u(x) \ge u(y) + 1, \\
x \, I \, y &\iff -1 < u(x) - u(y) < 1.
\end{aligned}
\tag{2}
$$

In the finite case, it is well know that strict and nonstrict unit representations are equivalent (see Pirlot and Vincke [34], Chap. 3 and 4 or Roberts [37], Chap. 6). The same is true in the countably infinite case (Beja and Gilboa [4], Th. 3.8, p. 436). In the general case, the two types of representations are distinct. For instance, the canonical nonstrict semiorder on $\mathbb{R}$ ($x \, P \, y \iff x \ge y + 1, x \, I \, y \iff |x - y| < 1$) has a trivial nonstrict unit representation but has no strict unit representation.[2] A similar phenomenon occurs with biorders [14] and, hence, interval orders [17, 18]).

We shall only consider strict unit representations that assign the same value to equivalent elements of X w.r.t. the trace $\succsim$. Consequently, we shall assume henceforth that the equivalence class of each element of X w.r.t. the trace of the semiorder is reduced to a singleton. In other words, for all $x, y \in X$, $x \succsim y$ and $y \succsim x$ imply $x = y$. Therefore, the trace $\succsim$ is a linear order on X. Its asymmetric part is denoted by $\succ$ and its symmetric part by $\sim$. This is not restrictive (see Candeal and Induráin [9], Lemma 3.2).

---

[2]Indeed, in this semiorder, for all $x \in \mathbb{R}$, the ordered pair $(x + 1, x)$ is a nose, as defined in Definition 4. Hence, we have an uncountable number of noses, while the existence of a strict numerical representation implies that the number of noses must be finite or countably infinite. See Remark 2.

2.4 Axioms and Previous Results

2.4.1 Bounded P-Chain Condition

A necessary condition for the existence of a strict unit representation of a semiorder is the bounded P-chain condition. It says that every bounded P-chain is finite.

This condition was introduced by Manders [27] and Beja and Gilboa [4] under slightly different forms (we use here the version in Candeal and Induráin [9]). It is simple to check ([27], Prop. 8, p. 237) that if there is an I-chain joining any two elements of X, then the bounded P-chain condition holds. This condition is called *regularity* in Candeal and Induráin [9].

The bounded P-chain condition has the flavor of an Archimedean axiom. It sounds like "Every bounded standard sequence is finite" ([25], p. 25). Here, the sequence of pairs of objects in P plays a role that resembles that of equally spaced preference intervals used in standard sequences (see, in particular, the *strong standard sequences* defined in Gonzales [24], p. 51). Such properties are required for enabling representations using real numbers.

The bounded P-chain condition is clearly necessary for the existence of a strict unit representation as well as for a nonstrict unit representation. Consider, e.g., the case of strict representation. Suppose it has an infinite increasing chain $(x_i, x_i \in X, i \in J)$, indexed by the set of consecutive integers $\mathbb{N}$, such that $(x_{i+1} \ P \ x_i$, for all $i \in \mathbb{N}$ and such that $\omega \ P \ x_i$, for all $i \in \mathbb{N})$. This would imply

$$\omega \ P \ldots P \ x_{j+1} \ P \ x_j \ P \ldots P \ x_2 \ P \ x_1$$

for all $j \in \mathbb{N}$, implying $u(x_j) > u(x_1) + j$, so that $u(\omega) - u(x_1) > n$, for all $n \in \mathbb{N}$. This is clearly impossible.

2.4.2 s-Separability

Candeal and Induráin [9] introduce the following condition that they call semiorder-separability, ("s-separability", for short).

Definition 2 A semiorder $S = (P, I)$ on X is semiorder-separable if there is a denumerable set E, $E \subseteq X$, such that, for all $a, b \in X$ with $a \ P \ b$, there are

$$c \in E \text{ such that } a \ P \ c \succsim b \text{ and}$$
$$d \in E \text{ such that } a \succsim d \ P \ b$$

The fact that this condition is necessary for the existence of a strict unit representation is shown in Candeal and Induráin ([9], p. 488, 2nd col).

2.4.3   The main result in Candeal and Induráin [9]

The main result in [9] can be rephrased as follows.

Theorem 1 (Candeal and Induráin [9], Theorem 3.6) *A semiorder $S = (P, I)$ on a set X has a strict unit representation iff it satisfies the bounded P-chain condition and is s-separable.*

Notice that the trace of an s-separable semiorder is Debreu-separable (see Candeal and Induráin [9], Lemma 3.4. We often say d-separable instead of Debreu-separable, for short). This is a condition guaranteeing the existence of a numerical representation of the trace (i.e., the existence of a function $v : X \to \mathbb{R}$ such that $x \succsim y$ iff $v(x) \geq v(y)$). We recall the definition of Debreu-separability below. We refer to Bridges and Mehta [8] for a detailed analysis of this condition and several equivalent formulations found in the literature.

Definition 3 A semiorder $S = (P, I)$ is d-separable if its trace $\succsim$ is d-separable. The trace is d-separable if it has a denumerable order-dense set, i.e., there is a denumerable set D, $D \subseteq X$, such that, for all $a, b \in X$ with $a \succ b$, there is $d \in D$, such that $a \succsim d \succsim b$.

3   Results

We revisit the s-separability condition. Our aim is to factorize it into d-separability and another condition. The latter is expressed in terms of the *noses* of the semiorder.

3.1   Noses and Half-Noses

The notion of "nose" of a semiorder has been introduced in Pirlot [32, 33]. When X is finite, it is instrumental to build synthetic representations of a semiorder [33] as well as proving that it has a minimal representation and building it [32]. It was shown in Doignon and Falmagne [13] that the noses of a semiorder are exactly the ordered pairs of the *inner fringe* of the semiorder $S = (P, I)$. The inner fringe of semiorder consists in the set of ordered pairs belonging to P that can be removed from P and turned into I, while remaining in the set of semiorders.

Definition 4 The ordered pair $(a, b) \in X^2$ is a *nose* of the semiorder $S = (P, I)$ if $a \, P \, b$ and there is no $c \in X$ such that $a \, P \, c \succ b$ and there is no $d \in X$ such that $a \succ d \, P \, b$.

Noses play a special role w.r.t. s-separability as shown by the following lemma.

Lemma 1 *If the semiorder $S = (P, I)$ on X (for which $\succsim$ is antisymmetric) is s-separable by the denumerable set E, then a and b belong to E whenever (a, b) is a nose.*

Proof Let (a, b) be a nose, so that $a \, P \, b$. By the s-separability property, there is $c \in E$ such that $a \, P \, c \succsim b$. By definition of a nose, we have $c = b$ and therefore, $b \in E$. Using s-separability, there is also $d \in E$ such that $a \succsim d \, P \, b$, which implies $a = d$ and $a \in E$ since (a, b) is a nose. □

Remark 1 For later use, let us observe that if (a, b) is a nose then there cannot exist a nose (a, c) with $b \neq c$. Indeed, if $b \succ c$, we have $a P c$, $b \neq c$ and $a P b$, violating the definition of a nose. Hence if a is the left endpoint of a nose there is unique right endpoint b associated to it, so that (a, b) is a nose. A similar observation holds in the opposite direction: if b is the right endpoint of a nose there is unique left endpoint a associated to it, so that (a, b) is a nose.

We will also need to care about *half-noses*, as defined below.

Definition 5 The ordered pair $(a, b) \in X^2$ is a *lower half-nose* (l-h-nose) of the semiorder $S = (P, I)$ if $a \, P \, b$ and there is no $c \in X$ such that $a \, P \, c \succ b$. The ordered pair (a, b) is a *proper l-h-nose* if it is an l-h-nose that is not a nose, i.e., there is $d \in X$ such that $a \succ d \, P \, b$.

The ordered pair $(a, b) \in X^2$ is an *upper half-nose* (u-h-nose) of the semiorder $S = (P, I)$ if $a \, P \, b$ and there is no $d \in X$ such that $a \succ d \, P \, b$. The pair (a, b) is a *proper u-h-nose* if it is a u-h-nose that is not a nose, i.e., there is $c \in X$ such that $a \, P \, c \succ b$.

We denote by $\mathcal{N}_{plh}$ (resp. $\mathcal{N}_{puh}$) the set of right endpoints b (resp. left endpoints a) of proper l-h-noses (resp. u-h-noses) (a, b).

We have the following result.

Lemma 2 *If the semiorder $S = (P, I)$ is d-separable, then the sets $\mathcal{N}_{plh}$ and $\mathcal{N}_{puh}$ are denumerable.*

Proof We give the proof for $\mathcal{N}_{plh}$, the case of $\mathcal{N}_{puh}$ being similar.

Let (a, b) be a proper l-h-nose. Hence, we know that:

(i) $a \, P \, b$,
(ii) there is no $c \in X$ such that $a \, P \, c$ and $c \succ b$,
(iii) there is a $d \in X$ such that $a \succ d$ and $d \, P \, b$.

We define the set $N(b) = \{x \in X : x \, P \, b,$ and there is no $c \in X$ such that $x \, P \, c$ and $c \succ b\}$. In other words, for all $x \in N(b)$, (x, b) is a l-h-nose.

By hypothesis, $a \in N(b)$. We know that there is an element $d \in X$ such that $a \succ d \, P \, b$. We claim that d belongs to $N(b)$. Indeed, we have $d \, P \, b$. Suppose that there is $c \in X$ such that $d \, P \, c$ and $c \succ b$. Since $a \succ d$ and $d \, P \, c$, we obtain $a \, P \, c$. So that we have $a \, P \, c$ and $c \succ b$, a contradiction. Hence, the claim is proved and $N(b)$ contains at least two elements a and d.

We claim that $N(b)$ is an interval w.r.t. $\succ$. To prove this, let $x, y \in N(b)$. Take any z such that $x \succ z \succ y$. Let us show that we have $z \in N(b)$, which will prove the claim. Because, $z \succ y$ and $y \ P \ b$, we have $z \ P \ b$. Suppose that there is $c \in X$ such that $z \ P \ c$ and $c \succ b$. Because $x \succ z, z \ P \ c$ implies $x \ P \ c$. Hence, we have $x \ P \ c$ and $c \succ b$, contradicting the fact that $x \in \mathcal{N}_{plh}$. Hence, for all $b \in \mathcal{N}_{plh}$, $N(b)$ is a nondegenerate interval of $\succsim$.

Now, let (a, b) and (e, f) be two proper l-h-noses, so that $b, f \in \mathcal{N}_{plh}$. Suppose that $b \neq f$, We claim that the associated intervals $N(b)$ and $N(f)$ are disjoint. Indeed, $x \in N(b)$ implies $x \ P \ b$, and there is no $c \in X$ such that $x \ P \ c$ and $c \succ b$. Similarly, $x \in N(f)$ implies $x \ P \ f$, and there is no $d \in X$ such that $x \ P \ d$ and $d \succ f$. Because, $b \neq f$ we have either $b \succ f$ or $f \succ b$. Suppose that $f \succ b$, the other case being similar. We have $x \ P \ f$ and $f \succ b$, violating the fact that $x \in N(b)$.

Now each of these intervals contains at least two distinct points and therefore at least an element from the denumerable set D that d-separates $S = (P, I)$. Consequently, the set $\mathcal{N}_{plh}$ is denumerable.   □

3.2   Main Results

We are now in position to propose the announced factorization for the s-separability condition.

Proposition 1 *A semiorder $S = (P, I)$ on X is s-separable iff it is d-separable and its set of noses is denumerable.*

Proof [$\Rightarrow$] By Lemma 1, the set of noses is denumerable. The s-separability property implies that $\succsim$ is d-separable (see Candeal and Induráin [9], Lemma 3.4). We include the proof for completeness. Let $x, y \in X$ be such that $x \succ y$. There is $z \in X$ such that $x \ P \ z$ and $z \ S \ y$ and/or $w \in X$ such that $w \ P \ y$ and $x \ S \ w$. In the former case, s-separability entails that there is $d \in E$ such that $x \succsim d \ P \ z$ and, since $z \ S \ y$, we have $x \succsim d \succ y$. In the latter case, there is $c \in E$ such that $w \ P \ c \succsim y$ and, since $x \ S \ w$, we have $x \succ c \succsim y$.

[$\Leftarrow$] Let D be a denumerable set that d-separates $\succsim$. Let $x, y \in X$ be such that $x \ P \ y$. If (x, y) is not a nose, there are two cases:

1. either there is $y' \succ y$ such that $x \ P \ y'$,
2. or there is $x' \prec x$ such that $x' \ P \ y$.

In the first case, by the d-separability of $\succ$, there is $c \in D$ such that $y' \succsim c \succsim y$. Therefore we have $x \ P \ c \succsim y$. Further, either there is $x' \prec x$ such that $x' \ P \ y$ or, for all $x' \prec x$, we have $Not[x' \ P \ y]$. In the former case, d-separability implies that there is $d \in D$ such that $x' \precsim d \precsim x$. Otherwise, (x, y) is a proper u-h-nose. In order to have $d \in E$ such that $x \succsim d \ P \ y$, we set $d = x$ and include, using Lemma 2, the denumerable set $\mathcal{N}_{puh}$ of left endpoints of the proper u-h-noses in E.

In the second case, by the d-separability of $\succ$, there is $d \in D$ such that $x' \precsim d \precsim x$. Therefore we have $x \succsim d \ P \ y$. Further, there are two cases. Either there is $y' \succ y$ such that $x \, P \, y'$ or, for all $y' \succ y$, we have $Not[x \, P \, y']$. In the former case, d-separability implies that there is $c \in D$ such that $y' \succsim c \succsim y$. Then, we have $x \, P \, c \succsim y$. Otherwise, (x, y) is a proper l-h-nose. In order to have $c \in E$ such that $x \, P \, c \succsim y$, we set $c = y$ and include, using Lemma 2, the denumerable set $\mathcal{N}_{plh}$ of right endpoints of the proper l-h-noses in E.

Finally, by considering E as the union of D, $\mathcal{N}_{plh}$, $\mathcal{N}_{puh}$ and the set of elements a, b such that (a, b) is a nose, which is denumerable by hypothesis, we obtain a denumerable set E, which s-separates the semiorder (P, I). $\square$

Remark 2 It is easy to show that having a denumerable set of noses is a necessary condition for a semiorder to have a strict unit representation. Indeed, assume that f is a unit representation of the semiorder $S = (P, I)$ (and, consequently is a numerical representation of $\succsim$, since we have supposed $\succsim$ to be antisymmetric). Suppose that (a, b) is a nose of S. Since $a \, P \, b$, we have $f(a) > f(b) + 1$. Let ε_{ab} be the positive number $f(a) - f(b) - 1$. By definition of a nose, there is no element $c \neq b$ such that $a \, P \, c \succ b$ and therefore, there is no c such that $f(c) \in (f(b), f(a) - 1]$, an interval of length $\varepsilon_{ab} > 0$. To each nose (a, b) is associated such an interval of positive length and all these intervals are disjoint. Since there is only a denumerable number of disjoint intervals of positive length in $\mathbb{R}$, the number of noses is denumerable.

Combining Remark 2 and Proposition 1 with Theorem 1 leads to our main result.

Theorem 2 *A semiorder $S = (P, I)$ on a set X has a strict unit representation iff it satisfies the bounded P-chain condition, is d-separable and has a set of noses that is denumerable. These three conditions are independent.*

Proof The first part immediately follows from Remark 2 and Proposition 1, in view of Theorem 1. To prove the second part, we need three examples.

Example 1 Let $X = \mathbb{R}^2$. Consider the binary relation S such that $S = P \cup I$ with $(x_1, x_2) \, P \, (y_1, y_2)$ if $x_1 > y_1 + 1$ or $[x_1 = y_1 + 1$ and $x_2 > y_2]$, while I is the symmetric complement of P (i.e., $x \, I \, y \iff Not[x \, P \, y]$ and $Not[y \, P \, x]$).

It is not difficult to show that S is a semiorder (that is not a complete preorder). It is clear that for all $x, y \in X$, there is an I-chain joining them, so that the bounded P-chain condition holds. The set of noses of S is easily seen to be empty. The trace of S is the lexicographic preorder on $\mathbb{R}^2$. Hence, d-separability is violated (see [3, 8]).

Example 2 Let $X = \mathbb{R}$. We consider the binary relation S on X such that $x \, S \, y \iff x \geq y + 1$. It is clear that this relation is a semiorder. For all $x, y \in X$, there is an I-chain joining them, so that the bounded P-chain condition holds. The trace $\succsim$ of S is $\geq$, so that S is d-separable. All ordered pairs $(x, y) \in \mathbb{R}^2$ such that $x = y + 1$ are noses.

Example 3 Let $X = \mathbb{N} \cup \{\omega\}$. Consider the binary relation S such that $\omega \, P \, x$, for all $x \in \mathbb{N}$ and $x \, P \, y$ iff $x > y + 1$, for all $x, y \in \mathbb{N}$, while I is the symmetric complement of P. Since X is denumerable, d-separability and the condition on noses trivially hold. The bounded P-chain condition is violated. $\square$

Remark 3 In Bouyssou and Pirlot [7], for proving the existence of a numerical representation, we use *d*-separability and the condition that the number of noses is denumerable, instead of *s*-separability. In this proof, we only use the denumerable set *D* that is dense in the trace $\succsim$ and the denumerable set of noses endpoints. We do not need to add the proper half-noses as in the proof of Proposition 1. In other words, we do not use all the points in the set *E* involved in the *s*-separability condition.

4   Discussion

We have exhibited a set of three independent conditions that are necessary and sufficient for the existence of a strict unit representation of a semiorder. This has been achieved by factorizing *s*-separability into *d*-separability and the condition that the set of noses is denumerable. We feel that these three conditions have a clear interpretation.

The bounded *P*-chain condition deals with the fact that the threshold is constant and positive. As already noted, it resembles an Archimedean condition. It applies as soon as the set *X* is infinite, even countably infinite. It is not specific to strict unit representations. It is easy to check that it is also a necessary condition for nonstrict unit representations.

The *d*-separability condition ensures that the trace of the semiorder, which is a complete preorder, has a numerical representation. This is clearly necessary for strict unit representations but is not specific to them. As can be easily checked, *d*-separability is also necessary for nonstrict unit representations.

Our final condition states that the set of noses is denumerable. It is specific to strict unit representations. We show in Bouyssou and Pirlot [6, 7] how to deal with the case of nonstrict representations. This involves replacing our condition on "noses" by a condition on the dual notion of "hollows" [32, 33]. We do not develop this point here.

Our results are also linked to the discussion in Candeal and Induráin ([9], Sect. 4, p. 489) of Beja and Gilboa [4], Th. 4.5(a), p. 439). This theorem asserts that a "Generalized Numerical Representation" (GNR) with $\mathscr{S}$ open exists iff *S* is a semiorder (for which $\succsim$ is antisymmetric) satisfying *d*-separability and the bounded *P*-chain condition and such that the set of *P*-gap-edge-points is denumerable. An element $x \in X$ is a *P*-gap-edge-point if there is a $y \in X$ such that $y \, P \, x$ and there is no $z \in X$ for which $y \, P \, z \succ x$.

As noted by Candeal and Induráin [9], the proof of this result (see Beja and Gilboa [4] p. 446–448) refers to "positive threshold GNR in which $\mathscr{S}$ is open". This is tantamount to what we have called a strict unit representation. Hence, Candeal and Induráin [9] wonder whether Beja and Gilboa [4] were the first to characterize semiorders having a strict unit representation. They state (p. 489, last par. of 2nd col.) that the result in Beja and Gilboa [4] should be amended by the addition of a condition stating, in our terms, that the set of all right endpoints of lower-half noses should be denumerable.

Our results allow to be more specific. It is clear that if x is a P-gap-edge-point, there is a y such that (y, x) is a nose or a proper lower-half nose (see Definitions 4 and 5). In other terms, x is the right endpoint of a l-h-nose. A l-h-nose, is either a proper l-h nose or a nose. Whenever d-separability is in force, we do not have to ensure the fact that the set of right endpoints of proper l-h noses is denumerable (Lemma 2). We only have to require that the set of noses is denumerable, which is clearly implied by the requirement that the set of all right endpoints of l-h noses is denumerable: requiring that the set of P-gap-edge-points is denumerable therefore implies that the set of right endpoints of noses as well as the set of right endpoints of proper l-h noses are denumerable. Proposition 1 and Theorem 2 show that this is sufficient to guarantee the existence of a strict unit representation. This condition can be weakened however since, as shown in Lemma 2, d-separability implies that the set of proper l-h-noses is denumerable. Hence, our result sharpens the result of Beja and Gilboa [4] discussed in Candeal and Induráin [9], Sect. 4 while ensuring its correctness. To bring our result closer to the one of Beja and Gilboa [4], we could require that the set of all right endpoints of noses is denumerable instead of requiring that the set of all noses is denumerable. Clearly, these two conditions are equivalent: to the right endpoint of a nose corresponds a unique nose (see Remark 1).

Let us finally mention two directions for future research.

The first is to relate our analysis to the several equivalent formulations of s-separability analyzed in Candeal et al. [11]. Candeal et al. ([11], Th. 4.11, p. 449) state that, for a semiorder, s-separability is equivalent to any of the conditions ensuring separability for interval orders, as introduced in Oloriz et al. [31] and detailed in [5] see also the analysis of separability conditions for biorders in Doignon et al. [14] and Nakamura [29]. It would be useful to investigate if one of these conditions, equivalent to s-separability for semiorders, allows us to obtain sharper results.

The second and more difficult question is to tackle the case in which representations are neither strict nor nonstrict, as suggested in [29]. The related problem of characterizing interval graphs using mixed intervals, i.e., using intervals that are not necessarily all closed or all open, has recently attracted attention in the Graph Theory community [15, 35, 36], while the classic result of Frankl and Maehara [20] may be consider as a dual to Beja and Gilboa ([4], Th. 3.8, p. 436) stating the equivalence of strict and nonstrict representations for denumerable sets.

Acknowledgements We would like to thank two referees for their useful comments on an earlier draft of this text.

References

1. Abrísqueta, F.J., Candeal, J.C., Induráin, E., Zudaire, M.: Scott-Suppes representability of semiorders: internal conditions. Math. Soc. Sci. **57**(2), 245–261 (2009)
2. Aleskerov, F., Bouyssou, D., Monjardet, B.: Utility Maximization, Choice and Preference, 2nd edn. Springer, Berlin (2007)

3. Beardon, A.F., Candeal, J.C., Herden, G., Induráin, E., Mehta, G.B.: The non-existence of a utility function and the structure of non-representable preference relations. J. Math. Econ. **37**(1), 17–38 (2002)
4. Beja, A., Gilboa, I.: Numerical representations of imperfectly ordered preferences (a unified geometric exposition). J. Math. Psychol. **36**(3), 426–449 (1992)
5. Bosi, G., Candeal, J.C., Induráin, E., Oloriz, E., Zudaire, M.: Numerical representations of interval orders. Order **18**(2), 171–190 (2001)
6. Bouyssou, D., Pirlot, M.: Unit representation of semiorders on countable sets. Working paper (2019)
7. Bouyssou, D., Pirlot, D.: Revisiting the representation of semiorders: the uncountable case. Working paper (2019)
8. Bridges, D.S., Mehta, D.S.: Representations of preferences orderings. Number 422 in Lecture Notes in Economics and Mathematical Systems, 1st edn. Springer, Berlin Heidelberg (1995)
9. Candeal, J.C., Induráin, E.: Semiorders and thresholds of utility discrimination: solving the Scott-Suppes representability problem. J. Math. Psychol. **54**(6), 485–490 (2010)
10. Candeal, J.C., Induráin, E., Zudaire, M.: Numerical representability of semiorders. Math. Soc. Sci. **43**(1), 61–77 (2002)
11. Candeal, J.C., Estevan, A., Gutiérrez García, J., Induráin, E.: Semiorders with separability properties. J. Math. Psychol. **56**, 444–451 (2012). ISSN 0022-2496,1096-0880. https://doi.org/10.1016/j.jmp.2013.01.003
12. Caspard, N., Leclerc, B., Monjardet, B.: Finite Ordered Sets: Concepts, Results and Uses. Number 144 in Encyclopedia of Mathematics and its Applications. Cambridge University Press (2012). ISBN 978-1-107-01369-8
13. Doignon, J.-P., Falmagne, J.-C.: Well-graded families of relations. Discret. Math. **173**(1–3), 35–44 (1997)
14. Doignon, J.-P., Ducamp, A., Falmagne, J.-C.: On realizable biorders and the biorder dimension of a relation. J. Math. Psychol. **28**(1), z 73–109 (1984)
15. Dourado, M.C., Le, V.B., Protti, F., Rautenbach, D., Szwarcfiter, J.L.: Mixed unit interval graphs. Discret. Math. **312**, 3357–3363 (2012). ISSN 0012–365X. https://doi.org/10.1016/j.disc.2012.07.037
16. Estevan, A., Gutiérrez García, J., Induráin, E.. Numerical representation of semiorders. Order **30**(2), 455–462 (2013). ISSN 1572-9273. https://doi.org/10.1007/s11083-012-9255-3
17. Fishburn, P.C.: Interval representations for interval orders and semiorders. J. Math. Psychol. **10**(1), 91–105 (1973)
18. Fishburn, P.C.: Interval Orders and Intervals Graphs. Wiley, New York (1985)
19. Fishburn, P.C., Monjardet, B.: Norbert Wiener on the theory of measurement (1914, 1915, 1921). J. Math. Psychol. **36**(2), 165–184 (1992)
20. Frankl, P., Maehara, H.: Open-interval graphs versus closed-interval graphs. Discret. Math. **63**, 97–100 (1987). ISSN 0012-365X. https://doi.org/10.1016/0012-365x(87)90156-7
21. Gensemer, S.H.: On relationships between numerical representations of interval orders and semiorders. J. Econ. Theory **43**, 157–169 (1987a)
22. Gensemer, S.H.: Continuous semiorder representations. J. Math. Econ. **16**, 275–289 (1987b)
23. Gensemer, S.H.: On numerical representations of semiorders. Math. Soc. Sci. **15**, 277–286 (1988)
24. Gonzales, Ch.: Additive utility without restricted solvability on every component. J. Math. Psychol. **47**(1), 47–65 (2003)
25. Krantz, D.H., Luce, R.D., Suppes, P., Tversky, A.: Foundations of Measurement: Additive and Polynomial Representations, vol. 1. Academic Press, New York (1971)
26. Luce, R.D.: Semiorders and a theory of utility discrimination. Econometrica **24**(2), 178–191 (1956)
27. Manders, K.L.: On JND representations of semiorders. J. Math. Psychol. **24**(3), 224–248 (1981)
28. Monjardet, B.: Axiomatiques et propriétés des quasi-ordres. Mathématiques et Sciences Humaines **63**, 51–82 (1978)
29. Nakamura, Y.: Real interval representations. J. Math. Psychol. **46**(2), 140–177 (2002)

30. Narens, L.: The measurement theory of dense threshold structures. J. Math. Psychol. **38**(3), 301–321 (1994)
31. Oloriz, E., Candeal, J.C., Induráin, E.: Representability of interval orders. J. Econ. Theory **78**(1), 219–227 (1998)
32. Pirlot, M.: Minimal representation of a semiorder. Theory Decis. **28**, 109–141 (1990)
33. Pirlot, M.: Synthetic description of a semiorder. Discret. Appl. Math. **31**, 299–308 (1991)
34. Pirlot, M., Vincke, Ph.: Semiorders. Properties, Representations, Applications. Kluwer, Dordrecht (1997)
35. Rautenbach, D., Szwarcfiter, J.L.: Unit interval graphs: a story with open ends. Electron. Notes Discret. Math. **38**, 737–742 (2011). ISSN 1571-0653. https://doi.org/10.1016/j.endm.2011.10.023
36. Rautenbach, D., Szwarcfiter, J.L.: Unit interval graphs of open and closed intervals. J. Graph Theory **72**, 418–429 (2013). ISSN 0364-9024,1097-0118. https://doi.org/10.1002/jgt.21650
37. Roberts, F.S.: Measurement Theory with Applications to Decision Making, Utility and the Social Sciences. Addison-Wesley, Reading (1979)
38. Roubens, M., Vincke, Ph.: Preference Modelling. Number 250 in Lecture Notes in Economics and Mathematical Systems, 1st edn. Springer, Berlin (1985)
39. Schröder, B.: Ordered Sets: An Introduction with Connections from Combinatorics to Topology, 2nd edn. Birkhäuser (2016). ISBN 978-3-319-29788-0
40. Scott, D.: Measurement structures and linear inequalities. J. Math. Psychol. **1**(2), 233–247 (1964)
41. Scott, D., Suppes, P.: Foundational aspects of theories of measurement. J. Symb. Log. **23**(2), 113–128 (1958)
42. Suppes, P., Krantz, D.H., Luce, R.D., Tversky, A.: Foundations of Measurement: Geometrical, Threshold, and Probabilistic Representations, vol. 2. Academic Press, New York (1989)

Chain Representations of Nested Families of Biorders

Denis Bouyssou and Jean-Paul Doignon

Abstract Among the real-valued representations of nested families of biorders some representations reflect the nestedness of the family in a simple way. Calling them chain representations, we prove their existence in the finite and countably infinite cases. For the general case, we obtain chain representations in a well-chosen linearly ordered set. Although the existence of real-valued representations in general remains an open problem, our analysis answers questions left pending in the literature. It also leads to new proofs of classical theorems on the existence of a real representation for a single biorder, as well as for a single interval order. A combinatorial property of the set of all biorders from a finite set to another finite set plays a central role in the new proof; called weak gradedness, it is a particularization of well-gradedness which derives from a simpler argument.

Keywords Biorders · Nested relations · Interval orders · Semiorders · Numerical representations · Chain representations · Well-gradedness · Weak gradedness

MSC (2010) 06A75

This work has benefited from many helpful discussions with Thierry Marchant.

D. Bouyssou (✉)
LAMSADE, UMR 7243, CNRS, Université Paris-Dauphine, PSL Research University, 75016 Paris, France
e-mail: bouyssou@lamsade.dauphine.fr

J.-P. Doignon
Département de Mathématique, Université Libre de Bruxelles, c.p. 216 Boulevard du Triomphe, 1050 Bruxelles, Belgique
e-mail: doignon@ulb.ac.be

1 Introduction

This chapter deals with an apparently new problem: the existence of chain representations, possibly using real numbers, of families of biorders. It connects three themes
present in the literature. The first is the existence of real (numerical) representations
of binary relations. The second is the consideration of binary relations from one set to
another set, instead of binary relations on a single set. The third is the consideration
of nested families of relations and of real representations reflecting this nestedness.

Understanding which types of binary relations admit a specific form of real representation has a long history. For instance, let us postulate here that a relation $\mathcal{R}$
on a set X admits a representation by real numbers when for a, b in X, there holds
$a \mathcal{R} b$ if and only if the real value attributed to a is greater or equal than the real value
attributed to b. When X is finite, such a representation exists if and only if the relation
$\mathcal{R}$ is a weak order (or a total preorder) on X. For infinite sets X, characterizing the
existence of a representation is a more involved problem whose solution essentially
goes back to Cantor (for an English version, see [7]) and in its modern form to
Debreu [10]. Debreu's result has important implications in Economics: it singles out
the preference relations that are faithfully described by a utility function assigning
real numbers to the items compared (see Bridges and Mehta [6] and Fishburn [18],
for thorough reviews).

Another example comes from the literature in Psychology. A Guttman scale [23,
24] amounts to a relation from one set to a second set which is captured by the
comparison of real values assigned to the items of each set. Ducamp and Falmagne
[17] provide a mathematical characterization of such a relation (at least in the finite
case). Here we use the term "biorder" coined by Doignon et al. [14] in a paper which
also handles the infinite case by relying on the results of Cantor and Debreu we just
alluded to. Biorders contain as particular cases interval orders ([19, 21, 22]; they are
biorders from one set to itself that are irreflexive) and semiorders ([25, 37]; they are
semitransitive interval orders).

The study of nested families of relations is a classic theme in the literature on
probabilistic consistency ([2, 20, 26–28, 34], [35, Ch. 6], [36, Ch. 5], [38, Ch. 17]).
The usual interpretation is that one observes the frequency $p(x, y)$ with which a
subject chooses between the two objects x and y. A family of nested relations is then
obtained by cutting the probabilistic relation $p(x, y)$ at different levels, usually taken
above 1/2. Indeed, if $\lambda_1 > \lambda_2 \geq 1/2$, letting $x \ R_1 \ y$ iff $p(x, y) \geq \lambda_1$ and $x \ R_2 \ y$ iff
$p(x, y) \geq \lambda_2$, we obtain a nested family of two relations $R_1 \subseteq R_2$.

A similar interpretation holds with biorders. Classically, the set A is a set of
"subjects" and the set Z is a set of "problems" that subjects are asked to solve.
Recording the frequency $p(a, x)$ with which subject $a \in A$ solves problem $x \in Z$,
we obtain a probabilistic relation from A to Z that can be used, as above, to define
a nested family of biorders [16].

In this chapter we extend representation results from single biorders to nested
families of biorders, using representations which reflect in a simple way the nest-

edness of the family. This leads to our main problem: *establishing the existence of chain representations of nested families of biorders*.

We give fairly complete results characterizing the existence of chain representations in some arbitrary linearly ordered set $(E, \geq)$. The case of real chain-representations (E is taken to be $\mathbb{R}$) is more delicate outside the denumerable case (denumerable means finite or countably infinite).

We have two main results. The first says that all nested families of biorders from one finite set to another finite set have a real chain-representation. The second says that the same is true for *all* chains of biorders from an arbitrary set to another arbitrary set if the chain representations are sought in an arbitrary linearly ordered set. Moreover, our results offer some new proofs of results characterizing the existence of representations for a *single* relation.

In the finite case, our sufficiency proof uses combinatorial properties of biorders. A weakening of the property of *well-gradedness*, called *weak gradedness*, plays a central role; both properties concern the family of all biorders from a finite set to a finite set (we refer to Doignon and Falmagne [12], for a study of well-gradedness).

The rest of the paper is organized as follows. Section 2 recalls a number of useful facts on biorders. Section 3 introduces weak gradedness and compares it to the stronger property of well-gradedness. Section 4 presents our main results, in the finite case, on chain representations of nested families of biorders. We also relate them to the existing literature. Turning to the general case in Sect. 5, we analyse the question of the existence of chain representations. We illustrate some of the difficulties of this more delicate problem on a number of examples. A final section discusses our results and lists several open problems that we see as interesting opportunities for future research.

We use a standard vocabulary for binary relations. To avoid any ambiguity, we define the main properties and structures that we use in Appendix A.

2 Biorders

2.1 Notation and Framework

Let A and Z be two disjoint sets. A binary relation $\mathcal{R}$ from A to Z is a subset of $A \times Z$.

Remark 1 The disjointness hypothesis on A and Z, which we make throughout the chapter, may seem quite restrictive. This is not so. If A and Z are not disjoint, we build duplications A' of A and Z' of Z with A' and Z' disjoint, and next a new relation $\mathcal{R}'$ from A' to Z' which faithfully encodes $\mathcal{R}$; see details in Doignon et al. [14, Def. 4, p. 79]. All properties and concepts used below are better understood if we work with $\mathcal{R}'$ rather than with $\mathcal{R}$. A similar remark holds for any relation defined on a set X: we replace it, w.l.o.g., with a relation from a duplication of X to another, disjoint duplication of X. ●

For a relation $\mathcal{R}$ from A to Z, we denote by $\overline{\mathcal{R}} = (A \times Z) \setminus \mathcal{R}$ its *complement*, by $\mathcal{R}^{-1} = \{(z, a) \in Z \times A : (a, z) \in \mathcal{R}\}$ its *converse* and by $\overline{\mathcal{R}}^{-1} = \{(z, a) \in Z \times A : (a, z) \notin \mathcal{R}\}$ its *dual*. Moreover, if $\mathcal{R}$ is a relation from A to Z and $\mathcal{T}$ is a relation from Z to K, we define the *product* $\mathcal{R}\mathcal{T}$ of $\mathcal{R}$ and $\mathcal{T}$ by letting $a \mathcal{R}\mathcal{T} \ell$, where $a \in A$ and $\ell \in K$, when $a \mathcal{R} x$ and $x \mathcal{T} \ell$ for some $x \in Z$.

Remark 2 Clearly the above definition of the complement $\overline{\mathcal{R}}$ depends on the specification of A and Z. In any case, the context will make clear which complement is denoted by $\overline{\mathcal{R}}$: when $\mathcal{R}$ is a relation from A to Z, the complement is taken with respect to $A \times Z$.   ●

2.2 Biorders

A binary relation $\mathcal{R}$ from A to Z is a *biorder* if it is Ferrers, that is, for all $a, b \in A$ and all $x, y \in Z$, we have:

$$[a \mathcal{R} x \text{ and } b \mathcal{R} y] \Rightarrow [a \mathcal{R} y \text{ or } b \mathcal{R} x].$$

More compactly, $\mathcal{R}$ is a biorder when $\mathcal{R}\overline{\mathcal{R}}^{-1}\mathcal{R} \subseteq \mathcal{R}$.

The following notation and concepts are central in this chapter. Let $\mathcal{R}$ be a relation from A to Z, with A and Z disjoint sets. For a in A, we let $a\mathcal{R} = \{z \in Z : a \mathcal{R} z\}$ and $\mathcal{R}z = \{a \in A : a \mathcal{R} z\}$. The *left trace* of $\mathcal{R}$ is the binary relation $\succsim_{\mathcal{R}}^{A}$ on A defined by letting, for all $a, b \in A$ (the second equivalence is trivial),

$$a \succsim_{\mathcal{R}}^{A} b \iff [b \mathcal{R} x \Rightarrow a \mathcal{R} x, \text{ for all } x \in Z] \iff a\mathcal{R} \supseteq b\mathcal{R}. \qquad (1)$$

Similarly, the *right trace* of $\mathcal{R}$ is the binary relation $\succsim_{\mathcal{R}}^{Z}$ on Z defined by letting, for all $x, y \in Z$,

$$x \succsim_{\mathcal{R}}^{Z} y \iff [a \mathcal{R} x \Rightarrow a \mathcal{R} y, \text{ for all } a \in A] \iff \mathcal{R}x \subseteq \mathcal{R}y. \qquad (2)$$

Whatever $\mathcal{R}$, the relations $\succsim_{\mathcal{R}}^{A}$ and $\succsim_{\mathcal{R}}^{Z}$ are, by construction, quasi orders (that is, reflexive and transitive relations).

Here are easy characterizations of biorders (see for instance [14, Proposition 2, p. 78]; [29, Theorem 1, p. 60]; [33, Theorem 2, p. 52]).

Proposition 1 *The following assertions about a relation $\mathcal{R}$ from A to Z are equivalent:*

   (i) $\mathcal{R}$ is a biorder;
  (ii) the sets $a\mathcal{R}$, for $a \in A$, form a chain;
 (iii) the sets $\mathcal{R}x$, for $x \in Z$, form a chain;
 (iv) the left trace $\succsim_{\mathcal{R}}^{A}$ of $\mathcal{R}$ is complete;
  (v) the right trace $\succsim_{\mathcal{R}}^{Z}$ of $\mathcal{R}$ is complete.

For the record let us spell out the following result from Ducamp and Falmagne [17, Th. 3, finite case] and Doignon et al. [14, Prop. 4, p. 79], which is the *representation theorem of biorders* in the denumerable case (meaning finite or countably infinite case).

Proposition 2 *Let $\mathcal{R}$ be a binary relation from A to Z. When each of A and Z is denumerable, the following statements are equivalent:*

(i) *the relation $\mathcal{R}$ is a biorder;*
(ii) *there are a real-valued function f on A and a real-valued function g on Z such that, for all $a \in A$ and $x \in Z$,*

$$a \,\mathcal{R}\, x \iff f(a) > g(x). \tag{3}$$

Furthermore, the functions f and g can always be chosen in such a way that, for all $a, b \in A$ and $x, y \in Z$,

$$a \succsim_{\mathcal{R}}^{A} b \iff f(a) \geq f(b),$$
$$x \succsim_{\mathcal{R}}^{Z} y \iff g(x) \geq g(y). \tag{4}$$

Let $\mathcal{R}$ be any relation from A to Z. In Proposition 2, it is tempting to replace the real-valued mappings f and g with mappings to some linearly ordered set $(E, \geq)$. As we will see, the equivalence in the above proposition then holds for *all* sets A and Z. We now give a name and a notation for the pair of mappings.

Definition 1 (*Representations*) A $>$-*representation in* $(E, \geq)$ of the relation $\mathcal{R}$ from A to Z consists in a linearly ordered set $(E, \geq)$ and in two mappings $f : A \to E$ and $g : Z \to E$ such that, for all a in A and z in Z:

$$a \,\mathcal{R}\, z \iff f(a) > g(z). \tag{5}$$

We then also say that (f, g) is a $>$-representation (also called a *strict representation*) of $\mathcal{R}$ in $(E, \geq)$. The representation is *trace-compatible*, or *respects the traces*, when moreover the following holds, for all $a, b \in A$ and $x, y \in Z$,

$$a \succsim_{\mathcal{R}}^{A} b \iff f(a) \geq f(b),$$
$$x \succsim_{\mathcal{R}}^{Z} y \iff g(x) \geq g(y). \tag{6}$$

The representation is *special* if, for all $a \in A$ and all $x \in Z$, we have $f(a) \neq g(x)$.

The definition of $\geq$-*representations* results from changing only the order sign in Eq. (5) from $>$ to $\geq$. Such representations are also called *nonstrict representations*.

As it is easily checked, a relation $\mathcal{R}$ from A to Z admits a $>$-representation in the ordered set $(E, \geq)$ exactly if the relation $\overline{\mathcal{R}}^{-1}$ from Z to A admits a $\leq$-representation in the same linearly ordered set $(E, \geq)$.

A general characterization of the existence of a $>$-representation in some linearly ordered set follows. It is a slight variation on Doignon et al. [15, Prop. 1, p. 4].

Proposition 3 *A relation $\mathcal{R}$ from A to Z has a $>$-representation in some linearly ordered set $(E, \geq)$ if and only if $\mathcal{R}$ is a biorder. For a biorder, there always exists a representation which is both special and trace-compatible.*

Proof (Sketch) Necessity is easy. We sketch sufficiency. Define a relation $\mathcal{Q}_m$ on $A \cup Z$ by the following table:

$$
\begin{array}{c|cc}
\mathcal{Q}_m & A & Z \\
\hline
A & \succsim_{\mathcal{R}}^{A} & \mathcal{R} \\
Z & \overline{\mathcal{R}}^{-1} & \succsim_{\mathcal{R}}^{Z}
\end{array}
$$

Using Eqs. (1) and (2), it is routine to show that $\mathcal{Q}_m$ is a weak order. This weak order is such that for all $a \in A$ and all $z \in Z$, it is never true that $a \ \mathcal{Q}_m \ z$ and $z \ \mathcal{Q}_m \ a$. The relation $\mathcal{Q}_m \cap \mathcal{Q}_m^{-1}$ is an equivalence (a reflexive, symmetric and transitive relation) on $A \cup Z$. We build the set E as the quotient of $A \cup Z$ by $\mathcal{Q}_m \cap \mathcal{Q}_m^{-1}$. This set is linearly ordered by the relation induced by $\mathcal{Q}_m$ on the quotient set. The function f associates to each $a \in A$ the equivalence class of $\succsim_{\mathcal{R}}^{A}$ to which it belongs. Similarly, the function g associates to each $x \in Z$ the equivalence class of $\succsim_{\mathcal{R}}^{Z}$ to which it belongs. It is clear that (f, g) is a $\geq$-representation of $\mathcal{R}$ in $(E, \geq)$.

By definition of $\mathcal{Q}_m$, the representation respects the traces. It is also special, so that it is at the same time a $>$-representation and a $\geq$-representation. $\square$

Remark 3 Notice that any representation in a denumerable linearly ordered set $(E, \geq)$ is easily turned into a real representation: it suffices to compose the mappings f and g with any embedding of $(E, \geq)$ into $(\mathbb{R}, \geq)$. Clearly, this is no more true in the general case.   ●

The study of nested families of biorders led us to new proofs of Propositions 2 and 3 (see Sect. 4.4 and after Corollary 1). The next section introduces a tool useful to investigate the combinatorial properties of biorders on finite sets.

3   Well-Gradedness and Weak Gradedness

3.1   Well-Gradedness

Consider two relations $\mathcal{R}$ and $\mathcal{S}$ from A to Z, both sets being finite (and disjoint). The distance between these two relations is $d(\mathcal{R}, \mathcal{S}) = |\mathcal{R} \triangle \mathcal{S}|$, with $\mathcal{R} \triangle \mathcal{S} = (\mathcal{R} \setminus \mathcal{S}) \cup (\mathcal{S} \setminus \mathcal{R})$, and $|U|$ denoting the cardinality of the set U.

Let $BO(A, Z)$ (or simply BO, when there is no ambiguity on the underlying sets) be the collection of all biorders from A to Z. Consider two relations $\mathcal{R}, \mathcal{S} \in BO$ such that $d(\mathcal{R}, \mathcal{S}) = \ell$. Doignon and Falmagne [12] show that there are $\mathcal{F}_0, \mathcal{F}_1, \ldots, \mathcal{F}_\ell$ in

BO such that $\mathcal{F}_0 = \mathcal{R}$, $\mathcal{F}_\ell = \mathcal{S}$, and $d(\mathcal{F}_{i-1}, \mathcal{F}_i) = 1$, for $i = 1, 2, \ldots, \ell$. In words, for any two biorders at distance ℓ from one another, some sequence of exactly ℓ elementary steps, each consisting in the *addition* or the *removal* of a single (ordered) pair, transforms the first biorder into the second one without never leaving BO. Collections, like BO, having this property are called *well-graded* in Doignon and Falmagne [12].

The previous paragraph points, for a given biorder, to exceptional pairs (a, x) whose addition to, or deletion from, the biorder produces again a biorder. The following definition and lemma capture the two resulting collections of pairs.

Definition 2 (*Inner and outer fringes*) The *inner fringe* and *outer fringe* of a biorder $\mathcal{R}$ are defined respectively by

$$\mathcal{R}^{\mathcal{I}} = \mathcal{R} \setminus \mathcal{R}\overline{\mathcal{R}}^{-1}\mathcal{R}, \qquad \mathcal{R}^{\mathcal{O}} = \overline{\mathcal{R}}\mathcal{R}^{-1}\overline{\mathcal{R}} \setminus \mathcal{R}. \tag{7}$$

Doignon and Falmagne [12] prove the following result.

Lemma 1 *We have* $\mathcal{R}^{\mathcal{I}} = \mathcal{R}\overline{\mathcal{R}}^{-1}\mathcal{R} \subseteq \mathcal{R}$ *and* $\mathcal{R}^{\mathcal{O}} = \overline{\mathcal{R}}\mathcal{R}^{-1}\overline{\mathcal{R}} \subseteq \overline{\mathcal{R}}$. *Moreover:*

$$\mathcal{R}^{\mathcal{I}} = \{p \in \mathcal{R} : \mathcal{R} \setminus \{p\} \text{ is a biorder}\}, \tag{8}$$
$$\mathcal{R}^{\mathcal{O}} = \{q \in (A \times Z) \setminus \mathcal{R} : \mathcal{R} \cup \{q\} \text{ is a biorder}\}. \tag{9}$$

Observe, in particular, that if $\mathcal{R}, \mathcal{S} \in BO$ are such that $\mathcal{R} \subseteq \mathcal{S}$ and $d(\mathcal{R}, \mathcal{S}) = \ell$, then there are $\mathcal{T}_0, \mathcal{T}_1, \ldots, \mathcal{T}_\ell \in BO$ such that $\mathcal{T}_0 = \mathcal{R}$, $\mathcal{T}_\ell = \mathcal{S}$, $d(\mathcal{T}_{i-1}, \mathcal{T}_i) = 1$, for $i = 1, 2, \ldots, \ell$, and $\mathcal{R} = \mathcal{T}_0 \subsetneq \mathcal{T}_1 \subsetneq \cdots \subsetneq \mathcal{T}_\ell = \mathcal{S}$. In words, some sequence of ℓ elementary steps, each one consisting in the addition of a single pair, transforms $\mathcal{R}$ into $\mathcal{S}$ without ever leaving BO (no pair removal is applied). This property, implied by well-gradedness (compare with Proposition 4 in Doignon and Falmagne [12]), is called here "weak gradedness". It is all we need to establish our results on nested families of biorders in the finite case.

3.2 Weak Gradedness

This subsection deals only with the finite case. We first establish a lemma which implies the property of "weak gradedness" from previous paragraph.

Lemma 2 *Suppose that the sets A and Z are finite. For any two biorders $\mathcal{R}$ and $\mathcal{S}$ from A to Z with $\mathcal{R} \subsetneq \mathcal{S}$, there exists some pair p in $\mathcal{S} \setminus \mathcal{R}$ such that $\mathcal{R} \cup \{p\}$ is again a biorder.*

Proof According to Proposition 1, the subsets $(a\mathcal{R})_{a \in A}$ form a chain. We may thus list the elements of A as $a_1, a_2, \ldots, a_n$ in such a way that

$$a_1\mathcal{R} \subseteq a_2\mathcal{R} \subseteq \cdots \subseteq a_n\mathcal{R} \tag{10}$$

(many or even all of the inclusions may be equalities). Notice that for any z in $a_{i+1} \mathcal{R} \setminus a_i \mathcal{R}$ where $i \in \{1, 2, \ldots, n-1\}$, the new relation $\mathcal{R}' = \mathcal{R} \cup \{(a_i, z)\}$ is again a biorder (the subsets $a_i \mathcal{R}'$, for $i = 1, 2, \ldots, n$, still form a chain).

For $i = 1, 2, \ldots, n$, we have by assumption

$$a_1 \mathcal{R} \subseteq a_1 \mathcal{S}, \qquad a_2 \mathcal{R} \subseteq a_2 \mathcal{S}, \qquad \ldots \qquad a_n \mathcal{R} \subseteq a_n \mathcal{S}. \tag{11}$$

Although the subsets $a_i \mathcal{S}$, for $i = 1, 2, \ldots, n$, also form a chain, we do not necessarily have $a_i \mathcal{S} \subseteq a_{i+1} \mathcal{S}$. We may however permute, if necessary, the elements of A in order to have both (10) and the following: for all i, j in $\{1, 2, \ldots, n\}$,

$$[i < j \text{ and } a_i \mathcal{R} = a_j \mathcal{R}] \implies a_i \mathcal{S} \subseteq a_j \mathcal{S}. \tag{12}$$

Now if there exists some i in $\{1, 2, \ldots, n-1\}$ such that $(a_{i+1} \mathcal{R} \setminus a_i \mathcal{R}) \cap a_i \mathcal{S} \neq \varnothing$, then for any z in the intersection the pair $p = (a_i, z)$ makes the thesis true. Moreover, if $a_n \mathcal{R} \subset a_n \mathcal{S}$, then for any element z in $a_n \mathcal{S} \setminus a_n \mathcal{R}$ the pair $p = (a_n, z)$ makes the thesis true.

We are thus left with the situation where we have for any i in $\{1, 2, \ldots, n-1\}$

$$a_{i+1} \mathcal{R} \setminus a_i \mathcal{R} \subseteq Z \setminus a_i \mathcal{S} \tag{13}$$

and moreover

$$a_n \mathcal{R} = a_n \mathcal{S}. \tag{14}$$

We next derive from Eqs. (13) and (14) the contradiction $\mathcal{S} = \mathcal{R}$. It suffices to prove by induction $a_i \mathcal{S} = a_i \mathcal{R}$, meaning here is $a_i \mathcal{S} \subseteq a_i \mathcal{R}$ for $i = n, n-1, \ldots, 1$. The case $i = n$ is Eq. (14). Assuming $a_{i+1} \mathcal{R} = a_{i+1} \mathcal{S}$ for some i with $n-1 \geq i \geq 1$, we show $a_i \mathcal{R} = a_i \mathcal{S}$ by working out two cases.

If $a_i \mathcal{R} = a_{i+1} \mathcal{R}$, then from Eq. (12) we have $a_i \mathcal{S} \subseteq a_{i+1} \mathcal{S} = a_{i+1} \mathcal{R} = a_i \mathcal{R}$ and we are done (remember $a_i \mathcal{R} \subseteq a_i \mathcal{S}$).

If $a_i \mathcal{R} \subsetneq a_{i+1} \mathcal{R}$, take some t in $a_{i+1} \mathcal{R} \setminus a_i \mathcal{R}$. By the induction assumption, we have also $a_{i+1} \mathcal{S} t$, and by (13) we have $(a_i, t) \notin \mathcal{S}$. Take any y in $a_i \mathcal{S}$. We deduce $a_{i+1} \mathcal{R} y$ because $\mathcal{R}$ is a biorder. Then $a_i \mathcal{R} y$ follows from the induction assumption. Now if the pair (a_i, y) would not be in $\mathcal{R}$, then y would contradict (13). $\qquad \square$

We have seen above that if $\mathcal{R}$ is a biorder from A to Z and (a, x) is a pair in $\overline{\mathcal{R}} = (A \times Z) \setminus \mathcal{R}$, then $\mathcal{R} \cup \{(a, x)\}$ is again a biorder if and only if it belongs to the outer fringe of $\mathcal{R}$. Hence, Lemma 2 asserts that if $\mathcal{R} \subsetneq \mathcal{S}$ and $d(\mathcal{R}, \mathcal{S}) = 1$, then the outer fringe of $\mathcal{R}$ has at least one of its pairs in $\mathcal{S}$. Alternatively, it says that if $\mathcal{R} \subsetneq \mathcal{S}$ and $d(\mathcal{R}, \mathcal{S}) = 1$, that the inner fringe of $\mathcal{S}$ has at least one of its pairs that does not belong to $\mathcal{R}$.

It is worth taking a more general point of view on Lemma 2. Biorders from A to Z, being relations, are subsets of $A \times Z$. Their collection is thus a collection of subsets of a ground set (here $A \times Z$). Lemma 2 implies a property of this collection, that we now designate in a more general setting.

Definition 3 Let $\mathcal{F}$ be a collection of subsets of a finite ground set E. We say that $\mathcal{F}$ is *weakly graded* when for any two elements F and G of $\mathcal{F}$ with $F \subsetneq G$, there is a sequence $F_0, F_1, ..., F_k$ of elements of $\mathcal{E}$ such that $F = F_0$, $F_k = G$ and for each $i = 1, 2, ..., k$ there hold both $F_{i-1} \subsetneq F_i$ and $|F_i \setminus F_{i-1}| = 1$.

In Definition 3, we necessarily have $k = |G \setminus F|$. Notice that the collection $\mathcal{F}$ is weakly graded as soon as for any two of its elements, say J and K, with $J \subsetneq K$, there exists a third element L such that $J \subseteq L \subseteq K$ and moreover $|L \setminus J| = 1$. This is the property obtained for the collection of biorders from A to Z in Lemma 2.

Let us compare "well-gradedness" and "weak gradedness" of the collection BO of biorders from A to Z. We know from Doignon and Falmagne [12] that BO is well-graded. This clearly implies that it is weakly graded. However, we prefer to stick here with weak gradedness, which is a weaker property. Indeed, the proof of well-gradedness of BO needs more elaborate arguments than those in the proof of Lemma 2. To see why, let $A = \{a, b\}$ and $Z = \{y, z\}$. Consider the two biorders $\mathcal{R} = \{(a, y)\}$ and $\mathcal{S} = \{(b, z)\}$, which are at distance 2 from one another. To transform $\mathcal{R}$ into $\mathcal{S}$ in 2 steps, we cannot start by adding some pair to $\mathcal{R}$. The first step has to consist in the deletion of (a, y). Notice that $\mathcal{R}$ and $\mathcal{S}$ form a collection of relations which is (trivially) weakly graded but not well-graded.

We are now fully equipped to tackle the case of nested families of biorders.

4 Nested Families of Biorders on Finite Sets

This section establishes results on nested families of biorders from A to Z, with A and Z finite sets. We deal with the extension of these results to the infinite case in the next section.

4.1 Relation to the Literature

Doignon et al. [16] have proposed a detailed study of (non-necessarily nested) families of biorders defined on finite sets and their real representations. Their findings consolidate and extend results due to Roberts [34], Fishburn [20], Monjardet [30], Monjardet [31], Roubens and Vincke [36, Ch. 5], and Doignon [11]. Most of these results deal with situations in which the intersection of the left traces (or of the right traces) of all biorders in the family is complete. This gives rise to what [16] have called right (or left) homogeneous families of biorders. The real representation of such families involves using a single function either on A (for right-homogeneous families) or on Z (for left-homogeneous families).[1] This function is a real representation of the intersection of the traces on this set that is supposed to be complete.

---

[1]This is no mistake since our use of left and right trace does not conform to the terminology of Doignon et al. [16], as explained in Bouyssou and Marchant [5].

Doignon et al. [16] also particularize their results to the case of nested families of biorders (see Sect. IV of their paper) through the study of the cuts of a valued relation. Nested families of biorders are the subject of the present text. We study real-valued representations of such families without supposing that they are left or right homogeneous.

4.2 Results

Definition 4 A *family of relations* from A to Z consists in an integer $k \geq 1$ and k relations $(\mathcal{R}_1, \mathcal{R}_2, \ldots, \mathcal{R}_k)$. It is *nested* if

$$\mathcal{R}_1 \subseteq \mathcal{R}_2 \subseteq \cdots \subseteq \mathcal{R}_k.$$

If all the relations $\mathcal{R}_i$, $i = 1, 2, \ldots, k$ are biorders, we speak of a *nested family of biorders*.

Doignon et al. [16, Proposition 1, p. 460–461] state that, if $(\mathcal{R}_1, \mathcal{R}_2, \ldots, \mathcal{R}_k)$ is a nested family of biorders between the finite sets A and Z, then there are real-valued functions f_i on A and g_i on Z, for $i = 1, 2, \ldots, k$, such that, for all $a \in A$, $x \in Z$, and $i = 1, 2, \ldots, k$,

$$a \, \mathcal{R}_i \, x \iff f_i(a) > g_i(x),$$

and, for all $i, j \in \{1, 2, \ldots, k\}$ with $i > j$,

$$f_i(a) > g_i(x) \Rightarrow f_j(a) > g_j(x).$$

We use a different notion of representation, called a "chain representation", that makes obvious the nested character of the family (see Fig. 1).

Definition 5 A *real chain->-representation* of a family of relations $(\mathcal{R}_1, \mathcal{R}_2, \ldots, \mathcal{R}_k)$ consists in real-valued functions $f_1, f_2, \ldots, f_k$ on A and $g_1, g_2, \ldots, g_k$ on Z such that, for all $a \in A$, $x \in Z$ and $i \in \{1, 2, \ldots, k\}$,

$$a \, \mathcal{R}_i \, x \iff f_i(a) > g_i(x),$$
$$f_k(a) \geq f_{k-1}(a) \geq \cdots \geq f_1(a), \tag{15}$$
$$g_1(x) \geq g_2(x) \geq \cdots \geq g_k(x).$$

The chain->-representation *respects the traces*, or is *trace-compatible*, when for all $a, b \in A$, $x, y \in Z$ and $i \in \{1, 2, \ldots, k\}$

$$a \succsim^A_{\mathcal{R}_i} b \iff f_i(a) \geq f_i(b), \tag{16}$$
$$x \succsim^Z_{\mathcal{R}_i} y \iff g_i(x) \geq g_i(y). \tag{17}$$

It is *special* when for all $a \in A$, $x \in Z$ and $i \in \{1, 2, \ldots, k\}$

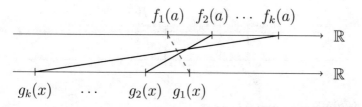

Fig. 1 Illustration of the real values assigned to a and x in a chain representation: here, $a\, \mathcal{R}_2\, x$, $a\, \mathcal{R}_3\, x$, ..., $a\, \mathcal{R}_k\, x$ hold but $a\, \mathcal{R}_1\, x$ does not hold. We have $\mathcal{R}_1 \subseteq \mathcal{R}_2 \subseteq \cdots \subseteq \mathcal{R}_k$, $f_k \geq f_{k-1} \geq \cdots \geq f_1$, and $g_1 \geq g_2 \geq \cdots \geq g_k$

$$f_i(a) \neq g_i(z). \tag{18}$$

We illustrate a chain representation in Fig. 1.

Our main purpose in this section is to prove the following proposition that gives necessary and sufficient conditions for the existence of real chain->-representations. This clearly tightens the result from Doignon et al. [16] recalled above.

Proposition 4 *Let $(\mathcal{R}_1, \mathcal{R}_2, \ldots, \mathcal{R}_k)$ be a family of relations from the finite set A to the finite set Z. This family has a real chain->-representation $f_1, f_2, \ldots, f_k$ and $g_1, g_2, \ldots, g_k$ if and only if it is a nested family of biorders.*

Any nested family of biorders admits some real >-representation which is both special and trace-compatible.

It is clear that the desired real representation in Proposition 4 implies that $(\mathcal{R}_1, \mathcal{R}_2, \ldots, \mathcal{R}_k)$ is a nested family of biorders. The proof of the converse implication in Sect. 4.3 relies on the following lemma.

Lemma 3 *Let $\mathcal{R}$ and $\mathcal{S}$ be two biorders from the finite set A to the finite set Z such that $\mathcal{S} \setminus \{p\} = \mathcal{R}$ for some pair p in $\mathcal{S} \setminus \mathcal{R}$. Assume the real-valued functions f on A and g on Z are such that for all b in A, y in Z,*

$$b\, \mathcal{R}\, y \iff f(b) > g(y), \tag{19}$$

and suppose furthermore that the >-representation f, g is special and trace-compatible.

Then there are also real-valued functions f^* on A and g^* on Z such that, for all b in A, y in Z,

$$b\, \mathcal{S}\, y \iff f^*(b) > g^*(y), \tag{20}$$

$$f^*(b) \geq f(b), \tag{21}$$

$$g(y) \geq g^*(y), \tag{22}$$

and moreover the >-representation f^*, g^* of $\mathcal{S}$ is special and trace-compatible.

The following subsection collects the proofs.

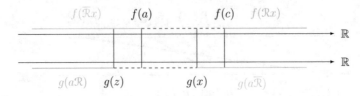

Fig. 2 Illustration of the proof of Lemma 3

4.3 Proofs

The proof of Lemma 3 which we now give focuses on the representations themselves. Some additional comments of combinatorial nature appear after the proof.

Proof of Lemma 3 Assume $p = (a, x)$, with thus $\mathcal{R} \cup \{(a, x)\} = \mathcal{S}$. We are given a special representation (f, g) of $\mathcal{R}$ which respects the traces of $\mathcal{R}$. If $b \in A \setminus \{a\}$ and $y \in Z \setminus \{z\}$, then $(b, y) \in \mathcal{R}$ if and only if $(b, y) \in \mathcal{S}$; we are thus tempted to set $f^*(b) = f(b)$ and $g^*(y) = g(y)$, which we do. In view on the assumptions on f and g, the required inequalities (20)–(22) are satisfied for all $b \in A \setminus \{a\}$ and $y \in Z \setminus \{z\}$.

There remains only to assign values to $f^*(a)$ and $g^*(x)$. To this aim, select some c in A and some z in Z such that

$$f(c) = \min\{f(b) : b \in A \text{ and } f(b) > g(x)\} = \min f(\mathcal{R}x), \qquad (23)$$
$$g(z) = \max\{g(y) : z \in Z \text{ and } f(a) > g(z)\} = \max g(a\mathcal{R}). \qquad (24)$$

Notice that c is well defined in A except if $\mathcal{R}x$ is empty, and similarly z is well defined in Z except if $a\mathcal{R}$ is empty. We leave to the reader the two cases $\mathcal{R}x = \varnothing$ and $a\mathcal{R} = \varnothing$, and in the sequel assume that c and z are well defined. Notice $g(z) < f(a) < g(x) < f(c)$. There is more to be said on the values of f and g (compare with Fig. 2).

No value $f(b)$, for $b \in A$, can be in the interval $]\,f(a), f(c)\,[$. Indeed: (i) if we had $f(a) < f(b) < g(x)$, there would exist some y in Z such that $f(a) \le g(y) < f(b)$ (because f respects the trace on A we must have $a \succ_{\mathcal{R}}^A b$). Then $a \,\overline{\mathcal{R}}\, y \,\mathcal{R}^{-1}\, b \,\overline{\mathcal{R}}\, x$ contradicts the assumption that $\mathcal{R} \cup \{(a, x)\}$ is a biorder; (ii) $f(b) = g(x)$ contradicts that f, g is a special $>$-representation; (iii) $g(x) < f(b) < f(c)$ would give $b \in \mathcal{R}x$ in contradiction with the definition of $f(c)$. Because the $>$-representation f, g respects the traces, we derive that $\{b \in A : f(b) = f(c)\}$ is the equivalence class just above the one of a in A w.r.t. the trace $\succsim_{\mathcal{R}}^A$ (the class of a is *just above* the class of b when $a \succ_{\mathcal{R}}^A b$ and for no element c in A do we have $a \succ_{\mathcal{R}}^A c \succ_{\mathcal{R}}^A b$).

Similarly, no value $g(y)$, for $y \in Z$, can be in $]\,g(z), g(x)\,[$. Indeed, (i) $g(z) < g(y) < f(a)$ would contradict the definition of z; (ii) $g(y) = f(a)$ cannot hold because the $>$-representation f, g is special; (iii) $f(a) < g(y) < g(x)$ would imply the existence of b in A such that $y \,\mathcal{R}^{-1}\, b \,\overline{\mathcal{R}}\, x$ which together with $a \,\overline{\mathcal{R}}\, y$ contradicts that $\mathcal{R} \cup \{(a, x)\}$ is a biorder. Consequently, $\{y \in Z : g(y) = g(z)\}$ is the equivalence class just below the one of x in Z (w.r.t. the trace $\succsim_{\mathcal{R}}^Z$).

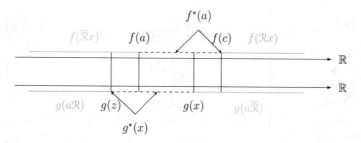

Fig. 3 Assignations of $f^*(a)$ and $g^*(x)$ in the proof of Lemma 3

Because S differs from R only by the addition of the pair (a, x), we must assign values to $f^*(a)$ and $g^*(x)$ in such a way that $g(z) \leq g(x) < f(a) \leq f(c)$; moreover, all pairs (a, y) and (b, x), for $b \in A$ and $y \in Z$, are then correctly represented. Hence any such assignment of $f^*(a)$ and $g^*(x)$ delivers a $>$-representation f^*, g^* of S. However, to make sure that the representation respects the traces of S, we must take care. Remember that the trace $\succsim_R^A$ reflects comparisons among themselves of the subsets bR of Z—see Eq. (1); also, $\succsim_R^Z$ reflects comparisons among themselves of the subsets Ry of A. The addition of (a, x) to R modifies only two such sets, namely aR to which x is added, and Rx to which a is added. For the element c defined in Eq. (23), we derive that $aS = aR \cup \{x\}$ forms a subset of $cR = cS$, but we can have either $aS = cS$ or $aS \subsetneq cS$.

Let us now define $f^*(a)$ (see Fig. 3): if $aS = cS$, then a and c become equivalent in the trace $\succsim_S^A$ of S, and we set $f^*(a) = f(c)$; otherwise, we set $f^*(a) = (f(a) + 2\, g(x))/3$. In a similar way, if $Sx = Sz$, we set $g^*(x) = g(z)$, otherwise $g^*(x) = (2\, f(a) + g(x))/3$.

With the above assignments of $f^*(a)$ and of $g^*(x)$, the $>$-representation f^*, g^* respects the traces of S. Moreover, the representation is special in view of the assumption that the $>$-representation f, g is special and the use of strict inequalities in the definitions of $f^*(a)$ and $g^*(x)$. □

The above proof takes care of the evolution of the collection of equivalence classes when the biorder R changes into the biorder $S = R \cup\{(a, x)\}$. Note that only four classes are susceptible of modification: the class of a and the class just above it in A, the class of x and the class just below it in Z (the meaning of "just above" is explained in the proof). The next example illustrates various ways in which these four classes can evolve.

Example 1 On the upper part of Fig. 4 there are three biorders. The addition of the single pair (a, x) produces the corresponding biorders on the lower part. ◇

Proof of Proposition 4
Let $R_1, R_2, \ldots, R_k$ be a nested family of biorders from a finite set A to a finite set Z. We may assume $R_1 = \varnothing$, because otherwise we may add $\varnothing$ as the very first biorder (and renumber the other biorders). If $|R_{i+1} \setminus R_i| > 1$, we apply Lemma 2 and insert

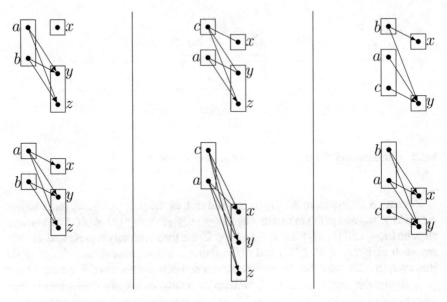

Fig. 4 Three examples of evolution of the equivalence classes of the traces, when the addition of the single pair (a, x) to the upper biorder produces the lower biorder (Example 1)

a new biorder in between $\mathcal{R}_i$ and $\mathcal{R}_{i+1}$. Repeating the construction while it remains possible, we end up with a nested family of biorders such that two consecutive biorders differ by exactly one pair (finiteness comes into play here); all biorders $\mathcal{R}_1 = \varnothing, \mathcal{R}_2, ..., \mathcal{R}_k$ belong to the constructed family.

Clearly, there exists a trivial real $>$-representation of the empty biorder that is special and trace-compatible (for an example, take $f_1(b) = 0$ for all $b \in A$, and $g_1(y) = 1$ for all $y \in Z$). Then repeated applications of Lemma 3 produces a real chain-$>$-representation of the constructed family, thus also a real chain-$>$-representation of the given family $\mathcal{R}_1, \mathcal{R}_2, ..., \mathcal{R}_k$. Moreover, the resulting representation is special and trace-compatible.   □

4.4 Remarks and Extensions

As announced above, we observe that Proposition 4 offers an alternative proof of the fact that any biorder $\mathcal{R}$ from a finite set A to a finite set Z has a real representation which is special and trace-compatible. This is obvious observing that the single biorder $\mathcal{R}$ forms a nested family of biorder(s), to which Proposition 4 applies. As a matter of fact, we do not even need Lemma 2 here: it suffices to check that each nonempty biorder $\mathcal{R}$ from A to Z contains at least one pair (a, x) such that $\mathcal{R} \setminus \{(a, x)\}$ is again a biorder (in other words, the inner fringe of $\mathcal{R}$ is nonempty). To get such a pair (a, x), first select in A any element a with $a\mathcal{R}$ minimal for the left trace among the nonempty sets $b\mathcal{R}$, where $b \in A$ (minimal for the trace means minimal for the inclusion), and next pick any element x in $a\mathcal{R}$: that $\mathcal{R} \setminus \{(a, x)\}$ is again a biorder follows at once from Proposition 1(ii).

This new proof for the existence of a real representation is quite different from the previous ones given in the literature: it is in some sense "algorithmic", the representation with the desired property being built up step by step, with the number of steps equal to the number of pairs in $\mathcal{R}$.

Notice that, because our method of proof starts with the empty biorder to end up with the given biorder by addition of a single pair at each step, it also applies to interval orders. Indeed, an interval order $\mathcal{P}$ is nothing but a biorder from a set X to the same set X that is irreflexive. Our proof builds the real representation of the interval order $\mathcal{P}$ by forging the real representation of the biorder $\mathcal{R}$ from X' to X'' which is the duplication of $\mathcal{P}$ (remember Remark 1). Because the pairs (x', x''), for $x \in X$, are outside of both biorders $\varnothing$ and $\mathcal{R}$, they are never added to the current biorder (in terms of X, the pairs (x, x) are never added and the proof works only with irreflexive biorders).

For the record, we spell out the following:

Proposition 5 *Let $(\mathcal{P}_1, \mathcal{P}_2, \ldots, \mathcal{P}_k)$ be a family of relations on the finite set X. There are real-valued functions $f_1, f_2, \ldots, f_k$ on A and $g_1, g_2, \ldots, g_k$ on Z such that, for all $x, y \in X$ and $i \in \{1, 2, \ldots, k\}$,*

$$x \, \mathcal{P}_i \, y \iff f_i(x) > g_i(y),$$
$$f_k(x) \geq f_{k-1}(x) \geq \cdots \geq f_1(x),$$
$$g_1(x) \geq g_2(x) \geq \cdots \geq g_k(x),$$
$$f_i(x) \leq g_i(x),$$

if and only if $(\mathcal{P}_1, \mathcal{P}_2, \ldots, \mathcal{P}_k)$ is a nested family of interval orders on X.

For a nested family of interval orders, there always exists a real $>$-representation which is special and trace-compatible.

Semiorders are the particular interval orders which are semitransitive, or equivalently the interval orders for which the left and right traces are never contradictory [1, 29]. The following proposition immediately follows from Proposition 5.

Proposition 6 *Let $(\mathcal{S}_1, \mathcal{S}_2, \ldots, \mathcal{S}_k)$ be a family of relations on the finite set X. There are real-valued functions $f_1, f_2, \ldots, f_k$ on A and $g_1, g_2, \ldots, g_k$ on Z such that, for all $x, y \in X$ and $i \in \{1, 2, \ldots, k\}$,*

$$x \, \mathcal{S}_i \, y \iff f_i(x) > g_i(y),$$
$$f_k(x) \geq f_{k-1}(x) \geq \cdots \geq f_1(x),$$
$$g_1(x) \geq g_2(x) \geq \cdots \geq g_k(x),$$
$$f_i(x) \leq g_i(x),$$
$$f_i(x) > f_i(y) \Rightarrow g_i(x) \geq g_i(y)$$

if and only if $(S_1, S_2, \ldots, S_k)$ *is a nested family of semiorders.*

For any nested family of semiorders, the mappings $f_1, f_2, \ldots, f_k$ *and* $g_1, g_2, \ldots, g_k$ *can be selected in order to moreover form a special* $>$-*representation which respects the traces.*

A more difficult question asks whether any nested family of semiorders on a finite set admits some representation 'with no nesting' [1, 18]. The latter means a representation as in Proposition 6 which furthermore satisfies $f_i(x) \geq f_i(y) \Rightarrow g_i(x) \geq g_i(y)$, for all $x, y \in X$ and $i \in \{1, 2, \ldots, k\}$. An even more advanced question asks for the existence of some representation with constant thresholds, in the sense that $g_i = f_i + \tau_i$ for some positive constant τ_i, for $i = 1, 2, \ldots, k$ [37]. While we know that any semiorder defined on a denumerable set admits a representation with no nesting, and also that a semiorder defined on a finite set admits a constant threshold representation, extending the results to the case of nested families of semiorders looks as a delicate problem. We leave the latter questions on semiorders for further study. Crucial steps leading to positive answers would be proofs of lemmas similar to Lemma 3.

5 The Infinite Case

5.1 *Definitions*

When the sets A and Z are not restricted to be finite as in the preceding section, the situation becomes more difficult. Observe first that biorders can form finite or infinite families, which we now call 'chains of biorders' to emphasize the fact that they generalize the finite families studied until here. The notation BO designates again the collection of all biorders from A to Z.

Definition 6 (*Chain of biorders*) A chain of biorders from A to Z consists in a nonempty index set I and a mapping from I to the collection BO, which we denote as $(\mathcal{R}_i)_{i \in I}$, such that for all i, j in I, either $\mathcal{R}_i \supset \mathcal{R}_j$ or $\mathcal{R}_j \supset \mathcal{R}_i$.

Clearly, when I is finite, a chain a biorders is nothing more than a nested family of biorders (up to an adequate renumbering of the biorders).

Definition 7 (*Chain* $>$-*Representations*) A chain representation $(f_i, g_i)_{i \in I}$ of a chain $(\mathcal{R}_i)_{i \in I}$ of biorders consists in a linearly ordered set $(E, \geq)$ and, for each i in I, in two mappings $f_i : A \to E$ and $g_i : Z \to E$ such that

1. for all i in I, a in A and z in Z:

$$a \, \mathcal{R}_i \, z \iff f_i(a) > g_i(z);$$

2. for all i, j in I,

$$\mathcal{R}_i \subseteq \mathcal{R}_j \Rightarrow f_j \geq f_i \text{ and } g_i \geq g_j.$$

The representation is *real* when $(E, \geq) = (\mathbb{R}, \geq)$. The definition of *chain $\geq$-representations* is similar (with $\geq$ instead of $>$).

We first characterize chains of biorders that admit a chain $>$-representation in some linearly ordered set $(E, \geq)$. For real representations we have more questions than answers. From now on we concentrate on chain $>$-representations (results on $\geq$-representations are easily derived).

5.2 Chain Representations in a Linearly Ordered Set

According to Proposition 3, any biorder admits a representation in some linearly ordered set $(E, \geq)$. Hence, it is natural to ask whether the same holds for any chain of biorders. To derive a positive answer in Proposition 7 below, we introduce techniques different from the ones in Sect. 4.3 (in the finite case, we handled nested families of biorders with the crucial tool of weak gradedness). The need for a new technique can be grasped from the following counterexample showing that several statements on finite biorders in Doignon and Falmagne [12] do not extend to the infinite setting.

Example 2 Here is a biorder whose inner and outer fringes are both empty. Let $A = Z = \mathbb{Q}$. Take any three real numbers which are linearly independent over $\mathbb{Q}$, for instance 1, $\sqrt{2}$ and $\sqrt{3}$. Setting $f(a) = \sqrt{2}\,a + \sqrt{3}$ for $a \in A$, and $g(z) = z$ for $z \in Z$, we obtain the biorder $\mathcal{R} = \{(a, z) \in A \times Z : \sqrt{2}\,a + \sqrt{3} > z\}$ for which (f, g) is a $>$-representation in $(\mathbb{Q}, \geq)$. Let us prove that the inner fringe $\mathcal{R}^{\mathcal{I}}$ is empty. For any pair (a, z) in $\mathcal{R}$, we have $\sqrt{2}\,a + \sqrt{3} > z$. There thus exist some rational number y such that $\sqrt{2}\,a + \sqrt{3} > y > z$ and then some rational number b such that $y \geq \sqrt{2}\,b + \sqrt{3} > z$. This gives $(a, y) \in \mathcal{R}$, $(b, y) \notin \mathcal{R}$, $(b, z) \in \mathcal{R}$. We obtain $(a, z) \in \mathcal{R}\overline{\mathcal{R}}^{-1}\mathcal{R}$, and thus $\mathcal{R}^{\mathcal{I}} = \varnothing$. One proves in a similar way $\mathcal{R}^{\mathcal{O}} = \varnothing$. $\diamond$

Proposition 7 *Any chain of biorders has a chain $>$-representation in some linearly ordered set.*

Proof Let $(\mathcal{R}_i)_{i \in I}$ be a chain of biorders from A to Z. To build a chain $>$-representation $(f_i, g_i)_{i \in I}$ in the linearly ordered set $(E, \geq)$, we first specify a set E and, for each i in I, two mappings $f_i : A \to E$ and $g_i : Z \to E$ by letting

$$E = (A \times I) \cup (Z \times I),$$
$$f_i(a) = (a, i), \quad \text{for } a \in A,$$
$$g_i(z) = (z, i), \quad \text{for } z \in Z.$$

It remains to equip E with an adequate linear ordering. To this end, consider first the following set of pairs of E:

$$\mathfrak{X} = \{((a, j), (a, i)) \ : \ a \in A, \ \mathcal{R}_i \subsetneq \mathcal{R}_j\} \cup$$
$$\{((z, i), (z, j)) \ : \ z \in Z, \ \mathcal{R}_i \subsetneq \mathcal{R}_j\} \cup$$
$$\{((a, i), (z, i)) \ : \ a \ \mathcal{R}_i \ z\} \cup$$
$$\{((z, i), (a, i)) \ : \ a \ \overline{\mathcal{R}_i}^{-1} \ z\}.$$

Remark that $\mathfrak{X}$ consists exactly of the pairs forced in the linear ordering $\geq$ of E by any representation of the given chain in $(E, \geq)$. Now we check that $\mathfrak{X}$ is acyclic, that is, $\mathfrak{X}$ has no cycle. Here we view a cycle of $\mathfrak{X}$ as a finite sequence $e_1, e_2, ..., e_k$ of elements in E such that $(e_1, e_2), (e_2, e_3), ..., (e_{k-1}, e_k), (e_k, e_1)$ are all in $\mathfrak{X}$. The length of the latter cycle is its number k of elements.

Here are properties of such a cycle $\mathcal{C}$, where we always assume $a, b \in A$ and $y, z \in Z$:

- (i) no four successive elements of $\mathcal{C}$ can have a same second component i: if not, we would meet, say, either $(a, i), (y, i), (b, i)$ and (z, i) along the cycle, or $(y, i), (a, i), (z, i)$ and (b, i) along the cycle. In the first case we have $a \ \mathcal{R}_i \ y \ \overline{\mathcal{R}}^{-1} \ b \ \mathcal{R}_i \ z$ which implies $a \ \mathcal{R}_i \ z$ and we could shorten the cycle by replacing $(a, i), (y, i), (b, i)$ and (z, i) with $(a, i), (z, i)$, a contradiction. A similar contradiction occurs in the second case.
- (ii) no three successive elements of C can have a same second component i: if not, we would meet, say, either $(a, i), (z, i)$ and (b, i) along the cycle, or (y, i), (a, i) and (z, i) along the cycle. In the first case, by (i), we must have some (a, j), with $\mathcal{R}_j \supset \mathcal{R}_i$, before (a, i) in the cycle and some (b, h) with $\mathcal{R}_i \supset \mathcal{R}_h$ after (b, i) in the cycle. Then replacing $(a, j), (a, i), (z, i), (b, i) \ (b, h)$ with $(a, i), (z, j), (z, h), (b, h)$ shortens the cycle, a contradiction. The second case also leads to a contradiction.
- (iii) if (a, i) and (z, i) are successive elements of the cycle $\mathcal{C}$, then by (ii) the cycle contains successive elements $(a, j), (a, i), (z, i)$ and (z, k) with $\mathcal{R}_j \supset \mathcal{R}_i$ and $\mathcal{R}_k \supset \mathcal{R}_i$. We must then have $j = k$, otherwise the cycle could be shortened (use (a, k) or (z, j)).
- (iv) similarly, if (z, i) and (b, i) are successive elements in $\mathcal{C}$, then we have also successive elements $(z, h), (z, i), (b, i)$ and (b, h) in the cycle.

Properties (i)–(iv) of $\mathcal{C}$ imply that $\mathcal{C}$ must be of the form $(a, j), (a, i), (z, i), (z, k)$, (a, j) with $\mathcal{R}_k \supset \mathcal{R}_i$, which is impossible. So we have proved that $\mathfrak{X}$ is an acyclic relation on E. By the following lemma, there exists a strict linear ordering $>$ on E extending $\mathfrak{X}$. After taking the reflexive closure of $>$, we obtain a linear order $\geq$. Then $(f_i, g_i)_{i \in I}$ is a chain $>$-representation of the chain $(\mathcal{R}_i)_{i \in I}$ of biorders in $(E, \geq)$. $\qquad \square$

The following lemma is a well-known fact. We sketch its proof for completeness.

Lemma 4 *Let $\mathfrak{X}$ be an irreflexive relation on an arbitrary nonempty set E. There exists a strict linear ordering $>$ on E extending $\mathfrak{X}$ (that is, $x \mathfrak{X} y$ implies $x > y$) if and only if the relation $\mathfrak{X}$ is acyclic.*

Proof Necessity being obvious, we prove only sufficiency. First define a relation $\mathfrak{R}$ (the transitive closure of $\mathfrak{X}$) by letting, for a, b in E,

$$a \, \mathfrak{R} \, b \iff \text{there exist } n \text{ in } \mathbb{N} \text{ and } c_1, c_2, \ldots, c_n \text{ in } E \text{ such that}$$
$$a \, \mathfrak{X} \, c_1, \; c_1 \, \mathfrak{X} \, c_2, \; c_2 \, \mathfrak{X} \, c_3, \; \ldots, \; c_{n-1} \, \mathfrak{X} \, c_n, \; c_n \, \mathfrak{X} \, b.$$

The acyclicity of $\mathfrak{X}$ implies that $\mathfrak{R}$ is a strict partial order on E. By the main result of Szpilrajn [39], $\mathfrak{R}$ is contained in some strict linear ordering of E. $\qquad\square$

Here is an easy reinforcement of Proposition 7.

Corollary 1 *Any chain $(\mathfrak{R}_i)_{i \in I}$ of biorders from A to Z has a chain $>$-representation in some linearly ordered set, such that the representation of any biorder $\mathfrak{R}_i$ in the chain is special and trace-compatible.*

Proof In the proof of Proposition 7, we modify the definition of E. Using the equivalence classes and quotient sets of the equivalence relations $\sim_{\mathfrak{R}_i}^{A}$ and $\sim_{\mathfrak{R}_i}^{Z}$ for each biorder $\mathfrak{R}_i$ in the chain, we let

$$E = \bigcup_{i \in I} (A/\sim_{\mathfrak{R}_i}^{A}, i) \cup (Z/\sim_{\mathfrak{R}_i}^{Z} \times I),$$
$$f_i(a) = (\sim_{\mathfrak{R}_i}^{A} a, i), \quad \text{for } a \in A,$$
$$g_i(z) = (\sim_{\mathfrak{R}_i}^{Z} z, i), \quad \text{for } z \in Z.$$

The definition of the relation $\mathfrak{X}$ on E is now

$$\mathfrak{X} = \{((\sim_{\mathfrak{R}_j}^{A} a, j), (\sim_{\mathfrak{R}_i}^{A} a, i)) \, : \, a \in A, \, \mathfrak{R}_i \subsetneq \mathfrak{R}_j\} \cup$$
$$\{((\sim_{\mathfrak{R}_i}^{Z} z, i), (\sim_{\mathfrak{R}_j}^{Z} z, j)) \, : \, z \in Z, \, \mathfrak{R}_i \subsetneq \mathfrak{R}_j\} \cup$$
$$\{((\sim_{\mathfrak{R}_i}^{A} a, i), (\sim_{\mathfrak{R}_i}^{Z} z, i)) \, : \, a \, \mathfrak{R}_i \, z\} \cup$$
$$\{((\sim_{\mathfrak{R}_i}^{Z} z, i), (\sim_{\mathfrak{R}_i}^{A} a, i)) \, : \, a \, \overline{\mathfrak{R}_i}^{-1} \, z\}.$$

The rest of the proof is similar to the one of Proposition 7. $\qquad\square$

Proposition 7 is quite general since it covers any chain of biorders (the index set I may be finite, countably infinite or uncountable), defined from any set A to any set Z. Now, if we restrict attention to denumerable chains of nested biorders with A and Z denumerable sets, it is clear that the set E built in the proof of Proposition 7 is denumerable. Then in the statement we may replace E with $\mathbb{R}$.

Corollary 2 *Any denumerable chain of biorders from a denumerable set to another denumerable set has a real chain $>$-representation which is special and trace-compatible.*

Proof The domain of the linearly ordered set $(E, \geq)$ built in the proof of Corollary 1 is the disjoint union of copies of $A \cup Z$, and the number of copies is just the number of biorders in the actual chain. Hence for any denumerable chain of biorders on denumerable sets A and Z, the set E is itself denumerable. Consequently, there is an order embedding of $(E, \geq)$ in $(\mathbb{R}, \geq)$ and by composing the $>$-representation from Proposition 7 with the embedding we get a real $>$-representation.   □

Notice that the way we obtained Corollary 2 provides another proof of Proposition 4, a proof which moreover applies to the countable case. Also, Corollary 1 does not extend to all chains of biorders on countable domains. The reason lies in the existence of uncountable chains of biorders on infinite countable domains, as the following example shows.

Example 3 Here is an uncountable chain of biorders on countable domains (a similar example thus exists for any pair of countable domains). As in Example 2, let $A = Z = \mathbb{Q}$. Then, for any r in $\mathbb{R}$, set $f_r(a) = a$ and $g_r(z) = z + r$ (where $a \in A$ and $z \in Z$). The pair (f_r, g_r) of real-valued mappings defines the biorder $\mathcal{R}_r = \{(a, z) \in A \times Z : a > z + r\}$. Notice that if the real numbers r and s differ, then $\mathcal{R}_r \neq \mathcal{R}_s$ (in geometric terms: given two parallel lines in the real plane, there are points with rational coordinates lying strictly in between the two lines). We conclude that the biorders $\mathcal{R}_r$, for $r \in \mathbb{R}$, form an uncountable chain.   ◇

5.3 Real Chain Representations

We now turn to the study of real chain $>$-representations of chains of biorders in the general case. Having few results, we mainly offer examples and open problems. It is useful to examine first the existence of real representations of a single biorder in the general case.

5.3.1 Single Biorders

Proposition 3 shows that any biorder has a special $>$-representation respecting the traces in some linearly ordered set $(E, \geq)$. This is clearly no more true when we take E to be $\mathbb{R}$. Indeed, a special $>$-representation is also a $\geq$-representation. But there are biorders on $\mathbb{R}$ having a $>$-representation and no $\geq$-representation and, hence, no special representation.

Example 4 Let $A = Z = \mathbb{R}$, after taking adequate disjoint duplications, and $\mathcal{R} = \{(a, z) \in A \times Z : a \geq z\}$. The relation $\mathcal{R}$ is a biorder which does not admit any real $>$-representation. That $\mathcal{R}$ is a biorder is clear because it admits a $\geq$-representation. Now suppose that $\mathcal{R}$ admits some real $>$-representation in $(\mathbb{R}, \geq)$ using functions f and g, that is $a \geq z \iff f(a) > g(z)$ (for all a in A and z in Z). Notice that with $a = z$ we get $f(a) > g(a)$. Moreover, two open intervals $]f(a), g(a)[$ and

$]f(b), g(b)[$, where $a, b \in \mathbb{R}$ with $a \neq b$, are disjoint. Indeed, if $a < b$, we get $a \,\overline{\mathcal{R}}\, b$, and so $f(a) \leq g(b)$. Now selecting some rational number q_a in $]f(a), g(a)[$, we form a collection $(q_a)_{a \in \mathbb{R}}$ of distinct rational numbers, in contradiction with the countability of $\mathbb{Q}$. $\diamond$

Here is a necessary condition for real $\geq$-representability. It is taken from Doignon et al. [14] (other equivalent conditions can be found in [32]).

Definition 8 Let $\mathcal{R}$ be a relation from A to Z. A countable subset M^* of $A \cup Z$ is *widely dense* for $\mathcal{R}$ when, for all a in A and x in Z, there follows from $a \,\mathcal{R}\, x$ the existence of some $m^* \in M^*$ such that

$$m^* \in Z, \ a \,\mathcal{R}\, m^*, \ \text{and} \ m^* \succsim_{\mathcal{R}}^{Z} x,$$

or

$$m^* \in A, \ a \succsim_{\mathcal{R}}^{A} m^*, \ \text{and} \ m^* \,\mathcal{R}\, d.$$

The following statement is Proposition 9 in Doignon et al. [14].

Proposition 8 *A biorder $\mathcal{R}$ has a real $\geq$-representation if and only if there is a countable subset M^* of $A \cup Z$ that is widely dense for $\mathcal{R}$. This representation can always be chosen so as to respect the traces.*

The proof of Proposition 8 in Doignon et al. [14] uses a construction that differs from (but resembles) the one used in Proposition 3 with the relation $\mathcal{Q}_m$ on $A \cup Z$. This is because the relation $\mathcal{Q}_m$ always produces a special representation, and any real special representation is at the same time a $>$-representation and a $\geq$-representation. But there are biorders having a $>$-representation while having no $\geq$-representation (take $\overline{\mathcal{R}}^{-1}$ with $\mathcal{R}$ as in Example 4). Now the proof of Doignon et al. [14] uses a different construction dating back to Bouchet [4] and Cogis [8, 9]. Let $\mathcal{R}$ be a relation (not necessarily a biorder here) from A to Z. Among all quasi orders (that is, reflexive and transitive relations) $\mathcal{W}$ on $A \cup Z$ such that $\mathcal{W} \cap (A \times Z) = \mathcal{R}$, there is one which includes all the other ones; it is the *maximum quasi order* $\mathcal{Q}_M$ attached to $\mathcal{R}$. Then $\mathcal{Q}_M$ is the union of the four relations in the cells of Table 1. Moreover, the quasi order $\mathcal{Q}_M$ is complete if and only if the relation $\mathcal{R}$ is a biorder. The proof of Proposition 8 by Doignon et al. [14] then consists in showing that the existence of a widely dense subset for $\mathcal{R}$ implies that $\mathcal{Q}_M$ has a real representation $h : A \times Z \to \mathbb{R}$ as a weak order (meaning $\alpha \,\mathcal{Q}_M\, \beta \iff h(\alpha) \geq h(\beta)$). This directly leads to a real

Table 1 The four parts of the maximum quasi order $\mathcal{Q}_M$ attached to a relation $\mathcal{R}$

$\mathcal{Q}_M$	A	Z
A	$\overline{\mathcal{R}}\ \mathcal{R}^{-1}$	$\mathcal{R}$
Z	$\mathcal{R}^{-1}\ \overline{\mathcal{R}}\ \mathcal{R}^{-1}$	$\mathcal{R}^{-1}\ \overline{\mathcal{R}}$

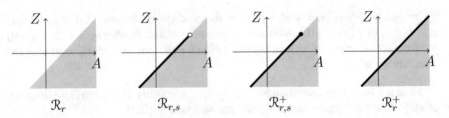

Fig. 5 Some biorders from the chain in Example 6

$\geq$-representation f, g of $\mathcal{R}$ respecting the traces (just take for f and g the restrictions of h to respectively A and Z). Observe that the representation f, g is in general not special: we may have $f(a) = g(x)$, which happens exactly when the pair (a, x) is in the outer fringe of $\overline{\mathcal{R}}^{-1}$ as is easily checked. To obtain a real $>$-representation, it suffices to apply the same process to $\overline{\mathcal{R}}^{-1}$ instead of $\mathcal{R}$.

Bosi et al. [3] compare methods for establishing real representations of interval orders (the latter are particular cases of biorders).

5.3.2   Chains of Biorders on Infinite Sets: Open Problems

We now turn to the investigation of real chain representations of nested families of biorders in the infinite case. Notice that this means not only that the sets A and Z may be infinite, but also that the chain may contain infinitely many biorders.

We have almost no result here, even under the assumption that the chain of biorders is finite (in other words, it is a nested family of biorders). We will simply illustrate the difficulty of the problem by giving two (related) examples. Our first example is a chain of biorders for which all relations in the chain have empty fringes.

Example 5 Here is a chain of biorders which all have empty inner fringes. Let $A = Z = \mathbb{R}$ and, for $r \in \mathbb{R}$, set $f_r(a) = a$ and $g_r(z) = z + r$ (where $a \in A$ and $z \in Z$). The pair (f_r, g_r) of real-valued mappings defines the biorder $\mathcal{R}_r = \{(a, z) \in A \times Z : a > z + r\}$. For any pair (a, z) in $\mathcal{R}_r$, there are real numbers y and b such that $a > y + r > z + r$, and then $y + r \geq b > z + r$, and so $(a, y) \in \mathcal{R}_r$, $(b, y) \notin \mathcal{R}_r$, $(b, z) \in \mathcal{R}_r$. We obtain $(a, z) \in \mathcal{R}_r \overline{\mathcal{R}_r}^{-1} \mathcal{R}_r$, thus $\mathcal{R}_r^{\mathcal{I}} = \varnothing$. For $r, s \in \mathbb{R}$, notice $\mathcal{R}_r \subsetneq \mathcal{R}_s$ if and only if $r > s$. Thus $(\mathcal{R}_r)_{r \in \mathbb{R}}$ is a chain of biorders, all satisfying $\mathcal{R}_r = \cup\{\mathcal{R}_s : s \in \mathbb{R}, r > s\}$. However $(\mathcal{R}_r)_{r \in \mathbb{R}}$, even after the addition of the empty and the full biorders, is not a maximal chain (see next example).   ◇

It is not difficult to show that any chain of biorders is included in some maximal chain of biorders (refering to the Axiom of Choice). Working with maximal chains families does not change the picture revealed by the preceding example, as shown below.

Example 6 We build one of the (infinitely many) maximal chains of biorders containing the family $(\mathcal{R}_r)_{r \in \mathbb{R}}$ from the preceding example. A geometric viewpoint is helpful here. Each biorder $\mathcal{R}_r$, consisting of pairs (a, z) from $A \times Z$, is a subset of $\mathbb{R}^2$ (remember $A = Z = \mathbb{R}$); more precisely, $\mathcal{R}_r$ is the open half-plane defined by the inequality $a > z + r$ (we keep the letters a and z to denote the two coordinates of the point (a, z) in $\mathbb{R}^2$). Now consider the additional following subsets, where $(r, s) \in \mathbb{R}^2$ (see Fig. 5 for an illustration):

$$\mathcal{R}_{r,s} = \mathcal{R}_r \cup \{(a + r, a) : a < s\},$$
$$\mathcal{R}_{r,s}^+ = \mathcal{R}_{r,s} \cup \{(s + r, s)\},$$
$$\mathcal{R}_r^+ = \mathcal{R}_r \cup \{(a + r, a) : a \in \mathbb{R}\}.$$

It is easily checked that the whole collection $\{\mathcal{R}_r, \mathcal{R}_{r,s}, \mathcal{R}_{r,s}^+, \mathcal{R}_r^+ : (r, s) \in \mathbb{R}^2\}$ is a chain of biorders. Moreover, augmented by the empty and the full biorders, the chain becomes a maximal chain. $\diamond$

Although all biorders forming the chain in Example 5 have a real representation, this is not true for all biorders in the extended, maximal chain in Example 6 (remember Example 4). Because of the last fact, it seems difficult to find conditions for the real representability of infinite chains of biorders—even for those chains of biorders which all admit individually a real representation.

Also for the apparently simpler question of characterizing finite nested chains of biorders having a real representation, we have no clear answer. An important open problem is the following: *given a nested finite family of biorders all having a real >-representation, what are the conditions allowing to build a real chain representation of this nested family?*

This will eventually lead us to work out a notion of weak gradedness (or well-gradedness) in more details on infinite sets. Some results on a type of well-gradedness on infinite sets are available in Doignon and Falmagne [13, Ch. 4].

6 Discussion

We summarize the (apparently) new results established above and list open problems for further investigation. All results concern the existence of chain representations of nested families, or more generally chains, of biorders; they fall into two types.

Any chain of biorders from a set A to a set Z has a chain >-representation in some adequately chosen linearly ordered set $(E, \geq)$ (Proposition 7). The proof builds an acyclic relation on the union of disjoint duplications of A and Z, as many duplication as there are biorders in the chain. Applied to the special family $(\varnothing, \mathcal{R}, A \times Z)$ of biorders, it delivers a new argument for the existence of a >-representation (or $\geq$-representation) of any biorder $\mathcal{R}$ in some linearly ordered set $(E, \geq)$ (Proposition 3, established in Sect. 4.4); the argument is quite different from the ones existing in the literature (as for instance in [14, 15]).

Assume now that the target $(E, \geq)$ is the linearly ordered set of the reals, and that biorders are from a finite set A to a finite set Z. Each nested family of biorders has a real representation (Proposition 4). The proof is based on the notion of weak gradedness (a weakening of well-gradedness, [12]). Weak gradedness asserts that for any two biorders $\mathcal{R}$ and $\mathcal{S}$ from A to Z such that $\mathcal{R} \subsetneq \mathcal{S}$, there is a sequence of elementary transformations, each one consisting of the addition of a single pair, which transforms $\mathcal{R}$ into $\mathcal{S}$ while producing a biorder at each step (weak gradedness follows from Lemma 3). For a single biorder from A to Z, the argument for the existence of a real representation in Sect. 4.4 is quite different from those in the literature [14, 17]. Exactly as for a nested family of biorders, it is a constructive argument. The proof directly extends to the construction of a chain representation for any nested family of interval orders or semiorders (Propositions 5 and 6) again on finite sets; in the case of semiorders, the representations obtained are without proper nesting [1].

Quite a few problems are left unsolved. Here are the most intriguing ones in our view.

(i) Working with chain representations of chains of biorders benefits from a better understanding of the collection of all representations of a *single* biorder. We obtained several results on the structure of the collection, whether or not the representations are required to be special and/or to respect the traces. More remains to be done however.

(ii) Among the chain representations of a nested family of semiorders (as in Proposition 6), do there exist representations which, for each of the semiorders, avoid proper nesting, and even are constant threshold representations? See [1, 15] for the latter notions in the case of a single semiorder.

(iii) Finally, we do not have much knowledge about real chain representations in the general case (that is, with no restrictions on the cardinalities of the sets A and Z), even for finite nested families of biorders. We repeat an important open question: let $(\mathcal{R}_1, \mathcal{R}_2, \dots, \mathcal{R}_k)$ be a nested family of $k \geq 1$ biorders from the set A to the set Z. Suppose furthermore that each of these biorders has a real $>$-representation (or $\geq$-representation). When is it true that this finite family of nested biorders has a chain representation?

As far as we know, papers on nested families of relations assume that the left or the right traces are compatible (meaning that the intersection of the traces is complete, see for instance [16]). Working without any compatibility assumption proves to be more difficult but also quite rewarding: for instance, new proofs for the existence of a representation for a single relation are important by-products of our investigations.

Appendix

A Binary Relations on a Set

A binary relation S on a set X is a subset of $X \times X$. For $x, y \in X$, we often write, as is usual, $x \, S \, y$ instead of $(x, y) \in S$. Whenever the symbol $\succsim$ denotes a binary relation, $\succ$ stands for its asymmetric part ($x \succ y \iff x \succsim y$ and $Not[y \succsim x]$) and $\sim$ stands for its symmetric part ($x \sim y \iff x \succsim y$ and $y \succsim x$). A similar convention holds when subscripts or superscripts appear to $\succsim$.

A binary relation S on X is

   (i) *reflexive* if $x \, S \, y$,
  (ii) *irreflexive* if $Not[x \, S \, y]$,
 (iii) *complete* if $x \, S \, y$ or $y \, S \, x$,
 (iv) *symmetric* if $x \, S \, y$ implies $y \, S \, x$,
  (v) *asymmetric* if $x \, S \, y$ implies $Not[y \, S \, x]$,
 (vi) *antisymmetric* if $x \, S \, y$ and $y \, S \, x$ imply $x = y$,
(vii) *transitive* if $x \, S \, y$ and $y \, S \, z$ imply $x \, S \, z$,
(viii) *Ferrers* if $[x \, S \, y$ and $z \, S \, w] \Rightarrow [x \, S \, w$ or $z \, S \, y]$,
 (ix) *semitransitive* if $[x \, S \, y$ and $y \, S \, z] \Rightarrow [x \, S \, w$ or $w \, S \, z]$,

for all $x, y, z, w \in X$.

We list below a number of remarkable structures. A binary relation S on X is

   (i) a *quasi order* if it is reflexive and transitive,
  (ii) a *weak order* or a *complete preorder* if it is complete and transitive,
 (iii) a *linear order* if it is an antisymmetric weak order,
 (iv) a *strict linear order* if it is the asymmetric part of a linear order,
  (v) an *equivalence* if it is reflexive, symmetric, and transitive,
 (vi) a *strict partial order* if it is irreflexive and transitive,
(vii) *interval order* if it is irreflexive and Ferrers;
(viii) *semiorder* if it is irreflexive, Ferrers and semitransitive.

It is well known that an equivalence relation S partitions the set X into equivalence classes. The equivalence class of an element x of X is denoted Sx. The set X/S of all equivalence classes of X under S is the *quotient* of X by S. Any weak order S on X leads to the equivalence relation $S \cap S^{-1}$ on X, whose classes are also the *classes* of S.

A relation S generates a *left trace* $\succsim_S^\ell$ and a *right trace* $\succsim_S^r$ on X, which are the binary relations on X such that

$$x \succsim_S^\ell y \iff [y \, S \, z \Rightarrow x \, S \, z],$$
$$x \succsim_S^r y \iff [z \, S \, x \Rightarrow z \, S \, y]$$

for all $x, y, z \in X$. It is clear that $\succsim_S^\ell$ and $\succsim_S^r$ are always quasi orders.

A characteristic feature of interval orders combines their irreflexivity with the facts that their left and right traces are complete (and therefore weak orders).

A characteristic feature of semiorders is that they are interval orders with noncontradictory left and right traces. The last condition requires that $x \succ_S^\ell y$ and $y \succ_S^r x$ never occur together, or equivalently that the relation defined as $\succsim_S = \succsim_S^\ell \cap \succsim_S^r$ is complete. The relation $\succsim_S$ is thus a weak order, called the *trace* of the semiorder.

A characteristic feature of weak orders is that they are identical to their trace.

References

1. Aleskerov, F., Bouyssou, D., Monjardet, B.: Utility Maximization, Choice and Preference, 2nd edn. Springer, Berlin (2007)
2. Block, H.D., Marschak, J.: Random orderings and stochastic theories of responses. In: Olkin, I., Ghurye, S.G., Hoeffding, W., Madow, W.G., Mann, H.B. (eds.) Contibutions to Probability and Statistics. Stanford Studies in Mathematics and Statistics, pp. 97–132. Stanford University Press, Stanford (1960)
3. Bosi, B., Candeal, J.C., Induráin, E., Oloriz, E., Zudaire, M.: Numerical representations of interval orders. Order **18**(2), 171–190 (2001)
4. Bouchet, A.: Étude combinatoire des ordonnés finis. Applications. Thèse, Université Scientifique et Médicale de Grenoble, Grenoble, France (1981)
5. Bouyssou, D., Marchant, T.: Biorders with frontier. Order **28**(1), 53–87 (2011)
6. Bridges, D.S., Mehta, G.B.: Representations of Preferences Orderings. Lecture Notes in Economics and Mathematical Systems, vol. 422, 1st edn. Springer, Berlin (1995)
7. Cantor, G.: Contributions to the Founding of the Theory of Transfinite Numbers. Dover Publications, United States (1955). Beiträge zur Begründung der transfiniten Mengenlehre, Translation of the original text in German dated 1895–1897
8. Cogis, O.: On the Ferrers dimension of a digraph. Discret. Math. **38**(1), 47–52 (1982a)
9. Cogis, O.: Ferrers digraphs and threshold graphs. Discret. Math. **38**(1), 33–46 (1982b)
10. Debreu, G.: Representation of a preference ordering by a numerical function. In: Thrall, R., Coombs, C.H., Davies, R. (eds.) Decision Processes, pp. 159–175. Wiley, New York (1954)
11. Doignon, J.-P.: Partial structures of preference. In: Kacprzyk, J., Roubens, M. (eds.) Nonconventional Preference Relations in Decision Making. Lecture Notes in Economics and Mathematical Systems, vol. 301, pp. 22–35. Springer, Berlin (1988)
12. Doignon, J.-P., Falmagne, J.-C.: Well-graded families of relations. Discret. Math. **173**(1–3), 35–44 (1997)
13. Doignon, J.-P., Falmagne, J.-C.: Learning Spaces. Springer, Berlin (2011)
14. Doignon, J.-P., Ducamp, A., Falmagne, J.-C.: On realizable biorders and the biorder dimension of a relation. J. Math. Psychol. **28**(1), 73–109 (1984)
15. Doignon, J.-P., Ducamp, A., Falmagne, J.-C.: On the separation of two relations by a biorder or a semiorder. Math. Soc. Sci. **13**(1), 1–18 (1987)
16. Doignon, J.-P., Monjardet, B., Roubens, M., Vincke, Ph: Biorder families, valued relations and preference modelling. J. Math. Psychol. **30**(1), 435–480 (1988)
17. Ducamp, A., Falmagne, J.-C.: Composite measurement. J. Math. Psychol. **6**(3), 359–390 (1969)
18. Fishburn, P.C.: Utility Theory for Decision-Making. Wiley, New York (1970a)
19. Fishburn, P.C.: Intransitive indifference with unequal indifference intervals. J. Math. Psychol. **7**(1), 144–149 (1970b)
20. Fishburn, P.C.: Binary choice probabilities: on the varieties of stochastic transitivity. J. Math. Psychol. **10**(4), 327–352 (1973a)
21. Fishburn, P.C.: Interval representations for interval orders and semiorders. J. Math. Psychol. **10**(1), 91–105 (1973b)

22. Fishburn, P.C.: Interval Orders and Intervals Graphs. Wiley, New York (1985)
23. Guttman, L.: A basis for scaling qualitative data. Am. Sociol. Rev. **9**(2), 139–150 (1944)
24. Guttman, L.: The basis for scalogram analysis. In: Stouffer, S.A., Guttman, L., Suchman, E.A., Lazarsfeld, P.F., Star, S.A., Clausen, J.A. (eds.) Measurement and Prediction. Studies in Social Psychology in World War II, vol. IV, pp. 60–90. Princeton University Press, Princeton (1950)
25. Luce, R.D.: Semiorders and a theory of utility discrimination. Econometrica **24**(2), 178–191 (1956)
26. Luce, R.D., Suppes, P.: Preference, utility, and subjective probability. In: Luce, R.D., Bush, R.R., Galanter, E. (eds.) Handbook of Mathematical Psychology, vol. 3. Wiley, New York (1965)
27. Marley, A.A.J.: Some probabilistic models of simple choice and ranking. J. Math. Psychol. **5**(2), 311–332 (1968)
28. Marschak, J.: Binary-choice constraints and random utility indicators. In: Arrow, K.J., Karlin, S., Suppes, P. (eds.) Mathematical Methods in the Social Sciences, pp. 312–329. Stanford University Press, Stanford (1960)
29. Monjardet, B.: Axiomatiques et propriétés des quasi-ordres. Mathématiques Sci. Hum. **63**, 51–82 (1978)
30. Monjardet, B.: Probabilistic consistency, homogeneous families of relations and linear Λ-relations. In: Degreef, E., van Buggenhaut, J. (eds.) Trends in Mathematical Psychology, pp. 271–281. North-Holland, Amsterdam (1984)
31. Monjardet, B.: A generalization of probabilistic consistency: linearity conditions for valued preference relation. In: Kacprzyk, J., Roubens, M. (eds.) Non-conventional Preference Relations in Decision Making. LNEMS, vol. 301, pp. 36–53. Springer, Berlin (1988)
32. Nakamura, Y.: Real interval representations. J. Math. Psychol. **46**(2), 140–177 (2002)
33. Rabinovitch, I.: The dimension of semiorders. J. Comb. Theory **25**(1), 50–61 (1978)
34. Roberts, F.S.: Homogeneous families of semiorders and theory of probabilistic consistency. J. Math. Psychol. **8**(2), 248–263 (1971)
35. Roberts, F.S.: Measurement Theory with Applications to Decision Making, Utility and the Social Sciences. Encyclopedia of Mathematics and its Applications, vol. 7. Addison-Wesley, Reading (1979)
36. Roubens, M., Vincke, P.: Preference Modelling. LNEMS, vol. 250. Springer, Berlin (1985)
37. Scott, D., Suppes, P.: Foundational aspects of theories of measurement. J. Symb. Log. **23**(2), 113–128 (1958)
38. Suppes, P., Krantz, D.H., Luce, R.D., Tversky, A.: Foundations of measurement, Volume 2: Geometrical, Threshold, and Probabilistic Representations. Academic Press, New York (1989)
39. Szpilrajn, E.: Sur l'extension de l'ordre partiel. Fundam. Math. **16**(1), 386–389 (1930)

A Note on Representable Group Topologies

C. Chis, H.-P. A. Künzi and M. Sanchis

Abstract We study natural topologies in the sense of Debreu in the poset of topological group topologies on a topological group. We call this kind of topologies g-topologies. To be precise, groups admitting a non-totally disconnected g-natural topology as well as the non-totally disconnected g-topologies are identified. Moreover, the non-totally disconnected g-representable topologies as well as the total orders inducing non-totally disconnected group topologies are characterized. It is worth noting that our framework is more general than the usual one in representation theory: we assume no translation invariant properties. We also deal with some questions concerning order and topological algebra related to the semicontinuous representation property (SRP): we establish some results related to the Sorgenfrey line and SRP (some of them connected to the Proper Forcing Axiom (PFA)) and, we characterize σ-compact and (locally) precompact groups which satisfy SRP.

Keywords Order preserving real-valued functions · Continuous and additive order preserving real-valued functions · Representability of an ordered structure · (Locally pseudocompact) Topological group · Linear order · Totally ordered

---

During the first stages of the writing of the paper the second named author was visiting the Mathematics Department of the University Jaume I of Castellón. He takes the opportunity to thank his hosts for the generous hospitality and support. He also thanks the National Research Foundation of South Africa for partial financial support (Grant Number: 118517).

C. Chis
Departament de Matemàtiques, Institut Universitari de Matemàtiques i Aplicacions de Castelló (IMAC), Universitat Jaume I, Campus del Riu Sec. s/n, 12071 Castellón de la Plana, Spain
e-mail: chis@mat.uji.es

H.-P. A. Künzi
Department of Mathematics and Applied Mathematics, University of Cape Town, Rondebosch, Cape Town 7701, South Africa
e-mail: hans-peter.kunzi@uct.ac.za

M. Sanchis (✉)
Departament de Matemàtiques, Universitat Jaume I, Campus del Riu Sec. s/n, 12071 Castellón de la Plana, Spain
e-mail: sanchis@mat.uji.es

group · Algebraically orderable group · Topologically orderable group · Non-Archimedean metric · The Sorgenfrey line · Cardinal function · Locally pseudocompact group

2010 Mathematics Subject Classification Primary 54F05 · 22A05; Secondary 54F50 · 54G10 · 06A10 · 06A55

1   Introduction

An important tool and theoretical support in mathematical social sciences, decision theory, majorization and order statistics, etc. is the theory of order-preserving real-valued functions. For example, major concern in decision analysis has been with an inquiry on the existence and the properties of this kind of functions, usually defined on a probabilistic discrimination structure. It is worth noting that the theory blends order, algebraic and topological aspects.

The theory goes back to Cantor [12, 13] and since then the theory has undergone a great development and it has diversified in many different branches of research. In the beginning of the theory, the emphasis in the literature was on topological or algebraic aspects of ordered structures but not in topological algebra. We must cite early papers by Eilenberg [19] and Nachbin [35] in the framework of order and topology and by Birkhoff [5] and Fuchs [21] in the framework of order and algebra.

Seminal papers in order and topological algebra are [7, 9, 10, 32, 33, 37]. Among other, the interested reader might also consult [11, 22–24, 26, 31].

Although some are well known, we need to give some definitions. A preorder $\preceq$ on an arbitrary nonempty set X is a binary relation on X which is reflexive and transitive. Antisymmetric preorders are called orders. A total preorder $\preceq$ on a set X is a preorder such that any pair of elements are comparable, i.e., for every $x, y \in X$ either $x \preceq y$ or $x \preceq y$. The notion of a total order is self-explanatory. If $\preceq$ is a (total) (pre)order on a nonempty set X, the pair $(X, \preceq)$ is called a (totally) (pre)ordered set.

Let $(X, \preceq)$ be a preordered set. Then the sets

$$L(x) = \{a \in X \mid a < x\}$$

and

$$G(x) = \{a \in X \mid a > x\},$$

where x runs over X, form a subbasis for a topology $\tau_{\preceq}$ on X, the so-called *order topology*. Given a topological space (X, τ), a total preorder $\preceq$ on X is said to be τ-continuous if it is coarser than the τ-topology. A topological space (X, τ) is said to be (pre)orderable if $\tau = \tau_{\preceq}$ for some total (pre)order $\tau = \tau_{\preceq}$.

Given a totally preordered set (X, τ), an injective real-valued function $u : X \to \mathbb{R}$ is said to be an *isotony* if for every $x, y \in X$, the inequality $x \preceq y$ implies $u(x) \le$

$u(y)$ (respectively, the inequality $x < y$ implies $u(x) < u(y)$). If such a function u does exist, then $\preceq$ is called *representable*. We also refer to this function as an order-preserving real-valued function. An *order isomorphism* is a surjective isotony. For a topological space (X, τ), a total preorder $\preceq$ on X is said to be *continuously representable* if there exists an order-preserving function for $\preceq$ that is continuous with respect to the topology τ and the usual (Euclidean) topology on the real line $\mathbb{R}$.

Definition 1 A topological space (X, τ) satisfies the continuous representability property (in short, CRP) if every continuous total preorder $\preceq$ defined on X is continuously representable.

A total preorder is semicontinuous with respect to a given topology on a set X if either the upper or the lower topology induced by the total preorder is coarser than the given topology. Below the following concept will be crucial.

Definition 2 A topology defined on a nonempty set satisfies the semicontinuous representability property (in short, SRP) if every lower (upper) semicontinuous total preorder defined on the given set admits a numerical representation by means of a lower (upper) semicontinuous real-valued order-monomorphism.

One of the specially relevant problems in the theory deals with the representability of totally ordered groups. In this context, a totally ordered group $(G, \circ, \preceq)$ means an algebraic group $(G, \circ)$ equipped with a total order $\preceq$ which is translation-invariant, that is, $(G, \circ)$ is an algebraically ordered group. In other words, $x \preceq y$ implies $x + z \preceq y + z$ and $z + x \preceq z + y$ for every $x, y, z \in G$.

When dealing with totally ordered groups $(G, \circ, \preceq)$, an outstanding theorem of Hölder [28] states that the Archimedean property is a sufficient (and necessary) condition for the existence of an order-preserving group homomorphism h from $(G, \circ, \preceq)$ into $(\mathbb{R}, +, \leq)$. Although no explicit topological considerations are made in Hölder's theorem, it is interesting to remark that the homomorphism h is continuous when we consider $(G, \circ, \preceq)$ and $(\mathbb{R}, +, \leq)$ endowed with the order topology (see, for example, [8, Theorem 1]). In this framework the question of finding topological properties on an ordered group that force the order structure to be representable by means of a continuous (additive) order-preserving real-valued function arises in a natural way (for a similar question on semigroups the reader might consult [9]). For instance, Iseki [29] shows that if a non-trivial totally ordered group $(G, \circ, \leq)$ is connected with regards to the order topology, then it is a non-compact, locally compact, one-dimensional group which implies by a theorem due to Montgomery [34] that it is isomorphic to the additive group of the real numbers. For results on representability of algebraically (pre)ordered groups, the interested reader can consult [11].

In Sect. 2 we address the problem of representability of linearly topologically ordered groups. Usually, an orderable topological space (X, τ) is called *linearly ordered* provided that $\tau = \tau_{\preceq}$ for a total order $\preceq$. A topological group (G, τ) will be called *topologically orderable* if it is orderable in this sense. Recall that a topological group (G, τ) is an abstract group $(G, \circ)$ endowed with a topology τ such that the functions $\phi: G \times G \to G$ and $\rho: G \to G$ defined, respectively, as

$$\phi(x, y) = x \circ y \quad \text{and} \quad \rho(x) = x^{-1}$$

are continuous for all $x, y \in G$ (where, as usual, x^{-1} stands for the inverse element of x). All topological groups are assumed to be Hausdorff. It is worth noting that, as we explain below, this kind of groups includes as a proper subclass the class of totally ordered groups.

Let $\preceq$ be a (total) order on a nonempty set X. Following Debreu [16, 17], a topology τ on X is said to be *natural* if τ is finer than the order topology $\tau_{\preceq}$ on X. The next definition adapts this notion in the realm of topological groups.

Definition 3 Let $(G, \tau_{\preceq})$ be a topologically orderable group. A topology τ on G is called *g-natural* if τ is finer than $\tau_{\preceq}$ and (G, τ) is a topological group.

So, *g*-natural means *natural* in the poset of group topologies. In the same spirit as the definition of g-natural topology we have

Definition 4 A group topology τ on a group G is said to be *g-representable* if every total order $\preceq$ such that $(G, \tau_{\preceq})$ is a topological group and $\tau_{\preceq} \subseteq \tau$ is representable by means of a order-preserving real-valued function.

The reader should note that although we consider groups endowed with total orders the order-preserving real-valued functions that we study do not necessarily preserve the algebraic structure (that is, they may fail to be group homomorphisms on, say, the additive real line). Therefore utility representations considered in this paper need not be group homomorphisms. Indeed we carefully distinguish between "representability" and "algebraic representability" (see remarks below).

The aim of Sect. 2 is twofold. First, the groups admitting a non-totally disconnected *g*-natural topology as well as the non-totally disconnected *g*-topologies are identified. Secondly, the non-totally disconnected *g*-representable topologies as well as the total orders inducing non-totally disconnected group topologies are characterized. Notice that, as we say above, our framework is more general than the usual one in representation theory: we assume no translation invariant properties. Actually, topological ordering of a group is quite different from algebraic ordering: it is easy to show that the only algebraic ordering of the additive group $\mathbb{Z}$ of integers are the usual one and its dual, both of which turn $\mathbb{Z}$ into a discrete space (in particular, a discrete topological group). Nevertheless $\mathbb{Z}$ admits many strongly zero-dimensional metrizable group topologies (for instance, the p-adic topology for a given prime p) and, by a theorem of Herrlich [27], with these topologies $\mathbb{Z}$ is a topologically orderable group. In Sect. 3 we address some questions related to order and topological algebra related to the SRP property. We present some results on the Sorgenfrey line and SRP (some of them related to the (PFA) axiom). We also characterize σ-compact and (locally) precompact groups which satisfy SRP.

Our notation and terminology is standard. For example, $\mathbb{R}$ stands for the reals endowed with its usual topology and $\aleph_0$ for the cardinal of the first countable ordinal. As usual, $\aleph_1$ stands for the cardinality of the set of all countable ordinal numbers.

Recall that in this paper topological groups are assumed Hausdorff (and, consequently, they are Tychonoff spaces). For concepts on order-preserving real-valued
functions (respectively, on topological groups) not defined here the reader can consult
[6] (respectively, [2]).

Finally we formulate the (PFA) axiom:

Definition 5 The (PFA) axiom holds if, given an proper poset $\mathbb{M}$ and a collection
$\mathcal{D}$ of dense subsets of $\mathbb{M}$, with cardinal of $\mathcal{D}$ less or equal to $\aleph_1$, then there is a filter
$G \subseteq \mathbb{M}$ such that $G \cap D \neq \emptyset$ for all $D \in \mathcal{D}$.

2 g-Natural Non-totally Disconnected Group Topologies

We start with the first aforementioned question in the introduction. A topological
space X is totally disconnected if the connected components in X are the one-point
sets. In the remainder of this section $c(G)$ will stand for the connected component
of the identity of a topological group G. Recall that $c(G)$ is the largest connected
subset of the topological group G that contains the identity element. The component
$c(G)$ is a closed normal subgroup of G; the cosets with respect to $c(G)$ coincide with
the connected components of G. The quotient group $G/c(G)$ is totally disconnected
(and Hausdorff). Moreover $c(G)$ is the smallest among the normal subgroups $H \subseteq G$
such that G/H is totally disconnected. If G is locally connected (for example, if G
is a Lie group), then $c(G)$ is open in G and $G/c(G)$ is discrete.

As one might expect, the structure of topologically orderable groups plays an
important role in our results. Let $\mathbb{R}_D$ denote the product topological group $\mathbb{R} \times D$
where $\mathbb{R}$ is the additive topological group of the real numbers and D is a discrete
group. Our first result states the following.

Theorem 1 *If τ is a non-totally disconnected g-natural topology on a group G,
then G is algebraically isomorphic to a product of the additive group $\mathbb{R}$ of the real
numbers and a topological group D. Moreover, the connected component $c(G)$ is an
open subgroup which is either algebraically isomorphic to $\mathbb{R}$ or trivial, and $\tau_{|\{0\} \times D}$
is the discrete topology.*

Proof Since τ is a g-natural topology, there exists a topology σ on G such that
(G, σ) is a topologically orderable group. As τ is non-totally disconnected, so is
σ. Hence, by Theorem 2.4 in [40], (G, σ) contains an open subgroup isomorphic
to the additive group of the real numbers. Since $\mathbb{R}$ is divisible we get that (G, σ)
is topologically isomorphic to an $\mathbb{R}_D$. Thus, $c(G)$ is an open subgroup and G is
algebraically isomorphic to $\mathbb{R} \times D$ with $\tau_{|\{0\} \times D}$ the discrete topology.

To establish the algebraic structure of $c(G)$, notice that $c(G)$ is connected as a
subgroup of $\mathbb{R}_D$ because $\sigma \subseteq \tau$. The result now follows from the fact that the only
connected proper subgroup of $\mathbb{R}$ is the trivial one. This completes the proof. □

The previous theorem tells us that for characterizing non-totally disconnected
g-natural topologies one needs to know the group topologies on $\mathbb{R}$ which are finer

than the usual order topology, the so-called Euclidean topology (on $\mathbb{R}$). The key to the description of these topologies is the following lemma which is of interest in itself. From now on, $\leq_r$ will denote the usual order on the reals.

Theorem 2 *For a group topology σ on the reals, the following conditions are equivalent:*

(1) σ is finer than the Euclidean topology $\tau_{\leq_r}$.
(2) σ satisfies property $(\star\star)$:

> $(\star\star)$ *For each $x \in \mathbb{R}$ with $0 <_r x$, there is an open neighborhood V of 0 in $(\mathbb{R}, \sigma)$ such that $v <_r x$ for every $v \in V$.*

Proof $(1) \Longrightarrow (2)$ Consider $x \in \mathbb{R}$ with $0 <_r x$. If $0 <_r y <_r x$, then $] - y, y [$ is a σ-open subset (because $\tau_\leq \subseteq \sigma$) which satisfies condition (2).

$(2) \Longrightarrow (1)$ Let us choose $x <_r 0$. We claim that there exists a symmetric σ-open neighborhood W of 0 such that $v <_r 0$ for every $v \in x + W$. In fact, since $0 <_r -x$, we can choose a symmetric σ-open neighborhood W of 0 such that $w <_r -x$ for every $w \in W$. It is an easy matter to see that $x + W$ enjoys the desirable property which proves the claim.

Now we shall show that an open interval $] y, +\infty [$ ($y \in \mathbb{R}$) is σ-open. To do so, consider $y <_r x$. By our claim there is a symmetric σ-open neighborhood W of $y - x$ such that $y - x + w <_r 0$ for every $w \in W$, so that $y <_r x + w$ for every $w \in W$. This means that $x + W$ is a σ-open neighborhood of x contained in $] y, +\infty [$. Thus, $] y, +\infty [$ is σ-open.

A similar argument to the previous one shows that $] - \infty, y [$ is a σ-open set for every $y \in \mathbb{R}$ which implies that σ is finer than the Euclidean topology on the reals. $\square$

Gathering Theorems 1 and 2 we can obtain the following characterization of non-totally disconnected g-natural topologies.

Theorem 3 *If (G, σ) is a topological group, then the following assertions are equivalent:*

(1) σ is a g-natural non-totally disconnected topology.
(2) There exists a topology τ on the reals satisfying condition $(\star\star)$ and an ordinal γ such that $\sigma = \bigoplus_{\alpha < \gamma} \tau_\alpha$ where $\tau_\alpha = \tau$ for every $\alpha < \gamma$.
(3) $c(G)$ is an open subgroup algebraically isomorphic to the additive group of the reals and $\sigma_{|c(G)}$ satisfies property $(\star\star)$.

Proof $(1) \Longrightarrow (2)$ By Theorem 1, (G, σ) is topologically isomorphic to a product group $\mathbb{R} \times D$ with $\sigma_{|\{0\} \times D}$ the discrete topology and the topology $\tau = \sigma_{|\mathbb{R}}$ finer than the usual topology on $\mathbb{R}$. Thus, if γ denotes the first ordinal of cardinality $|D|$, then $\sigma = \bigoplus_{\alpha < \gamma} \tau_\alpha$ where each $\tau_\alpha = \tau$. Moreover, since τ is finer than the Euclidean topology, Theorem 2 tells us that τ enjoys property $(\star\star)$.

$(2) \Longrightarrow (3)$ It is clear.

(3)$\Longrightarrow$(1) Consider on G a topology τ defined by the requirement that $c(G)$ must be an open subgroup of G endowed with the Euclidean topology. Since σ satisfies property ($\star\star$), σ is finer than τ. We shall conclude the proof by showing that (G, τ) is a topologically orderable group. To see this, enumerate the family $\{C_\alpha \mid \alpha < \gamma\}$ of connected components of (G, τ). Define an order $\preceq$ on G by letting $g_1 \preceq g_2$ whenever g_1, g_2 belong to the same component and $g_1 \leq_r g_2$ or $\alpha \leq \beta$ with $g_1 \in C_\alpha$ and $g_2 \in C_\beta$. If $\tau_{\preceq}$ is the order topology on G, it is clear that $c(G)$ is an open subgroup of $(G, \tau_{\preceq})$ and $\tau_{|c(G)} = \tau_{\preceq|c(G)}$. Thus, $\tau = \tau_{\preceq}$ ([18, Lemma 1]) and the proof is complete. $\qquad\square$

In essence Theorem 3 tells us that we must consider connected group topologies on $\mathbb{R}$ finer than the Euclidean one when we work with g-natural topologies. It seems interesting to recall that such topologies exist: the paper [1] contains a construction of a second countable connected group topology on the additive group of the reals which is strictly finer than the usual one, and the main result in [38] states that every separable connected Abelian torsion-free group admits a strictly finer separable connected group topology.

We look now at the non-totally disconnected topologies which are g-representable.

Theorem 4 *If (G, σ) is a topological group, then the following assertions are equivalent:*

(1) σ is a non-totally disconnected g-representable topology.
(2) There exist an ordinal $\gamma \leq \omega$ and a topology τ on the reals satisfying condition ($\star\star$) such that $\sigma = \bigoplus_{\alpha < \gamma} \tau_\alpha$ where $\tau_\alpha = \tau$ for every $\alpha < \gamma$.
(3) (G, σ) has either finite or countably many connected components and $c(G)$ is an open subgroup algebraically isomorphic to the additive group of the reals such that $\sigma_{|c(G)}$ satisfies property ($\star\star$).

Proof (1)$\Longrightarrow$(2) Let $\preceq$ be an order on G such that $\tau_{\preceq} \subseteq \sigma$ and $(G, \tau_{\preceq})$ is a topologically orderable group. By Theorem 1, $(G, \tau_{\preceq})$ is isomorphic to an $\mathbb{R}_D$ group. Then, since $\tau_{\preceq}$ is g-representable, the cardinal of D is either finite or countable. Now the proof follows a pattern similar to (1)$\Longrightarrow$(2) in Theorem 3.

(2)$\Longrightarrow$(3) It is clear.

(3)$\Longrightarrow$(1) Since $c(G)$ is an open subgroup algebraically isomorphic to the additive group of the reals, $c(G)$ is an open divisible subgroup of (G, σ) and, consequently, (G, σ) is topologically isomorphic to an $\mathbb{R}_D$ group. Obviously, D is not uncountable because (G, σ) does not have uncountably many connected components (and $c(G)$ is open), so $|D| < \omega$. Since $\sigma_{|c(G)}$ satisfies property ($\star\star$), we can consider an order $\preceq$ on G such that the group $\mathbb{R}_D$ is a topologically orderable group with $\tau_{\preceq} \subseteq \sigma$. Since D is countable, [40, Theorem 2.8] tells us that $(\mathbb{R}_D, \tau_{\preceq})$ is a separable metrizable group. Thus, $(\mathbb{R}_D, \tau_{\preceq})$ is second countable which implies that $\tau_{\preceq}$ is g-representable [6, Theorem 3.2.9]. $\qquad\square$

Now we turn to the description of linear orders which induce a non-totally disconnected group topology. By [40, Theorem 2.8] (see Theorem 1) a non-totally disconnected topologically orderable group is isomorphic to an $\mathbb{R}_D$ group. So our problem can be written out in the following way: let D be an algebraic group, what orders on $\mathbb{R} \times D$ induce a product topology τ with $\tau_{|\mathbb{R}}$ the Euclidean topology and $\tau_{|D}$ the discrete topology? Our next theorem answers this question. First we need to introduce some notation and to prove a lemma. Let us recall that if $\preceq$ is an order on a set X, the dual order $\preceq^d$ is defined by letting $y \preceq^d x$ whenever $x \preceq y$. It is a well-known fact that the usual order $\leq_r$ on the reals and its dual $\leq_r^d$ are the unique orders on the reals inducing the Euclidean topology (see for instance [19, Theorem II]). Consider now a product set $D \times \mathbb{R}$ with $(D, \preceq)$ a totally ordered set and let $\mathbb{R}_d$ denote the set

$$\mathbb{R}_d = \{(d, r) \mid r \in \mathbb{R}\}.$$

For each $d \in D$, $(\mathbb{R}_d, \leq)$ (in short, $\mathbb{R}_{d_\leq}$) stands either for the ordered set $(\mathbb{R}_d, \leq_r)$ or for the ordered set $(\mathbb{R}_d, \leq_r^d)$. If confusion is not possible, in the following we shall identify (d, r) with r. Given a family $\mathcal{R} = \{\mathbb{R}_{d_\leq} \mid d \in D\}$ of ordered sets, the *lexicographic order associated with* $\mathcal{R}$ on $D \times \mathbb{R}$ (in short $\leq_{lex_\mathcal{R}}$) is defined by letting $(d_1, r_1) <_{lex_\mathcal{R}} (d_2, r_2)$ whenever $d_1 \not\succeq d_2$ or $d_1 = d_2$ and $r_1 < r_2$. Notice that in the case $d_1 = d_2$, $r_1 < r_2$ means $r_1 <_r r_2$ or $r_1 <_r^d r_2$ depending on the order on $\mathbb{R}_{d_\leq}$. From now on, τ_d stands for the discrete topology on a set X and τ_e for the Euclidean topology on $\mathbb{R}$. By $\langle r, s \rangle$ (respectively, $\langle r, \rightarrow \rangle$, $\langle \leftarrow, r \rangle$) we denote the open interval $]r, s[$ (respectively, the open intervals $]r, \rightarrow [,] \leftarrow, r[$) either in $(\mathbb{R}, \leq_r)$ or in $(\mathbb{R}, \leq_r^d)$.

Lemma 1 *Let $(D, \preceq)$ be an ordered set. If $\leq_{lex_\mathcal{R}}$ is the lexicographic order on $D \times \mathbb{R}$ associated with a family $\mathcal{R} = \{\mathbb{R}_{d_\leq} \mid d \in D\}$, then the order topology $\tau_{\leq_{lex_\mathcal{R}}}$ on $D \times \mathbb{R}$ is $\tau_d \times \tau_e$.*

Proof If $d \in D$ and $] r, s [$ is a basic open set in $\mathbb{R}$, then

$$\{d\} \times] r, s [=] (d, m), (d, n) [$$

where $m = r$ and $n = s$ if the order on $\mathbb{R}_{d_\leq}$ is $\leq_r$ or $m = s$ and $n = r$ if the order on $\mathbb{R}_{d_\leq}$ is $\leq_r^d$. Thus, $\tau_d \times \tau_e \subseteq \tau_{\leq_{lex_\mathcal{R}}}$. To see the inclusion $\tau_{\leq_{lex_\mathcal{R}}} \subseteq \tau_d \times \tau_e$, first notice that $(D \times \mathbb{R}, \leq_{lex_\mathcal{R}})$ has neither a smallest nor a largest element. In fact, if $(d, r) \leq_{lex_\mathcal{R}} (d_1, r_1)$ (respectively, $(d_1, r_1) \leq_{lex_\mathcal{R}} (d, r)$) for every $(d_1, r_1) \in D \times \mathbb{R}$, then $r \leq s$ (respectively, $s \leq r$) for every $s \in \mathbb{R}$ which is a contradiction. Hence, the family of intervals of the form $](d_1, r_1), (d_2, r_2)[$ is a base for the topology $\tau_{\leq_{lex_\mathcal{R}}}$. Now an easy computation shows that, for each $(d_1, r), (d_2, s) \in D \times \mathbb{R}$,

$$](d_1, r), (d_2, s)[= (\{d_1\} \times \langle r, \rightarrow \rangle) \cup (\{d_2\} \times \langle \leftarrow, s \rangle) \cup \left(\bigcup_{d_1 < d < d_2} \{d\} \times \mathbb{R} \right),$$

whenever $d_1 < d_2$ (notice that $\bigcup_{d_1 < d < d_2} \{d\} \times \mathbb{R}$ can be the empty set if d_2 is the successor of d_1). Since

$$](d_1, r), (d_2, s)[= \{d_1\} \times \langle r, s \rangle$$

if $d_1 = d_2$, we have just proved that every basic open set $](d_1, r), (d_2, s)[$ is the union of $\tau_d \times \tau_e$-open sets which completes the proof. $\qquad\qquad\square$

Now we can show the promised characterization of the orders on non-totally disconnected topologically orderable groups.

Theorem 5 *Let D be a group. For an order $\preceq$ on $D \times \mathbb{R}$, the following assertions are equivalent:*

(1) $\tau_{\preceq}$ *is a non-totally disconnected group topology.*
(2) There is an order on D and a family $\mathcal{R} = \{\mathbb{R}_{d_{\preceq}} \mid d \in D\}$ such that $\preceq$ coincides with the lexicographic order $\leq_{lex_{\mathcal{R}}}$ associated with $\mathcal{R}$.

Proof $(2) \Longrightarrow (1)$ follows from Lemma 1. So, we only need to prove $(1) \Longrightarrow (2)$. To see this, Theorem 1 tells us that the restriction of $\tau_{\preceq}$ to each $\{d\} \times \mathbb{R}$ is the usual Euclidean topology (in particular, every $\mathbb{R}_{d_{\preceq}}$ is connected). Therefore $\tau_{\preceq|\{d\}\times\mathbb{R}}$ is either the usual order $\leq_r$ or its dual $\leq_r^d$. So, if $\mathbb{R}_{d_{\preceq}}$ stands for the ordered set $(\mathbb{R}_d, \preceq_{|\mathbb{R}_d})$, we can consider the family $\mathcal{R} = \{\mathbb{R}_{d_{\preceq}} \mid d \in D\}$ in the spirit of the definition of a lexicographic order associated with a family of ordered spaces as defined above. In order to describe $\preceq$ as a lexicographic order associated with the family $\mathcal{R}$ we need the following claim:

Claim: If $d_1 \neq d_2$, then there is no $x, y, z \in \mathbb{R}$ such that

$$(\star) \qquad (d_2, x) \npreceq (d_1, y) \npreceq (d_2, z).$$

In fact, suppose, by contradiction, that $(\star)$ holds. Then $] \leftarrow, (d_1, y)[\cap \mathbb{R}_{d_2}$ and $](d_1, y), \rightarrow [\cap \mathbb{R}_{d_2}$ are nonempty proper clopen subsets of $\mathbb{R}_{d_2}$ which is impossible because $\mathbb{R}_{d_2}$ is connected. Thus, our claim remains valid.

Define now an order $\lesssim$ on D by letting $d_1 \lesssim d_2$ if $(d_1, 0) \preceq (d_2, 0)$. Given the ordered set $(D, \lesssim)$, we shall prove that $\preceq$ is the lexicographic order on $D \times \mathbb{R}$ associated with the family $\mathcal{R}$. By the definition of the family $\mathcal{R}$, we only need to prove that, if $d_1 \neq d_2$, then $(d_1, z_1) \preceq (d_2, z_2)$ implies $(d_1, x) \preceq (d_2, y)$ for every $x, y \in \mathbb{R}$. To see this, notice that $(d_2, z_2) \preceq (d_1, x)$ is impossible because then $(d_1, z_1) \preceq (d_2, z_2) \preceq (d_1, x)$ which contradicts our claim. Hence $(d_1, x) \preceq (d_2, z_2)$. Therefore, by our claim again, $(d_1, x) \preceq (d_2, y)$. The proof is complete. $\qquad\square$

The previous theorem provides a description of all orders on a topological group which induce a non-totally disconnected group topology. Each such order may be constructed from an arbitrary order on a group D by considering on each $\mathbb{R}_d$ ($d \in D$) either the usual order on $\mathbb{R}$ or its dual and defining a lexicographic order on $D \times \mathbb{R}$. A natural way to do this is by means of a topologically discrete order on D: an order on a space X is named *topologically discrete* if it induces the discrete topology. By a theorem of Herrlich [27] for every strongly zero-dimensional metrizable space X there exists a linear order $<$ on the space X which induces the original topology of

X. Therefore for every set X there is a topologically discrete order on X. (An easy way to build a topologically discrete order is the following: if the cardinal of X is $\aleph_\alpha$ and ω_α is the first ordinal of cardinality $\aleph_\alpha$, then consider the lexicographic order on $[1, \omega_\alpha[\times \mathbb{Z}.)$

3   Some Other Results on Order and Topological Algebra

We begin with the notion of circular order on the reals. It is the order $\leqslant_c$ defined by letting $x \leqslant_c y$ whenever either $|x| < |y|$ or $x \leqslant y$ whenever $|x| = |y|$ (here $\leqslant$ stands for the usual order on $\mathbb{R}$). Circular orders are closely related to faithful group actions of countable groups on the circle (see [3]). An easy result is the following

Theorem 6 *The circular order on the reals is not representable.*

Proof For every $x, y > 0$ the following properties hold:

(1) If $x < y$, then $[-x, x] \cap [-y, y] = \varnothing$, and
(2) $[-x, x] = \{-x, x\}$.

   Suppose now that there exists a function $f : \mathbb{R} \longrightarrow \mathbb{R}$ such that $x \leqslant_c y$ if and only if $f(x) \leqslant f(y)$. Then the family $\{[f(-x), f(x)] : x \in \mathbb{R}, x > 0\}$ is an uncountable family of pairwise disjoint intervals which leads us to a contradiction.   □

   We now turn to the circular order on the Sorgenfrey line $\mathbb{S}$. Its relevance lies not only in itself but also in its applications. By the Sorgenfrey line it is understood the topological space which underlying set is the reals and the topology is generated by the family of open sets $\mathcal{B} = \{(x, y] \mid x < y\}$. In particular, since the intervals are closed on the right, $\mathbb{S}$ is known as the right Sorgenfrey topology. If instead we use intervals closed on the left, then it would be called the left Sorgenfrey topology. Notice that the Sorgenfrey line is a paratopological group which is not a topological group, that is, the addition operation is continuous but the function $\rho \colon \mathbb{S} \to \mathbb{S}$ defined as $\rho(x) = -x$ fails to be continuous.
   The first result is tedious but straightforward. We leave the proof to the reader.

Theorem 7 *The circular order on the Sorgenfrey line is semicontinuous.*

As a consequence of the previous results we have

Corollary 1 *The circular order on the Sorgenfrey line does not satisfy CRP.*

Corollary 2 *The circular order on the Sorgenfrey line does not satisfy SRP.*

Proof It follows from the fact that semicontinuous representability implies continuous representability (see Corollary 1).   □

In particular, we have

Corollary 3 *If X satisfies SRP, then X contains no copies of the Sorgenfrey line.*

The previous result allows us to obtain examples of spaces that do not satisfy SRP. For instance, the well-known fact that $\mathbb{R}_L$, the space $\mathbb{R} \times [0, 1]$ with the lexicographic order does not satisfy SRP. Indeed, the subset $\mathbb{R} \times \{0\}$ of $\mathbb{R}_L$ is homeomorphic to the right Sorgenfrey line (and the subset $\mathbb{R} \times \{1\}$ is homeomorphic to the left Sorgenfrey line).

In the same way, we have

Corollary 4 *The (Abelian) free group over the Sorgenfrey line does not satisfy SRP.*

Proof It suffices to note that the (Abelian) free group over the Sorgenfrey line contains a closed copy of $\mathbb{S}$. □

Corollary 5 *An arbitrary product of Sorgenfrey lines does not satisfy SRP. In particular, the Sorgenfrey plane $\mathbb{S} \times \mathbb{S}$ does not satisfy SRP.*

Another application is related to the celebrated Helly space which is defined as the subspace X of the cube $\prod_{t \in [0,1]}$ where $I_t = [0, 1]$ for every $t \in [0, 1]$, consisting of all nondecreasing functions from $[0, 1]$ to $[0, 1]$. The Helly space contains a copy of the Sorgenfrey line [20, 3.2.E(b)] so that

Corollary 6 *The Helly space does not satisfy SRP.*

Under the (PFA) axiom we can state a stronger result.

Given a cardinal κ, a set $X \subseteq \mathbb{R}$ is said to be κ-dense, if every interval meets X in κ points. Among the implications of the (PFA) axiom, we need the following celebrated theorem by Baumgartner [4] which states that, under the (PFA) axiom, it is consistent with ZFC that every pair of $\aleph_1$-dense subsets of $\mathbb{R}$ are order isomorphic.

Theorem 8 *Under (PFA) axiom, if X satisfies SRP, then X contains no uncountable subspaces of the Sorgenfrey line.*

Proof Let A be an uncountable subspace of the Sorgenfrey line. By Baumgartner's theorem, all $\aleph_1$-dense subsets of the reals are order isomorphic. Therefore we may assume that $x \in \mathbb{S}$ implies $-x \in \mathbb{S}$. An argument similar to the one used in Theorem 6 shows that the circular order on A is not representable for every uncountable subset A of the reals. Then the result follows as in Theorem 7. □

For the following corollary, we need the concept of a *network*. A family $\mathcal{B}$ of subsets of a topological space X is a network for X if for every point $x \in X$ and any neighborhood V of x there exists an $A \in \mathcal{B}$ such that $x \in A \subseteq V$. Clearly, any base for X is a network for X: it is a network of a special kind, namely, all members of which are open. Recall that a space X is cometrizable if X has a coarser metric topology such that each point of X has a (not necessarily open) neighborhood base of metric closed sets. It is easy to check that the Sorgenfrey line is a cometrizable space. More interesting examples of cometrizable spaces can be found in [25].

Theorem 9 *Let X be a cometrizable space with no uncountable discrete subspaces. Under the (PFA) axiom the following assertions are equivalent:*

(1) X satisfies SRP.
(2) X contains no copies of an uncountable subspace of the Sorgenfrey line.
(3) X has a countable network.

Proof (1) ⇒ (2) is Theorem 8 and (3) ⇒ (1) follows from the fact that every space with a countable network satisfies SRP. Thus, we only need to prove (2) ⇒ (3). Since X has no uncountable discrete subspaces and contains no copies of an uncountable subspace of the Sorgenfrey line, an outstanding theorem by Gruenhage [25] says us that X is a continuous image of a separable metric space and, consequently, X has a countable network.   □

In the realm of topological groups, (locally) precompact groups play an important role. In the sequel, we will study the relationship between SRP and (locally) precompact groups. Recall that a subset B of a topological group G is said to be precompact (in G) if for every neighborhood U of the identity, there is a finite subset F_U of G such that $B \subseteq \bigcup_{x \in F_U} x\, U$. Precompact groups are topological groups which are precompact in themselves. A topological group G is called locally precompact if the identity has a base of precompact neighborhoods.

(Locally) precompact groups are characterized as dense groups of (locally) compact groups. Indeed, a topological group G is (locally) precompact if, and only if, there is a unique (locally) compact group $\widehat{G}$ (up to topological isomorphisms leaving G fixed) such that G is dense in $\widehat{G}$. The topological group $\widehat{G}$ is called the Weil completion of G (see, for example, [2]).

A topological group G is said to be ω-bounded if for each open neighborhood V of the identity, there is a countable set F such that $G = FV$. The ω-bounded groups are a helpful tool in the sequel. The reason is apparent: since every space X which satisfies SRP is hereditarily Lindelöf space, a topological group G enjoying SRP is Lindelöf and, consequently ω-bounded (see [2]). The converse fails to be true.

Example 1 There exists an ω-bounded topological group that does not satisfy SRP.

Proof Uspenskij provides in [39] an example of an ω-bounded group G which has a pairwise disjoint family of nonempty open sets such that its cardinality is the continuum. Then G does not satisfy SRP. It is worth noting that G is a subgroup of a product of continuum many copies of the discrete group $\mathbb{Z}$ of the integers.   □

Related to the notion of ω-bounded groups we consider σ-compact topological groups, that is, topological groups which are the union of countable many compact subsets. The network weight of a space X is defined as the smallest cardinal number of the form $|\mathcal{N}|$ where $\mathcal{N}$ is a network for X; this cardinal number is denoted by $nw(X)$. In order to characterize σ-compact topological groups which satisfy SRP, we need the following result concerning the pseudocharacter of a space which satisfies SRP. Recall that the pseudocharacter of a point x in a space X is defined as the smallest cardinal number of the form $|\mathcal{O}|$ where $\mathcal{O}$ is a family of open subsets of X such that

$\bigcap \mathcal{O} = \{x\}$; this cardinal number is denoted by $\psi(x, X)$. The pseudocharacter of space X is defined as the supremum of all numbers $\psi(x, X)$ for $x \in X$. This cardinal number is denoted by $\psi(X)$. It is clear that, for a topological group G, the equality $\psi(G) = \psi(e, G)$ holds where e is the identity of the group G.

Theorem 10 *If X satisfies SRP, then $\psi(X) = \aleph_0$.*

Proof Let $L^*(X)$ denote the cardinal function

$$L^*(X) = \sup\{L(Y) : Y \leqslant X\}$$

where $L(Y)$ stands for the Lindelöf-degree of Y. It is known that $L^*(X) = h(X)$ where $h(X)$ denotes the height of X (see [30, Theorem 2.9]) where, by definition, $h(X)$ is the supremum taken over $|S| + \aleph_0$ where S is a right-separated subset of X. Since X is hereditarily Lindelöf, we have

$$h(X) = L^*(X) = \aleph_0.$$

The result now follows from the fact that $\psi(X) \leqslant h(X)$ in the realm of Hausdorff spaces (see [30, Theorem 2.10]). $\qquad\square$

The previous theorem will now be used to prove the following

Theorem 11 *A σ-compact topological group G satisfies SRP if and only if $nw(G) = \aleph_0$.*

Proof Suppose that G satisfies SRP. By Theorem 10 we have $\psi(G) = \aleph_0$. Since for σ-compact groups the equality $\Psi(G) = nw(G)$ holds we are done. The converse holds in general. $\qquad\square$

We now move on to locally precompact groups. The character of a group G is the smallest cardinality of a base of neighborhoods of the identity of the group. It is denoted by $\xi(G)$. We can obtain the following

Theorem 12 *A (locally) precompact bounded group G satisfies SRP if and only if G is a second countable metrizable group.*

Proof Second countable metrizable groups satisfy SRP so that we only need to show the converse. For this, suppose that G satisfies SRP. Then Theorem 10 tells us that $\psi(G) = \aleph_0$. Let $\widehat{G}$ be the Weil completion of G and let e denote the identity of G. Since $\{cl_{\widehat{G}}U\}$ where U runs over a base of neighborhoods of e is a base for the identity in $\widehat{G}$, we have $\psi(G) = \psi(\widehat{G})$. Now since G satisfies SRP, G is ω-bounded which implies that so is $\widehat{G}$ (groups with a dense ω-bounded subgroup are ω-bounded). $\widehat{G}$ being locally compact, $\psi(\widehat{G}) = \chi(\widehat{G}) = \aleph_0$. To close the proof, notice that the weight of a ω-bounded topological group coincides with $\chi(\widehat{G})$ and, consequently, G is second countable. Metrizability now follows from the classical Birkhoff–Kakutani theorem characterizing metrizable groups. $\qquad\square$

A topological space X is said to be pseudocompact if every continuous real-valued function on X is bounded. If every point of a space X has a pseudocompact neighborhood, then it is called locally pseudocompact. (Locally) pseudocompact groups are (locally) precompact (see [14, 15]). Since metrizable pseudocompact spaces are compact, we have

Corollary 7 *A (locally) pseudocompact group G satisfies SRP if and only if G is a second countable metrizable compact group.*

Remark A non metrizable σ-compact group G with $\psi(G) = \aleph_0$ provides an example of a topological group G that satisfies SRP but is not metrizable.

4   Open Problems

In [36], the authors show that a (totally disconnected) topologically ordered group is productively non-Archimedean, that is, its product with some non-discrete space Y is non-Archimedean. This fact motivates the following question which appears in [36].

(Q1) Is every productively non-Archimedean space linearly orderable?

Theorem 9 in [36] states that every locally compact, totally disconnected, topologically orderable group G is either discrete or has an open subgroup homeomorphic to the Cantor set. The following question arises in a natural way:

(Q2) Prove or disapprove: A locally compact, totally disconnected, topologically orderable group G admits a g-representable topology if and only if the cardinal of G is the continuum.

The previous question is a particular case of the following

(Q3) Characterize the totally disconnected groups which have a g-topology (respectively, a g-representable topology).

Our last question is motivated by the results obtained in Sect. 3.

(Q4) Characterize the topological groups which satisfy (SRP).

In the Abelian case, the structure theorems will be essential for (Q4).

5   Conclusion

Our framework is the realm of topologically orderable groups, that is, topological groups (G, τ) equipped with a topology induced by a total order. This class includes as a proper subclass the class of algebraically orderable groups. We generalize to the poset of group topologies the notion of a natural topology introduced by Debreu which we call a g-topology. In Sect. 2 we focus our attention on the class of non-totally disconnected groups. Among other things we show that if a group G admits a non-totally disconnected g-natural topology, then G is algebraically isomorphic to a product of the additive group of the reals and a discrete group. Moreover, we characterize the non-totally disconnected group topologies which are g-natural and also the non-totally disconnected g-representable topologies.

Section 3 is devoted to present some results related to the Sorgenfrey line and the (PFA) axiom, and also some conclusions for topological groups. We prove that if a space X satisfies the semicontinuous representability property, then X contains no copies of the Sorgenfrey line. Some consequences of this fact are obtained. For example, the Helly space does not satisfy the semicontinuous representability property. We also reveal the relationship between the semicontinuous representability property and the Sorgenfrey line. To be precise, under the (PFA) axiom, if a space X satisfies the semicontinuous representability property, then X contains no uncountable subspaces of the Sorgenfrey line. This allows us to characterize, (under the (PFA) axiom), the cometrizable spaces which satisfy the semicontinuous representability property. To close the section we characterize σ-compact topological groups and (locally) precompact groups (in particular, (locally) pseudocompact groups) which satisfy the semicontinuous representability property. In Sect. 4 we offer some questions for further research.

Acknowledgements The authors would like to thank the referees for many useful suggestions that improved the presentation of their results.

References

1. Alas, O.T., Tkachenko, M.G., Tkachuk, V.V., Wilson, R.G.: Connectedness and local connectedness of topological groups and extensions. Comment. Math. Univ. Carol. **40**, 735–753 (1999)
2. Arhangel'skii, A., Tkachenko, M.: Topological Groups and Related Structures. Atlantis Studies in Mathematics, vol. 1. Atlantis Press, Paris; World Scientific Publishing Co., Pte. Ltd., Hackensack (2008)
3. Baik, H., Hyungryul, S., Samperton, E.: Spaces of invariant circular orders of groups. Groups Geom. Dyn. **12**(2), 721–763 (2018)
4. Baumgartner, J.E.: All $\aleph_1$-dense sets of reals can be isomorphic. Fundam. Math. **79**(2), 101–106 (1973)
5. Birkhoff, G.: Lattice Theory, Revised. American Mathematical Society Colloquium Publications, vol. 25. American Mathematical Society, New York (1948)
6. Bridges, D.S., Metha, G.B.: Representations of Preferences Orderings. Lecture Notes in Economics and Mathematical Systems. Springer, Berlin (1995)
7. Candeal, J.C., Indurain, E., Mehta, G.B.: Some utility theorems on inductive limits of preordered topological spaces. Bull. Aust. Math. Soc. **52**, 235–246 (1995)
8. Candeal, J.C., De Miguel, J.R., Induráin, E.: Extensive measurement: continuous additive utility functions on semigroups. J. Math. Psychol. **40**(4), 281–286 (1996)
9. Candeal, J.C., De Miguel, J.R., Induráin, E.: Topological additively representable semigroups. J. Math. Anal. Appl. **210**, 375–389 (1997)
10. Candeal, J.C., Indurain, E., Mehta, G.B.: Order preserving functions on ordered topological vector spaces. Bull. Aust. Math. Soc. **60**, 55–65 (1999)
11. Candeal, J.C., Indurain, E., Sanchis, M.: Order representability in groups and vector spaces. Expo. Math. **30**, 103–123 (2012)

12. Cantor, G.: Beiträge zur Begründung der transfiniten Mengenlehre I. Math. Ann. **46**, 481–512 (1895)
13. Cantor, G.: Beiträge zur Begründung der transfiniten Mengenlehre II. Math. Ann. **49**, 207–246 (1897)
14. Comfort, W.W., Ross, K.A.: Pseudocompactness and uniform continuity in topological groups. Pac. J. Math. **16**, 483–496 (1966)
15. Comfort, W.W., Trigos-Arrieta, F.J.: Locally pseudocompact topological groups. Topol. Appl. **62**, 263–280 (1995)
16. Debreu, G.: Representation of a preference ordering by a numerical function. In: Thrall, R., Coombs, C., Davies, R. (eds.) Decision Processes. Wiley, New York (1954)
17. Debreu, G.: Continuity properties of Paretian utility. Int. Econ. Rev. **5**, 285–293 (1964)
18. Dierolf, S., Schwanengel, U.: Examples of locally compact non-compact minimal topological groups. Pac. J. Math. **82**, 349–355 (1979)
19. Eilenberg, S.: Ordered topological spaces. Am. J. Math. **63**, 39–45 (1941)
20. Engelking, R.: General Topology. Sigma Series in Pure Mathematics, 2nd edn., vol. 6. Heldermann Verlag, Berlin (1989)
21. Fuchs, L.: Partially Ordered Algebraical Systems. Pergamon Press, Oxford (1963)
22. Glass, A.M.W.: Partially Ordered Groups. Series in Algebra, vol. 7. World Scientific, Singapore (1999)
23. Glass, A.M.W., Marra, V.: Embedding finitely generated Abelian lattice-ordered groups: Highman's theorem and a realisation of π. J. Lond. Math. Soc. **68**, 545–562 (2003)
24. Glass, A.M.W., Macintyre, A., Point, F.: Free Abelian lattice-ordered groups. Ann. Pure Appl. Logic **134**, 265–283 (2005)
25. Gruenhage, G.: Cosmicity of cometrizable spaces. Trans. Am. Math. Soc. **313**, 301–315 (1989)
26. Hahn, H.: Über die nichtarchimedischen Größensysteme. Sitzungber. K. Akad. der Wiss. Wien. Math. Nat. Kl. (Wien. Ber.) **116**, 601–655 (1907)
27. Herrlich, H.: Ordnungsfähigkeit total-diskontinuierlicher Räume. Math. Ann. **159**, 77–80 (1965)
28. Hölder, O.: Die Axiome der Quantität und die Lehre vom Mass. Leipz. Berichte Math. Phys. **C1**(53), 1–64 (1901)
29. Iseki, K.: On simple ordered groups. Port. Math. **10**(2), 85–88 (1951)
30. Juhász, I.: Cardinal Functions in Topology–Ten Years Later. Mathematical Centre Tracts, 2nd edn., vol. 123. Mathematisch Centrum, Amsterdam (1980)
31. Manara, C., Marra, V., Mundici, D.: Lattice-ordered Abelian groups and Schauder bases of unimodular fans. Trans. Am. Math. Soc. **359**, 1593–1604 (2007)
32. Mehta, G.B., Monteiro, P.K.: Infinite dimensional utility representation theorems. Econ. Lett. **53**, 169–173 (1996)
33. Monteiro, P.K.: Some results on the existence of utility functions on path connected spaces. J. Math. Econ. **16**, 147–156 (1987)
34. Montgomery, D.: Connected one dimensional groups. Ann. Math. **40**(1), 195–204 (1948)
35. Nachbin, L.: Topology and Order. Van Nostrand Reinhold, New York (1965)
36. Nyikos, P.J., Reichel, H.-C.: Topologically orderable groups. Gen. Topol. Appl. **5**(3), 195–204 (1975)
37. Shafer, W.: Representations of Preorders on Normed Spaces. University of Southern California (1984)
38. Tkachenko, M.G., Villegas-Silva, L.: Refining connected topological group topologies on Abelian torsion groups. Topol. Appl. **84**, 77–90 (1998)
39. Uspenskij, V.V.: On the Suslin number of subgroups of products of countable groups, 23rd winter school on abstract analysis (Lhota and Rohanovem, 1995; Poděbrady, 1995). Acta Univ. Carol. Math. Phys. **36**, 85–87 (1995)
40. Venkataraman, M., Rajagopalan, M., Soundararajan, T.: Orderable topological spaces. Gen. Topol. Appl. **2**, 1–10 (1972)

Preferences in Abstract Convex Structures

Marta Cardin

Abstract Convexity of preferences is a canonical assumption in economic theory. In this paper we study a generalized definition of convex preferences that relies on the notion of a convex space, that is an abstraction of the standard notion of convexity in a linear space. We introduce also betweenness relations that characterize convex spaces. First we consider a ternary betweenness relation that gives rise to an interval space structure and then we propose a more general definition of betweenness.

Keywords Convex space · Interval space · Betweenness · Convex preference

1 Introduction

In Economic Theory it is often assumed for analytical convenience, but also in accordance with common intuition that consumer preferences are convex. Moreover convexity of preference is a condition considered in other fields (see for example [3]).

The standard Euclidean notion of convex preferences is an algebraic property used to express the notion that agents exhibit an inclination for diversification and so they prefer a more balanced bundle to bundles with a more extreme composition.

Then a preference in a subset of Euclidean space X is convex if whenever $x, y \in X$ and $x \succeq y$ then

$$tx + (1 - t)y \succeq y \text{ for all } t, \ 0 \le t \le 1.$$

In the present paper we consider abstract convex structures that are combinatorial objects studied in various areas of mathematics and we propose a general definition of convex preferences.

The paper is organized as follows. In Sect. 2, we collect definitions and basic results about abstract convex spaces and we propose some examples of convex spaces.

M. Cardin (✉)
Department of Economics, Ca' Foscari University of Venice, Venezia, VE, Italy
e-mail: mcardin@unive.it

© Springer Nature Switzerland AG 2020
G. Bosi et al. (eds.), *Mathematical Topics on Representations of Ordered Structures and Utility Theory*, Studies in Systems, Decision and Control 263,
https://doi.org/10.1007/978-3-030-34226-5_9

In Sect. 3 we introduce betweenness relations while in Sect. 4 we propose a definition of a convex preference in our framework and we study convex preferences in a class of interval spaces.

2 Abstract Convex Structures

The general notion of an abstract convexity structure studied in [13] is considered.

A family $\mathcal{C}$ of subsets of a set X is a convexity on a set X if $\emptyset$ and X belong to $\mathcal{C}$ and $\mathcal{C}$ is closed under arbitrary intersections and closed under unions of chains (totally ordered subsets). The elements of $\mathcal{C}$ are called convex sets of X and the pair $(X, \mathcal{C})$ is called a convex space. A convex set with a convex complement is called a half-space.

The convexity notion allows us to define the notion of the convex hull operator, which is similar to that of the closure operator in topology. If X is a set with a convexity $\mathcal{C}$ and A is a subset of X, then the convex hull of $A \subseteq X$ is the set

$$co(A) = \bigcap \{C \in \mathcal{C} : A \subseteq C\}.$$

A convex structure is completely determined by its hull operator, or even by its effect on finite sets (see Proposition 2.1 of [13]). This operator enjoys certain properties that are identical to those of usual convexity: for instance $co(A)$ is the smallest convex set that contains set A. It is easy to prove that C is convex if and only if $co(C) = C$.

The convex hull of a set $\{x_1, ..., x_n\}$ is called an n-polytope and is denoted by $[x_1, ..., x_n]$. A 2-polytope $[a, b]$ is called the segment joining a, b. A convexity C is called N-ary ($N \in \mathbb{N}$) if $A \subseteq \mathcal{C}$ whenever $co(F) \subseteq A$ for all $F \subseteq A$ where F has at most N elements. For a general theory of convexity we refer to [7, 8, 13].

2.1 Some Examples

We present some examples and classes of convex spaces. First of all we note that every real vector space together with the collection of all convex sets in the usual meaning, is a 2-arity convex space.

Ordered spaces The usual convexity on $\mathbb{R}$ can be defined in terms of ordering as follows: a set C is convex if and only if $a, b \in C$ and $a \leq x \leq b$ then $x \in C$. We can define in the same way a convexity on a partially ordered set (see [13], p. 6). Such a convexity is called the order convexity.

Convexity defined by orderings Let X be a non empty set and $\mathcal{P}$ a set of total preorders (complete and transitive relations) on X. We refer to [13] p. 10 for a definition of a base and a subbase of a convexity. Then we define the convexity generated by the sets $\{x \in X : x \succeq_i z\}$ where $z \in X$ and $\succeq_i$ is an element of $\mathcal{P}$ and the complement sets of these sets that are the sets $\{x \in X : x_i \prec z\}$ where $z \in X$ and $\succeq_i$ is an element of $\mathcal{P}$. So this convexity is generated by half-spaces.

Median spaces A median space is a convexity space X with a 2-arity convexity such that for each $a, b, c \in X$ there exists a unique point in $[a, b] \cap [a, c] \cap [b, c]$. We call it the median of a, b, c and denote by $m(a, b, c)$. This defines a map $m : X^3 \to X$, called the median operator on X. In any convexity space, every point in $[a, b] \cap [a, c] \cap [b, c]$ is called a median of a, b, c. There is a natural way to define the structure of a median space by means of the median operator (see [13]).

Property-based domains A property-based domain (as defined in [2]) is a pair $(X, \mathcal{H})$ where X is a non-empty set and $\mathcal{H}$ is a collection of non-empty subsets of X and if $x, y \in X$ and $x \neq y$ there exists $H \in \mathcal{H}$ such that $x \in H$ and $y \notin H$. The elements of $\mathcal{H}$ are referred to as properties and if $x \in H$ we say that x has the property represented by the subset H. This definition is slightly more general than that of [5], [9] and of [10], in fact it is not assumed that the set X is finite and we do not consider that the set H^c is a property if H is a property.

The "property space" model provides a very general framework for representing preferences, and then for aggregation of preferences. In every property-based domain we can define a convexity defined as follows. A subset $S \subseteq X$ is said to be convex if it is intersection of properties.

2.2 Lattices as Convexity Spaces

A partially ordered set L is a *lattice* if every pair of elements x, y has

(i) a *least upper bound* $x \vee y$ (called *join*), and
(ii) a *greatest lower bound* $x \wedge y$ (called *meet*);

that is

$$z \geq x \vee y \iff z \geq x \text{ and } z \geq y$$
$$z \leq x \wedge y \iff z \leq x \text{ and } z \leq y.$$

An alternative way to define a lattice is as an algebraic structure $\langle L; \wedge, \vee \rangle$ where L is a nonempty set, called *universe*, and where $\wedge$ and $\vee$ are two binary operations, called *meet* and *join*, respectively, which satisfy the following axioms:

(i) (idempotency) for every $a \in L$, $a \vee a = a \wedge a = a$;
(ii) (commutativity) for every $a, b \in L$, $a \vee b = b \vee a$ and $a \wedge b = b \wedge a$;
(iii) (associativity) for every $a, b, c \in L$, $a \vee (b \vee c) = (a \vee b) \vee c$ and $a \wedge (b \wedge c) = (a \wedge b) \wedge c$;
(iv) (absorption): for every $a, b \in L$, $a \wedge (a \vee b) = a$ and $a \vee (a \wedge b) = a$.

With no danger of ambiguity, we will denote lattices by their universes. As it is well-known, every lattice L constitutes a partially ordered set endowed with the partial order $\leq$ given by: for every $x, y \in L$, write $x \leqslant y$ if $x \wedge y = x$ or, equivalently, if

$x \vee y = y$. If for every $a, b \in L$, we have $a \leqslant b$ or $b \leqslant a$, then L is said to be a *chain*. A lattice L is said to be *bounded* if it has a least and a greatest element, denoted by 0 and 1, respectively.

A *filter* of a lattice L is a nonempty subset F such that

(i) if $x \in F$ and $x \leq y$ then $y \in F$,
(ii) $x, y \in F$ then $x \wedge y \in F$.

Sets satisfying Condition (i) of a filter are called upsets. The dual notation is that of an *ideal*. If $a \in L$ we define the *principal filter* generated by x as $\uparrow x = \{y \in L : y \geq x\}$. It is easy to prove that $\uparrow x$ is a filter for every $x \in L$. It can be proved that in a finite lattice each filter and each ideal are principal.

A *proper filter* is a filter that is neither empty nor the whole lattice while a *prime filter* is a proper filter P such that if $x \vee y \in P$ then $x \in P$ or $y \in P$. If $\langle L, \wedge, \vee \rangle$ is a lattice we denote by $\mathcal{L}$ and $\mathcal{U}$ the collections of all ideals and all filters respectively (the empty set and the whole lattice are treated as (non-proper) ideals and filters). Since the union of a chain of filters (ideals) is a filter (ideal), these are two convexities on L that will be called the lower and the upper lattice convexity respectively. Moreover there exists a convexity $\mathcal{C}$ generated by $\mathcal{L} \bigcup \mathcal{U}$ the least convexity containing all ideals and filters. This convexity will be called the lattice convexity on L.

Note that if L is linearly ordered then G equals the order convexity. The convexity of the dual lattice is the same as the original one.

2.3 Separation Axioms

Two nonempty sets are separated by a set A if one is contained in A and the other one is disjoint from A.

We shall consider the following separation axioms (see [13] p. 53):

S_0: For every two distinct points there exists a convex set which contains exactly one of them.
S_1: Every one-point subset is convex.
S_2: Distinct points are separated by half-spaces
S_3: Every convex set is an intersection of half-spaces
S_4: For every two disjoint convex sets there exists an half-space which contains exactly one of them.

A S_1 convex space is called point convex while axiom S_4 is called the Kakutani separation property. In fact the classical theorem of Kakutani says that each two disjoint convex sets in a real vector space can separated by a half- space. It can be proved that a median space satisfies the Kakutani property and that a lattice satisfies the Kakutani property if and only if it is a distributive lattice.

It is also clear that $S_4 \implies S_3$, $S_2 \implies S_1 \implies S_0$, and $S_1 + S_3 \implies S_2$.

We denote by $\mathcal{H} \subseteq \mathcal{C}$ the set of half spaces of the convex space $(X, \mathcal{C})$.

3 Betweenness

3.1 Ternary Betweenness and Interval Spaces

The notion of a point lying between two given points on a geometric line or a totally ordered set has strong intuitive appeal and has been generalized in a number of directions. In all of these, betweenness is taken to be a ternary relation that satisfies certain conditions.

The ternary relation of betweenness comes up in different structures on a given set, reflecting intuitions that range from order-theoretic to the geometrical and topological settings with different meanings.

A modern axiomatic definition of betweenness is due to Hedlíková [6] who introduced the ternary representation of betweenness. These relations have been introduced in the context of abstract convexity in [13], in the context of property spaces (see for example [9, 10]) as well as in graph theory (see [11]).

In a lattice L is defined a ternary betweenness relation

$$B = \{(x, z, y) \in L^3 : x \wedge y \leq z \leq x \vee y\}.$$

This ternary relation satisfies the following properties:

[B1] (Reflexivity) If $z \in \{x, y\}$ then $B(x, z, y)$

[B1] (Symmetry) If $B(x, z, y)$ then $B(y, z, x)$

[B1] (Transitivity) If $B(x, x', y)$, $B(x, y', y)$ and $B(x', z, y')$ then $B(x, z, y)$.

Let us assume that these properties characterize a ternary betweenness relation. There is a close link between interval spaces defined below and the relation of betweenness. We introduce interval spaces $(X, I(x, y))$ (see [14] and the references therein) namely a set X and a function $I : X^2 \to \mathcal{P}(X)$ that satisfy the following properties:

[I1] (Extension) $\{x, y\} \subseteq I(x, y)$.

[I2] (Symmetry) $I(x, y) = I(y, x)$.

[I3] (Convexity) If $\{x', y'\} \subseteq I(x, y)$ then $I(x', y') \subseteq I(x, y)$.

We could consider interval spaces that also satisfy the following property:

[I4] (Idempotence) $I(x, x) = x$.

The following proposition can be easily proved.

Proposition 1 *If it is defined a ternary betweenness on the set X, then the function*

$$I(x, y) = \{z \in X \mid B(x, z, y)\}$$

defines an interval space $(X, I(x, y))$.

If $(X, I(x, y))$ is an interval space, then the ternary relation

$$B(x, z, y) \quad \text{if and only if} \quad z \in I(x, y)$$

defines a ternary betweenness in X.

Note that an interval space $(X, I(x, y))$ is a convex space such that a set $C \in X$ is convex if and only if $x, y \in C$ implies that $I(x, y) \subseteq C$ and then interval spaces are 2-ary convexities.

3.2 Betweenness in Convex Structures

Let us consider a more general definition of betweenness as a binary relation involving points and finite subsets of a non empty set X.

We introduce this relation by considering for every $k \in \mathbb{N}$ a k-ary interval function $I_k \colon X^k \to \mathcal{P}(X)$ such that

F1 $x_i \in I_k(x_1, \ldots, x_k)$ for every k, $1 \leq i \leq k$.
F2 I_k is a symmetric function for every $k \in \mathbb{N}$.
F3 if $y_1, \ldots, y_h \in I_k(x_1, \ldots, x_k)$ then $I_h(y_1, \ldots, y_h) \subseteq I_k(x_1, \ldots, x_k)$.

The set $I_k(x_1, \ldots, x_k)$ is called a generalized k-interval and we say that an element x in X is between the elements $x_1, \ldots, x_k$ if $x \in I_k(x_1, \ldots, x_k)$. Then we assume that no point is between the empty set and we can describe also betweenness as binary relation involving points and finite subsets of a given set (see [13]). We say that an element x is between a finite set $\{x_1, \ldots, x_k\}$ of elements of X if $x \in I_k(x_1, \ldots, x_k)$.

A nonempty set X endowed with a family of k-ary functions $I_k \colon X^k \to \mathcal{P}(X)$ that satisfy properties F1–F3 is called a betweenness space.

The following result proves that a convex space is completely characterized by its betweenness relation.

Proposition 2 *Let X a nonempty set and $\{I_k : k \in \mathbb{N}\}$ a family of functions $I_k \colon X^k \to \mathcal{P}(X)$ that satisfy the properties F_1, F_2 and F_3.*

Then there exists a convexity $\mathcal{C}$ on X such that

$$x \in I_k(x_1, \ldots, x_k) \iff \left[\text{for all} \ \ C \in \mathcal{C} : x_1, \ldots, x_k \in C \implies x \in C \right]. \quad (1)$$

Conversely, if $(X, \mathcal{C})$ is a convex space then the functions $I_k \colon X^k \to \mathcal{P}(X)$ defined by

$$I_k(x_1, \ldots, x_k) = co(\{x_1, \ldots, x_k\}) \quad (2)$$

satisfy the properties F_1, F_2 and F_3.

Proof Let $\{I_k : k \in \mathbb{N}\}$ be a family of functions $I_k \colon X^k \to \mathcal{P}(X)$ that satisfy the properties F_1, F_2, and F_3. Then we define a convexity $\mathcal{C}$ on X where $C \in \mathcal{C}$ if and only if for every k $x \in I_k(x_1, \ldots, x_k)$ and $x_1, \ldots, x_k \in C$ then it holds that $x \in C$. We can easily prove that $\emptyset$ and X belong to $\mathcal{C}$ and that the intersection of elements of $\mathcal{C}$ is an element of $\mathcal{C}$.

Moreover, if we consider a chain $C_1 \subseteq C_2 \ldots \subseteq C_i \ldots$ of elements of $\mathcal{C}$ such that $\{x_1, \ldots, x_k\}$ is contained in $\bigcup_i C_i$, there exists j such that $\{x_1, \ldots, x_k\} \subseteq C_j$. Hence we can get that $I_k(x_1, \ldots, x_k) \subseteq C_j \subseteq \bigcup_i C_i$.

Now we have to prove that if for all $C \in \mathcal{C}$ if $x_1, \ldots, x_k \in C$ implies that $x \in C$ then $x \in I_k(x_1, \ldots, x_k)$. This can be proved by just observing that by property F3 $I_k(x_1, \ldots, x_k)$ is an element of the convexity $\mathcal{C}$. The second part can be easily verified. □

4 Convex Preferences

4.1 Basic Definitions

In this section we consider convex spaces $(X, \mathcal{C})$ that satisfy axioms S_1 and S_4. We propose a definition of convex preferences in abstract convex structures and we do not restrict ourselves to consider only complete relations as in [12].

In this paper a preference on a nonempty set X is a transitive and reflexive binary relations on X. A preference relation on a convex space $(X, \mathcal{C})$ is $\succeq$ on X is said to be a *convex preference* if $\{x \in X : x \succeq z\}$ is a convex set for every $z \in X$.

Example 1 If the convex space is an Euclidean space our definition is the well known definition of convexity of a preference.

Example 2 If $(X, \mathcal{C})$ is a convex space we consider the relation $\succeq$ in X such that $x \succeq y$ if and only if

$$\{H \in \mathcal{H}' : x \in H\} \supseteq \{H \in \mathcal{H}' : y \in H\},$$

where $\mathcal{H}' \subseteq \mathcal{H}$. This relation is transitive and complete and it can be proved that the set $\{x \in X : x \succeq z\}$ is an intersection of half-spaces and so is a convex set for every $z \in X$. Hence the considered relation is a convex preference.

The following result generalizes in our framework the well known property of convex preferences in Euclidean spaces.

Proposition 3 *Let $(X, \mathcal{C})$ be a convex space. If $\succeq$ is a convex preference in X then for every $x, y \in X$*

$$if \quad x \succeq y \quad and \quad z \in co\{x, y\} \quad then \quad z \succeq y.$$

Proof If we consider a convex preference $\succeq$ in $(X, \mathcal{C})$ and two elements x, y of X such that $x \succeq y$ then obviously x, y belong to the convex set $\{t \in X : t \succeq y\}$. Then if $z \in co\{x, y\}$ that is the smallest convex set that contains x and y. Note that z is an element of the convex set $\{t \in X : t \succeq y\}$ and then we get $z \succeq y$. □

4.2   Convex Preferences in Interval Spaces

The following result considers complete preference in interval spaces.

Proposition 4 *Let $(X, I(x, y))$ be an interval space. A transitive and complete relation $\succeq$ in X is a convex preference if an only if for every $x, y \in X$*

$$\text{if } x \succeq y \text{ and } B(x, z, y) \text{ then } z \succeq y.$$

Proof Let $\succeq$ be a a a convex preference in $(X, I(x, y))$. If x, y and two elements of X such that $x \succeq y$ then obviously x, y belong to the convex set $\{t \in X : t \succeq y\}$. Moreover if $B(x, z, y)$ holds, and z is element of X then we have that $z \in I(x, y)$ that is the smallest convex set that contains x and y.

Then if $B(x, z, y)$ is satisfied, z is an element of the convex set $\{t \in X : t \succeq y\}$ and then we get $z \succeq y$.

Conversely let $(X, I(x, y))$ an interval space and $\succeq$ a transitive and reflexive relation in X such that for every $x, y \in X$, if $x \succeq y$ and $B(x, z, y)$ then $z \succeq y$. Then if x_1, x_2 are elements of X such that $x_1 \succeq y$ and $x_2 \succeq y$ we can suppose that $x_1 \succeq x_2$. Hence $I(x_1, x_2)$ is contained in the set $\{t \in X : t \succeq y\}$ since if $z \in I(x_1, x_2)$ then $z \succeq x_2 \succeq x$.   □

We are now ready to state the main result of this paper concerning the representation of convex preferences in interval spaces by means of a set of complete preferences. Note that in standard convexity the orderings are represented by linear functions. In abstract convexity linear orderings are criteria used to evaluate alternatives.

We consider interval spaces that satisfy some additional properties, and that are named convex geometries in [4].

Proposition 5 *Let $(X, I(x, y))$ be an idempotent interval space that satisfies the following properties:*

(i) for all x, y, z, z_1, z_2 elements of X if $y \in I(z_1, z_2)$ and $z \in I(y, z_2)$ then there exists $t \in I(x, z_1)$ such that $z \in I(x, t)$;

(ii) for all x, y, z, t, z elements of X such that $x \neq y$ if $x \in I(y, t)$ and $y \in I(x, z)$ then $x \in I(t, z)$.

If $\succeq$ in X is a convex preference there exists a set $\{\succeq_i : i \in I\}$ of complete preferences such that if $z \in X$, for every $i \in I' \supseteq I$ there exists $y_i \in X$ such that

$$x \succeq z \quad \text{if and only if} \quad x \succeq_i y_i \quad \text{for every} \quad i \in I'. \tag{3}$$

Proof Note that properties (i) and (ii) are properties (C) and (L2) in [4] respectively. Then by Proposition II.10 in [4] $(X, I(x, y))$ is a convex space that satisfies the anti-exchange property (see [1, 4] for a definition of anti-exchange property). By Theorem 23 of [1] the interval convexity of $(X, I(x, y))$ is a convexity defined by orderings. Then this convexity is generated by half-spaces and satisfies separation axiom S_3.

If $z \in X$ the set $\{x \in X : x \succeq z\}$ being a convex set it is an intersection of half-spaces. Hence if we consider an element $z \in X$ is an intersection of half-spaces of type $\{x \in X : x \succeq_i y_i\}$, $i \in I$ and then Eq. (3) is satisfied. $\qquad\Box$

5 Concluding Remarks

This paper is a first step toward the study of a general notion of convex preferences in abstract structures. We consider an abstract notion of convexity in a base set and we study this notion when the convexity is an interval convexity or a lattice convexity.

Then we propose a general definition of a convex preferences. Note that there are many examples of economic models that consider convex spaces and convex preferences (see [12]). We prove that if the convex structure satisfies a set of properties a convex preference is defined by a set of orderings that in some sense play a role of linear functions in Euclidean spaces.

We plan to study convex preferences in general convex spaces, and to find more applications of our results in future work.

References

1. Adaricheva, K.: Algebraic convex geometries revisited. arXiv:1406.3721
2. M. Cardin, Sugeno integral on property-based preference domains. In: Kacprzyk, J., Szmidt, E., Zadrożny, S., Atanassov, K., Krawczak, M. (eds.) Advances in Fuzzy Logic and Technology 2017, EUSFLAT 2017. Advances in Intelligent Systems and Computing, vol. 641 (2018)
3. Chateauneuf, A., Tallon, J.-M.: Diversification, convex preferences, and non-empty core in the Choquet expected utility model. Econ. Theory **19**, 509–523 (2002)
4. Coppel, W.A.: Foundations of Convex Geometry. Cambridge University Press, Cambridge (1998)
5. Gordon, S.: Unanimity in attribute-based preference domains. Soc. Choice Welf. **44**, 13–29 (2015)
6. Hedlíková, J.: Ternary spaces, media and Chebyshev sets. Czechoslovak Math. J. **33**, 373–389 (1983)
7. Kubis, W.: Abstract convex structures in topology and set theory. Ph.D. thesis, University of Silesia, Katowice, Poland (1999)
8. Llinares, J.V.: Abstract convexity, some relations and applications. Optimization **51**(6), 797–818 (2002)
9. Nehring, K., Puppe, C.: The structure of strategy-proof social choice—part I: general characterization and possibility results on median spaces. J. Econ. Theory **135**(1), 269–305 (2007)
10. Nehring, K., Puppe, C.: Abstract Arrowian aggregation. J. Econ. Theory **145**, 467–494 (2010)
11. Rautenbach, D., Schäfer, P.M.: Strict betweennesses induced by posets as well as by graphs. Order **28**, 89–97 (2011)
12. Richter, M., Rubinstein, A.: "Convex preferences": a new definition. mimeo (2018)
13. van de Vel, M.L.J.: Theory of Convex Structures North-Holland Mathematical Library, vol. 50. Elsevier, Amsterdam (1993)
14. Vannucci, S.: Weakly unimodal domains, anti-exchange properties, and coalitional strategy-proofness of aggregation rules. Math. Soc. Sci **84**, 50–67 (2016)

Strictly Monotonic Preferences

Carlos Hervés-Beloso and Paulo K. Monteiro

Abstract Monotonicity assumptions of preferences are natural and useful. A strictly monotonic preference is such that an increase in even only one commodity consumption is always strictly preferred. However, when we consider a continuum of commodities, it is not easy to find examples of strictly monotonic preferences. We survey some previous results in order to show that purely strictly monotonic preferences always exist but, if the commodity space is rich enough, they cannot be continuous in any linear topology defined on the consumption set and they cannot be represented by a utility function.

Keywords Strictly monotonic preferences · Pure monotonicity · Continuum of commodities · Continuity of preferences · Utility representation

JEL Classifications D11 D50

1 Introduction

In this Chapter we elaborate on some previous results about preferences defined in a set X, that represent the set of alternatives of a decision maker, the strategies of a player in a game or the consumption set of any agent in an economy. We will address existence, continuity and utility representation of such preferences. Depending on the properties of the preference we consider, we will assume that X is a topological space, a metric space, a subset of a linear space or a subset of a partially ordered space.

C. Hervés-Beloso (✉)
RGEAF-ECOBAS, Universidad de Vigo, Vigo, Spain
e-mail: cherves@uvigo.es

P. K. Monteiro
FGV EPGE, Rio de Janeiro, Brazil
e-mail: Paulo.Klinger@fgv.br

© Springer Nature Switzerland AG 2020
G. Bosi et al. (eds.), *Mathematical Topics on Representations of Ordered Structures and Utility Theory*, Studies in Systems, Decision and Control 263,
https://doi.org/10.1007/978-3-030-34226-5_10

The existence of a utility function representing a preference is closely related with the properties of the set of alternatives to which we often refer as the consumption set. As we focus on continuous preferences, the topology of the consumption set will play a relevant role. In this scenario, the existence of utility functions is guaranteed by Eilenberg-Debreu Theorem, as soon as the underlying topological space is connected and separable or second countable.

Several authors have elaborated on generalizations of the Eilenberg–Debreu Theorem providing conditions for the existence of utility representation. We refer to Fleischer [13], Nachbin [26], Peleg [28], Jaffray [19], Mehta [24–26], Richter [29], Herden [14, 15]. See Bridges and Mehta [4] for a complete summary of this contributions and Stigler [33, 34] for a historical analysis of utility theory.

However, separability of the consumption set play a relevant role. As Estévez and Hervés [12] have shown, in every non-separable metric space, there are continuous preferences that are not representable by a utility function. Yet, these non-representable preferences fail to have some properties frequently assumed in the literature. In particular, if the consumption set is a convex subset of an ordered linear space as the positive cone of a Banach lattice, they are neither convex nor monotonic.

The preference relation of a consumer or a decision maker, defined in a (partially) ordered set, is monotonic if more is better and it is strictly monotonic if an increase in even only one commodity[1] consumption is always strictly preferred. In the case of finite or countably many commodities, it is easy to find examples of continuous and strictly monotonic preferences. Yet, it is not an easy task to give examples of strictly monotonic preferences defined on general commodity spaces involving uncountably many commodities. Moreover, in this situation, a representative commodity space, as $B([0, 1])$, the space of bounded functions defined on [0, 1], fails to be separable and thus, the classical utility representation results do not apply.

In Sect. 2.1 we set the notations and provide the background. The positive and negative results on preference representation are presented in Sect. 2.2. Since our objective focuses on the case of uncountably many commodities, we survey the findings by Monteiro [25] and the generalization provided by Candeal et al. [6] which give necessary and sufficient conditions for a continuous preference defined on a non-separable topological space to be representable by a utility function.

In Sect. 3, we provide examples of strictly monotonic and continuous utility functions defined on particular subspaces of the space of functions defined on the set of commodities $K = [0, 1]$. Moreover we stress that, for any set of commodities K, and any consumption set, there are strictly monotonic preferences.

---

[1]Following Debreu [9], a commodity is a good or a service completely specified physically, temporally, and spatially. The same good or service in different dates or locations is a different commodity. The date, the location or the quality of commodities could be treated as continuous variables. A consumption plan, or a bundle, is a specification for each commodity of the quantity that she will make available or that will be made available to her, i.e, a complete listing of the quantities of her inputs and of her outputs. The commodity space is the vector space that contains all possible consumption plans. The consumption set, for a given consumer, is the set of consumption plans that are available for that consumer.

Our final results are stated in Sect. 4. We show that strictly monotonic preferences on the positive cone of spaces like the Banach space of all bounded function defined in $[0, 1]$ are neither representable by utility functions nor continuous in any linear topology. We revisit Hervés-Beloso and Monteiro [17] to extend this result to *rich-enough* consumption subsets of the space of functions defined on any non-countable set of commodities.

2 Notation, Definitions and Background

2.1 Definitions and Notation

Let X be a set. A subset $R \subset X \times X$ is a binary relation. We write xRy if $(x, y) \in R$.

Definition 2.1 The binary relation, R on X is:

(a) reflexive if xRx for all $x \in X$;
(b) transitive if xRy and yRz implies xRz;
(c) antisymmetric if xRy and yRx implies $x = y$;
(d) complete (or total) if for every x, y in X either xRy or yRx;
(e) a partial order if is reflexive, transitive and antisymmetric.

An order is a complete partial order.

Definition 2.2 The ordered set $(K, \geqslant)$ is well ordered if for every non-empty subset $A \subset K$ there is $a \in A$ such that $x \geqslant a$ for every $x \in A$.

That every set can be well ordered is Zermelo's theorem.[2]

A reflexive and transitive binary relation $\succsim$ on X is a preference relation on X. If $(f, g) \in \succsim$, we write $f \succsim g$. Thus:

Reflexivity $f \succsim f$ for all $f \in X$;
Transitivity   If $f \succsim g$ and $g \succsim h$, then $f \succsim h$.

We read $f \succsim g$ as f is at least as preferred as g. Moreover, for $f, g \in X$ we write $f \succ g$, and we read f is more preferred than g, if $f \succsim g$ but $g \not\succsim f$ and $f \sim g$, and we read f is indifferent to g, if $f \succsim g$ and $g \succsim f$.

A preference $\succsim$ defined on a convex subset X of a linear vector space is convex if, for any two points $f, g \in X$ and $\lambda \in (0, 1)$, $f \succsim g$ implies $\lambda f + (1 - \lambda)g \succsim g$, and strictly convex if $f \succsim g$ and $\lambda \in (0, 1)$ imply $\lambda f + (1 - \lambda)g \succ g$.

Let $(X, \geqslant)$ be a partially ordered set. A preference $\succsim$ defined on X, is monotonic if, for $f, g \in X$, $f \geqslant g$ implies $f \succsim g$. The preference $\succsim$ is strictly monotonic if $f \geqslant g, f \neq g$, implies $f \succ g$.

---

[2]Or Zermelo's axiom as suggested by a referee. It is also called the well ordering principle. It is equivalent to the axiom of choice. See Kelley [20] page 33 or Chap. 4 of Ciesielski [7].

The topological space (X, τ) is said separable if it contains a countable subset whose closure is X. That is, there exists a sequence $Q = \{q_1, \ldots, q_n, \ldots\}$ such that $Q \cap V \neq \emptyset$ for every non-empty open set $V \subset X$. The topological space (X, τ) is second countable or perfectly separable if τ admits a countable basis of open sets. Every perfectly separable topological space is separable, and every separable metric space is perfectly separable.[3] A topological space (X, τ) is connected if there is no partition of X into two disjoint, non-empty closed sets. Also, X is path-connected if for all $f, g \in X$ there exists a continuous function $f : [0, 1] \rightarrow X$ with $f(0) = f$ and $f(1) = g$. Note that every path-connected space is connected and every convex set in a linear topological space is path-connected.

A preference $\succsim$ defined on a topological space (X, τ) is continuous if, for all $f \in X$, the sets $L_f = \{g \in X : f \succsim g\}$ and $U_f = \{g \in X : g \succsim f\}$ are τ-closed in X and the sets $\mathring{L}_f = \{g \in X : f \succ g\}$ and $\mathring{U}_f = \{g \in X : g \succ f\}$ are open sets for all $f \in X$. Note that if $\succsim$ is complete, then, for all $f \in X$, the sets U_f and L_f are closed if and only if the sets $\mathring{U}_f$ and $\mathring{L}_f$ are open. However, for example, in the Pareto order in $\mathbb{R}^n$, given by $a \succsim_P b$ if and only if $a_i \geqslant b_i$ for all $i = 1, \ldots, n$ we have that U_a and L_a are closed sets for all $a \in \mathbb{R}^n$, but $\succsim_P$ is not continuous since neither $\mathring{U}_a$ nor $\mathring{L}_a$ are open sets.

The following theorem shows that continuous preferences defined on connected topological spaces are automatically complete:

Theorem 2.1 (Schmeidler [30]) *A continuous preference, $\succsim$, defined on a connected space X, and non trivial (there are f and $g \in X$, such that $f \succ g$), is complete.*

Thus, in order to consider non-trivial continuous preferences, i.e., reflexive, transitive and continuous binary relations, we must assume that they are complete.

If $U : X \rightarrow \mathbb{R}$ is a function then $\succsim_U := \{(f, g) \in X^2 : U(f) \geq U(g)\}$, that is, $f \succsim_U g$ if and only if $U(f) \geqslant U(g)$, is a complete preference relation on X. Also $f \succ_U g$ if and only if $U(f) > U(g)$. A preference relation is representable by a utility function if there is a $U = U_{\succsim} : X \rightarrow \mathbb{R}$ such that $f \succsim g$ if and only if $U(f) \geq U(g)$.

2.2 Existence of Utility Representation

The representability of a preference relation by a utility function was taken first as a cardinal concept to measure a consumer's well-being by Pareto [27]. However, its relevance was recognized much later by Slutsky [31] and especially by Wold [36], who listed a number of axioms (or conditions) that preference must meet in order to guarantee the existence of a real-valued utility representation.

One basic requirement of a utility function in applications to consumer theory is that the utility function be continuous. For continuous preferences, the positive

---

[3]See Kelley [20], pp. 48–49.

results are very general. The **Eilenberg-Debreu Theorem**, establish that a continuous preference defined on a connected and separable, or second countable topological space (X, τ), has a continuous utility representation.

Theorem 2.2 (Debreu [8])

(a) *Let X be connected and separable. Then any continuous preference relation on X has a utility representation.*
(b) *Let X be a perfectly separable space. Then any continuous preference relation on X has a utility representation.*

The Debreu [8], (see also [10]), contribution on the existence of continuous utility representation of preferences is based on an earlier work by Eilenberg [11], that shows the existence of utility for a continuous total order in connected and separable spaces. For connected and separable spaces Debreu extended to preference relations the Eilenberg's result for total orders. Without requiring connectedness, but strengthening separability, Debreu's result shows existence of a continuous utility for continuous preferences defined on second countable topological spaces.

Note that for any utility function U representing a preference $\succsim$, if ψ is a strictly increasing function $\psi : \mathbb{R} \to \mathbb{R}$, the function $V = \psi \circ U$ is also a utility function for $\succsim$. Thus, a continuous preference can be represented by a utility function V that fails to be continuous but, using the gap theorem [10], it is shown the existence of a continuous function U representing the same preference.

Many authors elaborated on the problem of preference representability. We refer to Bridges and Mehta [4], for a survey of different approaches and also to Herden [16] for the connections between these approaches.

The most commonly known non representable preference is the lexicographic order $\succsim_L$ defined, for example, in $X = [0, 1] \times [0, 1]$ which is, in fact, a total order. Note that $\succsim_L$ is not continuous when one considers the Euclidean topology on X. However, it is continuous in the order space $(X, \succsim_L)$, but this topological space is non-separable.

Thus, connectedness and separability or second countability guarantee the existence of a utility representation of a continuous preference. Note that second countability is a hereditary property that, in particular, implies that any continuous preference defined on an arbitrary subset of any separable metric space is representable. Yet, in some interesting economic applications the set of alternatives could be non-separable. For example, in a situation where alternatives are renewable natural resources in continuous time or with an infinite temporal horizon, the decision-maker deals with infinitely many commodities in the commodity space L_∞ or l_∞ (see Bewley [3], Araujo [1]). In the case where one consider an infinite degree of commodity differentiation, the space of alternatives, following Mas-Colell [21], is $ca(K)$, the space of countably additive signed measures over the compact space of commodities K. The respective positive cones of the Banach spaces $ca(K)$, L_∞, l_∞, that usually play the role of consumption set, are non-separable metric spaces.

Moreover, given any non-separable metric space (X, d), Estévez and Hervés [12] show that there are continuous preference relations defined on X which cannot be represented by a utility function. To show this general non-existence result, it is used the characterizing property of non-separable metric spaces. In any non-separable metric space (X, d) there is an uncountable set $I \subset X$ and a real number $\epsilon > 0$ such that for every two different points $f, g \in I, d(f, g) > 2\epsilon$. They define a preference $\succsim$ such that each point $f \in I$ is the most desirable point in a ϵ neighborhood of f, and for any point g, outside the ϵ neighborhood of any point $f' \in I$ we have $f' \succ g$. By using the *long line*,[4] it is shown that $\succsim$ is non-representable by any utility function.

Yet, the class of non-representable preferences that guarantees the above result does not fulfill neither monotonicity, if X is a partially ordered metric space, nor convexity, when X is a convex set, nor local insatiability.[5] However, Monteiro [25] provides an example[6] of a convex, monotone and continuous preference defined on a closed convex subset of a Banach lattice that has no utility representation.

Given a preference $\succsim$ defined on X, a subset $F \subset X$ bounds $\succsim$ if for any $f \in X$, there are points $q_f, q^f \in F$ such that $q_f \succsim f \succsim q^f$. We say that $\succsim$ is countably bounded if there is a finite or countable set $F \subset X$ that bounds $\succsim$. It is easy to show that countable boundedness is a necessary condition for a preference to be representable by a utility function. Moreover, Monteiro [25] showed that it is also sufficient to guarantee the existence of a continuous utility representation for continuous preferences defined on the usual consumption sets. More precisely it is shown that:

If (X, τ) is path connected, any countably bounded continuous preference relation has a continuous utility function (Theorem 3, page 150).

In fact, this result is based on Theorem 1 in Monteiro [25] that shows that if there is $F \subset X$ connected and separable which bounds $\succsim$, then $\succsim$ has a continuous utility representation. Following this idea, Candeal et al. [6] consider the following definition:

A topological space (X, τ) is separably connected if and only if, given any two points $f, g \in X$ there is a connected and separable set, $F_{f,g} \subset X$ such that $f, g \in F_{f,g}$.

A separably connected space is connected and a path-connected topological space is separably connected because every path is connected and separable. However, not every separably connected space is path-connected.

Theorem 4 in Candeal et al. [6], shows that a continuous preference defined on a separably connected topological space is representable by a continuous utility function if and only if it is countably bounded. Countable boundedness is sometimes easy to get. For example a continuous preference on a countable union of compact sets is countably bounded. The weak* topology on the dual of a Banach space B, $\sigma(B', B)$, has this property. However the weak topology of a Banach space, $\sigma(B, B')$, is not weakly countably compact in general. Campión et al. [5] shown that even in this case, every weakly continuous preference relation is countably bounded.

---

[4]See Steen and Seebach [32], pp. 71–72.

[5]The preference $\succsim$ is locally insatiable if, for any point g in the consumption set X, and for any neighborhood V of f, there is another consumption $f \in V$ such that $f \succ g$.

[6]This is Theorem 6 on page 153 of Monteiro [25].

For examples of separably connected topological spaces that are neither path-connected nor separable we refer to Candeal et al. [6]. Actually, it was conjectured that any connected subset of a linear or metric space is separably connected, but Aron and Maestre [2] provide a counterexample to this conjecture. See also Wójcik [35].

On the other hand, if X is totally ordered by $\geqslant$ and we consider the order topology, then $(X, \geqslant)$ is separably connected if and only if it is path-connected.

Consider the lexicographic order $\succsim_L$ defined in $\mathbb{R}^n$. Note that $\succsim_L$ is countably bounded and if we consider the order topology in $\mathbb{R}^n$, $\succsim_L$ is a continuous preference. However the ordered space $(\mathbb{R}^n, \succsim_L)$ is not path-connected.

3 Strictly Monotonic Preferences

The property of monotonicity of preferences represents the idea that more is better. A preference $\succsim$ defined on a partially ordered set $(X, \geqslant)$ is monotonic if $f \geqslant g$ implies $f \succsim g$ and it is strictly monotonic if $f \geqslant g$ and $f \neq g$ implies $f \succ g$. In consumer theory, preferences are defined on the consumption set X. An element $f \in X$ represents a consumption plan that specifies the units, $f(k)$, of each commodity k that the consumer chooses to consume. Without loss of generality we assume that K is the set of commodities and $X \subset \{f : K \to \mathbb{R}_+\}$ is a set of non-negative functions defined on K. Thus, we can consider the standard partial order on X, that is $f \geqslant g$ if and only if $f(x) \geqslant g(x)$ for all $x \in K$. The following is Theorem 2 in Hervés-Beloso and Monteiro [17]:

Theorem 3.1 *Given a set K and any subset $X \subset \{f : K \to \mathbb{R}\}$, there are strictly monotonic preferences defined on X.*

Given any set K, let $\geqslant^*$ be a well-ordering of K. Let $f, g \in X$, $f \neq g$ and let $k_o = \min\{k \in K; f(k) \neq g(k)\}$. If $f(k_o) > g(k_o)$ we define $f \succ_L g$ and otherwise, we define $g \succ_L f$. Note that $f, g \in X$, $f \geqslant g$, $f \neq g$ implies $f \succ_L g$. The relation $\succsim_L$ defined by $f \succsim_L g$ if $f \succ_L g$ or $f = g$ is a strictly monotonic preference (in fact it is a strictly monotonic total order) on X.

When K is a finite set with n points we have that $X \subset \mathbb{R}^n$ and in case where K is a countable set, X is a subset of the space of sequences of real numbers. Note that $\succsim_L$ is the restriction of the lexicographic order in $\mathbb{R}^n$, respectively $\mathbb{R}^{\mathbb{N}}$ to X, which is neither continuous nor representable by a utility function.

It is very easy to find examples of continuous strictly monotonic preferences, representable by utility functions. In the countable case, the Banach spaces of bounded sequences, l_∞, and absolutely summable sequences, l_1, have been used to represent, respectively, economies with renewable and non-renewable resources. Consider $X \subset l_\infty^+$; given any sequence $\rho = (\rho_n)_{n \in \mathbb{N}}$, where $\rho_n > 0$ for all n, and $\sum_{n=1}^{\infty} \rho_n < +\infty$, the preference relation defined in X by $f \succsim_\rho g$ if and only if $u_\rho(f) \geqslant u_\rho(g)$, where $u_\rho(h) = \sum_{n=1}^{\infty} \rho_n h(n)$, is strictly monotone, continuous in both the norm and the weak* topology $\sigma(l_\infty, l_1)$, and, by definition, representable by the utility function u_ρ. Similarly, if the consumption set $X \subset l_1^+$, given any bounded sequence $b = (b_n)_{n \in \mathbb{N}}$, the preference relation defined in X by $f \succsim_b g$ if and only if $u_b(f) \geqslant u_b(g)$, where $u_b(h) = \sum_{n=1}^{\infty} b_n h(n)$, defines a continuous and strictly monotonic preference on X.

When K is an uncountable set, for instance $K = [0, 1]$, the commodity spaces like $C(K)$, the space of continuous functions defined on the compact set K, $L_p(K)$, with $1 \leqslant p < \infty$, the spaces of classes of functions for which the p-th power of the absolute value is Lebesgue integrable, $L_\infty(K)$, the space of classes of essentially bounded measurable functions defined on K, or the space $B(K)$ of bounded functions defined on K have been considered in the literature. For consumption sets contained in the positive cone of the spaces $C(K)$ or $L_p(K)$, it is easy to find examples of strictly monotonic preferences, representable by utilities. For example, $U(f) = \int f$, or respectively, $U(f) = \int f^p$ defines a strictly monotonic preference in $C(K)^+$ and $L_\infty(K)^+$, respectively in $L_p(K)^+$.

Moreover, we observe that given two continuous functions f, g defined, for example, in $K = [0, 1]$, such that $f \geqslant g$ and $f \neq g$, we have $f(k) > g(k)$ for uncountably many points $k \in K$. The same happens if f, g represent classes of integrable functions. Thus, when K is uncountable, the natural order in $C(K)$ or in the spaces $L_p(K)$, with $1 \leqslant p \leqslant \infty$, fails to represent, in a proper way, the idea that to consume more than just one commodity.

For this reason we focus on strictly monotonic preferences defined in $X \subset B(K)$ or, with more generality, $X \subset F(K) = \{f : K \to \mathbb{R}\}$. Note that any continuous and monotonic preference defined in the linear space of bounded functions $B(K)$ or in the positive cone $B(K)^+$ is countably bounded (by the set of constant functions with rational values) and, consequently, it has a continuous utility representation. However, unlike the case of $C(K^+)$ or $L_p(K)^+$, it is not so easy to find examples of strictly monotonic preferences in the positive cone of the space of all bounded functions $B(K)$.

The next section elaborates on this difficulty.

4　Strictly Monotonic Preferences on $F(K)$

Along this section K denotes an uncountable set and we consider first the linear space $F(K)$ of all functions defined on $K = [0, 1]$. Our main result is the following

Theorem 4.1 (Non-existence) *Suppose* $[0, 1]^K \subset X \subset F(K)$. *Every strictly monotonic preference relation on X is non-representable.*

Proof Let $\succsim$ be any strictly monotonic preference relation defined on X. Suppose $U : X \to \mathbb{R}$ represents $\succsim$. Define for $t \in [0, 1]$ the functions $g_t = \chi_{[0,t)}$ and $f_t = \chi_{[0,t]}$, where χ_A to denote the characteristic function of the set A; that is, $\chi_A(x) = 1$ if $x \in A$ and is 0 otherwise. Since $f_t > g_t$ we have that $U(f_t) > U(g_t)$. Now if $s > t$ we have that $g_s > f_t$ and therefore $U(g_s) > U(f_t)$. In particular we conclude that the family of intervals $\{I_t : t \in [0, 1]\}$, where $I_t = (U(g_t), U(f_t))$, is pairwise disjoint and uncountable. This impossibility shows the result.

The Corollary is Theorem 3 in Hervés-Beloso and Monteiro [17]:

Corollary 4.1 (Non-continuity) *Let τ be a linear topology on X. Then, no strictly monotonic preference on X can be continuous.*

Proof Let $\succsim$ be any strictly monotonic preference relation defined on X, and let us consider the restriction of $\succsim$ to $[0, 1]^K$. The constant functions $f(k) = 0$ and $g(k) = 1$ bounds $\succsim$. If $\succsim$ were continuous, since $[0, 1]^K$ is convex, it would have a (continuous) utility representation, which is forbidden by Theorem 4.1.

The negative results stated above refer to the standard consumption set of the commodity spaces $F([0, 1])$ or $B([0, 1])$ and, obviously, cannot be extended to any subset X, as the previous examples ($X \subset \mathbb{R}^n$, l_∞, l_1, $C[0, 1]$, $L_p[0, 1]$, with $1 \leqslant p \leqslant \infty$) show.

Finally, we will replicate Theorem 4.1 and Corollary 4.1 in the case where K is any uncountable set and $X \subset F(K)$ is a "rich enough" set of consumption plans.

Let be two consumption plans $f, g \in X$, $f \geqslant g$. It is natural to assume that if a consumer (a decision maker) can choose both f and g, she also can choose any consumption plan h with $f \geqslant h \geqslant g$. That means that if a consumer can consume an amount $f(k)$ or $g(k)$ of a commodity k, she also can consume any intermediate value. Note that this property is neither fulfilled when X is a set of continuous functions nor when X is a set of classes of integrable functions. Note also that to consider continuous functions on K implies a close relation among commodities; if the consumer chooses an amount $f(k)$, she is obliged to choose a similar amount of every commodity k' in a neighborhood of k.

In order to show the next results, we will require that the consumption set X be "rich enough". A set $X \subset F(K)$ is said "rich enough" if there are two functions $f, g \in X$ such that $f \geqslant g$, $f(k) > g(k)$ for all $k \in K_o \subset K$ with K_o uncountable, and such that the segment $[g, f] = \{h; g \leqslant h \leqslant f\} \subset X$.

Theorem 4.2 (Second Theorem of Non-existence) *Let K any uncountable set and let be $X \subset F(K)$ "rich enough". Every strictly monotonic preference relation on X is non-representable.*

The proof is, essentially, the same as in Theorem 1 in Hervés-Beloso and Monteiro [17]. See also Hervés-Beloso and del Valle-Inclán Cruces [18].

Let $\geqslant_K$ be a total order in K, the assumption guarantees that there are g and $f \geqslant g$[7] functions on K such that $[g, f] \subset X$ and $g(k) < f(k)$ for all k in the uncountable set K_o. For any $k \in K_o$, define $h_k(x) = f(x)$ if $k >_K x$ and $h_k(x) = g(x)$ otherwise, and $h^k(x) = f(x)$ if $k \geqslant_K x$ and $h^k(x) = g(x)$ otherwise. Note that $h^k > h_k$ for all $k \in K_o$ and thus, if U is a utility representing a strictly monotonic preference $\succsim$, for each $k \in K_o$, we would have that $I_k = (U(h_k), U(h^k))$ would be a non-empty open

---

[7]Note that $\geqslant_K$ is a total order on K, whereas $\geqslant$ represent the natural partial order in the set of functions X.

interval and for any other point $k' \in K_o, k' >_K k, h_{k'} > h^k$ and thus, $I_k \cap I_{k'} = \emptyset$. We would have uncountably many non-empty open and disjoint real intervals, an impossibility that shows the result.

Corollary 4.2 (Second Corollary of Non-continuity) *Let τ be a linear topology*[8] *on $X \subset F(K)$ "rich enough". Then, no strictly monotonic preference on X can be continuous.*

Proof Let $\succsim$ be any strictly monotonic preference relation defined on X, and let g, f as in the previous proof. Let us consider the restriction of $\succsim$ to the set $X' = [g, f] \subset X$. Note that $\succsim$ is bounded in the convex set X' by $\{g, f\}$. If $\succsim$ were continuous it would have a (continuous) utility representation in X' that is *"rich enough"*, but this contradicts Theorem 4.2.

Remark A relevant example of a *"rich enough"* consumption set is $B(K)^+$, the positive cone of $B(K)$, the Banach space of all bounded functions defined on K. We have shown that there are strictly monotonic preferences defined on $B(K)^+$, but none of them is continuous and none of them has a utility representation. Yet, there are infinite dimensional subspaces of $B(K)$ for which there are continuous and strictly monotonic preferences with utility representation defined on the positive cone. For it, consider $\ell^1([0, 1]) \subset B(K)$. If $f \in \ell^1_+([0, 1])$ then $\{t : f(t) \neq 0\}$ is finite or countable. Let $U(f) = \|f\|_1 = \sum_{t \in [0,1]} |f(t)| < \infty$. For $g, f \in \ell^1_+([0, 1])$, define $f \succsim g$ if and only if $U(f) \geqslant U(g)$. Observe that $\succsim$ is strictly monotonic, continuous, with the topology given by the norm $\|\ \|_1$ and representable by the utility U. This does not contradict the previous results since $\ell^1_+([0, 1])$ is not *"rich enough"*. In this consumption set no agent ever consumes an uncountable set of commodities.

5 Conclusion and Further Remarks

This work addresses the existence, continuity and representability of preferences defined on a set X that represents a consumption set or the set of available alternatives or strategies for an agent or a decision maker.

The Eilenberg-Debreu Theorem guarantees the existence of a continuous utility function for continuous preferences defined on a connected and separable or a second countable topological space X. On the other hand, Estévez and Hervés [12] showed that if X is any non-separable metric space, there are continuous preferences defined on X that are non-representable by a utility function.

In order to analyze positive results for continuous preferences defined on non-separable spaces, we have revisited some previous results by Monteiro [25] and Candeal et al. [6] to emphasize that, in very general consumption sets, namely separably connected sets, the property of countably boundedness is a necessary and sufficient condition for the representability of a continuous preference.

---

[8]Linearity guarantee that the path $t \to (1 - t)a + tb$ joining a to b is continuous.

Consider any good or service for which the quality, the date or the location where it will be make or that will be made available matters. Differences in quality, date or location give raise to different commodities. Consequently, if any of them, quality, date or location, is a continuous variable we will have a continuum of different commodities. Let assume that $K = [0, 1]$ is the set of commodities. A strategy or a consumption plan would be a specification for each commodity $x \in K$ of the quantity, let say, $f(x)$ that the agent is planing to buy or sell. Thus, it would be natural to assume that the commodity space is a subspace of $F(K)$, the space of all functions defined in K. Examples of such commodity spaces, already used in the literature, are the classical Banach spaces $B(K), C(K), L_p(K)$ with $1 \leqslant p \leqslant \infty$, or spaces like $\ell^p([0, 1]) \subset B(K)$.

If we consider a monotonic preference defined on a subset of $F(K)$, it could be easy to verify if the preference is countably bounded. This is the case where X is an order interval $X = [g, f]$, since the preference is bounded by $\{g, f\}$, or where X is a subset of the space of bounded functions, since the sequence of constant functions with rational values bounds the preference. In spite that X could be a non separable metric space, any monotonic and continuous preference defined in X is representable by a continuous utility function. Thus, we conclude that it is not possible to strengthen the general result by Estévez and Hervés [12], in the case of ordered non-separable metric spaces, by requiring monotonicity of the preference.

Finally, we addressed the case of strictly monotonic preferences. One can find different definitions of strictly monotonic preferences in the literature. A very weak definition of strict monotonicity is the following: the preference $\succsim$ defined on a subset $X \subset F(K)$ is *weakly strictly monotone* if and only if $f(k) > g(k)$ for all $k \in K$, implies $f \succ g$. For the case where K is a continuum, we can define: $\succsim$ is ω^*-*strictly monotone* if and only if $f \geqslant g$ and $f(k) > g(k)$ for uncountably many $k \in K$, implies $f \succ g$. In this work, $\succsim$ is *strictly monotone* if and only if $f \geqslant g$ and $f \neq g$, implies $f \succ g$. Note that if X is a subset of a Banach space as $C(K)$ or $L_p(K), 1 \leqslant p \leqslant \infty, f \geqslant g$ and $f \neq g$, implies $f(k) > g(k)$ for uncountably many $k \in K$, and we deal with ω^*-*strictly monotonicity*. Our strict monotonicity, let say, *purely strict monotonicity* appears in spaces like $F(K)$ or $B(K)$, where given any consumption plan g, one can consider the alternative f given by $f(k) = g(k)$ for all $k \neq k_o$ and $f(k_o) = g(k_o) + 1$, that represent just to consume one unit more of commodity k_o.

Importantly, we have shown (Theorem 3.1) that for any subset $X \subset F(K)$ there are *purely strict monotonic* preferences defined in X. Our final results explain that if X is " *rich enough*" these examples are neither representable nor continuous.

Acknowledgements We thank two anonymous referees for their carefully reading and helpful comments. Hervés-Beloso acknowledges the financial support of Research Grants ECO2016-75712-P (AEI/FEDER, UE) and RGEAF-ECOBAS (Xunta de Galicia). Monteiro acknowledges the financial support of CNPq–Brazil.

References

1. Araujo, A.P.: Lack of Pareto optimal allocations in economies with infinitely many commodities: the need for impatience. Econometrica **53**(2), 455–461 (1985)
2. Aron, R.M., Maestre, M.: A connected metric space that is not separably connected. Contemp. Math. **328**, 39–42 (2003)
3. Bewley, T.: Existence of equilibria with infinitely many commodities. J. Econ. Theory **4**, 514–540 (1972)
4. Bridges, D., Mehta, G.: Representations of Preferences Orderings. Lecture Notes in Economics and Mathematical Systems, vol. 422. Springer, Berlin (1995)
5. Campión, M.J., Candeal, J.C., Induraín, E.: The existence of utility functions for weakly continuous preferences on a Banach space. Math. Soc. Sci. **51**(2), 227–237 (2006)
6. Candeal, J.C., Hervés-Beloso, C., Induráin, E.: Some results on representation and extension of preferences. J. Math. Econ. **29**(1), 75–81 (1998)
7. Ciesielski, K.: Set Theory for the Working Mathematician. Cambridge University Press (1997)
8. Debreu, G.: Representation of a preference ordering by a numerical function. In: Thrall, R.M., Coombs, C.H., Davis, R.L. (eds.) Decision Processes, pp. 159–165 (Wiley, New York); also in Mathematical Economics: Twenty Papers of Gerard Debreu, pp. 105–110. Cambridge University Press, Cambridge (1954)
9. Debreu, G.: The Theory of Value: An Axiomatic Analysis of Economic Equilibrium. Wiley, New York (1959)
10. Debreu, G.: Continuity properties of Paretian utility. Int. Econ. Rev. **5**(3), 285–293 (1964)
11. Eilenberg, S.: Ordered topological spaces. Am. J. Math. **63**(1), 39–45 (1941)
12. Estévez, M., Hervés, C.: On the existence of continuous preference orderings without utility representations. J. Math. Econ. **24**, 305–309 (1995)
13. Fleischer, I.: Numerical representation of utility. J. Soc. Ind. Appl. Math. **9**(1), 48–50 (1961)
14. Herden, G.: On the existence of utility functions. Math. Soc. Sci. **17**, 297–313 (1989)
15. Herden, G.: On the existence of utility functions II. Math. Soc. Sci. **18**, 107–117 (1989)
16. Herden, G.: On some equivalent approaches to mathematical utility theory. Math. Soc. Sci. **29**(1), 19–31 (1995)
17. Hervés-Beloso, C., Monteiro, P.K.: Strictly monotonic preferences on continuum of goods commodity spaces. J. Math. Econ. **46**(5), 725–727 (2010)
18. Hervés-Beloso, C., del Valle-Inclán Cruces, H.: Continuous preference orderings representable by utility functions. J. Econ. Surv. **33**(1), 179–194 (2019)
19. Jaffray, J.-Y.: Existence of a continuous utility function: an elementary proof. Econometrica **43**, 981–983 (1975)
20. Kelley, J.: General Topology. Van Nostrand, New-York (1955)
21. Mas-Colell, A.: The price equilibrium existence problem in topological vector lattices. Econometrica **54**(5), 1039–1054 (1986)
22. Mehta, G.: Topological ordered spaces and utility functions. Int. Econ. Rev. **18**(3), 779–782 (1977)
23. Mehta, G.: Some general theorems on the existence of order preserving functions. Math. Soc. Sci. **15**, 135–143 (1988)
24. Mehta, G.B.: Preference and utility. In: Barberá, S., Hammond, P.J., Seidl, C. (eds.) Handbook of Utility Theory, Chapter 1, pp. 1–47. Kluwer Academic Publishers, Boston (1998)
25. Monteiro, P.K.: Some results on the existence of utility functions on path connected spaces. J. Math. Econ. **16**, 147–156 (1987)
26. Nachbin, L.: Topology and Order. D. Van Nostrand Company, Princeton, New Jersey (1965)
27. Pareto, V.: Course d' economie politique. Rouge, Lausanne (1986). See also Manuale di economia politica. Societa Editrice Libraria, Milan (1906). Reprinted as Manual of Political Economy. Augustus M. Kelley, New York (1971)
28. Peleg, B.: Utility functions for partially ordered topological spaces. Econometrica **38**, 93–96 (1970)

29. Richter, M.: Continuous and semi-continuous utility. Int. Econ. Rev. **21**(2), 293–299 (1980)
30. Schmeidler, D.: A condition for the completeness of partial preference relations. Econometrica **39**(2), 403–404 (1971)
31. Slutsky, E.: Sulla Teoria del Bilancio del Consumatore. Giornale Degli Economisti **51**(1), 1–26 (1915)
32. Steen, L.A., Seebach, J.A.J.: Counterexamples in Topology. Holt, Rinehart and Winston Inc., New York (1970)
33. Stigler, G.J.: The development of utility theory. I. J. Polit. Econ. **58**(4), 307–327 (1950)
34. Stigler, G.J.: The development of utility theory. II. J. Polit. Econ. **58**(5), 373–396 (1950)
35. Wójcik, R.M.: The generalized Aron-Maestre comb. Houston J. Math. **42**(2), 701–707 (2016)
36. Wold, H.: A synthesis of pure demand analysis; I, II and III. Scandinavian Actuarial J. **26**, 85–118, 220–263, 69–120 (1943)

Extensions of Utility

Non-total preorders; Multi-utility; Continuous representation; Subjective states; Flexibility; Constructive utility; The Arrow-Hahn construction; Fuzzy preferences; Utility in the fuzzy setting

Continuity and Continuous Multi-utility Representations of Nontotal Preorders: Some Considerations Concerning Restrictiveness

Gianni Bosi and Magalì Zuanon

Abstract A continuous multi-utility fully represents a not necessarily total preorder on a topological space by means of a family of continuous increasing functions. While it is very attractive for obvious reasons, and therefore it has been applied in different contexts, such as expected utility for example, it is nevertheless very restrictive. In this paper we first present some general characterizations of the existence of a continuous order-preserving function, and respectively a continuous multi-utility representation, for a preorder on a topological space. We then illustrate the restrictiveness associated to the existence of a continuous multi-utility representation, by referring both to appropriate continuity conditions which must be satisfied by a preorder admitting this kind of representation, and to the Hausdorff property of the quotient order topology corresponding to the equivalence relation induced by the preorder. We prove a very restrictive result, which may concisely described as follows: the continuous multi-utility representability of all closed (or equivalently weakly continuous) preorders on a topological space is equivalent to the requirement according to which the quotient topology with respect to the equivalence corresponding to the coincidence of all continuous functions is discrete.

1 Introduction

The necessity of considering *nontotal (incomplete) preferences* in order to deal with a more realistic framework dates back to the seminal paper of Aumann [3] published in 1962. Aumann pointed out that it is more appropriate not to assume that an individual may compare any two objects according with its own preferences, since

G. Bosi (✉)
DEAMS, University of Trieste, via Università 1, 34123 Trieste, Italy
e-mail: gianni.bosi@deams.units.it

M. Zuanon
DEM, University of Brescia, Contrada Santa Chiara 50, 25122 Brescia, Italy
e-mail: magali.zuanon@unibs.it

© Springer Nature Switzerland AG 2020   213
G. Bosi et al. (eds.), *Mathematical Topics on Representations of Ordered Structures and Utility Theory*, Studies in Systems, Decision and Control 263,
https://doi.org/10.1007/978-3-030-34226-5_11

"incomparability" may take place in some cases (see also Dubra et al. [26], Evren and Ok [29] and Ok [57]).

Clearly, when we deal with a nontotal binary relation, it is not possible to fully represent it by using only one function, as in the case of a total preorder. On the other hand, when the preference relation is defined on a topological space, continuity requirements of the representing functions naturally come into consideration. This is, needless to say, the spirit of the seminal famous papers by Debreu [24, 25] and Eilenberg [27], where general results about the existence of a *continuous utility function* for a *total preorder* on a topological space were presented.

Given a preorder $\precsim$ on a topological space (X, t), and the natural topology t_{nat} on the real line $\mathbb{R}$, we recall that a function $f : (X, \precsim, t) \rightarrow (\mathbb{R}, \leq, t_{nat})$ is said to be a continuous order-preserving function for $\precsim$ if f is continuous on the topological space (X, t) and, for all points $x, y \in X, x \precsim y$ implies that $f(x) \leq f(y)$, and $x \prec y$ implies that $f(x) < f(y)$. Clearly, $\prec$ is the *strict part* of the preorder $\precsim$.

Although an order-preserving function does not characterize a nontotal preorder, it is enough for many purposes, since it contains all the information concerning that preorder, which can be provided by a single real-valued function. So, for example, the maximization of an order-preserving function (when it is possible) for a preorder $\precsim$ on a set X leads to a *maximal element* x_0 of $(X, \precsim)$ (i.e., to a point $x_0 \in X$ such that $x_0 \prec z$ for no point $z \in X$).

It is worthwhile noticing that the mere existence of an order-preserving function for a binary relation $\precsim$ does not even imply transitivity of $\precsim$, but a weaker condition called *Suzumura consistency* (see Suzumura [61], Cato [22] and Bevilacqua et al. [8]).

Herden [33–35] introduced the concept of a *(decreasing) separable system* in order to derive very general conditions for the existence of a continuous order-preserving function for a nontotal preorder on a topological space. Actually, the notion of a decreasing separable system generalizes that of a *decreasing scale* in a *preordered topological space* (see Burgess and Fitzpatrick [20] and Johnson and Mandelker [39]).

Herden's efforts were addressed to the aim of unifying sparse but very significant results in the literature, concerning the existence of continuous order-preserving functions (*utility functions*) for total preorders. Classical examples of such results are the *Debreu's Theorem* (see Debreu [24, 25]), and respectively the *Eilenberg's Theorem* (see Eilenberg [27]), according to which every continuous total preorder on a second countable, and respectively on a connected and separable topological space admits a continuous utility representation.

An approach of this kind was actually initiated by Mehta [46–51], who explored the possibility of recovering deep results by following the general framework of Nachbin [56]. Indeed, Nachbin first investigated in a systematic way the connection between Order and Topology. Mehta was able to establish very general conditions for the existence of a continuous *order-preserving function* for a preorder that may fail to be total on a topological space. The reader may also consult the book by Bridges and Mehta [19] for a miscellanea of theorems concerning the existence of continuous order isomorphisms. Other general results were presented, for example, by Beardon and Mehta [6, 7], and Herden and Mehta [36].

Incidentally, the concept of a *complete separable system on a topological space* has been recently used in order to present a characterization of *useful topologies* on a set X (i.e., topologies on X with respect to which all the continuous total preorders are representable by a continuous utility function). Indeed, Bosi and Herden [16, Theorem 3.1] showed that a topology t on a set X is useful if and only if the topology $t_{\mathscr{E}}$ induced (generated) by every *complete separable system* $\mathscr{E}$ on (X, t) is second countable. Other authors prefer the terminology *continuously representable* topology instead of useful topology (see e.g., Campión et al. [21]).

We recall that, from Herden [33, 34], a family $\mathscr{E}$ of open subsets of a topological space (X, t) is said to be a *separable system on* (X, t) if there exist sets $E_1 \in \mathscr{E}$ and $E_2 \in \mathscr{E}$ such that $\overline{E_1} \subset E_2$, and for all sets $E_1 \in \mathscr{E}$ and $E_2 \in \mathscr{E}$ such that $\overline{E_1} \subset E_2$ there exists some set $E_3 \in \mathscr{E}$ such that $\overline{E_1} \subset E_3 \subset \overline{E_3} \subset E_2$. If, for all sets $E \in \mathscr{E}$ and $E' \in \mathscr{E}$, at least one of the following conditions $E = E'$ or $\overline{E} \subset E'$ or $\overline{E'} \subset E$ holds, then $\mathscr{E}$ is said to be *complete* (see Bosi and Herden [16, Definition 2.2]). If X is endowed with a preorder $\precsim$, then we get the concept of a *complete decreasing separable system on a preordered topological space* $(X, \precsim, t)$ by simply requiring every set $E \in \mathscr{E}$ in the previous definition to be *decreasing*.

We recall that a preorder $\precsim$ on a topological space (X, t) is said to admit a *continuous multi-utility representation* if there exists a family $\mathscr{F}$ of (continuous) increasing real functions on the preordered topological space $(X, \precsim, t)$ such that, for all $x, y \in X$, $x \precsim y$ is equivalent to $f(x) \leq f(y)$ for all functions $f \in \mathscr{F}$. This kind of representation, whose main feature is to fully characterize the preorder, was first introduced by Levin [42], who called *functionally closed* a preorder admitting a continuous multi-utility representation on a topological space. Levin's fundamental theorem [42, 43], that using the notation of Evren and Ok [29, Theorem 1] states that every *closed preorder* (i.e., every preorder which is a closed subset of $X \times X$ with respect to the product topology $t \times t$ on $X \times X$ that is induced by t) on a locally and σ-compact Hausdorff space has a continuous multi-utility representation, still belongs to the most quoted theorems in Mathematical Utility Theory (cf. the literature that has been quoted in Bosi and Herden [14]).

In the framework of our approach, the particular relevance of closed preorders $\precsim$ on (X, t) is based, on the one hand, on the observation that a preorder $\precsim$ on (X, t) that has a continuous multi-utility representation must be closed (cf. Bosi and Herden [15, Proposition 2.1]). On the other hand, it is beyond any doubt that closed preorders are of particular interest in Mathematical Economics (cf., for instance, the literature that has been quoted by Evren and Ok [29], Bosi and Herden [14, 15], Minguzzi [53, 54] and many others). Indeed, in some standard textbooks on microeconomics (such as Mas-Colell et al. [44, page 46]), the definitions of continuity of an (incomplete) preference relation and of a closed preference relation coincide. In combination with Proposition 2.1 in Bosi and Herden [15], these remarks on closed preorders suggest that the most fundamental problem in the theory of continuous multi-utility representations of preorders in some sense is the problem of precisely characterizing (determining) all topological spaces (Hausdorff spaces) (X, t) for which every closed preorder has a continuous multi-utility representation. A complete solution to this problem seems to be difficult. Indeed, since Levin's fundamental theorem [43], which we recalled above, no real progress towards a complete solution of the mentioned characterization problem has been made.

Levin's theorem only presents sufficient conditions for the existence of a continuous multi-utility representation. Therefore, in Bosi and Herden [15, Theorem 3.5] an effort has been made in order to also clarify up to which degree Levin's assumptions on the underlying topological space are really necessary.

Continuous multi-utility representations were first deeply studied in the framework of *Expected Utility* with incomplete preferences (see the seminal paper by Dubra et al. [26], followed by other papers like Evren [28], Galaabaatar and Karni [30] and Gorno [32]), and later they also appear in other branches of Applied Mathematics like Game Theory and Welfare Economics (see e.g., Baucells and Shapley [5], and Banerjee and Dubey [4]).

However, the first systematic study of multi-utility representations in the general case is due to Evren and Ok [29], who presented different conditions for the existence of semicontinuous and continuous multi-utility representations.

A sufficient condition for the existence of a continuous multi-utility representation is presented in Bosi and Zuanon [18], based on the concept of an *extremely continuous* preorder introduced by Mashburn [45]. Typical topologies with respect to which an upper (lower) semicontinuous multi-utility representation exists have been recently presented in Bosi et al. [10].

Ok [57] studied finite (continuous) multi-utility representation, as well as Kaminski [41] and Yilmaz [63].

Minguzzi [53, 54] introduced the concept of a *continuous Richter–Peleg multi-utility representation* $\mathscr{F}$ of a preorder $\precsim$, which is a particular kind of continuous multi-utility representation where every function $f \in \mathscr{F}$ is a *Richter–Peleg utility function* for $\precsim$ (i.e., every function $f \in \mathscr{F}$ is order-preserving). Richter–Peleg multi-utilities have been recently studied by Alcantud et al. [2], who in particular were concerned with the case of a countable representation (see also Bevilacqua et al. [9]). Although very restrictive, the case of a countable (upper semi) continuous multi-utility is particularly favorable, since it automatically implies the existence of a countable continuous Richter–Peleg multi-utility. Indeed, from Alcantud et al. [2], if there exists a countable continuous multi-utility, then there also exists a countable continuous Richter–Peleg multi-utility.

Conditions for the existence of countable multi-utilities are also found in Kabanov and Lépinette [40], and in Bevilacqua et al. [9].

Nishimura and Ok [55] generalized multi-utility, which necessarily implies transitivity, to the nontransitive case. Indeed, for a not necessarily transitive binary relation $\precsim$ on a set X, the following representation of $\precsim$ can be considered to hold for all points $x, y \in X$: $[x \precsim y \Leftrightarrow \sup_{\mathscr{F} \in \mathbb{F}} \inf_{f \in \mathscr{F}} (f(y) - f(x)) \geq 0,]$, where $\mathbb{F}$ is a set of sets $\mathscr{F}$ of real-valued functions f on X. While it is extremely nice, this kind of representation appears rather difficult to be modeled, at least according to our opinion.

Bosi et al. [11] generalized multi-utility by allowing *partial functions*, in order to also deal with nontransitive preferences. The idea of using partial functions is that of avoiding any unnecessary information, and to handle both incompleteness and intransitivity in a relatively easy way. Since transitivity is removed, typical nontransitive preference relations, like *interval orders* and *semiorders*, can be represented by using families of functions.

Let (X, t) be a topological space. In order for a preorder $\precsim$ on (X, t) to be representable by a continuous order-preserving real-valued function or else to admit a continuous multi-utility representation $\mathscr{F}$, it is necessary that for every pair $(x, y) \in \prec$ there exists a continuous increasing function $f_{xy} : (X, \precsim, t) \to (\mathbb{R}, \leq, t_{nat})$ such that $f_{xy}(x) < f_{xy}(y)$. Therefore, in this paper $\precsim$ is considered to be *weakly continuous* if it satisfies the just defined monotony behavior, that is obviously equivalent to requiring for every pair $(x, y) \in \prec$ to exist a complete decreasing separable system $\mathscr{E}$ on X such that for every pair $(x, y) \in \prec$ there exist sets $E \subset \overline{E} \subset E'$ in $\mathscr{E}$ such that $x \in E$ and $y \notin E'$.

We recall that *weak continuity* of a preorder on a topological space was introduced by Bosi and Herden [12, 13] in order to discuss the *continuous analogue of the Szpilrain Theorem*, i.e., the identification of conditions under which a weakly continuous preorder $\precsim$ on a topological space (X, t) admits a *refinement* $\lesssim$ by a total and continuous preorder (in the sense that $\precsim \subset \lesssim$ and $\prec \subset <$).

In this paper, we focus our attention on the *Hausdorff property* of the quotient *order topology* $t_{|\sim}^{\precsim}$, which is implied by the existence of a continuous multi-utility representation for the preorder $\precsim$ on the topological space (X, t). The quotient is considered with respect to the *equivalence relation* $\sim$.

Here, the order topology $t^{\precsim}$ induced by a preorder $\precsim$ on X is the *coarsest* topology on X with respect to which the *strict lower section* and respectively *upper section* $l_{\precsim}(x) := \{z \in X \mid z \prec x\}$ and $r_{\precsim}(x) := \{z \in X \mid x \prec z\}$ are open subsets of X for every point $x \in X$. We prove (see Proposition 2.3 below) that, when the quotient set $X_{|\sim}$ consists of at least two points, in order for $t_{|\sim}^{\precsim}$ to be a Hausdorff topology on $X_{|\sim}$, it is necessary that the sets $l_{\precsim}(x)$ and $r_{\precsim}(x)$, where x runs through X, constitute a subbasis of $t^{\precsim}$. In practice, this means that there is no point $x \in X$ which is at the same time a *minimal* and a *maximal* element for $\precsim$ on X.

Theorem 2.23 presents the equivalence of different concepts of continuity concerning a total preorder $\precsim$ on a topological space (X, t), also by using the concept of a complete decreasing separable system. The important role of weak continuity of a preorder on a topological space is illustrated. In particular, Theorem 2.27 shows that the existence of a continuous order-preserving function $f : (X, \precsim, t) \to (\mathbb{R}, \leq, t_{nat})$ is equivalent to the existence of a second countable topology t' on X, which is coarser than t and with respect to which $\precsim$ is weakly continuous.

Section 3 is devoted to the characterization of the existence of a continuous order-preserving function (see Theorem 3.1), and respectively the existence of a continuous multi-utility representation (see Theorem 3.2). The existence of a continuous Richter–Peleg multi-utility representation is considered in Theorem 3.3.

In Sect. 4 we present restrictive conditions, implied by the existence of a continuous multi-utility representation, which pose objective limitations to its applicability. Corollary 4.3 shows that if a preorder $\precsim$ on a topological space (X, t) admits a finite continuous Richter–Peleg multi-utility representation $\mathscr{F} = \{f_1, ..., f_n\}$, then the restriction of $\precsim$ to the components of (X, t) is total. This is a version of Proposition 5.2 in Alcantud et al. [2], who, based on a famous theorem by Schmeidler [60], showed that a nontrivial preorder on a connected topological space is total provided that it admits a finite continuous Richter–Peleg multi-utility representation.

Finally, Theorem 4.8 proves that the continuous multi-utility representability of all closed preorders on a topological space is equivalent to the continuous multi-utility representability of all weakly continuous preorders, and in turn to the requirement according to which the quotient topology with respect to the coincidence of all continuous functions is discrete. This very restrictive result is based on the fact that, given a topological space (X, t), the coarsest topology with respect to which every weakly continuous preorder $\precsim$ on (X, t) remains being continuous is actually the *weak topology* $\sigma(X, C(X, t, \mathbb{R}))$ of the real-valued continuous functions on (X, t), i.e., the coarsest topology with respect to which every real-valued continuous function $f \in C(X, t, \mathbb{R})$ remains being continuous (see Lemma 4.7).

2 Notation and Preliminary Results

2.1 Basic Definitions Concerning Preorders and Their Continuous Representability

Definition 2.1 A *preorder* $\precsim$ on a nonempty set X is a *reflexive* and *transitive* binary relation on X. Denote by $\prec$ and $\sim$ the *strict part* and respectively the *symmetric part* of a preorder $\precsim$ on X (i.e., for all $x, y \in X$, $x \prec y$ if and only if $(x \precsim y)$ *and not* $(y \precsim x)$, and respectively $x \sim y$ if and only if $(x \precsim y)$ *and* $(y \precsim x)$).

From time to time, we shall write "$(x, y) \in \prec$" instead of "$x \prec y$". A preorder $\precsim$ on X is said to be *nontrivial* if there exist $x, y \in X$ such that $x \prec y$.

Denote by $\bowtie$ the *incomparability relation* associated with a preorder $\precsim$ on a set X (i.e., for all $x, y \in X$, $x \bowtie y$ if and only if *not* $(x \precsim y)$ *and not* $(y \precsim x)$).

A preorder $\precsim$ is said to be *total* if, for all $x, y \in X$, either $x \precsim y$ or $y \precsim x$ (i.e., $\bowtie = \emptyset$).

Clearly, the symmetric part $\sim$ associated to any preorder $\precsim$ on X is an *equivalence relation* on X (i.e., $\sim$ is *reflexive*, *symmetric* and *transitive*).

We denote by $\precsim_{|\sim}$ the *quotient order* on the *quotient set* $X_{|\sim}$ (i.e., for all $x, y \in X$, $[x] \precsim_{|\sim} [y]$ if and only if $x \precsim y$, where $[x] = \{z \in X : z \sim x\}$ is the *indifference class* associated to $x \in X$).

Let $(X, \precsim)$ be an arbitrarily chosen *preordered set*. We define, for every point $x \in X$, the following subsets of X:

$$d_{\precsim}(x) := \{z \in X \mid z \precsim x\}, \quad i_{\precsim}(x) := \{z \in X \mid x \precsim z\},$$

$$l_{\precsim}(x) := \{z \in X \mid z \prec x\}, \quad r_{\precsim}(x) := \{z \in X \mid x \prec z\}.$$

For any pair $(x, y) \in X \times X$ such that $(x, y) \in \prec$, we shall denote by $]x, y[$ the (maybe empty) *open interval* defined as $]x, y[:= r_{\precsim}(x) \cap l_{\precsim}(y)$.

A pair $(x, y) \in \prec$ is said to be a *jump* in $(X, \precsim)$ if $]x, y[= \emptyset$.

Let us now present the basic definition of the *order topology* corresponding to a preorder $\precsim$ on a set X.

Definition 2.2 The *order topology* $t^{\precsim}$ on X associated with a preorder $\precsim$ on X is defined to be the coarsest topology on X for which the sets $l_{\precsim}(x)$ and $r_{\precsim}(x)$ are open.

In order to avoid artificial and superfluous considerations, we can assume for the moment that the *quotient order topology* $t_{|\sim}^{\precsim|\sim}$ (which in the sequel will be denoted by $t_{|\sim}^{\precsim}$ for the sake of convenience) to be a *Hausdorff topology* on $X_{|\sim}$. For underlining the importance of this assumption and for later use we still notice that in case that $X_{|\sim}$ contains at least two elements the following necessary condition for $t_{|\sim}^{\precsim}$ to be Hausdorff holds.

Indeed, the following proposition holds (see also Ward [62, Lemma 2]).

Proposition 2.3 *Let $\precsim$ be a preorder on X. Then the following assertion holds.*

SB*: In order for $t_{|\sim}^{\precsim}$ to be a Hausdorff topology on $X_{|\sim}$, it is necessary that the sets $l_{\precsim}(x)$ and $r_{\precsim}(x)$, where x runs through X, constitute a subbasis of $t^{\precsim}$.*

Proof In order to prove the above condition **SB**, it suffices to show that

$$\bigcup_{z \in X} (l_{\precsim}(z) \cup r_{\precsim}(z)) = X.$$

Consider any point $x \in X$. Since we assume that $X_{|\sim}$ contains at least two elements, there exists some point $y \in X$ such that $not\,(y \sim x)$. The fact that $t_{|\sim}^{\precsim}$ is a Hausdorff topology on $X_{|\sim}$ implies the existence of two points $y_1, x_1 \in X$ such that either $[y] \in l_{\precsim_{|\sim}}([y_1]), [x] \in r_{\precsim_{|\sim}}([x_1]), l_{\precsim_{|\sim}}([y_1]) \cap r_{\precsim_{|\sim}}([x_1]) = \emptyset$, or else $[x] \in l_{\precsim_{|\sim}}([x_1]), [y] \in r_{\precsim_{|\sim}}([y_1]), l_{\precsim_{|\sim}}([x_1]) \cap r_{\precsim_{|\sim}}([y_1]) = \emptyset$. Hence, either $x \in r_{\precsim}(x_1)$ or $x \in l_{\precsim}(x_1)$, and the thesis follows. $\qquad \square$

Definition 2.4 A point $x_0 \in X$ is said to be a *maximal* (*minimal*) element for a preorder $\precsim$ on a set X if $r_{\precsim}(x_0) = \emptyset$ ($l_{\precsim}(x_0) = \emptyset$).

Remark 2.5 It is immediate to check that the above condition **SB** is equivalent to the condition requiring that, for every point $x \in X$, either $l_{\precsim}(x)$ or $r_{\precsim}(x)$ is nonempty (i.e., no point $x \in X$ is at the same time a minimal and a maximal element for $\precsim$ on X).

Definition 2.6 A preorder $\precsim$ on a topological space (X, t) is said to be *continuous* if $l_\prec(x) = \{z \in X \mid z \prec x\}$ and $r_\prec(x) = \{z \in X \mid x \prec z\}$ are both open subsets of X for every $x \in X$.

Definition 2.7 A real-valued function u on a preordered set $(X, \precsim)$ is said to be

(i) *isotonic* or *increasing* if, for all $x, y \in X$,

$$x \precsim y \Rightarrow u(x) \le u(y);$$

(ii) a *weak utility for* $\prec$ if, for all $x, y \in X$,

$$x \prec y \Rightarrow u(x) < u(y);$$

(iii) *strictly isotonic* or *order-preserving* if it is both increasing and a weak utility for $\prec$.

Strictly isotonic functions on $(X, \precsim)$ are also called *Richter–Peleg representations* of $\precsim$ in the economic literature (see e.g. Peleg [58] and Richter [59]).

Definition 2.8 A preorder $\precsim$ on a topological space (X, t) is said to be

(i) *closed* if it is a closed subset of $X \times X$ with respect to the product topology $t \times t$ on $X \times X$ that is induced by t;

(ii) *semi-closed* if $d_\prec(x) = \{z \in X \mid z \precsim x\}$ and $i_\prec(x) = \{z \in X \mid x \precsim z\}$ are both closed subsets of X for every $x \in X$;

(iii) *weakly continuous* if for every pair $(x, y) \in \prec$ there exists a continuous and increasing real-valued function f_{xy} on X such that $f_{xy}(x) < f_{xy}(y)$.

It is immediate to check that every closed preorder is also semi-closed, while the converse is not true.

Remark 2.9 We notice that for other authors, actually, a preorder is continuous if it is semi-closed (see e.g. Bridges and Mehta [19, Definition 1.6.1]). Our choice is presently suggested by our definition of the order topology (see Definition 2.2 above).

Definition 2.10 A real-valued function u on a totally preordered set $(X, \precsim)$ is said to be a *utility function* for $\precsim$ if, for all $x, y \in X$,

$$x \precsim y \Leftrightarrow u(x) \le u(y).$$

An order-preserving function for a total preorder is necessarily a utility function. Clearly, a utility function characterizes a total preorder, while this is not the case of an order-preserving function for a nontotal preorder. On the other hand, in the general case of a nontotal preorder, an order-preserving function provides the greatest amount of information concerning the preorder which can be furnished by a real-valued function.

Definition 2.11 A preorder $\precsim$ on a topological space (X, t) is said to have a *continuous multi-utility representation* if there exists a family $\mathscr{F}$ of increasing and continuous functions $f : (X, \precsim, t) \longrightarrow (\mathbb{R}, \le, t_{nat})$ such that

$$\precsim = \{(x, y) \in X \times X \mid \forall f \in \mathscr{F} \, (f(x) \le f(y))\}$$

or, equivalently, if there exists for every pair $(x, y) \in X \times X$ such that $not\,(y \precsim x)$ some continuous and increasing function $f_{xy} : (X, \precsim, t) \longrightarrow (\mathbb{R}, \le, t_{nat})$ such that $f_{xy}(x) < f_{xy}(y)$.

The above fundamental Definition 2.8, (iii), of a weakly continuous preorder on a topological space is justified by the following proposition, whose immediate proof is left to the reader.

Proposition 2.12 *Let $\precsim$ be a preorder on a topological space (X, t). If either there exists a continuous order-preserving function $f : (X, \precsim, t) \longrightarrow (\mathbb{R}, \le, t_{nat})$ or $\precsim$ admits a continuous multi-utility representation $\mathscr{F}$, then $\precsim$ is weakly continuous.*

The following proposition appears as Proposition 2.1 in Bosi and Herden [15].

Proposition 2.13 *Let $\precsim$ be a preorder on (X, t) that admits a continuous multi-utility representation. Then $\precsim$ is a closed preorder on (X, t).*

The consideration of the Hausdorff property referred to the quotient order topology is justified by the following simple proposition. Needless to say, we assume that the quotient space $X_{|\sim}$ has at least two elements.

Proposition 2.14 *Let $\precsim$ be a preorder on a topological space (X, t) and assume that $\precsim$ admits a continuous multi-utility representation. Then $(X_{|\sim}, t_{|\sim}^{\precsim})$ is a Hausdorff space.*

Proof Let $\precsim$ have a continuous multi-utility representation and let $x \in X$ and $y \in X$ be arbitrarily chosen points such that $not\,(y \precsim x)$. Then there exists a continuous and increasing function $f_{xy} : (X, \precsim, t^{\precsim}) \longrightarrow (\mathbb{R}, \le, t_{nat})$ such that $f_{xy}(x) < f_{xy}(y)$. Therefore, for every real number $\alpha \in]f_{xy}(x), f_{xy}(y)[$, $f_{xy}^{-1}] - \infty, \alpha[$ and $f_{xy}^{-1}]\alpha, +\infty[$ are disjoint $t_{|\sim}^{\precsim}$-open sets containing $[x]$ and $[y]$, respectively. Hence, the desired conclusion follows. $\square$

By putting together Propositions 2.3 and 2.14, we obviously get the following result.

Proposition 2.15 *If a preorder $\precsim$ on a topological space (X, t) admits a continuous multi-utility representation $\mathscr{F}$, then the sets $l_{\precsim}(x)$ and $r_{\precsim}(x)$, where x runs through X, constitute a subbasis of $t^{\precsim}$.*

Definition 2.16 A preorder $\precsim$ on a topological space (X, t) is said to have a *continuous Richter–Peleg multi-utility representation* if there exists a continuous multi-utility representation $\mathscr{F}$ for $\precsim$ such that every function $f \in \mathscr{F}$ is order-preserving for $\precsim$.

Remark 2.17 It has been already observed (see Alcantud et al. [2, Remark 2.3]) that a (continuous) Richter–Peleg multi-utility representation for a preorder $\precsim$ characterizes the strict part $\prec$ of $\precsim$, i.e., for all $x, y \in X$,

$$x \prec y \Leftrightarrow \forall f \in \mathscr{F} \, (f(x) < f(y)).$$

The proof of the following proposition is contained in the proof of Proposition 5.2 in Alcantud et al. [2]. We include it separately for reader's convenience and for further use.

Proposition 2.18 *If a preorder $\precsim$ on a topological space (X, t) admits a finite continuous Richter–Peleg multi-utility representation $\mathscr{F} = \{f_1, ..., f_n\}$, then $\precsim$ is continuous.*

Proof Just consider that, due to Remark 2.17, the sets

$$l_{\precsim}(x) = \{z \in X \mid z \prec x\} = \{z \in X \mid f_i(z) < f_i(x), \text{ for all } i \in \{1, ..., n\}\} =$$
$$= \bigcap_{i=1}^{n} f_i^{-1}(] - \infty, f_i(x)[),$$
$$r_{\precsim}(x) = \{z \in X \mid x \prec z\} = \{z \in X \mid f_i(x) < f_i(z), \text{ for all } i \in \{1, ..., n\}\} =$$
$$= \bigcap_{i=1}^{n} f_i^{-1}(]f_i(x), +\infty[),$$

are open for every $x \in X$ as a consequence of the continuity of every function f_i $(i = 1, ..., n)$. $\square$

2.2 Decreasing Separable Systems and Continuity of Preorders

Definition 2.19 If $\precsim$ is a preorder on X, then a subset D of X is said to be *decreasing*, resp. *increasing*, if $d_{\precsim}(x) \subset D$, resp. $i_{\precsim}(x) \subset D$, for all $x \in D$.

Definition 2.20 (Herden [33, 34]) A family $\mathscr{E}$ of open decreasing subsets of X is said to be a *decreasing separable system on* $(X, \precsim, t)$ if it satisfies the following conditions:

DS1: There exist sets $E_1 \in \mathscr{E}$ and $E_2 \in \mathscr{E}$ such that $\overline{E_1} \subset E_2$.

DS2: For all sets $E_1 \in \mathscr{E}$ and $E_2 \in \mathscr{E}$ such that $\overline{E_1} \subset E_2$ there exists some set $E_3 \in \mathscr{E}$ such that $\overline{E_1} \subset E_3 \subset \overline{E_3} \subset E_2$.

If, for all sets $E \in \mathscr{E}$ and $E' \in \mathscr{E}$, at least one of the following conditions $E = E'$ or $\overline{E} \subset E'$ or $\overline{E'} \subset E$ holds, then $\mathscr{E}$ is said to be *complete*.

The concept of a complete separable system generalizes that of a *decreasing scale* in a preordered topological space (see Burgess and Fitzpatrick [20] and Johnson and Mandelker [39]). Decreasing scales, which have been widely used for providing characterizations of the existence of continuous real-valued order-preserving functions (see, e.g., Alcantud et al. [1] and Bosi and Mehta [17]) have the disadvantage to be countable.

The reader may notice that, given a complete decreasing separable system $\mathscr{E}$ on X, the inclusion $E \subsetneq E'$ for any two sets $E \in \mathscr{E}$ and $E' \in \mathscr{E}$ implies that $\overline{E} \subset E'$. In the remainder of this paper we shall always use this observation without extra hints.

Nevertheless, we must mention that in the arbitrary case, however, it cannot be concluded that the inclusion $E \subsetneq E'$ implies that $\overline{E} \subset E'$. In the arbitrary case (that will not be considered here) we, therefore, replace $\mathscr{E}$ by

$$\mathscr{E}^- := \mathscr{E} \setminus \{E' \in \mathscr{E} | \exists E \in \mathscr{E} \, (E \subsetneq E' \wedge \overline{E} \cap X \setminus E' \neq \emptyset)\}.$$

Because of the above considerations, the authors would like to take this opportunity of pointing out that it does not mean any loss of generality just to consider in the remainder of this paper complete decreasing separable systems.

Remark 2.21 The particular relevance of (complete) decreasing separable systems on $(X, \precsim, t)$ is given by the following two fundamental relations between continuous increasing real-valued functions on $(X, \precsim, t)$ and decreasing separable systems on $(X, \precsim, t)$ (cf. Herden [33] and Herden and Pallack [37, Lemma 3.6]).

1. Let f be a continuous increasing real-valued function on $(X, \precsim, t)$. Then

$$\mathscr{E} := \{f^{-1}(]-\infty, q[) \mid q \in \mathbb{Q}\}$$

   is a complete decreasing separable system on $(X, \precsim, t)$.
2. Let $\mathscr{E}$ be a (complete) decreasing separable system on $(X, \precsim, t)$. Then $\mathscr{E}$ induces a continuous increasing real-valued function on $(X, \precsim, t)$ by at first defining inductively a function $q \longrightarrow E_q$ from $[0, 1] \cap \mathbb{Q} \longrightarrow \mathscr{E}$ in such a way that $\overline{E_p} \subset E_q$ whenever $p < q$, in order to then define a continuous increasing real-valued function f on $(X, \precsim, t)$ by setting

$$f(x) := \begin{cases} \inf\{q \in [0, 1] \cap \mathbb{Q} \mid x \in E_q\}, & \text{if } x \in \bigcup_{q \in [0,1] \cap \mathbb{Q}} E_q, \\ 1, & \text{otherwise,} \end{cases}$$

   for all $x \in X$.

Remark 2.22 In case that $\precsim$ is the identity relation on X we merely speak of a separable system on X. Then the conditions **DS1** and **DS2** are abbreviated by **S1** and **S2**.

The following theorem holds, presenting different conditions all equivalent to the continuity of a total preorder on a topological space.

Theorem 2.23 *Let* $(X, \precsim, t)$ *be a totally preordered topological space. Then the following conditions are equivalent:*

1. $\precsim$ *is continuous;*
2. *The order topology* $t^{\precsim}$ *is coarser than* t;
3. $\precsim$ *is semi-closed;*
4. $\precsim$ *is closed;*
5. *For all* $x, y \in X$ *such that not* $(y \precsim x)$ *there exist an open decreasing subset* U_x *of* X *containing* x *and an open increasing subset* U_y *of* X *containing* y *such that* $U_x \cap U_y = \emptyset$;
6. $\prec$ *is an open subset of* $X \times X$ *considered with the product topology* $t \times t$;
7. $d_{\precsim}(x) = \{y \in X | y \precsim x\}$ *is a closed subset of* X *and* $l_{\prec}(x) = \{y \in X | y \prec x\}$ *is an open subset of* X *for every point* $x \in X$;
8. $i_{\precsim}(x) = \{z \in X | x \precsim z\}$ *is a closed subset of* X *and* $r_{\precsim}(x) = \{z \in X | x \prec z\}$ *is an open subset of* X *for every point* $x \in X$;
9. $\precsim$ *is weakly continuous;*
10. *For every pair* $(x, y) \in \prec$ *a decreasing separable system* $\mathscr{E}_{xy}$ *on* X *can be chosen in such a way that there exist sets* $E \subset \overline{E} \subset E'$ *in* $\mathscr{E}_{xy}$ *such that* $x \in E$ *and* $y \notin E'$;
11. *For every pair* $(x, y) \in \prec$ *a complete decreasing separable system* $\mathscr{E}_{xy}$ *on* X *can be chosen in such a way that there exist sets* $E \subset \overline{E} \subset E'$ *in* $\mathscr{E}_{xy}$ *such that* $x \in E$ *and* $y \notin E'$.

Proof The equivalence of conditions 1, 2, 3, 4, 6 was proved by Bridges and Mehta [19, Proposition 1.6.2]. The equivalence of conditions 4 and 5 appears in Ward [62, Lemma 1]. The equivalence of conditions 3, 7 and 8 is obvious. The equivalence of conditions 3 and 9 was proved by Herden and Pallack [38, Lemma 2.2]. Finally, the equivalence of conditions 9, 10 and 11 comes from Remark 2.21. □

Let $\mathbb{S}_C(X)$ be the set of all complete separable systems $\mathscr{E}$ on X that contain X (the reader may verify that the assumption X to be contained in $\mathscr{E}$ does not mean any loss of generality). In Bosi and Herden [16], $\mathbb{S}_C(X)$ has been endowed with the preorder $\precsim_s$ that is defined by considering for every complete separable system $\mathscr{E} \in \mathbb{S}_C(X)$ the topology $t_{\mathscr{E}}$ on X that is generated by $\mathscr{E}$ (i.e., $\mathscr{E}$ is a subbasis of $t_{\mathscr{E}}$), in order to then set

$$\mathscr{E} \precsim_s \mathscr{L} \Leftrightarrow t_{\mathscr{E}} \subset t_{\mathscr{L}}$$

for all complete separable systems $\mathscr{E} \in \mathbb{S}_C(X)$ and $\mathscr{L} \in \mathbb{S}_C(X)$.

Let $\mathbb{S}_C(X)_{\sim_s}$ be the set of indifference (equivalence) classes of $\precsim_s$ and let, in addition, $\mathbb{P}_{\lhd}(X)$ be the set of all total continuous preorders on (X, t).

Then the following fundamental proposition holds (cf. Bosi and Herden [16, Proposition 3.2]).

Proposition 2.24 *There exists a one-to-one correspondence between* $\mathbb{P}_{\unlhd}(X)$ *and* $\mathbb{S}_C(X)_{|_{\sim_{\mathscr{S}}}}$.

Proposition 2.24 leads us immediately to a first solution of the problem of characterizing in a simple way all *useful* (*continuously representable*) topologies on X, i.e., all topologies on X having the property according to which all their total continuous preorders are continuously representable by a utility function (cf. Bosi and Herden [16, Theorem 3.1] and Herden [35, Corollary 2.1]).

Proposition 2.25 *The following assertions are equivalent:*

(i) t *is useful.*

(ii) *For every complete separable system* $\mathscr{E} \in \mathbb{S}_C(X)$*, the topology* $t_{\mathscr{E}}$ *generated by* $\mathscr{E}$ *is second countable.*

One immediately verifies that Proposition 2.25 is a common generalization of **ET** (Eilenberg utility representation Theorem (Eilenberg [27])) and **DT** (Debreu utility representation Theorem (Debreu [24, 25])), which read as follows.

ET: *Every connected and separable topology t on X is useful.*
DT: *Every second countable topology t on X is useful.*

Herden and Pallack [38, Theorem 2.15] proved the following generalization of the Debreu utility representation Theorem.

Theorem 2.26 *Let* $\precsim$ *be a weakly continuous preorder on a second countable topological space* (X, t)*. Then there exists a continuous order-preserving function* $f : (X, \precsim, t) \to (\mathbb{R}, \leq, t_{nat})$.

Based on Theorem 2.26, we can present a characterization of the existence of a continuous order-preserving function, which utilizes weak continuity.

Theorem 2.27 *Let* $\precsim$ *be a preorder on a topological space* (X, t)*. Then the following conditions are equivalent:*

(i) *There exists a continuous order-preserving function* $f : (X, \precsim, t) \to (\mathbb{R}, \leq, t_{nat})$.
(ii) *There exists a second countable topology t' on X, which is coarser than t and with respect to which* $\precsim$ *is weakly continuous.*

Proof (i) $\Rightarrow$ (ii). Let $f : (X, \precsim, t) \to (\mathbb{R}, \leq, t_{nat})$ be a continuous order-preserving function, and consider the topology t' on X which is generated by the families $\{f^{-1}(]-\infty, q[)\}_{q \in \mathbb{Q}}$ and $\{f^{-1}(]q, +\infty[)\}_{q \in \mathbb{Q}}$, in order to immediately verify that t' is second countable, coarser than t and such that $\precsim$ is weakly continuous with respect to t'.

(ii) $\Rightarrow$ (i). Apply Theorem 2.26 to $\precsim$ on (X, t'), and just observe that a continuous order-preserving function $f : (X, \precsim, t') \to (\mathbb{R}, \leq, t_{nat})$ is continuous on (X, t). $\square$

Definition 2.28 A preorder $\precsim$ on a set X is said to be *Cantor-separable* if there exists a countable subset Z of X such that for all $(x, y) \in \prec$ there exists $z \in Z$ such that $x \prec z \prec y$.

The following theorem, which is closely related to the utility representation theorem of Peleg [58], illustrates the adequateness of the definition of weak continuity of a preorder (see the above Definition 2.8, (iii)), which was introduced by Bosi and Herden [12, 13]).

By the way, the reader may recall that Peleg [58], who was one of the first concerned with continuous representability of arbitrary preorders (actually, irreflexive and transitive binary relations) on (X, t) instead of only total preorders, when proving his continuous utility theorem, has taken advantage of the fact that a Cantor-separable preorder $\precsim$ on (X, t) is weakly continuous, provided that it satisfies the following properties:

P1: $l_{\precsim}(x)$ is open for every $x \in X$;
P2: $\overline{l_{\precsim}(x)} \subset l_{\precsim}(y)$ for every pair $(x, y) \in \prec$.

The reader may also consult Herden [33, Remark 4.1].

Proposition 2.29 *The following assertions hold:*

(i) *A Cantor-separable preorder $\precsim$ on (X, t) or, equivalently, a preorder $\precsim$ on (X, t) that has no jumps and which satisfies one of the conditions 7 or 8 of Theorem 2.23 is weakly continuous.*

(ii) *A preorder $\precsim$ on (X, t) is weakly continuous provided that it is both continuous and semi-closed.*

Proof A proof of assertion (i) is implicit in our proof of assertion (ii). In addition, the original proof of Peleg also can be applied in order to prove assertion (i). Hence, it suffices to concentrate on the proof of assertion (ii). Let, therefore, some pair $(x, y) \in \prec$ be arbitrarily chosen. Then the continuity of $\precsim$ will follow if we are able to prove the existence of a decreasing separable system $\mathscr{E}_{xy}$ on X such that $x \in E$ and $y \notin E$ for every set $E \in \mathscr{E}_{xy}$. We, thus, set $\mathscr{E}_{xy} := \{l_{\precsim}(z) | x \prec z \precsim y\}$ in order to then show that $\mathscr{E}_{xy}$ is the desired decreasing separable system on X satisfying the property that $x \in E$ and $y \notin E$ for every set $E \in \mathscr{E}_{xy}$. The definition of $\mathscr{E}_{xy}$ implies that we only must show $\mathscr{E}_{xy}$ to be a decreasing separable system on X. Hence, we distinguish between the following two cases.

Case 1: There exists some $z \in X$ such that $x \prec z \prec y$. Then we may conclude that $l_{\precsim}(z) \subset \overline{l_{\precsim}(z)} \subset d_{\precsim}(z) \subset l_{\precsim}(y)$ and the validity of condition **DS1** is guaranteed.

Case 2: There exists no $z \in X$ such that $x \prec z \prec y$. In this case $r_{\precsim}(x) \cup l_{\precsim}(y) = i_{\precsim}(x) \cup d_{\precsim}(y)$ is an open and closed subset of X (of course, the equation $r_{\precsim}(x) \cup l_{\precsim}(y) = i_{\precsim}(x) \cup d_{\precsim}(y)$ holds whenever $x \prec y$; the emptiness of the interval $]x, y[$ is not needed). But since $]x, y[= \emptyset$ we may conclude that $l_{\precsim}(y) = (i_{\precsim}(x) \cup d_{\precsim}(y)) \setminus r_{\precsim}(x)$ is open and closed, which also implies the validity of condition **DS1**.

In order to now finish the proof of the proposition take sets $l_{\precsim}(u) \in \mathscr{E}_{xy}$ and $l_{\precsim}(v) \in \mathscr{E}_{xy}$ such that $\overline{l_{\precsim}(u)} \subset l_{\precsim}(v)$ be arbitrarily chosen. Our arguments that have been used in order to do the second case allow us to assume without loss of generality that there exists some $w \in X$ such that $u \prec w \prec v$. Hence, our arguments that just have been used above apply. This statement already completes the proof of the proposition. $\qquad\square$

We must still mention that the concept of a continuous preorder on (X, t) up to now, however, has not completely been clarified in the literature. Indeed, some authors, in particular, Mas-Colell et al. [44] or Gerasimou [31] identify continuity of a preorder $\precsim$ on (X, t) with its closedness.

One immediately verifies (see the equivalence "4 $\Leftrightarrow$ 9" of Theorem 2.23), that in case that $\precsim$ is total, $\precsim$ is closed if and only if $\precsim$ is weakly continuous. But weak continuity of $\precsim$ and the property $\precsim$ to be a closed subset of the product space $(X \times X, t \times t)$ are not equivalent, in general, as the following example shows.

The following example of a closed and not weakly continuous preorder was presented by Herden and Pallack [38].

Example 2.30 Let $X := \mathbb{R}$. Then we consider the topology t on X that contains the empty set and the sets $X \setminus F$ where F runs through the empty set and all finite subsets of X. In addition, we choose the preorder

$$\precsim := \{(x, x) | x \in \mathbb{R}\} \cup \{1, 2\}.$$

Then $\precsim$ is obviously a closed subset of the product space $(X \times X, t \times t)$. Since the intersection of any two open set in nonempty, we have that every continuous real-valued function f on (X, t) is constant. Therefore, there cannot exist any continuous function $f_{12} : (X, \precsim, t)) \to (\mathbb{R}, \le, t_{nat})$ such that $f_{12}(1) < f_{12}(2)$, and this implies that $\precsim$ is not weakly continuous.

Remark 2.31 A slight extension of Schmeidler's proof in his well-known paper published in 1971 (see Schmeidler [60]) implies a variant of his theorem, which guarantees that a preorder $\precsim$ which is both continuous and semi-closed on a topological space (X, t) and which satisfies the additional property of the quotient topology $t_{\precsim}^{\precsim}$ being Hausdorf is such that the restriction of $\precsim$ to the components of (X, t) is total. The proof of this version of Schmeidler theorem will be completely presented in the last section of this paper, In opinion of the authors, this theorem underlines the relative strength of requiring both continuity and semi-closedness.

Despite Remark 2.31, we have that assertion (ii) of Proposition 2.29 cannot be improved. Indeed, we now present an example of a preorder $\precsim$ on (X, t) which satisfies, except for the assumption $r_{\precsim}(x)$ to be open for every point $x \in X$, any of the assumptions the validity of which is postulated by continuity and semi-closedness, but, nevertheless, fails to be continuous.

Example 2.32 Let $(X, t) := (\mathbb{R}, t_{nat})$ as underlying topological space. The preorder $\precsim$ to be considered on (X, t) is defined by setting

$$\precsim := \{(x, x) | x \in X\} \cup \{(x, y) \in X \times X | x \le y \le 0 \vee 1 \le x \le y \vee x \le 0 \wedge 1 \le y\}$$
$$\cup \{(z, y) \in X \times X | 0 < z < y \wedge 1 \le y\}.$$

A direct verification implies that for every $x \in X$ both sets $d_\prec(x)$ and $i_\prec(x)$ are closed. In addition, it follows that $l_\prec(x)$ is open for every $x \in X$. Let us now assume that there exists some complete decreasing separable system $\mathscr{E}_{01}$ on X that could be chosen in such a way that $0 \in E$ and $1 \notin E$ for every set $E \in \mathscr{E}_{01}$. Then the assumption every set $E \in \mathscr{E}_{01}$ to be the union of sets $l_\prec(z)$ ($0 < z \le 1$) does not mean any loss of generality. Since the interval $]0, 1[$ is empty our arguments that have been applied in the second case of the proof of Proposition 2.29 imply that in order for $\mathscr{E}_{01}$ to satisfy condition **DS1** it is necessary $l_\prec(1)$ to be closed. But $l_\prec(1) \subsetneq \overline{l_\prec(1)} = d_\prec(1)$ and we are done.

3 Characterization of Continuous Representations

Based on the concept of a complete decreasing separable system, the observations following Definition 2.20, and the above Remark 2.21, we can easily deduce the following variant of Theorem 3.2 in Herden and Pallack [38], providing a characterization of the existence of a continuous order-preserving function for a not necessarily total preorder on a topological space.

Theorem 3.1 *Let $\precsim$ be a preorder on a topological space (X, t). Then the following conditions are equivalent:*

(i) *There exists a continuous order-preserving function $f : (X, \precsim, t) \to (\mathbb{R}, \le, t_{nat})$.*
(ii) *There exists a countable complete decreasing separable system $\mathscr{E}$ on $(X, \precsim, t)$ such that for every pair $(x, y) \in \prec$ there exist sets $E, E' \in \mathscr{E}$ such that $\overline{E} \subset E'$, $x \in E$ and $y \in X \setminus E'$.*
(iii) *There exists a countable family $\{\mathscr{E}_n\}_{n \in \mathbb{N}}$ of complete decreasing separable systems on $(X, \precsim, t)$ such that for every pair $(x, y) \in \prec$ there exists some $n \in \mathbb{N}$ such that $x \in E$ and $y \in X \setminus E$ for every $E \in \mathscr{E}_n$.*

We are now interested in continuous multi-utility representation. For every $x \in X$ we denote by $\mathscr{E}(x)$ the family of all (complete) decreasing separable systems $\mathscr{E}$ on $(X, \precsim, t)$ such that $x \in \bigcap_{E \in \mathscr{E}} E$ for all $\mathscr{E} \in \mathscr{E}(x)$.

We are fully prepared for solving the problem of characterizing all preorders $\precsim$ on (X, t) that have a continuous multi-utility representation (cf. also Evren and Ok [29, Theorem 0 and Theorem 4], where the equivalence "(i) $\Leftrightarrow$ (ii)" has been proved).

Theorem 3.2 *Let $\precsim$ be a preorder on a topological space (X, t). Then the following assertions are equivalent:*

(i) $\precsim$ *has a continuous multi-utility representation.*

(ii) $x \precsim y$ *whenever for all* $\mathscr{E} \in \mathcal{E}(y)$ *and all* $E \in \mathscr{E}$ *the inclusion* $x \in E$ *holds.*

(iii) *For every pair* $(x, y) \in X \times X$ *such that* $not\,(y \precsim x)$ *there exists some* $\mathscr{E} \in$
$\mathcal{E}(x)$ *such that* $x \in E$ *and* $y \in X \setminus \overline{E}$ *for all* $E \in \mathscr{E}$.

(iv) $d_{\precsim}(x) = \bigcap\limits_{\mathscr{E} \in \mathcal{E}(x)} \bigcap\limits_{E \in \mathscr{E}} E$ *for all* $x \in X$.

Proof (i) $\Rightarrow$ (ii): Assertion (ii) means that $x \precsim y$, whenever $f(x) \le f(y)$ for all continuous increasing real-valued functions f on $(X, \precsim, t)$. This observation already guarantees the validity of the implication "(i) $\Rightarrow$ (ii)".

(ii) $\Rightarrow$ (iii): Since $not\,(y \precsim x)$, assertion (ii) implies the existence of some decreasing separable system $\mathscr{E} \in \mathcal{E}(x)$ for which there exists some $E \in \mathscr{E}$ such that $x \in E$ and $y \in X \setminus E$. Therefore, we distinguish between the following two cases.

Case 1: $E = \bigcap\limits_{E' \in \mathscr{E}} E'$. In this case the equation $\bigcap\limits_{E' \in \mathscr{E}} E' = \bigcap\limits_{E' \in \mathscr{E}} \overline{E'}$ implies that $E = \overline{E}$. Then $\{E\}$ is a separable system on $(X, \precsim, t)$ that belongs to $\mathcal{E}(x)$ and has the desired property that $x \in E$ and $y \in X \setminus \overline{E}$ for all $E \in \{E\}$.

Case 2: There exists some $E' \in \mathscr{E}$ such that $\overline{E'} \subsetneqq E$. Now $\mathscr{E}' = \{E'' \in \mathscr{E} \mid \overline{E''} \subset E'\}$ is a separable system on $(X, \precsim, t)$ that belongs to $\mathcal{E}(x)$ and has the desired property that $x \in E''$ and $y \in X \setminus \overline{E''}$ for all $E'' \in \mathscr{E}'$.

(iii) $\Rightarrow$ (iv): It is clear that, regardless the validity of condition (iii), we have that $d_{\precsim}(x) \subset \bigcap\limits_{\mathscr{E} \in \mathcal{E}(x)} \bigcap\limits_{E \in \mathscr{E}} E$ for all $x \in X$. Conversely, if for some $x \in X$ and $z \in X$ it happens that $z \notin d_{\precsim}(x) \Leftrightarrow not\,(z \precsim x)$, then from condition (iii) there exists some $\mathscr{E} \in \mathcal{E}(x)$ such that $x \in E$ and $z \in X \setminus \overline{E}$ for all $E \in \mathscr{E}$, so that $z \notin \bigcap\limits_{\mathscr{E} \in \mathcal{E}(x)} \bigcap\limits_{E \in \mathscr{E}} E$.

(iv) $\Rightarrow$ (i): In order to prove that $\precsim$ admits a continuous multi-utility representation, we may concentrate on the situation that points $x \in X$ and $y \in X$ such that $not\,(y \precsim x)$ or, equivalently, $y \notin d_{\precsim}(x)$ have been chosen. Then similar arguments as the ones that have been applied in the proof of the implication "(ii) $\Rightarrow$ (iii)" guarantee the existence of some separable system $\mathscr{E} = \{E_q\}_{q \in [0,1] \cap \mathbb{Q}}$ on $(X, \precsim, t)$ such that $x \in \bigcap\limits_{q \in [0,1] \cap \mathbb{Q}} E_q$ and $y \in X \setminus \bigcup\limits_{q \in [0,1] \cap \mathbb{Q}} E_q = X \setminus \bigcup\limits_{q \in [0,1] \cap \mathbb{Q}} \overline{E_q}$. With the help of these equations we may define a continuous decreasing function $f : (X, \precsim, t) \longrightarrow ([0, 1], \le, t_{nat})$ by setting

$$f(z) := \begin{cases} \inf\{q \in [0, 1] \cap \mathbb{Q} \mid z \in E_q\}, & \text{if } z \in \bigcup\limits_{q \in [0,1] \cap \mathbb{Q}} E_q, \\ 1, & \text{otherwise,} \end{cases}$$

for all $z \in X$. The definition of f implies that $f(x) = 0 < 1 = f(y)$. This consideration finishes the proof of the theorem. $\qquad\square$

Alcantud et al. [2, Proposition 3.2] proved that there exists a continuous Richter–Peleg multi-utility representation for a preorder on a topological space if and only if there exist both a continuous multi-utility representation and a continuous order-preserving function for the preorder. Therefore, from Theorem 3.1 and Theorem 3.2, we immediately get the following characterization of the existence of a continuous Richter–Peleg multi-utility representation.

Theorem 3.3 *Let $\precsim$ be a preorder on a topological space (X, t). Then the following assertions are equivalent:*

(i) $\precsim$ *has a continuous Richter–Peleg multi-utility representation.*
(ii) *The following conditions hold:*

  a. *For every pair $(x, y) \in X \times X$ such that $\mathrm{not}\,(y \precsim x)$ there exists some $\mathscr{E} \in \mathscr{E}(x)$ such that $x \in E$ and $y \in X \setminus \overline{E}$ for all $E \in \mathscr{E}$;*

  b. *There exists a countable complete decreasing separable system $\mathscr{E}$ on $(X, \precsim, t)$ such that for every pair $(x, y) \in \prec$ there exists sets $E, E' \in \mathscr{E}$ such that $\overline{E} \subset E'$, $x \in E$ and $y \in X \setminus E'$.*

4   Restrictive Results Concerning Continuous Multi-utility Representations

Schmeidler [60] proved the following famous theorem.

Theorem 4.1 (Schmeidler [60]) *Let $\precsim$ be a nontrivial preorder on a connected topological space (X, t). If, for every $x \in X$, the sets $d_{\precsim}(x)$ and $i_{\precsim}(x)$ are closed and the sets $l_{\precsim}(x)$ and $r_{\precsim}(x)$ are open, then the preorder $\precsim$ is total.*

In combination with the results of the preceding section, in this section we first want to present the following more general version of Schmeidler's Theorem.

Theorem 4.2 *Let $(X, \precsim, t)$ be a preordered topological space that satisfies the properties $t_{\widetilde{\precsim}}$ to be Hausdorff, and $\precsim$ to be both continuous and semi-closed. Then the restriction of $\precsim$ to the components of (X, t) is total.*

Proof Let $C \subset X$ be a component of (X, t) and let the point $x \in C$ be arbitrarily chosen. Since $t_{\widetilde{\precsim}}$ is Hausdorff it follows, with help of condition **SB** of Proposition 2.3, that at least one of the sets $l_{\precsim}(x)$ or $r_{\precsim}(x)$ is not empty. We, thus, must distinguish between the cases $l_{\precsim}(x) \neq \emptyset$ and $r_{\precsim}(x) = \emptyset$, $l_{\precsim}(x) = \emptyset$ and $r_{\precsim}(x) \neq \emptyset$ and $l_{\precsim}(x) \neq \emptyset$ and $r_{\precsim}(x) \neq \emptyset$. Since all these cases can be handled by analogous arguments it suffices to discuss the case that $l_{\precsim}(x)$ as well as $r_{\precsim}(x)$ is not empty. Let, therefore, $\mathscr{O}(x)$ be the collection of all open intervals $]y, z[$ that contain x. Then we set $O_x := \bigcup\limits_{]y,z[\,\in\mathscr{O}(x)}]y, z[$

in order to distinguish between the cases $C \cap O_x = \emptyset$ and $C \cap O_x \neq \emptyset$. In the first case we may conclude that $C = [x]$ and we are done. In the second case it follows that there exists some point $y \in C$ or some point $z \in C$ such that $y \prec x$ or $x \prec z$. Since $\precsim$ is assumed to be semi-closed these inequalities guarantee, however, that the sub-space $(C, \precsim_{|C}, t_{|C})$ of $(X, \precsim, t)$ satisfies the assumptions of Schmeidler's Theorem. Hence, the restriction $\precsim_{|C}$ of $\precsim$ to C is total. □

From Proposition 2.18 and Theorem 4.2, we immediately obtain the following restrictive result concerning the existence of finite continuous Richter–Peleg multi-utility representations.

Corollary 4.3 *Let $\precsim$ be a preorder on a topological space (X, t), which admits a finite continuous Richter–Peleg multi-utility representation $\mathscr{F} = \{f_1, ..., f_n\}$. Then the restriction of $\precsim$ to the components of (X, t) is total.*

A preorder $\precsim$ on (X, t) which has a continuous multi-utility representation must be both closed and continuous. The following three problems are therefore particularly important.

Problem 1 Determine all topological spaces (X, t) having the property that all their closed preorders are weakly continuous.

Problem 2 Determine all topological spaces (X, t) having the property that all their weakly continuous preorders are closed.

Problem 3 Determine all topological spaces (X, t) having the property that all their weakly continuous preorders admit a continuous multi-utility representation.

The first problem has been analyzed, at least partially, in Bosi and Herden [15]. The following definition is found in Nachbin [56].

Definition 4.4 A preorder $\precsim$ on a topological space (X, t) is said to be *normal* if for any two disjoint closed decreasing, respectively increasing subsets A and B of X there exist disjoint open decreasing, respectively increasing subsets U and V of X such that $A \subset U$ and $B \subset V$.

For example, the following results hold (see Bosi and Herden [15, Corollary 3.5 and Theorem 3.4]).

Theorem 4.5 *Let (X, t) be a connected metrizable space. Then the following assertions are equivalent:*

(i) *Every closed preorder $\precsim$ on (X, t) admits a continuous multi-utility representation.*

(ii) *(X, t) is locally compact and second countable.*

Theorem 4.6 *Let (X, t) be a Hausdorff space. Then the following assertions are equivalent:*

 (i) *Every closed preorder $\precsim$ on (X, t) admits a continuous multi-utility representa-
     tion.*
(ii) *Every closed preorder $\precsim$ on (X, t) is normal.*

As regards Problems 2 and 3 above, surprisingly, the common solution of both problems is possible in a very satisfactory and restrictive way.

Before stating the corresponding theorem, the reader may recall that, when we consider the space $C(X, t, \mathbb{R})$ of all continuous real-valued function on the topological space (X, t), the *weak topology* on X, $\sigma(X, C(X, t, \mathbb{R}))$, is the coarsest topology on X satisfying the property that every continuous real-valued function on (X, t) remains being continuous. Two points $x, y \in X$ are considered as being *equivalent* if $f(x) = f(y)$ for all functions $f \in C(X, t, \mathbb{R})$. For two equivalent points $x, y \in X$, we write $x \sim_{C(X,t,\mathbb{R})} y$ ($x \sim_C y$ for the sake of brevity).

It is well known that $(X_{|\sim_C}, \sigma(X, C(X, t, \mathbb{R}))_{|\sim_C})$ is a completely regular Hausdorff-space (cf., for instance, Cigler and Reichel [23, Satz 10, p. 101]). It is clear that $(X_{|\sim_C}, \sigma(X, C(X, t, \mathbb{R}))_{|\sim_C})$ is the quotient space of $\sigma(X, C(X, t, \mathbb{R}))$ that is induced by the equivalence relation $\sim_C$.

The following lemma holds true.

Lemma 4.7 *The coarsest topology on X satisfying the property that all weakly continuous preorders on (X, t) remain being continuous is $\sigma(X, C(X, t, \mathbb{R}))$. (Of course, this assertion is equivalent to the statement that a preorder $\precsim$ on (X, t) is weakly continuous if and only if it is weakly continuous with respect to $\sigma(X, C(X, t, \mathbb{R}))$).*

Proof Although the validity of this assertion appears somewhat surprisingly, its trueness is trivial. Indeed, since weak continuity of a preorder $\precsim$ on (X, t) is described by continuous (increasing) real-valued functions, its validity is immediate (the reader may notice that this phenomenon underlines once more the appropriateness of the concept of a weakly continuous preorder on (X, t)). □

We are now ready to prove the very restrictive result according to which the continuous multi-utility representability of all closed preorders on a topological space is equivalent to the continuous multi-utility representability of all weakly continuous preorders, and in turn to the requirement according to which the quotient topology with respect to the coincidence of all continuous functions is discrete

Theorem 4.8 *Let (X, t) be a topological space. The following assertions are equivalent:*

  (i) *Every weakly continuous preorder $\precsim$ on (X, t) has a continuous multi-utility
      representation.*
 (ii) *Every weakly continuous preorder $\precsim$ on (X, t) is closed.*
(iii) $t_{|\sim_C}$ *is the discrete topology on $X_{|\sim_C}$.*

Proof (i) $\Rightarrow$ (ii): We already know that a preorder $\precsim$ on (X, t) that has a continuous multi-utility representation must be closed. Hence, nothing remains to be shown.

(ii) $\Rightarrow$ (iii): Let (X, t) be a topological space for which every weakly continuous preorder $\precsim$ is closed. Then the properties of the defined equivalence relation $\sim_C$ on X imply that every weakly continuous preorder $\precsim$ on $X_{|\sim_C}$ is closed. Therefore, we may identify the topological spaces (X, t) and $(X_{|\sim_C}, t_{|\sim_C})$. This means that we may assume, in the remainder of the proof of the implication (ii) $\Rightarrow$ (iii), (X, t) to be a Hausdorff-space. In order to now prove the validity of the implication it, thus, suffices to show that there exists no point $y \in X$ such that the singleton $\{y\}$ is not an open subset of X. This will be done by contraposition. We, therefore, assume, in contrast, that there exists at least one point $y \in X$ such that $\{y\}$ is not an open subset of X and proceed by arbitrarily choosing some point $z \in X \setminus \{y\}$. Since (X, t) is a Hausdorff-space, it follows that $\{y\}$ as well as $\{z\}$ are closed subsets of X, which implies that $D := \{y, z\}$ is a closed subset of X. The inclusion $(X, \sigma(X, C(X, t, \mathbb{R})) \subset (X, t)$ allows to conclude, in addition, that every continuous function $f : (X, \sigma(X, C(X, t, \mathbb{R})) \to ([0, 1], t_{nat})$ is a continuous function $f : (X, t) \to ([0, 1], t_{nat})$. Since $(X, \sigma(X, C(X, t, \mathbb{R}))$ is completely regular, there exists for every point $x \in X \setminus D$ a continuous function $f_x : (X, t) \to ([0, 1], t_{nat})$ such that $f_x(x) = 0$ and $f_x(D) = \{1\}$. Hence, the preorder $\precsim$ on (X, t) that is defined by setting

$$\precsim := \{(v, v) | v \in X\} \cup \{(x, y) | x \in X \setminus D\} \cup \{(x, z) | x \in X \setminus D\}$$

is continuous. Assertion (ii), thus, implies that $\precsim$ is a closed subset of $X \times X$. Hence, it follows that there exist open subsets U and V of X such that $y \in U$ and $z \in V$ and, moreover, $U \times V \cap \precsim = \emptyset$. Since y is assumed to be not an open subset of X, we may conclude that U contains at least one point u that is different from y. Because of the definition of $\precsim$, this means, however, that $(u, z) \in U \times V \cap \precsim$. This contradiction proves assertion (iii).

(iii) $\Rightarrow$ (i): Assertion (iii) implies that $\sigma(X, C(X, t, \mathbb{R}))_{|\sim_C} = t_{|\sim_C}$. Hence, the space $(X_{\sim_C}, \sigma(X, C(X, t, \mathbb{R}))_{|\sim_C})$ is discrete, which allows us to conclude that every weakly continuous preorder $\precsim$ on $(X_{\sim_C}, \sigma(X, C(X, t, \mathbb{R}))_{|\sim_C})$ admits a continuous multi-utility representation. It, thus, follows that every weakly continuous preorder $\precsim$ on $(X, \sigma(X, C(X, t, \mathbb{R}))$ has a continuous multi-utility representation. This means that we may apply Lemma 4.7 in order to conclude that also every weakly continuous preorder $\precsim$ on (X, t) admits a continuous multi-utility representation, which finishes the proof of the implication. $\qquad \square$

Remark 4.9 Clearly, the equivalent assertions (i), (ii) and (iii) of Theorem 4.8 are also equivalent to any of the following (equivalent) assertions:

(iv) Every weakly continuous preorder $\precsim$ on $(X, \sigma(X, C(X, t, \mathbb{R}))$ has a continuous multi-utility representation;
(v) Every weakly continuous preorder $\precsim$ on $(X, \sigma(X, C(X, t, \mathbb{R}))$ is closed;
(vi) $\sigma(X, C(X, t, \mathbb{R}))_{\sim_C}$ is the discrete topology on $X_{\sim_C}$.

5 Conclusions

In this paper we have presented some general results concerning the existence of continuous representations of nontotal preorders on a topological space. The corresponding characterizations are mainly based on the concept of a *complete decreasing separable system* in a preordered topological space, which was introduced and widely studied by Herden [33–35].

We have focused our attention on continuity-like conditions which are necessary for the existence of a continuous order-preserving function and respectively a continuous multi-utility representation.

In particular, we have taken into consideration the property of *weak continuity*. Following the terminology introduced by Bosi and Herden [12, 13], a preorder on a topological space (X, t) is weakly continuous if for every pair $(x, y) \in \prec$ there exists a continuous and increasing real-valued function f_{xy} on X such that $f_{xy}(x) < f_{xy}(y)$.

We have presented some results which illustrate the restrictiveness of the continuous multi-utility representation, which nevertheless has been presented in the past as the best kind of continuous representation under incompleteness of the preference relation. To be precise, by using considerations according to which the quotient order topology is a Hausdorff topology, we have proven a variant of a famous theorem by Schmeidler [60]. Indeed, we have shown that if a continuous multi-utility representation exists for a preorder whose strict lower and upper sections are all open, then the preorder is total on each component. Further, we have proven that if a finite continuous Richter–Peleg multi-utility representation exists, then the preorder is total on every component.

Finally, using classical considerations concerning the *weak topology*, we have shown that the continuous multi-utility representability of all closed preorders (or equivalently weakly continuous preorders) on a topological space is equivalent to the requirement according to which the quotient topology with respect to the equivalence corresponding to the coincidence of all continuous functions is discrete.

References

1. Alcantud, J.C.R., Bosi, G., Campión, M.J., Candeal, J.C., Induráin, E., Rodríguez-Palmero, C.: Continuous utility functions through scales. Theory Decis. **64**, 479–494 (2008)
2. Alcantud, J.C.R., Bosi, G., Zuanon, M.: Richter-Peleg multi-utility representations of preorders. Theory Decis. **80**, 443–450 (2016)
3. Aumann, R.: Utility theory without the completeness axiom. Econometrica **30**, 445–462 (1962)
4. Banerjee, K., Dubey, R.S.: On multi-utility representation of equitable intergenerational preferences. In: Econophysics and Economics of Games, Social Choices and Quantitative Techniques. New Economic Windows, pp. 175–180. Springer, Berlin (2010)
5. Baucells, M., Shapley, L.S.: Multiperson utility. Games Econ. Behav. **62**, 329–347 (2008)
6. Beardon, A.F., Mehta, G.B.: The utility theorems of Wold, Debreu and Arrow-Hahn. Econometrica **62**, 181–186 (1994)
7. Beardon, A.F., Mehta, G.B.: Utility functions and the order type of the continuum. J. Math. Econ. **23**, 387–390 (1994)

8. Bevilacqua, P., Bosi, G., Zuanon, M.: Existence of order-preserving functions for nontotal fuzzy preference relations under decisiveness. Axioms **29**, 1–10 (2017)

9. Bevilacqua, P., Bosi, G., Zuanon, M.: Representation of a preorder on a topological space by a countable family of upper semicontinuous order-preserving functions. Adv. Appl. Math. Sci. **17**, 417–427 (2018)

10. Bosi, G., Estevan, A., Raventós-Pujol, A., Topologies for semicontinuous multi-utilities. Theory Decis., to appear

11. Bosi, G., Estevan, A., Zuanon, M.: Partial representations of orderings. Int. J. Uncertain., Fuzziness Knowl. Based Syst. **26**, 453–473 (2018)

12. Bosi, G., Herden, G.: On a strong continuous analogue of the Szpilrajn theorem and its strengthening by Dushnik and Miller. Order **22**, 329–342 (2005)

13. Bosi, G., Herden, G.: On a possible continuous analogue of the Szpilrajn theorem and its strengthening by Dushnik and Miller. Order **23**, 271–296 (2006)

14. Bosi, G., Herden, G.: Continuous multi-utility representations of preorders. J. Math. Econ. **48**, 212–218 (2012)

15. Bosi, G., Herden, G.: On continuous multi-utility representations of semi-closed and closed preorders. Math. Soc. Sci. **79**, 20–29 (2016)

16. Bosi, G., Herden, G.: The structure of useful topologies. J. Math. Econ. **82**, 69–73 (2019)

17. Bosi, G., Mehta, G.B.: Existence of a semicontinuous or continuous utility function: a unified approach and an elementary proof. J. Math. Econ. **38**, 311–328 (2002)

18. Bosi, G., Zuanon, M.: Continuous multi-utility for extremely continuous preorders. Int. J. Contemp. Math. Sci. **4**, 439–445 (2009)

19. Bridges, D.S., Mehta, G.B.: Representation of Preference Orderings. Springer, Berlin (1995)

20. Burgess, D.C.J., Fitzpatrick, M.: On separation axioms for certain types of ordered topological space. Math. Proc. Camb. Philos. Soc. **82**, 59–65 (1977)

21. Campión, M.J., Candeal, J.C., Induráin, E.: Preorderable topologies and order-representability of topological spaces. Topol. Its Appl. **156**, 2971–2978 (2009)

22. Cato, S.: Szpilrajn, Arrow and Suzumura: concise proofs of extensions theorems and an extension. Metroeconomica **63**, 235–249 (2012)

23. Cigler, J., Reichel, H.C.: Topologie. Bibliographisches Institut, Mannheim-Wien-Zürich (1978)

24. Debreu, G.: Representation of a preference ordering by a numerical function. In: Thrall, R., Coombs, C., Davies, R. (eds.) Decision Processes. Wiley, New York (1954)

25. Debreu, G.: Continuity properties of Paretian utility. Int. Econ. Rev. **5**, 285–293 (1964)

26. Dubra, J., Maccheroni, F., Ok, E.A., Expected utility theory without the completeness axiom. J. Econ. Theory **115**, 118–133 ((2004))

27. Eilenberg, S.: Ordered topological spaces. Am. J. Math. **63**, 39–45 (1941)

28. Evren, O.: On the existence of expected multi-utility representations. Econ. Theory **35**, 575–592 (2008)

29. Evren, O., Ok, E.A.: On the multi-utility representation of preference relations. J. Math. Econ. **47**, 554–563 (2011)

30. Galaabaatar, T., Karni, E.: Expected multi-utility representations. Math. Soc. Sci. **64**, 242–246 (2012)

31. Gerasimou, G.: On continuity of incomplete preferences. Soc. Choice Welf. **41**, 157–167 (2013)

32. Gorno, L.: A strict expected multi-utility theorem. J. Math. Econ. **71**, 92–95 (2017)

33. Herden, G.: On the existence of utility functions. Math. Soc. Sci. **17**, 297–313 (1989)

34. Herden, G.: On the existence of utility functions II. Math. Soc. Sci. **18**, 107–117 (1989)

35. Herden, G.: Topological spaces for which every continuous total preorder can be represented by a continuous utility function. Math. Soc. Sci. **22**, 123–136 (1991)

36. Herden, G., Mehta, G.B.: The Debreu gap lemma and some generalizations. J. Math. Econ. **40**, 747–769 (2004)

37. Herden, G., Pallack, A.: Useful topologies and separable systems. Appl. Gen. Topol. **1**, 61–82 (2000)

38. Herden, G., Pallack, A.: On the continuous analogue of the Szpilrajn theorem I. Mathematical Social Sciences **43**, 115–134 (2002)

39. Johnson, D.G., Mandelker, M., Separating chains in topological spaces. J. Lond. Math. Soc. **4**, 510–512 (1971/72)
40. Kabanov, Y., Lépinette, E.: Essential supremum and essential maximum with respect to random preference relations. J. Math. Econ. **49**, 488–495 (2013)
41. Kaminski, B.: On quasi-orderings and multi-objective functions. Eur. J. Oper. Res. **177**, 1591–1598 (2007)
42. Levin, V.: Measurable utility theorems for closed and lexicographic preference relations. Sov. Math. Dokl. **27**, 639–643 (1983)
43. Levin, V.L.: The Monge-Kantorovich problems and stochastic preference relation. Adv. Math. Econ. **3**, 97–124 (2001)
44. Mas-Colell, A., Whinston, M.D., Green, J.R.: Microeconomic Theory. Oxford University Press, Oxford (1995)
45. Mashburn, J.D.: A note on reordering ordered topological spaces and the existence of continuous, strictly increasing functions. Topol. Proc. **20**, 207–250 (1995)
46. Mehta, G.B.: Topological ordered spaces and utility functions. Int. Econ. Rev. **18**, 779–782 (1977)
47. Mehta, G.B.: A new extension procedure for the Arrow-Hahn theorem. Int. Econ. Rev. **22**, 113–118 (1981)
48. Mehta, G.B.: Continuous utility functions. Econ. Lett. **18**, 113–115 (1985)
49. Mehta, G.B.: Existence of an order preserving function on normally preordered spaces. Bull. Aust. Math. Soc. **34**, 141–147 (1986)
50. Mehta, G.B.: On a theorem of Fleischer. J. Aust. Math. Soc. **40**, 261–266 (1986)
51. Mehta, G.B.: Some general theorems on the existence of order-preserving functions. Math. Soc. Sci. **15**, 135–143 (1988)
52. Mehta, G.B., Preference and utility. In: Barberá, S., Hammond, P.J., Seidl, C. (eds.) Handbook of Utility Theory, pp. 1–47. Kluwer Academic Publishers, Dordrecht (1998)
53. Minguzzi, E.: Topological conditions for the representation of preorders by continuous utilities. Appl. Gen. Topol. **13**, 81–89 (2012)
54. Minguzzi, E.: Normally preordered spaces and utilities. Order **30**, 137–150 (2013)
55. Nishimura, H., Ok, E.A.: Utility representation of an incomplete and nontransitive preference relation. J. Econ. Theory **166**, 164–185 (2016)
56. Nachbin, L.: Topology and Order. D. Van Nostrand Company, New York (1965)
57. Ok, E.A.: Utility representation of an incomplete preference relation. J. Econ. Theory **104**, 429–449 (2002)
58. Peleg, B.: Utility functions for partially ordered topological spaces. Econometrica **38**, 93–96 (1970)
59. Richter, M.: Revealed preference theory. Econometrica **34**, 635–645 (1966)
60. Schmeidler, D.: A condition for the completeness of partial preference relations. Econometrica **3**, 403–404 (1971)
61. Suzumura, K.: Remarks on the theory of collective choice. Economica **43**, 381–390 (1976)
62. Ward Jr., L.E.: Partially ordered topologicl spaces. Proc. Am. Math. Soc. **5**, 144–161 (1954)
63. Yilmaz, O.: Utility representation of lower separable preferences. Math. Soc. Sci. **56**, 389–394 (2008)

Jointly Continuous Multi-utility Representations of Non-total Preorders

Alessandro Caterino, Rita Ceppitelli and Lubica Holá

Abstract In the present paper, dealing with non-total preorders, we are interested in the problem of the existence of multi-utility representations that are jointly continuous in both agents and commodities. Non-total preorders are more realistic to explain the behavior of an individual. Indeed, a decision maker may be incapable or unwilling to compare all feasible alternatives. We study the continuous multi-utility representation of a non-total preorder. This kind of representation is more functional than the classical Richter–Peleg utility representation as it allows a complete characterization of the preorder. Our representation theorems are proved in submetrizable k_ω-consumption spaces.

Keywords Non-total preorders · Submetrizable k_ω-spaces · Jointly continuous multi-utility functions · Fell topology

2010 Mathematics Subject Classification 54F05 · 91B16 · 54B20

1 Introduction

In Mathematical Economics, closed preorders are often interpreted as customer preference relations.

A. Caterino
Via dei Narcisi 41A, 06100 Perugia, Italy
e-mail: sandro.caterino@gmail.com

R. Ceppitelli (✉)
Department of Mathematics and Computer Sciences, University of Perugia, Via L. Vanvitelli 1, 06123 Perugia, Italy
e-mail: rita.ceppitelli@unipg.it

L. Holá
Institute of Mathematics, Academy of Science, Štefánikova 49, 81473 Bratislava, Slovakia
e-mail: hola@mat.savba.sk

© Springer Nature Switzerland AG 2020
G. Bosi et al. (eds.), *Mathematical Topics on Representations of Ordered Structures and Utility Theory*, Studies in Systems, Decision and Control 263,
https://doi.org/10.1007/978-3-030-34226-5_12

The problem concerning the existence of a continuous real-valued preorder-preserving map (a utility function) on a topological space endowed with a closed preorder is called a *continuous utility representation problem.*

In this paper we will suppose that instead of one preference relation there is a family $\mathcal{P}_X$ of preorders defined on a topological space X.

The study of the representability of a family of preorders is known in the literature as a *jointly continuous utility representation problem.* It consists in finding suitable topologies on $\mathcal{P}_X$ and X that ensure the existence of a continuous function defined on $\mathcal{P}_X \times X$ which is a continuous utility function for every $\preceq \in \mathcal{P}_X$.

Problems of this type are important in Mathematical Economics, where X is a commodity space and $\mathcal{P}_X$ can be considered as a family of preference relations $\{\preceq_i : i \in I\}$, where I is a set of economics agents.

Traditionally, the joint continuity has been widely studied in case all the preorders are total [3, 26, 33] (for a comprehensive reference see the last chapter of the book by Bridges and Mehta [10]).

Back in [3] studied the space $\mathcal{P}_{\mathcal{L}}$ of total closed preorders defined on closed subsets of X. Following Hildebrand [26] and Mass-Colell [33], Back [3] endowed $\mathcal{P}_{\mathcal{L}}$ with the Fell topology (*the topology of closed convergence*) which is a natural topology in Mathematical Economics. The space $\mathcal{U}$ of all continuous real partial maps defined on closed subsets of X (*utility functions*) was endowed with a topology that is a generalization of the compact-open topology. This topology has applications in Mathematical Economics as well as in Differential Equations and in the convergence of Dynamic Programming Models.

Back proved the existence of a continuous map from the space $\mathcal{P}_{\mathcal{L}}$ to the space $\mathcal{U}$ which is a utility function for every fixed preorder [3].

A classical result on the existence of the jointly continuous utility representation for a family $\mathcal{P}_X$ of closed not necessarily total preorders on locally compact and second countable spaces X was provided by Levin [28]. Levin's result can be extended to the space $\mathcal{P}$ of closed preorders defined on closed subsets of X (see Corollary 1, in [28]).

Note that Back's result uses the Corollary 1 of Levin [28].

In [13–17] the authors generalized Back's result in two ways. They considered a more general topological structure on X (a submetrizable k_ω-space) and studied the jointly continuous utility representation problem for non-total closed preorders defined on closed subsets of X.

In these studies they assumed the classical notion of a utility representation for a non-total preorder, referred as the *Richter–Peleg utility representation* (see also [5, 9, 17, 24, 25, 28, 35, 36, 39]).

Actually, the use of this representation is limited for it does not characterize the original non-total preorder. In fact, one cannot retrieve the original preorder from a Richter–Peleg utility function, which provides a strictly smaller information. A second and more recent approach to represent a non-total preorder is based on the concept of a *continuous multi-utility representation.* A non-total preorder $\preceq$ on a topological space (X, τ) admits a *continuous multi-utility representation* if there exists a family $\mathcal{F}_{\preceq}$ of continuous increasing real functions on X such that, for all $x, y \in X$, $x \preceq y$ is equivalent to $f(x) \leq f(y)$ for all $f \in \mathcal{F}_{\preceq}$.

After this, Minguzzi in [37] introduced the concept of a continuous Richter–Peleg multi-utility representation of a preorder $\preceq$, i.e. every function $f \in \mathcal{F}_{\preceq}$ is a Richter–Peleg utility function for $\preceq$.

With respect to this second approach, in the present paper we study the *jointly continuous multi-utility representation problem.* Precisely, we determine a multi-utility representation which is jointly continuous with respect to agents and commodities in a fixed submetrizable k_ω-consumption space X.

We recall that a k_ω-space (i.e. a hemicompact k-space) is submetrizable if and only if its compact subsets are metrizable [12]. Interesting examples of submetrizable k_ω-spaces that are not metrizable spaces are the inclusion inductive limit $lim_{\rightarrow}\mathbb{R}^n$ (see Definition 2.4 in [13]), and the space S' of tempered distributions ([12], example 4.4.1). These spaces have applications in Mathematical Economics as shown in [13, 16]. In these applications the existence of utilities representing families of preorders is important.

In the following let (X, τ) be a topological space, $C(X, \mathbb{R})$ be the space of all continuous real-valued functions defined on X and $CL(X \times X)$ be the family of all nonempty closed subsets of $X \times X$. Let

$$\mathcal{P}_X = \{\preceq : \preceq \text{ is a preorder on } X\} \subset CL(X \times X)$$

be the family of closed preorders defined on X.
For every $\preceq \in \mathcal{P}_X$ let

$$H_{\preceq} = \{u \in C(X, \mathbb{R}) : u \text{ is a utility function for } \preceq\}.$$

For every $\preceq \in \mathcal{P}_X$, $H_{\preceq}$ is an element of $2^{C(X,\mathbb{R})}$.

The jointly continuous multi-utility representation problem consists in finding suitable topologies on $\mathcal{P}_X$ and $2^{C(X,\mathbb{R})}$ ensuring the existence of a continuous function $L : \mathcal{P}_X \rightarrow 2^{C(X,\mathbb{R})}$ such that for every $\preceq \in \mathcal{P}_X$, $L(\preceq)$ is a continuous multi-utility representation for $\preceq$.

It is known that if X is a submetrizable k_ω-space, then $(C(X, \mathbb{R}), \tau_{UC})$, the space of continuous real-valued functions with the topology of uniform convergence on compacta, has a countable base and it is metrizable [34]. We will introduce in $2^{C(X,\mathbb{R})}$ the gap $D_\rho(A, B) = \inf\{\rho(f, g) : f \in A, g \in B\}$, where ρ stands for a compatible metric in $C(X, \mathbb{R})$ and A and B are two nonempty subsets of $(C(X, \mathbb{R}), \rho)$.

As regard the topology in $\mathcal{P}_X$, it is appropriate to require that it satisfies the following condition:

$$(*) \qquad x_\sigma \rightarrow x, \ y_\sigma \rightarrow y, \ \preceq_\sigma \rightarrow \preceq, \ x_\sigma \preceq_\sigma y_\sigma \Longrightarrow x \preceq y.$$

It is known that if X is a locally compact Hausdorff space, the classical topology of closed convergence (the Fell topology) on $\mathcal{P}_X$ satisfies the property $(*)$ (see [4] for a comprehensive reference on the Fell topology).

For a submetrizable k_ω-space X we will define a mapping $L : \mathcal{P}_X \to 2^{C(X, \mathbb{R})}$, such that for every $\preceq \in \mathcal{P}_X$, $L(\preceq)$ is a continuous multi-utility representation for $\preceq$. In Theorem 4.12 we will prove the following result: If $\{\preceq_\sigma : \sigma \in \Sigma\}$ converges to $\preceq$ in $(\mathcal{P}_X, \tau(\mathcal{G}))$, then $\{D_\rho(L(\preceq_\sigma), L(\preceq)) : \sigma \in \Sigma\}$ converges to 0. The topologies $\tau(\mathcal{G})$ were introduced in [16] using compatible uniform structures $\mathbf{U}$ on X. These topologies satisfy the property $(*)$.

We note that if X is a Hausdorff second countable locally compact space, then there is a compatible metric d such that (X, d) is a boundedly compact space [41]. Denote by $\mathcal{G}_d$ the uniformity generated by the metric d. It is easy to verify that the topology $\tau(\mathcal{G}_d)$ coincides with the Fell topology.

Notice that the mapping $L : \mathcal{P}_X \to 2^{C(X, \mathbb{R})}$ is a bijection between $\mathcal{P}_X$ and $L(\mathcal{P}_X)$. In Theorem 4.15 we will prove that the mapping $L^{-1} : L(\mathcal{P}_X) \to \mathcal{P}_X$ is continuous, if $L(\mathcal{P}_X)$ is equipped with the lower Vietoris topology and $\mathcal{P}_X$ is equipped with the upper Fell topology (see [19] for the definition and properties of the lower Vietoris topology).

2   Notation and Preliminaries

In Mathematical Economics a preference relation $\preceq$ on a set X of alternatives is often assumed to have the structure of a preorder, i.e. a binary relation which is *reflexive* and *transitive*. A preference relation $\preceq$ is *complete* or *total* if, for all $x; y \in X$, either $x \preceq y$ or $y \preceq x$. The literature has focused on two main problems: representability of preorders by means of real-valued order-preserving functions (utility representation of $\preceq$) and continuity of utility functions. Classical hypotheses for the existence of continuous representations of preorders are the continuity or the closedness of the preorder.

A preorder $\preceq$ on a topological space (X, τ) is *continuous* if for every $x \in X$ the sets $(-\infty, x]$ and $[x, +\infty)$ are closed in X.

A preorder $\preceq$ on a topological space (X, τ) is said to be *closed* if $\preceq$ is closed as a subset of $X \times X$ in the product topology $\tau \times \tau$.

It is well known that if the preorder is total, then it is closed iff it is continuous. In general a closed preorder is always continuous [38].

The representability of total preorders is defined as follows.

Definition 2.1 A total preorder $\preceq$ on a set X is representable in $\mathbb{R}$ if there exists a function $u : X \longrightarrow \mathbb{R}$ such that

$$\forall x, y \in X, \, x \preceq y \iff f(x) \le f(y).$$

The literature has mostly been studying the representability of a total preorder; an interesting and complete reference on the representation of total preorders is provided by the book of Bridges and Mehta [10].

Later on, the study of non-total (partial) preorders developed a growing interest. Peleg [40] was the first who presented sufficient conditions for the existence of a continuous utility function for a non-total order on a topological space. Peleg solved a problem which was posed by Aumann in the context of expected utility. Aumann [2] had observed that a rational decision-maker may express *indecisiveness* (or equivalently *incomparability*) between two alternatives, so that he is not forced to express *indifference*.

Non-total preorders are usually considered in the context of decision-making under uncertainty and risk and they are more realistic to explain the behavior of an agent, of an individual.

There are two main notions of utility representation for a non-total preorder on a topological space X. The first and most classical one is often referred as the *Richter–Peleg utility representation*:

Definition 2.2 A function $u : X \longrightarrow \mathbb{R}$ is a utility function representing a preference relation $\preceq$ if:

- $\forall x, y \in X, x \preceq y \Rightarrow f(x) \leq f(y)$;
- $\forall x, y \in X, x \prec y \Rightarrow f(x) < f(y)$.

Many authors have explored this notion of representation [5, 9, 17, 24, 25, 28, 39]. Mehta proved very general conditions for the existence of a continuous utility function for a not necessarily total preorder on a topological space (see e.g. Mehta [35] and the survey in Mehta [36]).

The Richter–Peleg utility representation extends to non-total preorders the usual representation of total preorders, but it does not characterize the original non-total preorder. Indeed, one cannot retrieve the original relation from a Richter–Peleg utility function, the information is strictly weaker than that of the preorder.

A second approach to represent $\preceq$ introduces the concept of a *continuous multi-utility representation*.

Definition 2.3 A preorder $\preceq$ on a topological space (X, τ) is said to satisfy the continuous multi-utility representation property if a family $\mathcal{F}_{\preceq}$ of continuous real functions f on (X, τ) can be chosen in such a way that $x \preceq y$ if and only if $f(x) \leq f(y)$ for every $f \in \mathcal{F}_{\preceq}$.

This kind of representation has the feature to fully characterize the preorder.

Continuous multi-utility representations of not necessarily total preorders were first introduced by Levin [29–32], who called functionally closed a preorder on a topological space admitting a representation of this kind. The author characterized functionally closed preorders on a completely regular topological space by means of an extension theorem and a separation theorem ([29], Proposition 1). As a consequence of this characterization, Levin showed that every closed preorder on a compact topological space admits a continuous multi-utility representation.

The first systematic study of multi-utility representations is due to Ok [39], who presented different conditions for the existence of continuous multi-utility representations. After, many authors have been interested in the existence of a continuous multi-utility representation and investigated the relation between such a representation and the concept of a *normally topological preordered space* introduced by Nachbin [38]. We refer the reader to [1, 6–8, 22, 37].

Authors in [6] proved the following result on limit point compact spaces.

We recall that a topological space (X, τ) is said to be *limit point compact* if every infinite subset L of X has an accumulation point.

Proposition 2.4 (Proposition 3.12 in [6]) *Let (X, τ) be a limit point compact space. If (X, τ) is a normal space then every closed preorder on (X, τ) has a continuous multi-utility representation.*

3 Utility and Multi-utility Representation on k_ω-Spaces

Definition 3.1 A topological space X is said to be a k_ω-space (i.e. a hemicompact k-space) if X is the inclusion inductive limit of a countable increasing family $(K_n)_n$ of Hausdorff compact subspaces. The family $(K_n)_n$ is called a k_ω-*decomposition* of X.

An interesting survey on k_ω-spaces can be found in [12, 23].

It is well known that if every K_n is metrizable then X is submetrizable. We recall that a k_ω-space is submetrizable if and only if its compact subsets are metrizable [12].

Several authors studied the continuous utility and multi-utility representation problem in these spaces [5, 6, 11, 13, 16, 17, 37].

We recall some results that we will use in the next sections.

The following lemma generalizes to k_ω-spaces an extension theorem (Lemma 2, [28]) proved by Levin for compact spaces.

Lemma 3.2 (Lemma 3.3 [17]) *Let X be a hemicompact k-space with a closed preorder $\preceq$, let S be a compact subset of X and let $u : S \to [0, 1]$ be a continuous increasing mapping. Then there is a continuous increasing function $f : X \to [0, 1]$ that extends u.*

Theorem 3.3 (Theorem 3.5 [17]) *Let X be a submetrizable k_ω-space and let $\preceq$ be a closed preorder on X. Then there is a continuous utility function on X.*

The proof of Theorem 3.3 is based on Lemma 3.2 and on the concept of a *network*. The same result was proved in [5] by using a different technique. The concepts of a network and a netweight have already been used in [18] to prove new continuous representation theorems for total preorders.

A family $\mathcal{N}$ of subsets of a topological space (X, τ) is called a *network* for X if for every point $x \in X$ and any neighbourhood U of x there exist $M \in \mathcal{N}$ such that $x \in M \subset U$ (see [21]).

The *network weight* (or *netweight*) of (X, τ) is defined by

$$nw(X, \tau) = \min\{|\mathcal{N}| : \mathcal{N} \text{ is a network for } (X, \tau)\} + \aleph_0.$$

Of course, the existence of a countable netweight generalizes the concept of second countability of a topology.

We remember that if a topological space X has a countable netweight then $X \times X$ is hereditarily Lindelöf.

The concept of a network allowed to generalize some results obtained by Minguzzi [37] in terms of a countable continuous multi-utility representation.

Minguzzi in [37] followed the terminology and notations of Nachbin [38].

A topological preordered space is a triple $(X, \tau, \preceq)$, i.e. a topological space (X, τ) endowed with a preorder $\preceq$.

Definition 3.4 A preordered topological space $(X, \tau, \preceq)$ is said to be
(i) regularly preordered if for every point $x \in X$ and for every increasing closed set $F \subset X$ such that $x \notin F$ there are two disjoint open subsets U and V, decreasing and increasing respectively, such that $x \in U$ and $F \subset V$. A dual property is required if $x \notin F$ and F is decreasing.
(ii) normally preordered if for any two disjoint closed decreasing, respectively increasing subsets A and B of X, there exist disjoint open decreasing, respectively increasing, subsets U and V of X such that $A \subset U$ and $B \subset V$.

Minguzzi combined the two approaches of the Richter–Peleg utility representation and a multi-utility representation.

Definition 3.5 A preorder $\preceq$ on a topological space (X, τ) admits a countable continuous multi-utility representation if there exists a countable set $\{f_n : n \in \mathbb{N}^+\}_{\preceq}$ of continuous utility functions $f_n : X \to \mathbb{R}$ such that

$$x \preceq y \Leftrightarrow f_n(x) \le f_n(y), \text{ for every } n \in \mathbb{N}^+.$$

Minguzzi ([37], Theorem 5.5) proved the following result.

Theorem 3.6 ([37], Theorem 5.5) *Every second countable regularly preordered space $(X, \tau, \preceq)$ admits a countable continuous utility representation, that is, there is a countable set $\{f_n : n \in \mathbb{N}^+\}_{\preceq}$ of continuous utility functions $f_n : X \to \mathbb{R}$ such that*

$$x \preceq y \Leftrightarrow f_n(x) \le f_n(y), \text{ for every } n \in \mathbb{N}^+.$$

In [6] the authors generalized Minguzzi's result by using the concept of a *network*.

Proposition 3.7 ([6] Proposition 4.1) *Let $(X, \tau, \preceq)$ be a regularly preordered space with a closed preorder. If the product topology $\tau \times \tau$ on $X \times X$ is hereditarily Lindelöf, then $\preceq$ has a countable continuous multi-utility representation.*

Corollary 3.8 ([6] Corollary 4.2) *Let $(X, \tau, \preceq)$ be a regularly preordered space with a countable netweight and assume that $\preceq$ is a closed preorder. Then $\preceq$ has a countable continuous multi-utility representation.*

In Minguzzi ([37], Theorem 2.7) it is proved that a k_ω-space is normally preordered with respect to every closed preorder.

Particularly, a submetrizable k_ω-space has a countable network, therefore the following result holds.

Corollary 3.9 ([6] Corollary 4.4) *Let $(X, \tau, \preceq)$ be a submetrizable k_ω-space, where $\preceq$ is a closed preorder. Then $\preceq$ has a countable continuous multi-utility representation.*

4 Jointly Continuous Multi-utility Representations

Let (X, τ) be a topological space. Let

$$\mathcal{P}_X = \{\preceq : \ \preceq \ \text{is a preorder on } X\} \subset CL(X \times X)$$

be the family of closed preorders defined on X.

For every $\preceq \in \mathcal{P}_X$ let

$$H_{\preceq} = \{u \in C(X, \mathbb{R}) : u \text{ is a utility function for } \preceq\}.$$

For every $\preceq \in \mathcal{P}_X$, $H_{\preceq}$ is an element of $2^{C(X,\mathbb{R})}$.

The jointly continuous multi-utility representation problem consists in finding suitable topologies on $\mathcal{P}_X$ and $2^{C(X,\mathbb{R})}$ ensuring the existence of a continuous function $L : \mathcal{P}_X \to 2^{C(X,\mathbb{R})}$ such that for every $\preceq \in \mathcal{P}_X$ $L(\preceq)$ is a continuous multi-utility representation for $\preceq$.

A natural topology (convergence) on the set $\mathcal{P}_X$ of preorders should satisfy the following condition:

$$(*) \qquad x_\sigma \to x, \ y_\sigma \to y, \ \preceq_\sigma \to \preceq, \ x_\sigma \preceq_\sigma y_\sigma \Longrightarrow x \preceq y.$$

The next theorem proved in [16] gives a necessary and sufficient condition for a topology defined on the set of preorders of an arbitrary topological space to satisfy the property $(*)$.

Theorem 4.1 *Let X be a topological space. Let η be a topology on $\mathcal{P}_X$. The following conditions are equivalent:*

(1) η satisfies the property $(*)$;
(2) whenever $\{\preceq_\sigma : \sigma \in \Sigma\}$ converges to $\preceq$ in $(\mathcal{P}_X, \eta)$, then $\mathrm{Ls} \preceq_\sigma \subseteq \preceq$.

4.1 Properties of Submetrizable k_ω-Spaces

In this section we suppose that X is a submetrizable k_ω-space.

Denote $C(X, \mathbb{R})$ the space of continuous real-valued functions and τ_{UC} the topology of uniform convergence on compacta on $C(X, \mathbb{R})$. It is known that if X is a submetrizable k_ω-space, then $(C(X, \mathbb{R}), \tau_{UC})$ has a countable base and is metrizable [34].

For every $\preceq \in \mathcal{P}_X$ we will define $\mathcal{H}_{\preceq}$ as follows.

Put $H_{\preceq} = \{u \in C(X, \mathbb{R}) : u$ is a utility function for $\preceq\}$.

By Theorem 3.3 $H_{\preceq} \neq \emptyset$. Since $(C(X, \mathbb{R}), \tau_{UC})$ has a countable base, there is a countable dense set $\{u_n : n \in \mathbb{N}^+\}$ in $(H_{\preceq}, \tau_{UC})$.

We will show

Proposition 4.2 $x \preceq y \Leftrightarrow u_n(x) \leq u_n(y)$ for every $n \in \mathbb{N}^+$.

Proof On the contrary, let $x, y \in X$ such that $u_n(x) \leq u_n(y)$ for every $n \in \mathbb{N}^+$, but not $(x \preceq y)$. Then, using Lemma 3.2, the function $u : \{x, y\} \to \mathbb{R}$ defined by $u(y) = 0, u(x) = 1$ can be extended to an increasing function $u^* : X \to [0, 1]$. Let $U : X \to [0, 1]$ be a utility function for $\preceq$. The function $g : X \to \mathbb{R}$ defined by

$$g(x) = u^*(x) + U(x)/4$$

is a utility function for $\preceq$. Thus $g \in H_{\preceq}$ and $g(x) \geq 1, g(y) \leq 1/4$. Since $\{u_n : n \in \mathbb{N}^+\}$ is a dense set in $(H_{\preceq}, \tau_{UC})$ there must exist $n \in \mathbb{N}^+$ such that

$$| u_n(x) - g(x) | < 1/4 \text{ and } | u_n(y) - g(y) | < 1/4.$$

Then $u_n(x) > g(x) - 1/4 \geq 3/4$ and $u_n(y) < g(y) + 1/4 \leq 2/4$. Thus $u_n(x) > u_n(y)$, a contradiction. $\qquad\square$

Put $\mathcal{H}_{\preceq} = \{u_n : n \in \mathbb{N}^+\}$.

Define the mapping $L : \mathcal{P}_X \to 2^{C(X, \mathbb{R})}$ as follows: $L(\preceq) = \mathcal{H}_{\preceq}$.

The space $(C(X, \mathbb{R}), \tau_{UC})$ is metrizable. We will define a compatible metric ρ on $C(X, \mathbb{R})$.

Let $\{K_n : n \in \mathbb{N}^+\}$ be a fixed countable cofinal subfamily in the family $K(X)$ of compact sets with respect to the inclusion, i.e., for every $K \in K(X)$ there exists $n \in \mathbb{N}^+$ such that $K \subset K_n$. We define a metric on the space $C(X, \mathbb{R})$.

For every $K \in K(X)$ let p_K be the pseudometric on $C(X, \mathbb{R})$ defined by

$$p_K(f, g) = \sup\{d(f(x), g(x)) : x \in K\}.$$

Then for every $K \in K(X)$ we have a real-valued pseudometric h_K defined as

$$h_K(f, g) = \min\{1, p_K(f, g)\}.$$

We define a function $\rho : C(X, \mathbb{R}) \times C(X, \mathbb{R}) \to \mathbb{R}$ as follows

$$\rho(f, g) = \sum_{n=1}^{\infty} \frac{1}{2^n} h_{K_n}(f, g).$$

It is easy to see that ρ is a metric on $C(X, \mathbb{R})$ and the topology τ_{UC} is generated by ρ.

Let A and B be two nonempty subsets of $(C(X, \mathbb{R}), \rho)$. The gap $D_\rho(A, B)$ between A and B is given by (see [4])

$$D_\rho(A, B) = \inf\{\rho(f, g) : f \in A, g \in B\} =$$
$$\inf\{\rho(f, B) : f \in A\} = \inf\{\rho(g, A) : g \in B\}.$$

4.2   Jointly Continuous Multi-utility Representation on Locally Compact Second Countable Spaces

In this subsection we suppose that (X, τ) is a Hausdorff second countable locally compact space and $CL(X)$ is the space of nonempty closed subsets of X.

Let

$$\mathcal{P} = \{\preceq : \ \preceq \ \text{is a preorder on } D(\preceq) \in CL(X)\} \subset CL(X \times X).$$

In this spaces, a currently applied topology on $\mathcal{P}$ is the topology of closed convergence (the Fell topology) (see [3, 26–28, 33]).

The Fell topology $F(\tau)$ on $CL(X, \tau)$ is the topology having as a subbase all sets of the form

$$U^- = \{B \in CL(X, \tau) : B \cap U \neq \emptyset\}, \ U \in \tau$$
$$(K^c)^+ = \{B \in CL(X, \tau) : B \cap K = \emptyset\}, \ K \text{ compact in } (X, \tau).$$

A comprehensive reference on the Fell topology is [4].

We recall the following results:

Proposition 4.3 ([4]) *Let (X, τ) be a Hausdorff space. The following are equivalent:*

(1) $(CL(X), F(\tau))$ is Hausdorff;
(2) $(CL(X), F(\tau))$ is regular;
(3) $(CL(X), F(\tau))$ is completely regular;
(4) (X, τ) is locally compact.

Proposition 4.4 ([4]) *Let (X, τ) be a Hausdorff space. The following are equivalent:*

(1) (X, τ) is locally compact and second countable;
(2) $(CL(X), F(\tau))$ is a Polish space (separable and metrizable with a complete metric);

(3) $(CL(X), F(\tau))$ is metrizable.

Levin [28] and Back [3] considered Richter–Peleg utility-functions which are jointly continuous in both consumption sets and preferences.

Theorem 4.5 ([28], Corollary 1) *Let $\mathcal{P}$ be the space of closed preorders defined on closed subsets $D \subset X$ and $\Phi = \{(\preceq, x) : \preceq \in \mathcal{P}, x \in D(\preceq)\}$.*

If $\mathcal{P}$ is metrizable, X is a Hausdorff locally compact and second countable space and

$$M = \{(\preceq, x, y) : \preceq \in \mathcal{P}, x, y \in D(\preceq), x \preceq y\}$$

is closed in $\mathcal{P} \times X \times X$, then there exists a continuous function $u : \Phi \to [0, 1]$ such that, for each $\preceq \in \mathcal{P}$, $u(\preceq, \cdot)$ is a continuous utility function.

Notice that by Proposition 4.4, Theorem 4.5 holds if $\mathcal{P}$ is equipped with the Fell topology $F(\tau \times \tau)$. Also, it is easy to verify that if (X, τ) is a Hausdorff locally compact space, then the Fell topology $F(\tau \times \tau)$ on $\mathcal{P}$ satisfies the property $(*)$.

In [16] the authors proved the following characterization of the local compactness.

Theorem 4.6 ([16]) *Let (X, τ) be a Hausdorff space. Let $\mathcal{P}$ be equipped with the Fell topology $F(\tau \times \tau)$. Then $F(\tau \times \tau)$ satisfies the property $(*)$ if and only if X is locally compact.*

Back used Levin's Theorem to prove a result in terms of partial maps and the Fell topology.

Let $\mathcal{U}$ be the set of continuous partial maps that is

$$\mathcal{U} = \{(D, u) : D \in CL(X), u \in C(D, \mathbb{R})\}$$

and let $\mathcal{P}_L$ be the space of total closed preorders.

Back endowed $\mathcal{U}$ with the topology τ_c.

The τ_c-topology on $\mathcal{U}$ is the topology that has as a subbase the sets

$$[G] = \{(D, u) \in \mathcal{U} : D \cap G \neq \emptyset\}$$

$$[K : I] = \{(D, u) \in \mathcal{U} : u(D \cap K) \subset I\}$$

where G is a τ-open subset of X, $K \subset X$ is τ-compact and $I \subset \mathbb{R}$ is open (possibly empty).

Theorem 4.7 (Back, [3]) *Let (X, τ) be a Hausdorff locally compact and second countable space. There exists a continuous map $\nu : (\mathcal{P}_L, F(\tau \times \tau)) \to (\mathcal{U}, \tau_c)$ such that $\nu(\preceq)$ is a utility function for every $\preceq \in \mathcal{P}_L$. Any such map ν is actually a homeomorphism of $\mathcal{P}_{lns}$ onto $\nu(\mathcal{P}_{lns})$, where $\mathcal{P}_{lns}$ is the family of total locally non-satiated preorders.*

We recall that a preorder $\preceq$ is called *locally non-satiated* if for each $x \in D(\preceq)$ and each neighbourhood U of x there is $y \in U$ such that $x \prec y$.

In [15] it was pointed out that the first part of Back's proof holds also in the case of closed preorders not necessarily total. By using the same technique it is possible to prove the existence of a continuous map $\nu : (\mathcal{P}, F(\tau \times \tau)) \to (\mathcal{U}, \tau_c)$.

We have the following theorem concerning a continuous multi-utility representation.

Theorem 4.8 *Let (X, τ) be a Hausdorff locally compact second countable space. If $\{\preceq_\sigma : \sigma \in \Sigma\}$ converges to $\preceq$ in $(\mathcal{P}_X, F(\tau \times \tau))$, then $\{D_\rho(L(\preceq_\sigma), L(\preceq)) : \sigma \in \Sigma\}$ converges to 0.*

Proof Since we consider only closed preorders on X, utility functions for such preorders are elements of $C(X, \mathbb{R})$. Notice that τ_c topology restricted on $C(X, \mathbb{R})$ is just the compact-open topology. It is a well-known fact that the compact-open topology and the topology of uniform convergence on compacta on $C(X, \mathbb{R})$ coincide. Thus by Back's result, there exists a continuous map $\nu : (\mathcal{P}_X, F(\tau \times \tau)) \to (C(X, \mathbb{R}), \tau_{UC})$ such that $\nu(\preceq)$ is a utility function for every $\preceq \in \mathcal{P}_X$. If $\{\preceq_\sigma : \sigma \in \Sigma\}$ converges to $\preceq$ in $(\mathcal{P}_X, F(\tau \times \tau))$, then $\{\nu(\preceq_\sigma) : \sigma \subset \Sigma\}$ converges to $\nu(\preceq)$ in $(C(X, \mathbb{R}), \tau_{UC})$. To prove that $\{D_\rho(L(\preceq_\sigma), L(\preceq)) : \sigma \in \Sigma\}$ converges to 0, let $\varepsilon > 0$. There is $\sigma_0 \in \Sigma$ such that

$$\rho(\nu(\preceq), \nu(\preceq_\sigma)) < \varepsilon/2 \text{ for every } \sigma \geq \sigma_0.$$

We claim that

$$D_\rho(L(\preceq_\sigma), L(\preceq)) < \varepsilon \text{ for every } \sigma \geq \sigma_0.$$

Let $\sigma \geq \sigma_0$. Since $\mathcal{H}_\preceq$ is a countable dense set in the set of continuous utility functions for $\preceq$ equipped with τ_{UC}, there is $u_\preceq \in \mathcal{H}_\preceq$ such that

$$\rho(\nu(\preceq), u_\preceq) < \varepsilon/4.$$

Also since $\mathcal{H}_{\preceq_\sigma}$ is a countable dense set in the set of continuous utility functions for $\preceq_\sigma$ equipped with τ_{UC}, there is $u_{\preceq_\sigma} \in \mathcal{H}_{\preceq_\sigma}$ such that

$$\rho(\nu(\preceq_\sigma), u_{\preceq_\sigma}) < \varepsilon/4.$$

Thus $\rho(u_\preceq, u_{\preceq_\sigma}) < \varepsilon$, i.e. $D_\rho(L(\preceq_\sigma), L(\preceq)) < \varepsilon$ for every $\sigma \geq \sigma_0$. $\qquad \square$

4.3 Jointly Continuous Multi-utility Representation on Submetrizable k_ω-Spaces

Let $(X_n, \tau_n)_n$ be a countable increasing family of locally compact second countable Hausdorff spaces such that each space X_n is closed in X_{n+1}. Such a family $X_1 \subset X_2 \subset \dots \subset X_n \subset \dots$ is called *a closed tower*.

In [16] the authors established the following result

Proposition 4.9 ([16], Proposition 2.2) *Let (X, τ) be a topological space. TFAE:*

(i) *X is a submetrizable k_ω-space;*

(ii) *X is the inclusion inductive limit of a countable increasing family of metric compact spaces;*

(iii) *X is the inclusion inductive limit of a closed tower of Hausdorff second countable locally compact spaces.*

In the following we assume that (X, τ) is a submetrizable k_ω-space, the inclusion inductive limit of a closed tower $(X_n, \tau_n)_n$ of Hausdorff second countable locally compact spaces.

Let $\mathcal{G}$ be a compatible uniformity on (X, τ) and $\mathcal{B}$ be a base of closed symmetric elements from $\mathcal{G}$.

A comprehensive reference on the uniformities and the uniform topologies is [21].

In [16] a new topology $\tau(\mathcal{G})$ on $\mathcal{P}$ was introduced. $\tau(\mathcal{G})$ is generated by all sets of the form

$$U^- = \{\preceq \in \mathcal{P} : \preceq \cap U \neq \emptyset\}, \ U \text{ open in } (X_n, \tau_n) \times (X_n, \tau_n) \text{ for some } n \in \mathbb{N}$$

$$[(B[x] \times B[y])^c]^+ = \{\preceq \in \mathcal{P} : \preceq \cap (B[x] \times B[y]) = \emptyset\}, \ B \in \mathcal{B}, \ x, y \in X.$$

Of course two different compatible uniformities on X can give two different topologies.

In [16] the following two theorems were proved.

Theorem 4.10 *Let (X, τ) be a submetrizable k_ω-space, the inclusion inductive limit of a closed tower $(X_n, \tau_n)_n$ of Hausdorff second countable locally compact spaces. Let $\mathcal{G}$ be a compatible uniformity on X. The topology $\tau(\mathcal{G})$ on $\mathcal{P}$ satisfies the condition $(*)$.*

Theorem 4.11 *Let (X, τ) be a submetrizable k_ω-space, the inclusion inductive limit of a closed tower $(X_n, \tau_n)_n$ of Hausdorff second countable locally compact spaces. Let $\mathcal{G}$ be a compatible uniformity on X. There exists a continuous map*

$$\nu : (\mathcal{P}, \tau(\mathcal{G})) \to (\mathcal{U}, \tau_c)$$

such that $\nu(\preceq)$ is a utility function for every $\preceq \in \mathcal{P}$.

Using Theorem 4.11 we can prove the following result for a family of preorders on a fixed consumption space X.

Theorem 4.12 *Let (X, τ) be a submetrizable k_ω-space, the inclusion inductive limit of a closed tower $(X_n, \tau_n)_n$ of Hausdorff second countable locally compact spaces. Let $\mathcal{G}$ be a compatible uniformity on X. If $\{\preceq_\sigma : \sigma \in \Sigma\}$ converges to $\preceq$ in $(\mathcal{P}_X, \tau(\mathcal{G}))$, then $\{D_\rho(L(\preceq_\sigma), L(\preceq)) : \sigma \in \Sigma\}$ converges to 0.*

Proof We can use the same ideas as in the proof of Theorem 4.8. Utility functions for $\preceq$ in $\mathcal{P}_X$ are elements of $C(X, \mathbb{R})$. The τ_c topology restricted on $C(X, \mathbb{R})$ is just the compact-open topology which coincides with the topology of uniform convergence on compacta.

Thus by Theorem 4.11 there exists a continuous map

$$\nu : (\mathcal{P}_X, \tau(\mathcal{G})) \to (C(X, \mathbb{R}), \tau_{UC})$$

such that $\nu(\preceq)$ is a utility function for every $\preceq \in \mathcal{P}_X$. Let $\{\preceq_\sigma : \sigma \in \Sigma\}$ converge to $\preceq$ in $(\mathcal{P}_X, \tau(\mathcal{G}))$. Thus $\{\nu(\preceq_\sigma) : \sigma \in \Sigma\}$ converges to $\nu(\preceq)$ in $(C(X, \mathbb{R}), \tau_{UC})$. To prove that $\{D_\rho(L(\preceq_\sigma), L(\preceq)) : \sigma \in \Sigma\}$ converges to 0, we will use the same argumentation as in the proof of Theorem 4.8. $\qquad\square$

Remark 4.13 The Theorem 4.8 can be proved as a consequence of the Theorem 4.12. Indeed any Hausdorff second countable locally compact space has a compatible metric d such that (X, d) is a boundedly compact space [41]. We say that a metric space (X, d) is boundedly compact [4] if every closed bounded subset is compact. Denote by $\mathcal{G}_d$ the uniformity generated by the metric d. It is easy to verify that $\tau(\mathcal{G}_d)$ coincides with the Fell topology.

Remark 4.14 Notice that the mapping $L : \mathcal{P}_X \to 2^{C(X,\mathbb{R})}$ defined by $L(\preceq) = \mathcal{H}_{\preceq}$ is a bijection between $\mathcal{P}_X$ and $L(\mathcal{P}_X)$.

In the following theorem we will use notions of the lower Vietoris topology [19] and the upper Fell topology. The upper Fell topology is known in the literature also as the co-compact topology [20].

Let (X, τ) be a topological space. The lower Vietoris topology $V(\tau)^-$ on 2^X is the topology which has as a subbase all sets of the form

$$U^- = \{B \in 2^X : B \cap U \neq \emptyset\}, \ U \in \tau$$

and the upper Fell topology $F(\tau)^+$ on 2^X is the topology having as a base all sets of the form

$(K^c)^+ = \{B \in 2^X : B \cap K = \emptyset\}$, where K runs over the compact subsets of (X, τ).

Theorem 4.15 *Let (X, τ) be a submetrizable k_ω-space, the inclusion inductive limit of a closed tower $(X_n, \tau_n)_n$ of Hausdorff second countable locally compact spaces. The mapping $L^{-1} : (L(\mathcal{P}_X), V(\tau)^-) \to (\mathcal{P}_X, F(\tau \times \tau)^+)$ is continuous.*

Proof Let $\{L(\preceq_\sigma) : \sigma \in \Sigma\}$ converge to $L(\preceq)$ in $V(\tau)^-$ topology. We will show that $\{\preceq_\sigma : \sigma \in \Sigma\}$ converges to $\preceq$ in the upper Fell topology $F(\tau \times \tau)^+$. Let K be a compact set in $X \times X$ such that

$$\preceq \cap K = \emptyset.$$

We claim that there is $\sigma_0 \in \Sigma$ such that

$$\preceq_\sigma \cap K = \emptyset \text{ for every } \sigma \geq \sigma_0.$$

Suppose that this is not true. Thus there is a cofinal directed subset $\Sigma_1 \subset \Sigma$ such that

$$\preceq_\eta \cap K \neq \emptyset \text{ for ever } \eta \in \Sigma_1.$$

Let $(x_\eta, y_\eta) \in \preceq_\eta \cap K$ for every $\eta \in \Sigma_1$. There is $(x, y) \in K$ such that (x, y) is a cluster point of $\{(x_\eta, y_\eta) : \eta \in \Sigma_1\}$. Without loss of generality we can suppose that $\{(x_\eta, y_\eta) : \eta \in \Sigma_1\}$ converges to (x, y).

Since $(x, y) \notin \preceq$ there must exist $u_n \in \mathcal{H}_\preceq$ such that $u_n(x) > u_n(y)$. Let $\varepsilon > 0$ be such that

$$u_n(y) + \varepsilon < u_n(x) - \varepsilon.$$

Let $\eta_0 \in \Sigma_1$ be such that

$$u_n(y_\eta) < u_n(y) + \varepsilon/4 \text{ and } u_n(x) - \varepsilon/4 < u_n(x_\eta) \text{ for every } \eta \geq \eta_0.$$

Put $C = \pi_1(K) \cup \pi_2(K)$. Put

$$G = \{f \in C(X, \mathbb{R}) : |f(z) - u_n(z)| < \varepsilon/4 \text{ for every } z \in C\}.$$

Then G is an open set in $(C(X, \mathbb{R}), \tau_{UC})$ and $L(\preceq) = \mathcal{H}_\preceq \in G^-$. $\{L(\preceq_\sigma) : \sigma \in \Sigma\}$ converges to $L(\preceq)$ in $V(\tau)^-$ topology. There is $\sigma_0 \in \Sigma$ such that

$$L(\preceq_\sigma) \cap G \neq \emptyset \text{ for every } \sigma \geq \sigma_0.$$

Let $\eta \in \Sigma_1$ be such that $\eta \geq \sigma_0$ and $\eta \geq \eta_0$. Let $u_{n(\eta)} \in \mathcal{H}_{\preceq_\eta} \cap G$. Then

$$u_{n(\eta)}(y_\eta) < u_n(y_\eta) + \varepsilon/4 < u_n(y) + \varepsilon/2,$$

$$u_n(y) + \varepsilon < u_n(x) - \varepsilon < u_n(x) - \varepsilon/2, and$$

$$u_n(x) - \varepsilon/2 < u_n(x_\eta) - \varepsilon/4 < u_{n(\eta)}(x_\eta),$$

a contradiction since $(x_\eta, y_\eta) \in \preceq_\eta$. $\qquad\square$

5 Conclusions

In our paper we study the jointly continuous multi-utility representation problem for a given topological space (X, τ), which consists in finding suitable topologies on $\mathcal{P}_X$, $2^{C(X,\mathbb{R})}$ and a continuous function $L : \mathcal{P}_X \to 2^{C(X,\mathbb{R})}$ such that for every $\preceq \in \mathcal{P}_X$, $L(\preceq)$ is a continuous multi-utility representation for $\preceq$. Concerning topologies on $\mathcal{P}_X$ we are interested in topologies which satisfy the property $(*)$.

For a submetrizable k_ω-space (X, τ) we defined the mapping L from $\mathcal{P}_X$ to $2^{C(X,\mathbb{R})}$, which assigned to every closed preorder $\preceq$ on X a countable set of continuous utility functions for $\preceq$, which is dense in the family of all continuous utility functions for $\preceq$ equipped with τ_{UC}, the topology of uniform convergence on compacta. The mapping $L : \mathcal{P}_X \to 2^{C(X,\mathbb{R})}$ is a bijection between $\mathcal{P}_X$ and $L(\mathcal{P}_X)$.

We proved the following results concerning the mapping L and its inverse L^{-1}:

- If $\mathcal{G}$ is a compatible uniformity on (X, τ) and $\{\preceq_\sigma : \sigma \in \Sigma\}$ converges to $\preceq$ in $(\mathcal{P}_X, \tau(\mathcal{G}))$, then $\{D_\rho(L(\preceq_\sigma), L(\preceq)) : \sigma \in \Sigma\}$ converges to 0.
- The mapping $L^{-1} : L(\mathcal{P}_X) \to \mathcal{P}_X$ is continuous, if $L(\mathcal{P}_X)$ is equipped with the lower Vietoris topology $V(\tau)^-$ and $\mathcal{P}_X$ is equipped with the upper Fell topology $F(\tau \times \tau)^+$.

Notice that if $\mathcal{G}$ is a compatible uniformity on (X, τ), then the topology $\tau(\mathcal{G})$ on $\mathcal{P}_X$ seems to be reasonable for our purposes, since it satisfies the property $(*)$. We would like to find a suitable topology on $L(\mathcal{P}_X)$ to guarantee the continuity of the mapping L.

We have the following conjecture.

Conjecture *Let (X, τ) be a submetrizable k_ω-space, the inclusion inductive limit of a closed tower $(X_n, \tau_n)_n$ of Hausdorff second countable locally compact spaces. Let $\mathcal{G}$ be a compatible uniformity on (X, τ). The mapping $L : (\mathcal{P}_X, \tau(\mathcal{G})) \to (L(\mathcal{P}_X), V(\tau)^-)$ is continuous.*

Acknowledgements This research was carried out within the Gruppo Nazionale per l'Analisi Matematica, la Probabilità e le loro Applicazioni, Istituto Nazionale di Alta Matematica (Italy). L. Holá would like to thank grant Vega 2/0006/16.

References

1. Alcantud, J.C.R., Bosi, G., Zuanon, M.: Richter-Peleg multi-utility representations of preorders. Theory Decis. **80**, 443–450 (2016)
2. Aumann, R.: Utility theory without the completeness axiom. Econometrica **30**, 445–462 (1962)
3. Back, K.: Concepts of similarity for utility functions. J. Math. Econ. **15**, 129–142 (1986)
4. Beer, G.: Topologies on Closed and Closed Convex Sets. Kluwer Academic Publishers, Dordrecht (1993)
5. Bosi, G., Caterino, A., Ceppitelli, R.: Existence of continuous utility functions for arbitrary binary relations: some sufficient conditions. Tatra Mt. Math. Publ. **46**, 15–27 (2010)
6. Bosi, G., Caterino, A., Ceppitelli, R.: Normally preordered spaces and continuous multi-utilities. Appl. Gen. Topol. **17**, 71–81 (2016)
7. Bosi, G., Herden, G.: Continuous multi-utility representations of preorders. J. Math. Econ. **48**, 212–218 (2012)
8. Bosi, G., Herden, G.: On continuous multi-utility representations of semi-closed and closed preorders. Math. Soc. Sci. **79**, 20–29 (2016)
9. Bosi, G., Mehta, G.B.: Existence of a semicontinuous or continuous utility function: a unified approach and an elementary proof. J. Math. Econ. **38**, 311–328 (2002)
10. Bridges, D.S., Mehta, G.B.: Representations of Preference Orderings. Springer, Berlin (1995)

11. Candeal, J.C., Indurain, E., Mehta, G.B.: Some utility theorems on inductive limits of pre-ordered topological spaces. Bull. Aust. Math. Soc. **52**, 235–246 (1995)
12. Castaing, C., de Fitte, P.R., Valadier, M.: Young Measures on Topological Spaces with Applications in Control Theory and Probability Theory, vol. 571. Kluwer Academic Publishers, Dordrecht (2004)
13. Caterino, A., Ceppitelli, R.: Jointly continuous utility functions on submetrizable k_ω-spaces. Topol. Its Appl. **190**, 109–118 (2015)
14. Caterino, A., Ceppitelli, R.: An application of Back's Theorem to an ordering of distributions of wellbeing. In: Proceedings of 14th Conference on Applied Mathemathics Aplimat 2015, Bratislava, Slovak Republic, pp. 191–198 (2015)
15. Caterino, A., Ceppitelli, R., Holá, Ľ.: Some generalizations of Back's Theorem. Topol. Its Appl. **160**, 2386–2395 (2013)
16. Caterino, A., Ceppitelli, R., Holá, Ľ.: On the jointly continuous utility representation problem. Topol. Its. Appl. **268** (2019). https://www.doi.org/10.1016/jtopol.2019.106919
17. Caterino, A., Ceppitelli, R., Maccarino, F.: Continuous utility functions on submetrizable hemi-compact k-spaces. Appl. Gen. Topol. **10**, 187–195 (2009)
18. Caterino, A., Ceppitelli, R., Mehta, G.B.: Representations of topological preordered spaces. Math. Slovaca **61**(1), 93–106 (2011)
19. Di Maio, G., Holá, Ľ., Meccariello, E.: Notes on hit-and-miss topologies. Rostock Math. Kolloq. **52**, 19–32 (1998)
20. Dolecki, S., Greco, G.H., Lechicki, A.: When do the upper Kuratowski topology (homeomorphically, Scott topology) and the co-compact topology coincide? Trans. Am. Math. Soc. **347**, 2869–2884 (1995)
21. Engelking, R.: General Topology. Heldermann Verlag, Berlin (1989)
22. Evren, O., Ok, E.A.: On the multi-utility representation of preference relations. J. Math. Econ. **47**, 554–563 (2011)
23. Franklin, S.P., Smith Thomas, B.V.: A survey of k $_\omega$-spaces. Topol. Proc. **2**, 111–124 (1977)
24. Herden, G.: On the existence of utility functions. Math. Soc. Sci. **17**, 297–313 (1989)
25. Herden, G., Pallack, A.: On the continuous analogue of the Szpilrajn Theorem I. Math. Soc. Sci. **43**, 115–134 (2002)
26. Hildebrand, W.: On economies with many agents. J. Econ. Theory **2**, 1–188 (1970)
27. Hildebrand, W.: Core and Equilibria of a Large Economy. Princeton University Press, Princeton (1974)
28. Levin, V.L.: A continuous utility theorem for closed preorders on a σ-compact metrizable space. Soviet Math. Dokl. **28**, 715–718 (1983)
29. Levin, V.L.: Functionally closed preorders and strong stochastic dominance. Soviet Math. Dokl. **32**, 22–26 (1985)
30. Levin, V.L.: Extremal problems with probability measures, functionally closed preorders and strong stochastic dominance. LN Control Inf. Sci. **81**, 435–447 (1986)
31. Levin, V.L.: General Monge-Kantorovich problem and its applications in measure theory and mathematical economics. In: Leifman, L.J. (ed.) Functional Analysis, Optimization and Mathematical Economics. A Collection of Papers Dedicated to the Memory of L.V. Kantorovich, pp. 141–176. Oxford University Press, Oxford (1990)
32. Levin, V.L.: The Monge-Kantorovich problems and stochastic preference relations. Adv. Math. Econ. **3**, 97–124 (2001)
33. Mas-Colell, A.: On the continuous representation of preorders. Int. Econ. Rev. **18**, 509–513 (1977)
34. McCoy, R.A., Ntantu, I.: Topological Properties of Spaces of Continuous Functions. Springer, Berlin (1980)
35. Mehta, G.B.: Some general theorems on the existence of order-preserving functions. Math. Soc. Sci. **15**, 135–146 (1988)
36. Mehta, G.B.: Preference and utility. In: Barberá, S., Hammond, P., Seidl, C. (eds.) Handbook of Utility Theory, vol. 1, pp. 1–47. Kluwer Academic Publishers, Dordrecht (1988)
37. Minguzzi, E.: Normally preordered spaces and utilities. Order **30**, 137–150 (2013)

38. Nachbin, L.: Topology and Order. Van Nostrand, Princeton (1965)
39. Ok, E.A.: Utility representation of an incomplete preference relation. J. Econ. Theory **104**, 429–449 (2002)
40. Peleg, B.: Utility functions for partially ordered topological spaces. Econometrica **38**, 93–96 (1970)
41. Vaughan, H.: On locally compact metrizable spaces. Bull. Amer. Math. Soc. **43**, 532–535 (1937)

Subjective States Without the Completeness Axiom

Asen Kochov

Abstract Existing work has shown how an agent's desire to defer choice may reveal the contingencies she considers relevant to her decision. We explore if this conclusion remains true when preferences are incomplete, that is, when the agent is unable to express a clear ranking among all alternatives. Remarkably we show that the incompleteness of preference does not preclude the existence of a subjective state space; rather it reflects an agent who cannot aggregate the various states she considers possible. An application to social choice and some general results concerning the representation of incomplete but transitive relations are also provided.

1 Introduction

In a seminal paper, Kreps [16] showed that preferences over opportunity sets can be represented using a subjective state space. Thus, the agent behaves *as if* she contemplates a list of contingencies that are relevant to her choice. Dekel, Lipman, and Rustichini [6], henceforth DLR, extended Kreps' analysis to opportunity sets over lotteries and showed that, under suitable conditions, the subjective state space is unique, strengthening its interpretation as "the agent's (implicit) view of future possibilities."

This paper explores how the above results depend on the completeness axiom. We draw motivation from the work of Bewley [3] who used incomplete preferences to capture an agent who in the face of uncertainty struggles to rank all available actions. Bewley's setting is one in which an exogenously given state space S specifies all the contingencies that are relevant to the agent's decision; the agent's difficulty comes from the need to assign likelihoods to the various states $s \in S$. In this setting, Bewley

An earlier draft of this paper was first completed in 2007 as part of my doctoral work. For their helpful comments, I am grateful to Larry Epstein, Özgür Evren, Bart Lipman, Efe Ok, and John Stovall.

A. Kochov (✉)
Department of Economics, University of Rochester, Rochester, NY 14627, USA
e-mail: asen.kochov@rochester.edu

© Springer Nature Switzerland AG 2020
G. Bosi et al. (eds.), *Mathematical Topics on Representations of Ordered Structures and Utility Theory*, Studies in Systems, Decision and Control 263,
https://doi.org/10.1007/978-3-030-34226-5_13

255

shows that preferences are incomplete if and only if the agent is unable to formulate a single belief μ over the state space S. Instead, she contemplates a set of beliefs μ and ranks alternative x higher than y only when x gives higher utility under every belief μ.

What happens if an exogenous state space is not given? Surely, this extra layer of uncertainty cannot make the agent's decision any easier. As Kreps [17] points out: *"I have little doubt that assumption 3 (completeness) is more questionable now than before; the added complications might cause you to waver in your determination to rank all available alternatives..."* Curiously, we show that incompleteness does not preclude the existence of a unique subjective state space. Instead, and in parallel with Bewley [3], it implies an agent who cannot aggregate the various states she considers possible.

The analysis allows us to address a natural question in the context of aggregating preferences. Consider a group of agents, each agent i having a unique subjective state space S_i. What state space S should a social planner use? Presumably, the answer to this question depends on how the social planner aggregates the agents' preferences. It turns out however that this is not the case. Assuming only that the social planner is strictly Paretian, which in particular allows for the social planner's ranking of alternatives to be incomplete, we show that S should be equal to the union of the S_i.

The paper begins with some preliminary results concerning the representation of incomplete but transitive relations in abstract domains. One of the results invokes Nachbin's [22] separation theorem, which has featured prominently in Professor Mehta's own work on utility representations. See, for instance, Mehta [18, 19] and Bridges and Mehta [4].

2 Representation of Preorders

A preorder $\succeq$ on a set X is a reflexive and transitive binary relation. When the set X is endowed with a topology τ, the preorder $\succeq$ is said to be **continuous** if its graph $\{(x, y) \in X \times X : x \succeq y\}$ is closed in the product topology on $X \times X$.[1] A preorder $\succeq$ on a set X is **representable** if there exists a set V of functions $v : X \to \mathbb{R}$ such that $x \succeq y$ if and only if $v(x) \geq v(y)$ for all $v \in V$. The preorder $\succeq$ is **continuously representable** if it is representable and every function $v \in V$ is continuous given the product topology on $\mathbb{R}^X$.

The next result shows that every preorder is representable.

Theorem 1 *Every preorder is representable.*

---

[1] Another common definition of continuity is to require that the strict upper and lower contour sets of $\succeq$ be open. Unfortunately, the two definitions are not equivalent when $\succeq$ is incomplete. Moreover, one must choose which definition to use since, as Schmeidler [25] shows, their conjunction forces $\succeq$ to be complete.

Proof Let $(X, \succeq)$ be a preordered set and for each $x \in X$, let $U_x = \{z \in X : z \succeq x\}$ be the upper contour set of x. Consider the set $V := \{I_{U_x} : x \in X\}$, where I_A denotes the indicator function for a set A. Clearly, $x \succeq y$ implies that $v(x) \geq v(y)$ for all $v \in V$. Suppose now that $x \succ y$ for some $x, y \in X$, which means that $x \succeq y$ but not $y \succeq x$. It suffices to show that $v(x) > v(y)$ for some $v \in V$. Taking $v \in V$ to be the indicator function of U_x does just that. ∎

Suppose now that X is a topological space and the preorder $\succeq$ is continuous. The proof of Theorem 1 suggests that a continuous representation would exist if the indicator functions in V can be approximated by continuous functions v that are also **increasing** in the sense that $x \succeq y$ implies $v(x) \geq v(y)$. Nachbin's [22] separation theorem shows that this is possible whenever X is a compact, Hausdorff space. Let $L_y = \{z \in X : y \succeq z\}$ be the lower contour set of y. More precisely, Nachbin shows that for every $x \succ y$, we can find a continuous increasing function $v : X \to \mathbb{R}$ such that $v(z) = 1$ for all $z \in U_x$ and $v(z) = 0$ for all $z \in L_y$.[2] This leads to the following result:

Theorem 2 *Let X be a compact, Hausdorff space. A preorder $\succeq$ on X is continuous if and only if it is continuously representable.*

Remark Evren and Ok [10] obtained Theorems 1 and 2 in independent and contemporaneous work.

3   Subjective State Spaces with Incomplete Preferences

Turn to our main question: Is the incompleteness of preference compatible with the existence of a unique subjective state space? To answer this question, we adopt the DLR model of choice among opportunity sets. Let $B = \{b_1, \dots b_K\}$ be a finite set of K prizes, ΔB the set of lotteries over B, and X the set of nonempty subsets of ΔB. The lotteries in ΔB are denoted as β, β'. The elements of X are called **menus** and denoted as x, x', y, etc. The set X of menus is endowed with the Hausdorff topology. Given $\lambda \in [0, 1]$, let

$$\lambda x + (1 - \lambda)x' := \{\lambda\beta + (1 - \lambda)\beta' : \beta \in x, \beta' \in x'\}$$

and refer to $\lambda x + (1 - \lambda)x'$ as a **mixture** of x and x'. The agent is assumed to choose among menus, with the implicit understanding that at some ex post stage she refines her choice by selecting an available alternative. The choice of a restaurant, for example, corresponds to the choice of a menu from which the agent selects a certain meal. Similarly, the choice of current savings determines the budget set from which the agent decides on next period consumption. With these examples in mind,

---

[2]Nachbin [22, Chap. 1, Theorem 4] states the separation theorem for compact, partially ordered spaces. It is easily verified that his arguments extend to preordered, compact, Hausdorff spaces.

it is intuitive that the agent's ex ante choice of a menu should reflect her conception of the possible future tastes she might have. Indeed, when preferences are complete, DLR show how a set of ex post preferences, or the agent's *subjective state space*, can be uniquely identified from behavior. This paper addresses the problem in the context of incomplete preferences.

3.1 Two Representation Theorems

Thus, let $\succeq$ be a preorder on the space X of menus. We begin with two representation theorems for such preorders. As in DLR, we assume that the option to randomize does not improve a menu.

Indifference to Randomization: Every menu $x \in X$ is indifferent to its convex hull $conv(x)$.

Combining Theorem 2 in Sect. 2 with insights from DLR delivers the following result.

Theorem 3 *A continuous preorder $\succeq$ on X satisfies Indifference to Randomization if and only if there exists a set S, a state-dependent utility function $U : \Delta B \times S \to \mathbb{R}$ and a set V of continuous functions $v : \mathbb{R}^S \to \mathbb{R}$ such that:*

(i) $x \succeq y$ if and only if

$$v((\sup_{\beta \in x} U(\beta, s))_{s \in S}) \geq v((\sup_{\beta \in x} U(\beta, s))_{s \in S}) \quad \text{for all } v \in V;$$

(ii) each $U(\cdot, s)$ is an expected-utility function in the sense that

$$U(\beta, s) = \sum_{b \in B} \beta(b) U(b, s).$$

To understand the representation, recall that we have in mind an agent who chooses a menu x facing uncertainty about her future tastes. As Theorem 3 shows, the agent's conception of this uncertainty can be elicited from choice behavior and represented by the set of ex post preferences $\{U(\cdot, s) : s \in S\}$. At some unmodeled stage, this uncertainty is resolved and the agent selects the best alternative in x according to the realized state $U(\cdot, s)$. The ex ante problem or the agent's evaluation of a menu therefore requires her to aggregate across the conflicting future tastes she considers possible. When V is a singleton or equivalently when preferences over menus are complete as in DLR, the agent has a unique way of trading off utility across states. In contrast, the representation in Theorem 3 departs from DLR by allowing for agents whose conflicting views of the future remain unresolved at the time they choose a menu.

Notably, Theorem 3 shows that weakening completeness does not preclude the existence of a subjective state space. This conclusion would be further strengthened by the analysis in Sect. 3.2 where we show that, under mild additional assumptions, the agent's state space is unique. Before doing so, we introduce another representation result which restricts the aggregators $v \in V$ to be probability measures over the state space S. The result is an analogue of Bewley [3] but with a subjective state space.

Some additional axioms are obviously required. The first is a standard independence condition.[3]

Independence: For all $\lambda \in (0, 1]$ and all $x, x', y \in X$,

$$x \succeq x' \quad \text{if and only if} \quad \lambda x + (1 - \lambda)y \succeq \lambda x' + (1 - \lambda)y.$$

The next axiom, first introduced by Kreps [16], requires that the agent, who is uncertain about her future tastes, values information and wants to preserve flexibility. The axiom rules out models of temptation, as in the pioneering work of Gul and Pesendorfer [13], in which agents may seek commitment and, hence, prefer smaller menus.

Monotonicity: For all menus $x, y \in X$, $x \supseteq y$ implies that $x \succeq y$.

The final axiom rules out degenerate cases.

Nontriviality: There are menus $x, x' \in X$ such that $x \succ x'$.

Theorem 4 *A continuous preorder $\succeq$ on X satisfies Independence, Monotonicity, and Nontriviality if and only if there exists a set S, a state-dependent utility function $U : \Delta B \times S \to \mathbb{R}$ and a closed and convex set Δ of probability measures over S such that:*
(i) $x \succeq y$ if and only if

$$\int_S \sup_{\beta \in x} U(\beta, s)d\mu \geq \int_S \sup_{\beta \in y} U(\beta, s)d\mu \quad \text{for all } \mu \in \Delta;$$

(ii) each $U(\cdot, s)$ is an expected-utility function in the sense that

$$U(\beta, s) = \sum_{b \in B} \beta(b)U(b, s).$$

A brief sketch of the proof might be helpful. Given Independence and the continuity of $\succeq$, each menu is indifferent to its closed convex hull. The insight of DLR is that the set of closed and convex menus x can be identified with the set of support functions

$$\sigma_x : s \mapsto (\max_{\beta \in x} \beta \cdot s),$$

---

[3]Independence implies Indifference to Randomization. See DLR for a proof as well as a careful explanation of Independence. Note as well that DLR assume a slightly weaker version of Independence. The axiom stated here is implied by the conjunction of Independence and Continuity in DLR.

where each s is a vector in $\mathbb{R}^K$ viewed as the von-Neumann–Morgenstern (vNM) index of an ex post expected-utility preference on ΔB.[4] Consequently, we can also identify the preorder $\succeq$ over menus with a preorder $\succeq^*$ over the set of support functions. Since the latter is a mixture space, a theorem of Shapley and Baucells [26, Theorem 1.8] shows that $\succeq^*$ can be represented by a set of linear functions whenever its **domination cone**

$$D := \{\lambda(\sigma - \sigma') : \lambda \geq 0, \sigma \succeq^* \sigma'\}$$

has nonempty interior. The gist of the present proof is to note that in our context this technical requirement is insured by Monotonicity. The latter also implies that the linear functions are continuous and increasing which, by Riesz Representation Theorem, insures that they can be represented as probability measures over the subjective state space.

3.2   Uniqueness of the State Space

In Theorems 3 and 4, each state $s \in S$ in the representation determines, through the function $U(\cdot, s)$, an ex post expected-utility preference over the set ΔB of lotteries. As DLR make clear, it is this set of ex post preferences that one hopes to pin down uniquely and that is behaviorally meaningful, not the cardinal functions $U(\cdot, s)$. To achieve that, let

$$S^K = \left\{ s \in \mathbb{R}^K : \sum_i s_i = 0, \sum_i s_i^2 = 1 \right\}$$

and note that by thinking of each $s \in S^K$ as a vNM utility index, we can identify every non-trivial expected-utility preference on ΔB with a vector $s \in S^K$.[5] Using the set S^K, we can also rewrite the representation obtained in Theorem 3 as a pair (S, V) where S is a subset of S^K and V is a set of continuous functions $v : \mathbb{R}^S \to \mathbb{R}$ such that

$$x \succeq y \quad \text{if and only if} \quad v\left((\sup_{\beta \in x} \beta \cdot s)_{s \in S} \right) \geq v\left((\sup_{\beta \in y} \beta \cdot s)_{s \in S} \right) \quad \text{for all } v \in V.$$

It remains to single out the representations (S, V) in which every state $s \in S$ matters in at least some choice situation. That is, we have to single out the states that the agent actually considers possible. To that end, say that (S, V) is a **proper representation**

---

[4] $\beta \cdot s$ denotes the scalar product of $\beta, s \in \mathbb{R}^K$.

[5] There are other ways one can choose the set S^K so that it maps one-to-one onto the space of non-trivial expected-utility preferences. As in Dekel, Lipman, Rustichini, and Sarver [7], this particular normalization is chosen for technical convenience.

if for every $s \in S$ and every open neighborhood O of s (in the relative topology on S), there exist menus $x, y \in X$ such that $\sup_{\beta \in x} \beta \cdot s = \sup_{\beta \in y} \beta \cdot s$ for all $s \in S \setminus O$ and $x \nsim y$ (which means that x, y are either strictly ranked or not comparable). In the words of DLR, (S, V) is proper if each state $s \in S$ is **relevant**. The next theorem summarizes our uniqueness results.

Theorem 5 *If a continuous preorder $\succeq$ satisfies Indifference to Randomization, Monotonicity, and Nontriviality, then $\succeq$ has a proper representation (S, V). Moreover, if (S', V') is another proper representation of $\succeq$, then $cl(S') = cl(S)$.*

4 Aggregation

Consider a group of agents, each agent i having a subjective state space S_i. What state space should a social planner use? The uniqueness result of the previous section allows us to address this question. In particular, suppose each agent $i = 1, 2, \ldots, n$ has a preference relation $\succeq_i$ on X satisfying the assumptions of Theorem 5 and having a proper representation (S_i, V_i). Say that a preorder $\succeq$ on X is **Paretian** if $x \succeq_i y$ for every i implies $x \succeq y$ and, in addition, if one of the rankings $x \succeq_i y$ is strict, then $x \succ y$. It is easy to see that any Paretian order $\succeq$ must satisfy Indifference to Randomization, Monotonicity, and Nontriviality. Assuming that the preorder $\succeq$ is continuous, we can apply Theorem 5 to deduce that $\succeq$ has a proper representation (S, V) where the state space S is unique up to closure. Our next and final theorem shows that S is equal to the union of the S_i.

Theorem 6 *If the preorder $\succeq$ is Paretian and continuous, then it has a proper representation (S, V). Moreover, $cl(S) = \cup_i cl(S_i)$.*

We highlight that the conclusion of Theorem 6 does not depend on whether $\succeq$ is the Pareto ordering itself or some extension thereof, or whether $\succeq$ satisfies some additional desiderata. All that matters is for $\succeq$ to respect the Pareto ordering and be continuous.

5 Concluding Remarks

The paper showed that the incompleteness of preference does not preclude the existence of a well-defined and unique subjective state space describing the agent's conceptualization of the future. As is known from DLR, the existence of a subjective state space is similarly robust to relaxations of Independence and Monotonicity. This brings up a question that has swirled around the literature on preference for flexibility for a long time: How should one model the plausible scenario in which an agent's conceptualization of the future is coarse or otherwise incomplete? Additionally, and perhaps more importantly, what behavior would distinguish an agent who believes that her conceptualization of the future is correct from one who doesn't? As the literature on preference for flexibility has struggled to give clear answers to these questions, recent work on the subject has sought other ways to address them.[6] My own view is that preference for flexibility should remain central to the analysis of these questions. In fact, it may be particularly beneficial to augment some of the alternative approaches by incorporating preference for flexibility. This is because, as Kreps [17, p.260] first argued, a demand for flexibility is the most natural response when one is confronted with uncertainty whose exact contours are hard to anticipate and contract on.

6 Appendix

6.1 Preliminary Results

Endow X with the semimetric Hausdorff topology. Let $\mathcal{K}(\Delta B) \subset X$ be the set of closed, convex, and nonempty subsets of ΔB. The relative topology on $\mathcal{K}(\Delta B)$ is metrizable by the Hausdorff distance

$$h(x, y) := \max\{\sup_{\beta \in x} \inf_{\beta' \in y} d(\beta, \beta'), \sup_{\beta \in y} \inf_{\beta' \in x} d(\beta, \beta')\},$$

where $d(\cdot, \cdot)$ denotes the Euclidean distance in $\mathbb{R}^K$. Let $C(S^K)$ be the space of continuous functions on S^K endowed with the sup-norm and the usual pointwise order. Following DLR, identify each $x \in \mathcal{K}(\Delta B)$ with its support function $\sigma_x : S^K \to \mathbb{R}$ defined as $\sigma_x(s) = \max_{\beta \in x} \beta \cdot s$. Note that $C := \{\sigma_x : x \in \mathcal{K}(\Delta B)\}$ is a compact and convex subset of $C(S^K)$. The following lemmas summarize mathematical results obtained in DLR and Dekel et al. [7].

Lemma 7 *The mapping $x \mapsto \sigma_x$ is an isometry from $\mathcal{K}(\Delta B)$ to C such that:*

(i) $\sigma_{\lambda x + (1-\lambda)y} = \lambda \sigma_x + (1 - \lambda)\sigma_y$;

(ii) $x \subset y$ if and only if $\sigma_x \leq \sigma_y$;

(iii) *For all $c \in [0, \frac{\sqrt{2}}{2K})$, there exists a menu $x \in \mathcal{K}(\Delta B)$ such that $\sigma_x \equiv c$. In particular, the function that is identically zero on S^K lies in C, i.e., $\mathbf{0} \in C \subset C(S^K)$.*

Lemma 8 *The set $H := \{\lambda(\sigma - \sigma') : \lambda \geq 0, \sigma, \sigma' \in C\}$ is a linear and lattice subspace of $C(S^K)$.*

---

[6]See, among others, Ahn and Ergin [1], Dietrich [8], Grant and Quiggin [12], Minardi and Savochkin [20], Karni and Vierø [14], Kochov [15], and Schipper [24], as well as the earlier work of Dekel, Lipman, and Rustichini [5], Epstein, Marinacci, and Seo [9], Ghirardato [11], and Mukerji [21].

6.2 Proof of Theorem 3

Write $x \sim y$ if $x \succeq y$ and $y \succeq x$.

Lemma 9 *For every menu $x \in X$, $x \sim cl(x)$.*

Proof The sequence $\{x, x, x, ldots\} \subset X$ converges to both x and $cl(x)$. Since $\succeq$ is continuous, $x \sim cl(x)$. ■

Define a preorder $\succeq^*$ on C by letting $\sigma_x \succeq^* \sigma_y$ if $x, y \in \mathcal{K}(\Delta B)$ are such that $x \succeq y$. By Lemma 7, $\succeq^*$ is continuous since $\succeq$ is continuous. By Theorem 2, $\succeq^*$ is continuously representable. Given Lemma 9 and Indifference to Randomization, the representation of $\succeq^*$ leads to the representation sought after in Theorem 3.

6.3 Proof of Theorem 4

Lemma 10 *For every closed subset x of ΔB, $x \sim conv(x)$.*

Proof See Dekel et al. [6, p. 922]. Though in the latter paper $\succeq$ is complete, one can verify that the arguments do not require completeness. ■

As before, we can define a continuous preorder $\succeq^*$ on C by letting $\sigma_x \succeq^* \sigma_y$ if $x \succeq y$. The proof of the following lemma is immediate.

Lemma 11 *The preorder $\succeq^*$ on C satisfies*
(i) (Independence) For all $\lambda \in (0, 1]$ and $\sigma, \sigma', \hat{\sigma} \in C$,

$$\sigma \succeq^* \sigma' \quad \text{if and only if} \quad \lambda\sigma + (1 - \lambda)\hat{\sigma} \succeq^* \lambda\sigma' + (1 - \lambda)\hat{\sigma};$$

(ii) (Monotonicity) $\sigma \geq \sigma'$ implies $\sigma \succeq^* \sigma'$.

By Lemmas 9 and 10, it suffices to find a representation for $\succeq^*$ on C. Consider the set

$$D := \{\lambda(\sigma - \sigma') : \lambda \geq 0, \sigma \succeq^* \sigma' \, \forall \sigma, \sigma' \in C\} \subset C(S^K)$$

known as the domination cone for the preorder $\succeq^*$ and let $span(D)$ be its linear span.

Lemma 12 *$span(D) = H$. In addition, D has nonempty topological interior in H.*

Proof By Monotonicity, the positive orthant $H_+ := \{f \in H : f \geq 0\}$ of H is a subset of D. By Lemma 8, $H = H_+ - H_+$ and so $span(D) = H$. Take $f \in H_+$ such that $M := \min\{f(s) : s \in S^K\} > 0$ and note that the ball $\{f' \in H : \|f' - f\| < \frac{M}{2}\}$ is contained in H_+. Since $H_+ \subset D$, D has nonempty topological interior in H. ■

To state the next result, due to Shapley and Baucells [26], say that a function $v : C \to \mathbb{R}$ is **affine** if $v(\lambda\sigma + (1 - \lambda)\sigma') = \lambda v(\sigma) + (1 - \lambda)v(\sigma')$ for all $\sigma, \sigma' \in C$ and $\lambda \in [0, 1]$; and **homogeneous** if $v(\mathbf{0}) = 0$.

Lemma 13 *There exists a nonempty set V of affine, homogeneous functions $v : C \to \mathbb{R}$ such that $\sigma \succeq^* \sigma'$ if and only if $v(\sigma) \geq v(\sigma')$ for all $v \in V$.*

Since $H = \cup_{\lambda \geq 0} \lambda(C - C)$, each $f \in H$ can be written as $f = \lambda(\sigma - \sigma')$. Extend each function $v \in V$ to a linear function on H by letting $v(f) = \lambda(v(\sigma) - v(\sigma'))$. Since $\succeq^*$ is monotone, V is a set of positive linear functions on H. Since H_+ has nonempty topological interior, all positive, linear functions are continuous. See, e.g., Schaefer [23, p. 225]. By Lemma 6.13 in Aliprantis and Border [2], each $v \in V$ can be extended to a continuous linear function on $C(S^K)$. By the Riesz Representation Theorem, each $v \in V$ can be represented as an integral with respect to a countably additive Borel probability measure μ_v on S^K. Letting $\Delta := \{\mu_v : v \in V\}$, it follows that for all $x, y \in \mathcal{K}(\Delta B)$,

$$x \succeq y \quad \text{if and only if} \quad \int_{S^K} \max_{\beta \in x}(\beta \cdot s)d\mu \geq \int_{S^K} \max_{\beta \in y}(\beta \cdot s)d\mu \quad \text{for all } \mu \in \Delta.$$

6.4 Proof of Theorem 5

Say that $s \in S^K$ is **strongly relevant** for $\succeq$ if for every open neighborhood O of s (in the relative topology on S^K), there exist menus $x, y \in \mathcal{K}(\Delta B)$ such that $\sigma_x(s') = \sigma_y(s')$ for all $s' \in S^K \setminus O$ and $x \nsim y$. Also, say that a subset $S \subset S^K$ is **sufficient** for the preorder $\succeq$ if for all $x, y \in \mathcal{K}(\Delta B)$, $\sigma_x(s) = \sigma_y(s)$ for all $s \in S$ implies $x \sim y$.

By Theorem 3, $\succeq$ has a representation (S^K, V). By Nontriviality, we can assume that none of the functions $v \in V$ are constant. By Dekel et al. [6, Theorem 1], each v defines a nonempty closed set S_v of strongly relevant states for the order on X induced by v.

Lemma 14 $S^* := cl(\cup_{v \in V} S_v)$ *is the set of all strongly relevant states for $\succeq$.*

Proof Fix some $s \in S^*$ and an open neighborhood $O \subset S^K$ of s. For some $v \in V$, $s' \in O \cap S_v$ is strongly relevant for v. Hence, there are menus $x, y \in \mathcal{K}(\Delta B)$ such that $\sigma_x|_{S^K \setminus O} = \sigma_y|_{S^K \setminus O}$ and $v(\sigma_x) \neq v(\sigma_y)$. By Monotonicity, $v(\sigma_{conv(x \cup y)}) > v(\sigma_x)$ and so $conv(x \cup y) \succ x$. Furthermore, $\sigma_{conv(x \cup y)} = \max\{\sigma_x, \sigma_y\}$ implies that $\sigma_x|_{S^K \setminus O} = \sigma_{conv(x \cup y)}|_{S^K \setminus O}$ and so $s \in S^*$ is strongly relevant for $\succeq$. Since S_v is sufficient for each $v \in V$ and (S^K, V) is a representation of $\succeq$, S^* is sufficient for $\succeq$. ∎

Lemma 15 S^* *is the smallest closed sufficient set for $\succeq$.*

Proof By way of contradiction, suppose there is a closed sufficient set $S \subset S^K$ such that $S^* \setminus S \neq \emptyset$. Take some $s \in S^* \setminus S$. Since S is closed, there is an open neighborhood $O \subset S^K$ of s such that $O \cap S = \emptyset$. Since $s \in S^*$ is strongly relevant, there are menus $x, y \in \mathcal{K}(\Delta B)$ such that $\sigma_x(s') = \sigma_y(s')$ for all $s' \in S^K \setminus O$ and $x \sim y$. But since $S \subset S^K \setminus O$, $x \sim y$ contradicts the sufficiency of S. ∎

Lemma 16 *Suppose S is a sufficient set such that each $s \in S$ is relevant. Then $cl(S) = S^*$.*

Proof By the previous lemma, we know that $S^* \subset cl(S)$ and so it suffices to show that $cl(S) \subset S^*$. By way of contradiction, suppose there is a state $s \in S \setminus S^*$. Since S^* is closed, there is a neighborhood O of s such that $O \cap S^* = \emptyset$. By the sufficiency of S^*, any two menus whose support functions differ only on O must be indifferent. But then the same property applies to $O \cap S$, which is a neighborhood of s in the relative topology on S, contradicting the fact that $s \in S$ is relevant. ∎

To complete the proof of Theorem 5, for each $x \in \mathcal{K}(\Delta B)$, define $\hat{\sigma}_x$ to be the restriction of σ_x to S^* and, for each $v \in V$, let $\hat{v}(\hat{\sigma}_x) := v(\sigma_x)$. Then $(S^*, \{\hat{v} : v \in V\})$ is a proper representation for $\succeq$.

6.5 Proof of Theorem 6

If $s \in \cup_{i=1}^n cl(S_i)$, then s is strongly relevant for some $\succeq_i$. Hence, for every open neighborhood O of s, there exist menus $x \supset y$ such that $x \succ_i y$ and $\sigma_x = \sigma_y$ on $S^K \setminus O$. By Monotonicity, $x \succeq_j y$ for all j. Since $\succeq$ is Paretian, we have $x \succ y$. Thus, s is strongly relevant for $\succeq$. Conclude that $\cup_{i=1}^n cl(S_i) \subset cl(S)$. To see the reverse inclusion, suppose by way of contradiction that there exists $s \in cl(S) \setminus \cup_{i=1}^n cl(S_i)$ and let O be an open neighborhood of s such that $O \cap (\cup_{i=1}^n cl(S_i)) = \emptyset$. Since s is strongly relevant for $\succeq$, there exist menus $x, y \in \mathcal{K}(\Delta B)$ such that $x \sim y$ and $\sigma_x|_{S^K \setminus O} = \sigma_y|_{S^K \setminus O}$. Since $S_j \subset S^K \setminus O$ for all j, the sufficiency of each S_j implies that $x \sim_j y$ for all j. But then, $x \sim y$ contradicts the assumption that $\succeq$ is Paretian.

References

1. Ahn, D., Ergin, H.: Framing contingencies. Econometrica **78**(2), 655–695 (2010)
2. Aliprantis, C., Border, K.: Infinite Dimensional Analysis, 2nd edn. Springer, Berlin (1999)
3. Bewley, T.: Knightian decision theory part I. Cowles foundation discussion papers, Yale University (1986)
4. Bridges, D.S., Mehta, G.B. Representations of Preference Orderings. Springer, Berlin (1995)
5. Dekel, E., Lipman, B., Rustichini, A.: Recent developments in modeling unforeseen contingencies. Eur. Econ. Rev. **42**, 523–542 (1998)
6. Dekel, E., Lipman, B., Rustichini, A.: Representing preferences with a unique subjective state space. Econometrica **69**, 891–934 (2001)
7. Dekel, E., Lipman, B., Rustichini, A., Sarver, T.: Representing preferences with a unique subjective state space: corrigendum. Econometrica **75**, 591–600 (2007)
8. Dietrich, F.: Savage's theorem under changing awareness. J. Econ. Theory **176**, 1–54 (2018)

9. Epstein, L.G., Marinacci, M., Seo, K.: Coarse contingencies and ambiguity. Theor. Econ. **2**(4), 355–394 (2007)
10. Evren, Ö., Ok, E.A.: On the multi-utility representation of preference relations. J. Math. Econ. **47**(4–5), 554–563 (2011)
11. Ghirardato, P.: Coping with ignorance: unforeseen contingencies and non-additive uncertainty. Econ. Theory **17**(2), 247–276 (2001)
12. Grant, S., Quiggin, J.: A preference model for choice subject to surprise. Theory Decis. **79**(2), 167–180 (2015)
13. Gul, F., Pesendorfer, W.: Temptation and self-control. Econometrica **69**, 1403–1435 (2001)
14. Karni, E., Vierø, M.-L.: Reverse Bayesianism: a choice-based theory of growing awareness. Am. Econ. Rev. **103**(7), 2790–2810 (2013)
15. Kochov, A.: A behavioral definition of unforeseen contingencies. J. Econ. Theory **175**, 265–290 (2018)
16. Kreps, D.: A representation theorem for preference for flexibility. Econometrica **47**, 565–576 (1979)
17. Kreps, M.: Static choice in the presence of unforeseen contingencies. In: Dasgupta, P., Gale, D., Hart, O., Maskin, E.(eds.) Economic Analysis of Markets and Games: Essays in Honor of Frank Hahn, pp. 259–281. MIT Press, Cambridge (1992)
18. Mehta, G.: Topological ordered spaces and utility functions. Int. Econ. Rev. **18**(3), 779–782 (1977)
19. Mehta, G.: Some general theorems on the existence of order-preserving functions. Math. Soc. Sci. **15**(2), 135–143 (1988)
20. Minardi, S., Savochkin, A.: Subjective contingencies and limited bayesian updating. J. Econ. Theory **183**, 1–45 (2019)
21. Mukerji, S.: Understanding the nonadditive probability decision model. Econ. Theory **9**, 23–46 (1997)
22. Nachbin, L.: Topology and Order. D. Van Nostrand Company Inc. (1965)
23. Schaefer, H.: Topological Vector Spaces. The Macmillan Company, New York (1966)
24. Schipper, B.C.: Awareness-dependent subjective expected utility. Int. J. Game Theory. **42**(3), 725–753 (2013)
25. Schmeidler, D.: A condition for the completeness of partial preference relations. Econometrica **39**(2), 403–404 (1971)
26. Shapley, L., Baucells, M.: Multiperson utility. UCLA Working paper (1998)

Preference for Flexibility: A Continuous Representation in an Ordinal Setup

Özgür Evren

Abstract In a menu-choice framework with a compact metric space of alternatives, I prove a representation of preference for flexibility that features continuous ex-post utility functions as well as a continuous aggregator over ex-post utility levels. Continuity of the ex-post utility functions ensures that each function attains its maximum on any compact menu of alternatives. This, in turn, supports the standard interpretation that relates the value of flexibility to the maximization of ex-post preferences. I also show that an incomplete menu-preference is characterized by the Pareto order induced by a collection of aggregators, in place of a single one.

Keywords Preference for flexibility · Subjective states · Freedom of choice · Continuity · Incompleteness

JEL classification D11 · D71 · D81 · D83

1 Introduction

In a menu-choice problem, the decision maker (DM) chooses a set of alternatives, such as a restaurant menu or a budget set, which determines her feasible options in a subsequent period. Following Kreps' [24] seminal work on representation of preference for flexibility, this menu-choice framework proved to be a powerful tool to model a variety of behavioral phenomena including preference for freedom of choice [2, 16, 29], temptation and self-control [9, 20, 21], anticipated regret [31],

---

I wrote the first draft of this paper in 2007, when I was a Ph.D. student at New York University, which circulated under the title "An Ordinal, Continuous Representation Theorem for Preference for Flexibility." I owe special thanks to Efe Ok, who was my advisor at the time. I also thank Barton Lipman, Pietro Ortoleva, Clemens Puppe, Debraj Ray, Gil Riella, two anonymous referees and the editors of this volume for useful discussions and suggestions.

Ö. Evren (✉)
New Economic School, Skolkovskoe Shosse 45, Moscow 121353, Russia
e-mail: oevren@nes.ru

© Springer Nature Switzerland AG 2020
G. Bosi et al. (eds.), *Mathematical Topics on Representations of Ordered Structures and Utility Theory*, Studies in Systems, Decision and Control 263,
https://doi.org/10.1007/978-3-030-34226-5_14

costly contemplation [12, 13], perfectionism [23], thinking aversion [27], and other-regarding behavior driven by social image concerns [10, 14, 30].

A large part of this literature focuses on menus of lotteries, which brings about a vector structure, and an arsenal of mathematical tools that can be used in that setting.[1] Compared to their ordinal counterparts, these cardinal models possess stronger uniqueness/identification properties [8], and tend to be more amenable to comparative statics exercises. Nevertheless, the ordinal framework, originally selected by Kreps, still appears to be worthwhile to study because objective lotteries are rarely present in real life choice problems. On a related note, the empirical content of an ordinal model and the associated axioms may well be more transparent than that of an analogous cardinal model based on von Neumann–Morgenstern independence axiom or its derivatives.[2]

In this paper, I revisit Kreps' [24] ordinal framework to provide a proof of a conjecture on representation of preference for flexibility that Kreps has left unproved. Specifically, I show that given a preference relation on closed menus in a compact metric space of alternatives $\mathcal{A}$, a standard continuity property and Kreps' axioms jointly imply the existence of a representation with a collection of continuous (state-dependent) utility functions on $\mathcal{A}$, and a further continuous function, referred to as an *aggregator*, that determines the overall value of a menu based on the state-dependent functions' maximum values on the menu. The interpretation is that the flexibility offered by a given menu allows the DM to wait and learn more about her tastes, so that ex-post she can select an alternative that maximizes the "correct" utility function over that menu.[3] The state-dependent utility functions do attain their maximum values because they are continuous. While Kreps were only able to find lower semi-continuous utility functions, he noted that a fully continuous version of the representation may be feasible.[4]

My construction of this continuous representation is inspired by Pavel Urysohn's characterization of normal topological spaces, and Nachbin's [26] application of this

---

[1]To mention a few examples, the proof of Gul and Pesendorfer's [20] representation appeals to the von Neumann–Morgenstern representation theorem, whereas Dekel, Lipman and Rustichini [8] build upon some classical results in functional analysis such as the Hahn-Banach extension theorem and the Riesz representation theorem.

[2]For instance, the independence axiom of Dekel, Lipman and Rustichini [8] combined with the monotonicity axiom implies Kreps' [24] submodularity property [8, Footnote 21]. While the independence axiom becomes irrelevant in an ordinal setup, submodularity is still indispensable, which arguably makes the latter more substantive as regards the notion of preference for flexibility.

[3]Representations of this sort arise in the analysis of preference for flexibility and related notions such as liquidity demand for money (see, e.g., [17, 18]). Multi-preference approach to the measurement of value of freedom of choice is another related area of research (see, e.g., [2]). Bleichrodt and Quiggin [3] provide a more recent application to health economics.

[4]"If ...$\succsim$ is represented ...with $U(\cdot, s)$ continuous for each s ..., then $\succsim$ is continuous on X^c. The converse seems reasonable: If $\succsim$ is continuous on X^c, then a representation with continuous $U(\cdot, s)$ and u is possible. But I am unable to supply a proof of the converse—U as constructed in the proof of the theorem will be lower semi-continuous only" (Kreps [24, p. 575]). Here, X^c stands for the collection of menus, $\succsim$ is the DM's preference relation over menus, u denotes the aggregator, and $U(\cdot, s)$ is a state-dependent utility function.

technique to topological order theory. Pioneered by Mehta [25], similar methods have been extensively used to prove representation theorems for preference relations on ordinary alternatives.[5] To the best of my knowledge, this is the first paper that illustrates potential uses of Urysohn–Nachbin type constructions in the context of menu-preferences.

In addition, my representation dispenses with the completeness axiom to let the DM remain indecisive on occasion. Incompleteness is characterized by the Pareto order induced by a collection of aggregators, in place of a single one. This provides an ordinal version of Kochov's [22] multi-prior representation for incomplete preferences over menus of lotteries.[6] It is also worth noting that indecisiveness may naturally lead to preference for flexibility [1, 6, 7, 28], which motivates relaxation of the completeness axiom in the present setting.

Following a bit of notation, in Sect. 3, I state and prove the representation theorem described above. Section 4 focuses on an alternative setup in which the DM simultaneously selects an alternative for immediate consumption and a budget set for future use.

2 Notation and Terminology

Let $\mathcal{X}$ be a compact metric space. $\mathcal{C}(\mathcal{X})$ stands for the set of all continuous, real valued functions on $\mathcal{X}$. For a nonempty set $U \subseteq \mathcal{C}(\mathcal{X})$, and a nonempty closed set $A \subseteq \mathcal{X}$, the vector $(\max_{a \in A} u(a))_{u \in U} \in \mathbb{R}^U$ is denoted as $\max_A U$. The space $\mathbb{R}^U$ is endowed with the product topology and the pointwise (partial) order $\geq$. The asymmetric and symmetric parts of $\geq$ are denoted as $>$ and $=$, respectively. I denote by $\mathcal{K}(\mathcal{X})$ the collection of all nonempty closed subsets of $\mathcal{X}$, equipped with the Hausdorff metric.[7]

Given a binary relation $\succsim$ on $\mathcal{X}$, I denote by $\succ$ and $\sim$ the asymmetric and symmetric parts of $\succsim$, respectively. As usual, $x \succsim y$ means $(x, y) \in \succsim$, and similarly for $\succ$ and $\sim$. In turn, if $x \succsim y$ is not true, I write $x \not\succsim y$. The relation $\succsim$ is said to be a *preorder* if it is reflexive and transitive. The incomplete part of $\succsim$ is denoted by $\bowtie$. That is, for any $x, y \in \mathcal{X}$, we have $x \bowtie y$ if and only if neither $x \succsim y$ nor $y \succsim x$.

A binary relation $\succsim$ on $\mathcal{K}(\mathcal{X})$ is said to be *monotonic* if for any $A, B \in \mathcal{K}(\mathcal{X})$ with $A \supseteq B$, we have $A \succsim B$.

---

[5]Recent reviews of this literature can be found in [4, 5] for representations with single and multiple utility functions, respectively.

[6]In Kochov's model, a prior refers to a probability distribution over subjective states, or state-dependent utility functions. Each prior defines an expectation operator over ex-post utility levels, which corresponds to an additive aggregator in the present setup (see Footnote 8).

[7]The Hausdorff distance between $A, B \in \mathcal{K}(\mathcal{X})$ equals the maximum of $\max_{a \in A} \min_{b \in B} \sigma(a, b)$ and $\max_{b \in B} \min_{a \in A} \sigma(a, b)$, where σ stands for the metric on $\mathcal{X}$.

3   A Basic Representation

Consider a two-stage choice problem, where the set of all alternatives is a compact metric space $\mathcal{A}$. In the first stage, the DM has to choose a *menu*, i.e., an element of $\mathcal{K}(\mathcal{A})$, which determines her feasible options in the second stage. The DM's first-stage preference is represented by a preorder $\succsim$ on $\mathcal{K}(\mathcal{A})$. The relation $\succsim$ is influenced by the DM's anticipated second-stage behavior. Nevertheless, following a standard practice, I do not model the second-stage behavior explicitly.

The basic representation of preference for flexibility requires the following axioms.

Monotonicity (M) $\succsim$ is monotonic.

Submodularity (SM) For any $A, B \in \mathcal{K}(\mathcal{A})$ with $A \sim A \cup B$, we have $A \cup C \sim A \cup C \cup B$ for every $C \in \mathcal{K}(\mathcal{A})$.

Continuity (C) $\succsim$ is closed in $\mathcal{K}(\mathcal{A}) \times \mathcal{K}(\mathcal{A})$.

If the DM cannot precisely predict which alternative she will select from a given menu A, she may strictly prefer A to any of its subsets because the former allows the DM to keep her options open. The monotonicity axiom accommodates such behavior, while ruling out potential instances of preference for commitment. The submodularity axiom posits that if adding a menu B to a menu A is of no value, then adding B to a superset of A should again be of no value. The closed-continuity axiom (C) is fairly standard, especially in the literature on representation of incomplete preferences (see, e.g., [11, 15, 22]). It should also be noted that this axiom holds trivially when $\mathcal{A}$ is finite.

For a complete preorder $\succsim$, the desired representation consists of a nonempty set $U \subseteq C(\mathcal{A})$ and a real valued function φ on the set $\{\max_A U : A \in \mathcal{K}(\mathcal{A})\} \subseteq \mathbb{R}^U$, strictly increasing with respect to the partial order $\geq$, such that, for any $A, B \in \mathcal{K}(\mathcal{A})$

$$A \succsim B \quad \Leftrightarrow \quad \varphi\left(\max_A U\right) \geq \varphi\left(\max_B U\right). \tag{1}$$

Each element u of the set U is interpreted as a utility function that represents a probable, future preference relation s_u on the set of alternatives, while the set $S := \{s_u : u \in U\}$ can be viewed as a subjective (i.e., endogenous) state space. Thus, the representation describes a DM who behaves as if she (a) will learn her future tastes at the beginning of the second stage; (b) will make her final choice to maximize her ex-post preference relation; (c) uses a monotonically increasing function φ to compute the ex-ante value of a menu based on her state-dependent indirect utility levels. In what follows, an *aggregator* refers to such a function φ.

Kreps [24] showed that when the set $\mathcal{A}$ is finite and $\succsim$ is complete, the axioms (M) and (SM) are equivalent to the existence of a representation of the form (1) with an additive aggregator φ.[8] For a compact metric space $\mathcal{A}$, he also noted that adding the continuity axiom (C) guarantees a representation of the form (1) with potentially

---

[8]For a finite set of utility functions U, a (monotonic) additive aggregator φ is defined by nonnegative numbers $\alpha_u (u \in U)$ such that $\varphi(\max_A U) = \sum_{u \in U} \alpha_u \max_{a \in A} u(a)$ for every $A \in \mathcal{K}(\mathcal{A})$. When U

discontinuous functions, $u \in U$. The next theorem shows that the continuity axiom is sufficient to obtain a fully continuous representation. Moreover, relaxation of the completeness axiom leads to a set of aggregators in place of a single one.

Theorem 1 *Let $\mathcal{A}$ be a compact metric space. A binary relation $\succsim$ on $\mathcal{K}(\mathcal{A})$ is a preorder that satisfies (M), (SM) and (C) if, and only if, there exist a nonempty set $U \subseteq \mathcal{C}(\mathcal{A})$ and a nonempty set Φ of strictly increasing, continuous, real valued functions on the set $\{\max_A U : A \in \mathcal{K}(\mathcal{A})\} \subseteq \mathbb{R}^U$ such that, for any $A, B \in \mathcal{K}(\mathcal{A})$,*

$$A \succsim B \quad \Leftrightarrow \quad \varphi\left(\max_A U\right) \geq \varphi\left(\max_B U\right) \text{ for every } \varphi \in \Phi. \tag{2}$$

Here, Φ can be chosen to be a singleton if, and only if, $\succsim$ is complete.

In Theorem 1, incompleteness of $\succsim$ is characterized by the multiplicity of aggregators, $\varphi \in \Phi$, which corresponds to a DM who is not able to form a firm belief about her probable future preference relations. It is also worth noting that if $\succsim$ is a preorder that satisfies (M), for any $a, b \in \mathcal{A}$ with $\{a\} \bowtie \{b\}$ we have $\{a, b\} \succ \{a\}$ and $\{a, b\} \succ \{b\}$. Thus, the model characterized in Theorem 1 embodies a close connection between indecisiveness and preference for flexibility, as mentioned in the introduction.

Given a preorder $\succsim$ on $\mathcal{K}(\mathcal{A})$ that satisfies the axioms in Theorem 1, we can define a dominance relation $\overset{\cdot}{\succsim}$ on $\mathcal{K}(\mathcal{A})$ as follows: For any $A, B \in \mathcal{K}(\mathcal{A})$,

$$A \overset{\cdot}{\succsim} B \quad \Leftrightarrow \quad A \sim A \cup B. \tag{3}$$

The next lemma will enable us to characterize this relation in terms of the Pareto order induced by a collection of indirect utility functions.

Lemma 1 *Given a compact metric space $\mathcal{X}$, let $\overset{\star}{\succsim}$ be a binary relation on $\mathcal{K}(\mathcal{X})$. The following two statements are equivalent.*

(a) $\overset{\star}{\succsim}$ is a closed and monotonic preorder such that for any $A, B \in \mathcal{K}(\mathcal{X})$,

$$A \overset{\star}{\succsim} B \quad \Rightarrow \quad A \overset{\star}{\succsim} A \cup B. \tag{4}$$

(b) There exists a nonempty set $U \subseteq \mathcal{C}(\mathcal{X})$ such that for any $A, B \in \mathcal{K}(\mathcal{X})$,

$$A \overset{\star}{\succsim} B \quad \Leftrightarrow \quad \max_A U \geq \max_B U.$$

Lemma 1 is a key step towards Theorem 1. Since it is somewhat involved, I postpone the proof of this lemma to the end of this section, which is based on a Urysohn–Nachbin type construction.

---

is infinite, the summation operator is replaced by the expectation operator induced by a probability measure on U.

The next lemma follows from Corollary 1 in [15]. This helps establish the existence of aggregators.

Lemma 2 *Let $\mathcal{X}$ be a compact metric space. A binary relation $\succsim$ on $\mathcal{X}$ is a closed preorder if, and only if, there exists a nonempty set $V \subseteq \mathcal{C}(\mathcal{X})$ that satisfies the following two properties for any $x, y \in \mathcal{X}$.*
(i) $x \succ y$ implies $v(x) > v(y)$ for every $v \in V$.
(ii) $x \succsim y$ if, and only if, $v(x) \geq v(y)$ for every $v \in V$.

We are now ready to prove Theorem 1.
Proof of Theorem 1. The proof of the "if" part is straightforward, and omitted. To prove the "only if" part, the first step is to show that the dominance relation $\dot{\succsim}$ in (3) satisfies the properties in Lemma 1(a), where $\mathcal{A}$ takes the role of $\mathcal{X}$.

It is clear that $\dot{\succsim}$ is reflexive, monotonic, and $A \dot{\succsim} B$ implies $A \dot{\succsim} A \cup B$ for any $A, B \in \mathcal{K}(\mathcal{A})$. Moreover, if $A \dot{\succsim} B$ and $B \dot{\succsim} C$ for some $A, B, C \in \mathcal{K}(\mathcal{A})$, by (SM) and the definition of $\dot{\succsim}$ we have $A \sim A \cup B \sim A \cup C \cup B$. From (M) it then follows that $A \sim A \cup C$, i.e., $A \dot{\succsim} C$. Thus $\dot{\succsim}$ is transitive.

Since $(A, B) \to A \cup B$ is a continuous function from $\mathcal{K}(\mathcal{A}) \times \mathcal{K}(\mathcal{A})$ into $\mathcal{K}(\mathcal{A})$, it is also a routine exercise to check that $\dot{\succsim}$ inherits the continuity property (C) from the relation $\succsim$.

By Lemma 1, we conclude that there exists a nonempty set $U \subseteq \mathcal{C}(\mathcal{A})$ such that $A \dot{\succsim} B$ iff $\max_A U \geq \max_B U$.

Also, Lemma 2 applied to the preorder $\dot{\succsim}$, where $\mathcal{K}(\mathcal{A})$ takes the role of $\mathcal{X}$, yields a nonempty set $V \subseteq \mathcal{C}(\mathcal{K}(\mathcal{A}))$ such that (i) $A \dot{\succ} B$ implies $v(A) > v(B)$ for every $v \in V$; and (ii) $A \dot{\succsim} B$ iff $v(A) \geq v(B)$ for every $v \in V$.

Note that $A \dot{\succ} (\dot{\sim}) B$ imply $A \succ (\sim) B$. In particular,

$$\max_A U = \max_B U \quad \Rightarrow \quad A \dot{\sim} B \quad \Rightarrow \quad A \sim B \quad \Rightarrow \quad v(A) = v(B).$$

Thus, for each $v \in V$ we can define a real valued function φ_v on the set $\{\max_A U : A \in \mathcal{K}(\mathcal{A})\}$ by the rule $\varphi_v(\max_A U) \equiv v(A)$. The function φ_v is strictly increasing with respect to $\geq$, for

$$\max_A U > \max_B U \quad \Rightarrow \quad A \dot{\succ} B \quad \Rightarrow \quad A \succ B \quad \Rightarrow \quad v(A) > v(B).$$

Put $\Phi := \{\varphi_v : v \in V\}$. By construction, (2) holds for every $A, B \in \mathcal{K}(\mathcal{A})$. We shall next show that for each $v \in V$ the function φ_v is continuous on its domain. Pick any $v \in V$, and suppose by contradiction that there exist a set $A \in \mathcal{K}(\mathcal{A})$, a neighborhood N of $\varphi_v(\max_A U)$, and a net (A_α) in $\mathcal{K}(\mathcal{A})$ such that the net $(\max_{A_\alpha} U)$ converges to $\max_A U$, while $\varphi_v(\max_{A_\alpha} U) \notin N$ for every α. Since $\mathcal{K}(\mathcal{A})$ is compact, a subnet of (A_α), say (A_β), converges to some $A' \in \mathcal{K}(\mathcal{A})$. Given that U consists of continuous functions on $\mathcal{A}$, it clearly follows that the net $(\max_{A_\beta} U)$ converges to $\max_{A'} U$. Then $\max_{A'} U = \max_A U$, so that $v(A') = v(A) = \varphi_v(\max_A U)$. Since v is continuous and (A_β) converges to A', it also follows that $v(A_\beta) = \varphi_v(\max_{A_\beta} U) \in N$ for all sufficiently large β, a contradiction.

Finally, it should be noted that when $\succsim$ is complete, we can select a single function v on $\mathcal{K}(\mathcal{A})$ to represent $\succsim$, and let $V = \{v\}$. ∎

3.1 Proof of Lemma 1

That statement (b) implies statement (a) is obvious. To prove the converse implication, let $\mathcal{X}$ be a compact metric space, and take a preorder $\overset{*}{\succsim}$ on $\mathcal{K}(\mathcal{X})$ that satisfies the properties in statement (a). Throughout the proof, generic elements of $\mathcal{K}(\mathcal{X})$ are denoted as A, B, C etc.

The first point to note is that since $\overset{*}{\succsim}$ is a monotonic preorder, the property (4) is equivalent to the following statement

$$ A \overset{*}{\succsim} B \quad \Leftrightarrow \quad A \overset{*}{\sim} A \cup B. $$

Let us proceed with a couple of claims.

Claim 1 $A \overset{*}{\succsim} B$ and $A \overset{*}{\succsim} C$ imply $A \overset{*}{\succsim} B \cup C$.

Proof Assume $A \overset{*}{\succsim} B$ and $A \overset{*}{\succsim} C$, or equivalently, $A \overset{*}{\sim} A \cup B$ and $A \overset{*}{\sim} A \cup C$. Then $A \cup B \overset{*}{\sim} A \cup C$ by transitivity of $\overset{*}{\succsim}$, which implies $A \cup B \overset{*}{\sim} (A \cup B) \cup (A \cup C) = A \cup B \cup C$. Since $\overset{*}{\succsim}$ is transitive, it also follows that $A \overset{*}{\sim} A \cup B \cup C$. Hence, $A \overset{*}{\succsim} B \cup C$, as we sought. ∎

Claim 2 *For each A in $\mathcal{K}(\mathcal{X})$, define $f(A) := \left\{ a \in \mathcal{X} : A \overset{*}{\succsim} \{a\} \right\}$. Then*

$$ A \overset{*}{\succsim} B \quad \Leftrightarrow \quad f(A) \supseteq f(B) \quad \Leftrightarrow \quad f(A) \supseteq B. \tag{5} $$

Thus, for any $A \in \mathcal{K}(\mathcal{X})$, we have $f(A) \supseteq A$, $f(A) \overset{*}{\sim} A$, and $f(f(A)) = f(A)$.

Proof If $A \overset{*}{\succsim} B$, by transitivity of $\overset{*}{\succsim}$ we have $f(A) \supseteq f(B)$. Moreover, the latter inclusion implies $f(A) \supseteq B$ because $f(B) \supseteq B$ by monotonicity of $\overset{*}{\succsim}$. Suppose now $f(A) \supseteq B$, i.e., $A \overset{*}{\succsim} \{b\}$ for every $b \in B$. Then, by Claim 1, $A \overset{*}{\succsim} B'$ for every finite subset B' of B. Since B is the limit of a sequence of such subsets, and $\overset{*}{\succsim}$ is closed, we must then have $A \overset{*}{\succsim} B$. This proves the statement (5).

To prove the remaining assertions, let $B = f(A)$, so that $f(A) \supseteq B$. The latter inclusion implies $f(A) \supseteq f(B) = f(f(A))$, and hence, $A \overset{*}{\succsim} f(A)$. Finally, $f(B) \supseteq B$ means $f(f(A)) \supseteq f(A)$, or equivalently, $f(A) \overset{*}{\succsim} A$. ∎

Claim 2 is analogous to Lemma 1(e) in Kreps [24]. Building upon this lemma, Kreps observed that the set $U := \{-\mathbf{1}_{f(A)} : A \in \mathcal{K}(\mathcal{X})\}$ leads to a representation of the form (1), under the completeness axiom. Since $\overset{*}{\succsim}$ is closed, each $f(A)$ is a closed subset of $\mathcal{X}$. Thus, the functions in U as defined by Kreps typically lack upper semi-continuity. This problem becomes especially severe when $\mathcal{X}$ is connected so that a nonempty proper subset, $f(A)$, cannot be open and closed simultaneously.

The next claim is the backbone of the present proof. In what follows, $\mathbb{Q}_0$ denotes the set of all dyadic rational numbers in $(0, 1)$, i.e., $\mathbb{Q}_0 := \{\frac{k}{2^n} : n \in \mathbb{N}, k \in \{1, \ldots, 2^n - 1\}\}$. We shall use $\mathbb{Q}_0$ as an index set to construct the functions in the set U.

Claim 3 *For any $A \in \mathcal{K}(\mathcal{X})$ and $b \in X$ with $A \overset{*}{\nsucceq} \{b\}$, there exists a collection of sets $\{X_r : r \in \mathbb{Q}_0\} \subseteq \mathcal{K}(\mathcal{X})$ such that for every $r \in \mathbb{Q}_0$ (i) $A \subseteq f(X_r)$; (ii) $b \notin f(X_r)$; and (iii) $f(X_r) \subseteq \text{int } f(X_{r'})$ whenever $r' > r$, where $\text{int } f(X_{r'})$ stands for the interior of $f(X_{r'})$.*

Let $\sigma(\cdot, \cdot)$ denote the metric on $\mathcal{X}$. For any $C \in \mathcal{K}(\mathcal{X})$ and $\gamma > 0$, the set $B_\gamma(C) := \bigcup_{c \in C} \{b \in \mathcal{X} : \sigma(b, c) < \gamma\}$ is an open neighborhood of C. The closure of $B_\gamma(C)$ will be denoted as $\overline{B}_\gamma(C)$. The following claim is the first step towards the proof of Claim 3.

Claim 4 *Suppose $f(C) = C$ for a set $C \in \mathcal{K}(\mathcal{X})$. Then for any $\varepsilon > 0$ there exists a $\delta > 0$ such that $f(\overline{B}_\delta(C)) \subseteq B_\varepsilon(C)$.*

Proof Given an $\varepsilon > 0$, suppose by contradiction that there exist a sequence of positive numbers (δ_n) that converges to 0, and a sequence (a_n) in $\mathcal{X}$ such that $\overline{B}_{\delta_n}(C) \overset{*}{\succsim} \{a_n\}$ and $\inf_{c \in C} \sigma(a_n, c) \geq \varepsilon$ for every n. By passing to a subsequence if necessary, assume (a_n) converges and put $a := \lim_n a_n$. Note that $C \subseteq \overline{B}_{\delta_n}(C) \subseteq \bigcup_{c \in C} \{b \in \mathcal{X} : \sigma(b, c) \leq \delta_n\}$ for every n. Thus, $(\overline{B}_{\delta_n}(C))$ converges to C in the Hausdorff metric. Since $\overset{*}{\succsim}$ is closed, it follows that $C \overset{*}{\succsim} \{a\}$, i.e., $a \in f(C) = C$. Then we must also have $\sigma(a_n, a) \geq \inf_{c \in C} \sigma(a_n, c) \geq \varepsilon$ for every n, which contradicts the assumption that (a_n) converges to a. ∎

We are now ready to prove Claim 3.

Proof of Claim 3. Pick any $A \in \mathcal{K}(\mathcal{X})$ and $b \in \mathcal{X}$ with $A \overset{*}{\nsucceq} \{b\}$, which means $b \notin f(A)$. Put $X_0 := A$. Since $f(X_0)$ is closed, there exists an $\varepsilon > 0$ such that $b \notin B_\varepsilon(f(X_0))$. By Claim 2, $f(f(X_0)) = f(X_0)$. Thus Claim 4 implies that there exists a $\delta > 0$ with $f(\overline{B}_\delta(f(X_0))) \subseteq B_\varepsilon(f(X_0))$.

Put $X_1 := \overline{B}_\delta(f(X_0))$. Then $A \subseteq f(X_0) \subseteq B_\delta(f(X_0)) \subseteq \text{int } X_1 \subseteq \text{int } f(X_1) \subseteq f(X_1)$, while $b \notin f(X_1)$. Inductively, given a nonnegative integer $\bar{n}$, suppose that we have found a collection $\{f(X_r) : r = \frac{k}{2^{\bar{n}}}, k \in \{0, \ldots, 2^{\bar{n}}\}\}$ such that (i) $A \subseteq f(X_0)$; (ii) $b \notin f(X_1)$; and (iii) $f(X_r) \subseteq \text{int } f(X_{r'})$ whenever $r < r'$.

Obviously, to complete the proof it suffices to show that for any $k \in \{0, \ldots, 2^{\bar{n}} - 1\}$, there exists a set $C \in \mathcal{K}(\mathcal{X})$ such that $f(X_r) \subseteq \text{int } f(C) \subseteq f(C) \subseteq$

int $f(X_{r'})$ where $r := \frac{k}{2^n}$ and $r' := \frac{k+1}{2^n}$. To this end, first note that since $f(X_r)$ is a compact subset of int $f(X_{r'})$, there exists an $\varepsilon > 0$ with $B_\varepsilon(f(X_r)) \subseteq$ int $f(X_{r'})$. Moreover, by Claim 4, there is a $\delta > 0$ such that $f(\overline{B}_\delta(f(X_r))) \subseteq B_\varepsilon(f(X_r))$. Let $C := \overline{B}_\delta(f(X_r))$. Then $f(X_r) \subseteq B_\delta(f(X_r)) \subseteq$ int $C \subseteq$ int $f(C) \subseteq f(C) \subseteq B_\varepsilon(f(X_r)) \subseteq$ int $f(X_{r'})$. In particular, $f(X_r) \subseteq$ int $f(C) \subseteq f(C) \subseteq$ int $f(X_{r'})$, as we sought. This completes the proof of Claim 3. ∎

Pick any $A, B \in \mathcal{K}(\mathcal{X})$ such that $A \overset{*}{\not\succsim} B$. Then, by Claim 2, B is not a subset of $f(A)$. That is, $A \overset{*}{\not\succsim} \{b\}$ for some $b \in B$. Let the collection $\{X_r : r \in \mathbb{Q}_0\}$ be as in Claim 3. Define a function $u : \mathcal{X} \to [0, 1]$ as follows: For any $c \in \mathcal{X}$,

$$u(c) := \begin{cases} \inf\{r \in \mathbb{Q}_0 : c \in f(X_r)\} & \text{if } c \in f(X_r) \text{ for some } r \in \mathbb{Q}_0, \\ 1 & \text{otherwise.} \end{cases}$$

In the remainder of the proof, the following observation will be used repeatedly. This is an immediate consequence of the definition of u.

Claim 5 *For any $c \in \mathcal{X}$ and any number γ with $u(c) < \gamma \leq 1$, there exists an $r \in \mathbb{Q}_0$ such that $r < \gamma$ and $c \in f(X_r)$.*

By definition of the collection $\{X_r : r \in \mathbb{Q}_0\}$, we have $A \subseteq f(X_r)$ and $b \notin f(X_r)$ for any $r \in \mathbb{Q}_0$. Consequently, $\sup_{a \in A} u(a) = 0$ while $u(b) = 1 = \sup_{b' \in B} u(b')$.

The next step is to show that for any $C, D \in \mathcal{K}(\mathcal{X})$,

$$C \overset{*}{\succsim} D \quad \Rightarrow \quad \sup_{c \in C} u(c) \geq \sup_{d \in D} u(d). \tag{6}$$

Suppose by contradiction that $C \overset{*}{\succsim} D$ and $\alpha := \sup_{c \in C} u(c) < \sup_{d \in D} u(d) =: \beta$. Then there exists an $r' \in \mathbb{Q}_0$ with $\alpha < r' < \beta$ because $\mathbb{Q}_0$ is dense in $[0, 1]$. Since $\alpha < r'$, Claim 5 implies that for each $c \in C$, there exists an $r \in \mathbb{Q}_0$ such that $r < r'$ and $c \in f(X_r)$. Then c also belongs to $f(X_{r'})$ because $f(X_r) \subseteq f(X_{r'})$. It follows that $C \subseteq f(X_{r'})$. By Claim 2, the latter inclusion implies $X_{r'} \overset{*}{\succsim} C$, and hence, $X_{r'} \overset{*}{\succsim} D$. Using Claim 2 once again, we see that $f(X_{r'}) \supseteq D$. But then $\beta = \sup_{d \in D} u(d) \leq r'$ by the definition of u, which is a contradiction. This proves (6).

It remains to show that u is continuous on $\mathcal{X}$. Suppose by contradiction that there is a convergent sequence (c_n) in $\mathcal{X}$ and a number $\varepsilon > 0$ such that $|u(c_n) - u(c)| > \varepsilon$ for every n, where $\lim_n c_n := c$. By passing to a subsequence if necessary, we can assume either (i) $u(c_n) > u(c) + \varepsilon$ for every n; or (ii) $u(c_n) + \varepsilon < u(c)$ for every n. Let us first assume that property (i) holds. Then $u(c) < u(c) + \varepsilon < 1$. By Claim 5, there is an $r \in \mathbb{Q}_0$ such that $r < u(c) + \varepsilon$ and $c \in f(X_r)$. Pick any $r' \in \mathbb{Q}_0$ with $r < r' < u(c) + \varepsilon$. Since $f(X_r) \subseteq$ int $f(X_{r'})$ and $\lim_n c_n = c$, we must have $c_n \in f(X_{r'})$, and hence, $u(c_n) \leq r'$ for all sufficiently large n. This contradicts the property (i). Suppose now that (ii) holds. Then, by Claim 5, for every n there exists a number $r_n \in \mathbb{Q}_0$ with $r_n < u(c_n) + \varepsilon/2$ and $c_n \in f(X_{r_n})$. Pick any $r' \in \mathbb{Q}_0$ with $u(c) - \varepsilon/2 < r' < u(c)$, and note that $r_n < r'$ for every n. Thus, $f(X_{r_n}) \subseteq f(X_{r'})$, which also implies $c_n \in f(X_{r'})$ for every n. Since $f(X_{r'})$ is a closed set, it follows that $c \in f(X_{r'})$, and hence, $u(c) \leq r'$, a contradiction. This proves continuity of u.

We have thereby proved that there exists a continuous function $u : \mathcal{X} \to [0, 1]$ such that $\max_{a \in A} u(a) = 0$, $\max_{b \in B} u(b) = 1$, and $\max_{c \in C} u(c) \geq \max_{d \in D} u(d)$ for any $C, D \in \mathcal{K}(\mathcal{X})$ with $C \overset{*}{\succsim} D$. Since A and B are arbitrarily selected elements of $\mathcal{K}(X)$ with $A \overset{*}{\not\succsim} B$, it easily follows that the set

$$U := \left\{ \tilde{u} \in \mathcal{C}(X) : \max_{c \in C} \tilde{u}(c) \geq \max_{d \in D} \tilde{u}(d) \; \forall C, D \in \mathcal{K}(X) \text{ with } C \overset{*}{\succsim} D \right\}$$

satisfies the properties in statement (b) of Lemma 1.   ∎

4 Consumption-Investment Budgeting

In dynamic choice problems, often the DM's investment or saving depends on her instantaneous consumption. Effectively, the DM is selecting both her instantaneous consumption and a set of alternatives that will become available in the future. The purpose of this section is to extend Theorem 1 along these lines.

Let $\mathcal{Z}$ and $\mathcal{A}$ be compact metric spaces, which represent alternatives for immediate and future consumption, respectively. The DM's preference relation is a preorder $\succsim$ on $\mathcal{Z} \times \mathcal{K}(\mathcal{A})$. As usual, $\mathcal{K}(\mathcal{A})$ is equipped with the Hausdorff metric, while the (metrizable) product topologies are imposed on $\mathcal{Z} \times \mathcal{A}$ and $\mathcal{Z} \times \mathcal{K}(\mathcal{A})$.

The following axioms are straightforward modifications of (M), (SM) and (C), respectively.

(M') For any $z \in \mathcal{Z}$ and $A, B \in \mathcal{K}(\mathcal{A})$ with $A \supseteq B$, we have $(z, A) \succsim (z, B)$.

(SM') For any $z \in \mathcal{Z}$ and $A, B \in \mathcal{K}(\mathcal{A})$ with $(z, A) \sim (z, A \cup B)$, we have $(z, A \cup C) \sim (z, A \cup C \cup B)$ for every $C \in \mathcal{K}(\mathcal{A})$.

(C') $\succsim$ is closed in $(\mathcal{Z} \times \mathcal{K}(\mathcal{A})) \times (\mathcal{Z} \times \mathcal{K}(\mathcal{A}))$.

The corresponding modification of Theorem 1 reads as follows.

Theorem 2 *Let $\mathcal{Z}$ and $\mathcal{A}$ be compact metric spaces. A binary relation $\succsim$ on $\mathcal{Z} \times \mathcal{K}(\mathcal{A})$ is a preorder that satisfies (M'), (SM') and (C') if, and only if, there exist a nonempty set $U \subseteq \mathcal{C}(\mathcal{Z} \times \mathcal{A})$ and a nonempty set Φ of strictly increasing, continuous, real valued functions on the set $\{\max_{\{z\} \times A} U : (z, A) \in \mathcal{Z} \times \mathcal{K}(\mathcal{A})\} \subseteq \mathbb{R}^U$ such that, for any $(z, A), (z', B) \in \mathcal{Z} \times \mathcal{K}(\mathcal{A})$,*

$$(z, A) \succsim (z', B) \quad \Leftrightarrow \quad \varphi\left(\max_{\{z\} \times A} U \right) \geq \varphi\left(\max_{\{z'\} \times B} U \right) \text{ for every } \varphi \in \Phi.$$

Here, Φ can be chosen to be a singleton if, and only if, $\succsim$ is complete.

Proof The proof of the "if" part is simple and omitted. To prove the "only if" part, let ρ be a product metric on $\mathcal{Z} \times \mathcal{A}$, and equip $\mathcal{K}(\mathcal{Z} \times \mathcal{A})$ with the Hausdorff metric induced by ρ. The projection of a set $K \in \mathcal{K}(\mathcal{Z} \times \mathcal{A})$ on $\mathcal{Z}$ is defined as $P_{\mathcal{Z}}(K) :=$ $\{z \in \mathcal{Z} : (z, a) \in K$ for some $a \in \mathcal{A}\}$. In turn, the cross section of $K \in \mathcal{K}(\mathcal{Z} \times \mathcal{A})$ at a point $z \in P_{\mathcal{Z}}(K)$ is the nonempty compact set $K_z := \{a \in \mathcal{A} : (z, a) \in K\}$.

Define a binary relation $\overset{*}{\succsim}$ on $\mathcal{K}(\mathcal{Z} \times \mathcal{A})$ as follows: For any $K, K' \in \mathcal{K}(\mathcal{Z} \times \mathcal{A})$,

$$K \overset{*}{\succsim} K' \quad \Leftrightarrow \quad P_{\mathcal{Z}}(K) \supseteq P_{\mathcal{Z}}(K') \text{ and } (z, K_z) \sim (z, K_z \cup K'_z) \, \forall z \in P_{\mathcal{Z}}(K').$$

For now, let us assume that the following claim holds true, which will be proved momentarily.

Claim 6 $\overset{*}{\succsim}$ *is a monotonic and closed preorder on* $\mathcal{K}(\mathcal{Z} \times \mathcal{A})$ *such that* $K \overset{*}{\succsim} K'$ *implies* $K \overset{*}{\succsim} K \cup K'$ *for any* $K, K' \in \mathcal{K}(\mathcal{Z} \times \mathcal{A})$.

Assuming Claim 6, we can apply Lemma 1 to the preorder $\overset{*}{\succsim}$, where $\mathcal{Z} \times \mathcal{A}$ takes the role of $\mathcal{X}$. Thereby, we find a nonempty set $U \subseteq \mathcal{C}(\mathcal{Z} \times \mathcal{A})$ such that $K \overset{*}{\succsim} K'$ iff $\max_K U \geq \max_{K'} U$ for any $K, K' \in \mathcal{K}(\mathcal{Z} \times \mathcal{A})$.

Further, applying Lemma 2 to the preorder $\succsim$, with $\mathcal{X} = \mathcal{Z} \times \mathcal{K}(\mathcal{A})$, yields a nonempty set $V \subseteq \mathcal{C}(\mathcal{Z} \times \mathcal{K}(\mathcal{A}))$ such that for any $(z, A), (z', B) \in \mathcal{Z} \times \mathcal{K}(\mathcal{A})$, (i) $(z, A) \succ (z', B)$ implies $v(z, A) > v(z', B)$ for every $v \in V$; and (ii) $(z, A) \succsim (z', B)$ iff $v(z, A) \geq v(z', B)$ for every $v \in V$.

Note that for any $(z, A), (z', B) \in \mathcal{Z} \times \mathcal{K}(\mathcal{A})$, we have $\{z\} \times A \overset{*}{\succsim} \{z'\} \times B$ iff $z = z'$ and $(z, A) \sim (z, A \cup B)$. Thus, $\max_{\{z\} \times A} U > (=) \max_{\{z'\} \times B} U$ imply $(z, A) \succ (\sim) (z', B)$. Consequently, for each $v \in V$ we can define a strictly increasing, real valued function φ_v on the set $\{\max_{\{z\} \times A} U : (z, A) \in \mathcal{Z} \times \mathcal{K}(\mathcal{A})\}$ by the rule $\varphi_v (\max_{\{z\} \times A} U) \equiv v(z, A)$, and proceed just as in the proof of Theorem 1.

It remains to prove Claim 6. To this end, for each $z \in \mathcal{Z}$ define a dominance relation $\overset{\cdot}{\succsim}_z$ on $\mathcal{K}(\mathcal{A})$ by $A \overset{\cdot}{\succsim}_z B$ iff $(z, A) \sim (z, A \cup B)$. As in Claim 1, it is easily checked that the relation $\overset{\cdot}{\succsim}_z$ is a monotonic preorder such that, for any $A, B, C \in \mathcal{K}(\mathcal{A})$,

$$A \overset{\cdot}{\succsim}_z B \text{ and } A \overset{\cdot}{\succsim}_z C \quad \Rightarrow \quad A \overset{\cdot}{\succsim}_z B \cup C, \tag{7}$$

where the property (7) follows from (M') and (SM'). Moreover, from (C'), it follows that for convergent sequences $(A_n), (B_n)$ in $\mathcal{K}(\mathcal{A})$, and (z_n) in $\mathcal{Z}$, with $z := \lim_n z_n$, we have

$$\lim_n A_n \overset{\cdot}{\succsim}_z \lim_n B_n \quad \text{whenever} \quad A_n \overset{\cdot}{\succsim}_{z_n} B_n \, \forall n \in \mathbb{N}. \tag{8}$$

Note also that the definition of $\overset{*}{\succsim}$ can be rewritten as follows: For any $K, K' \in \mathcal{K}(\mathcal{Z} \times \mathcal{A})$,

$$K \overset{*}{\succsim} K' \quad \Leftrightarrow \quad P_{\mathcal{Z}}(K) \supseteq P_{\mathcal{Z}}(K') \text{ and } K_z \overset{\cdot}{\succsim}_z K'_z \, \forall z \in P_{\mathcal{Z}}(K'). \tag{9}$$

If $K \supseteq K'$, then $P_{\mathcal{Z}}(K) \supseteq P_{\mathcal{Z}}(K')$ and $K_z \supseteq K'_z$ for each $z \in P_{\mathcal{Z}}(K')$. Since $\overset{\cdot}{\succsim}_z$ is monotonic, it follows that so is $\overset{*}{\succsim}$. This also implies that $\overset{*}{\succsim}$ is reflexive.

Suppose now that $K \overset{*}{\succsim} K'$ and $K' \overset{*}{\succsim} K''$ for some $K, K', K'' \in \mathcal{K}(\mathcal{Z} \times \mathcal{A})$. Then $P_{\mathcal{Z}}(K) \supseteq P_{\mathcal{Z}}(K') \supseteq P_{\mathcal{Z}}(K'')$, while $K_z \overset{\cdot}{\succsim}_z K'_z \overset{\cdot}{\succsim}_z K''_z$ for each $z \in P_{\mathcal{Z}}(K'')$. Thus, $\overset{*}{\succsim}$ inherits transitivity from the relations $\overset{\cdot}{\succsim}_z$.

Now, pick any $K, K' \in \mathcal{K}(\mathcal{Z} \times \mathcal{A})$ with $K \overset{*}{\succsim} K'$. Then $P_{\mathcal{Z}}(K) \supseteq P_{\mathcal{Z}}(K')$, and hence, $P_{\mathcal{Z}}(K \cup K') = P_{\mathcal{Z}}(K)$. Moreover, for any $z \in P_{\mathcal{Z}}(K \cup K')$ we have either (i) $z \notin P_{\mathcal{Z}}(K')$, or (ii) $z \in P_{\mathcal{Z}}(K')$. Case (i) implies $(K \cup K')_z = K_z$, so that $K_z \overset{\cdot}{\succsim}_z (K \cup K')_z$ by reflexivity of $\overset{\cdot}{\succsim}_z$. In case (ii), $(K \cup K')_z = K_z \cup K'_z$ while $K_z \overset{\cdot}{\succsim}_z K'_z$ by definition of $\overset{*}{\succsim}$. Since $\overset{\cdot}{\succsim}_z$ is reflexive, from (7) it follows that $K_z \overset{\cdot}{\succsim}_z (K \cup K')_z$. Hence, in both cases we have $K \overset{*}{\succsim} K \cup K'$ by (9).

The final step is to show that $\overset{*}{\succsim}$ is closed. Let (K^n) and (K'^n) be convergent sequences in $\mathcal{K}(\mathcal{Z} \times \mathcal{A})$ with $K^n \overset{*}{\succsim} K'^n$ for every n. Put $K := \lim_n K^n$, $K' := \lim_n K'^n$, and pick any $(z, a) \in K'$. Since (K'^n) converges to K', for each n we can select a point $(z_n, a_n) \in K'^n$ such that $(z, a) = \lim_n (z_n, a_n)$. Since $K^n \overset{*}{\succsim} K'^n$, we have $K^n_{z_n} \overset{\cdot}{\succsim}_{z_n} K'^n_{z_n}$ for each n. Moreover, a_n belongs to $K'^n_{z_n}$ by definitions, while monotonicity of $\overset{\cdot}{\succsim}_{z_n}$ implies $K'^n_{z_n} \overset{\cdot}{\succsim}_{z_n} \{a_n\}$. Since $\overset{\cdot}{\succsim}_{z_n}$ is transitive, it follows that $K^n_{z_n} \overset{\cdot}{\succsim}_{z_n} \{a_n\}$ for every n. By passing to a subsequence if necessary, assume that $(K^n_{z_n})$ converges to some $B \in \mathcal{K}(\mathcal{A})$. Then, by (8), we have $B \overset{\cdot}{\succsim}_z \{a\}$.

I now claim that $z \in P_{\mathcal{Z}}(K)$ and $B \subseteq K_z$. Take any $b \in B$. Since $\lim_n K^n_{z_n} = B$, there exists a sequence (b_n) such that $b = \lim_n b_n$ and $b_n \in K^n_{z_n}$ for every n. Given that $\lim_n K^n = K$ and $(z_n, b_n) \in K^n$, there exists a further sequence (z'_n, b'_n) in K such that $\lim_n \rho\big((z'_n, b'_n), (z_n, b_n)\big) = 0$. Also note that (z_n, b_n) converges to (z, b) by construction. Thus, (z'_n, b'_n) also converges to (z, b). Since (z'_n, b'_n) is contained in the closed set K, it follows that $(z, b) \in K$. That is, $z \in P_{\mathcal{Z}}(K)$ and $b \in K_z$. Since b is an arbitrary point in B, it also follows that $B \subseteq K_z$, as claimed.

Since $\overset{\cdot}{\succsim}_z$ is monotonic, $B \subseteq K_z$ implies $K_z \overset{\cdot}{\succsim}_z B$. Then $K_z \overset{\cdot}{\succsim}_z \{a\}$ by transitivity of $\overset{\cdot}{\succsim}_z$, for we have $B \overset{\cdot}{\succsim}_z \{a\}$ as noted earlier. Since (z, a) is an arbitrary point in K', we have shown that $P_{\mathcal{Z}}(K) \supseteq P_{\mathcal{Z}}(K')$, and $K_z \overset{\cdot}{\succsim}_z \{a\}$ for each $z \in P_{\mathcal{Z}}(K')$ and $a \in K'_z$. By (7), the latter property implies $K_z \overset{\cdot}{\succsim}_z F$ for every $z \in P_{\mathcal{Z}}(K')$ and finite nonempty subset F of K'_z. From (8), it follows that $K_z \overset{\cdot}{\succsim}_z K'_z$ for each $z \in P_{\mathcal{Z}}(K')$, because K'_z is the limit of a sequence of finite nonempty subsets of itself. Hence, $K \overset{*}{\succsim} K'$, as we sought. ∎

5 Concluding Remarks

In this paper, I have proved a continuous version of Kreps' [24] ordinal representation of preference for flexibility. It is also shown that the relaxation of the completeness axiom leads to multiple aggregators, and that the representation carries over to a consumption-investment budgeting framework. Lemma 1 is the key step in the proofs, which provides a multi-utility characterization of the dominance relation embodied in a preorder that exhibits preference for flexibility. The proof of Lemma 1 builds upon a Urysohn–Nachbin type construction centered on Claim 3.

Using Kreps' analysis, it can also be shown that when the space of alternatives, $\mathcal{A}$, is finite, then each aggregator in Theorem 1 can be selected to be additive.[9] Unfortunately, however, I have not been able to determine if an analogous conclusion holds when $\mathcal{A}$ is a compact metric space. Nevertheless, in relation to Theorem 1, it is possible to show that for a particular choice of the set U, each aggregator in the set Φ can be chosen to be a Choquet integral.[10] For brevity, I omit the proof of this assertion.

References

1. Arlegi, R., Nieto, J.: Incomplete preferences and the preference for flexibility. Math. Soc. Sci. **41**, 151–165 (2001)
2. Arrow, K.J.: A note on freedom and flexibility. In: Basu, K., Pattanaik, P., Suzumura, K. (eds.) Choice, Welfare, and Development. A Festschrift in Honour of Amartya K. Sen, pp. 7–16. Clarendon Press, Oxford (1995)
3. Bleichrodt, H., Quiggin, J.: Capabilities as menus: a non-welfarist basis for QALY evaluation. J. Health Econ. **32**, 128–137 (2013)
4. Bosi, G., Caterino, A., Ceppitelli, R.: Existence of continuous utility functions for arbitrary binary relations: some sufficient conditions. Tatra Mt. Math. Publ. **46**, 15–27 (2010)
5. Bosi, G., Herden, G.: On continuous multi-utility representations of semi-closed and closed preorders. Math. Soc. Sci. **79**, 20–29 (2016)
6. Danan, E.: A Behavioral Model of Individual Welfare. Université de Paris, Mimeo (2003)
7. Danan, E., Ziegelmeyer, A.: Are preferences complete? An experimental measurement of indecisiveness under risk. Max Planck Institute of Economics, Mimeo (2006)
8. Dekel, E., Lipman, B.L., Rustichini, A.: Representing preferences with a unique subjective state space. Econometrica **69**, 891–934 (2001)
9. Dekel, E., Lipman, B.L., Rustichini, A.: Temptation-driven preferences. Rev. Econ. Stud. **76**, 937–971 (2009)
10. Dillenberger, D., Sadowski, P.: Ashamed to be selfish. Theor. Econ. **7**, 99–124 (2012)

---

[9]More generally, Gorno [19] shows that any complete and transitive preference relation on nonempty subsets of a finite set can be represented with an additive aggregator over state-dependent utility levels. The distinctive feature of Kreps' version is that each state-dependent utility function has a nonnegative weight, in line with the monotonicity axiom (see Footnote 8).

[10]Specifically, for every $\varphi \in \Phi$, we can find a finite capacity η_φ on the set U such that for every $A \in \mathcal{K}(\mathcal{A})$, $\varphi\left(\max_A U\right) = \int_0^\infty \eta_\varphi\left(\{u \in U : \max_{a \in A} u(a) \geq \alpha\}\right) d\alpha$. (A *finite capacity* η on U is a real valued function on the collection of all subsets of U such that $\eta(\emptyset) = 0$, $\eta(U) < \infty$, and $\eta(V) \leq \eta(W)$ whenever $V \subseteq W \subseteq U$.)

11. Dubra, J., Maccheroni, F., Ok, E.A.: Expected utility theory without the completeness axiom. J. Econ. Theory **115**, 118–133 (2004)
12. Ergin, H.: Costly Contemplation. Mimeo, MIT (2003)
13. Ergin, H., Sarver, T.: A unique costly contemplation representation. Econometrica **78**, 1285–1339 (2010)
14. Evren, Ö., Minardi, S.: Warm-glow giving and freedom to be selfish. Econ. J. **127**, 1381–1409 (2017)
15. Evren, Ö., Ok, E.A.: On the multi-utility representation of preference relations. J. Math. Econ. **47**, 554–563 (2011)
16. Foster, J.E.: Freedom, opportunity, and well-being. In: Arrow, K.J., Sen, A.K., Suzumura, K. (eds.) Handbook of Social Choice and Welfare, vol. 2, pp. 687–728. Elsevier, Amsterdam (2011)
17. Goldman, S.M.: Flexibility and the demand for money. J. Econ. Theory **9**, 203–222 (1974)
18. Goldman, S.M.: Portfolio choice and flexibility: the precautionary motive. J. Monet. Econ. **4**, 263–280 (1978)
19. Gorno, L.: Additive representation for preferences over menus in finite choice settings. J. Math. Econ. **65**, 41–47 (2016)
20. Gul, F., Pesendorfer, W.: Temptation and self-control. Econometrica **69**, 1403–1435 (2001)
21. Gul, F., Pesendorfer, W.: A simple theory of temptation and self-control. Princeton University, Mimeo (2005)
22. Kochov, A.S.: Subjective states without the completeness axiom. University of Rochester, Mimeo (2007)
23. Kopylov, I.: Perfectionism and choice. Econometrica **80**, 1819–1843 (2012)
24. Kreps, D.M.: A representation theorem for "preference for flexibility,". Econometrica **47**, 565–577 (1979)
25. Mehta, G.: Topological ordered spaces and utility functions. Int. Econ. Rev. **18**, 779–782 (1977)
26. Nachbin, L.: Topology and Order. Van Nostrand, Princeton (1965)
27. Ortoleva, P.: The price of flexibility: towards a theory of thinking aversion. J. Econ. Theory **148**, 903–934 (2013)
28. Pejsachowicz, L., Toussaert, S.: Choice deferral, indecisiveness and preference for flexibility. J. Econ. Theory **170**, 417–425 (2017)
29. Puppe, C.: An axiomatic approach to "preference for freedom of choice,". J. Econ. Theory **68**, 174–199 (1996)
30. Saito, K.: Impure altruism and impure selfishness. J. Econ. Theory **158**, 336–370 (2015)
31. Sarver, T.: Anticipating regret: why fewer options may be better. Econometrica **76**, 263–305 (2008)

The Arrow-Hahn Construction
in a Locally Compact Metric Space

Douglas S. Bridges

Abstract The Arrow-Hahn existence theorem for representing a continuous, locally nonsatiated preference relation on a closed convex subset of N-dimensional Euclidean space is generalised, without requiring local nonsatiation, to the context of certain locally compact metric spaces. In order to accomplish this, convexity is replaced by a new geometrical condition, the uniform ball-centre distance property on compact sets. The paper is written within the framework of (Bishop's) constructive analysis, which requires the inclusion of several computationally empowered hypotheses.

1 Introduction

Among Ghanshyam Mehta's many outstanding contributions to the theory of preference and utility are some dealing with the Arrow-Hahn proof of the existence of a continuous utility function representing a preference relation on a closed convex subset X of $\mathbf{R}^N$ ([3], pages 82–87; [16, 17]). In this paper we consider aspects of that proof once more, briefly from a classical-logical viewpoint, and in more detail from a constructive-logical one.

The Arrow-Hahn construction requires the preference relation $\succ$ to be locally nonsatiated. Since a continuous function is bounded on each compact subset of the individual's consumption set X, local nonsatiation rules out the possibility that X is bounded. We note that Arrow and Hahn are incorrect in stating, on page 82 of [3], that their proof does not require local nonsatiation to prove the existence of the continuous utility function u representing $\succ$: in fact, they use local nonsatiation once, to establish that the set $\{x \in X : u(x) \geq \alpha\}$ is closed for each $\alpha \in \mathbf{R}$. Hence, while they do not need local nonsatiation to prove the existence of u, they do need it for their proof of the continuity of u. By implication, they are therefore dealing

For Ghanshyam Mehta on the occasion of his 75th birthday.

D. S. Bridges (✉)
School of Mathematics & Statistics, University of Canterbury, Christchurch 8140, New Zealand
e-mail: douglas.bridges@canterbury.ac.nz

© Springer Nature Switzerland AG 2020
G. Bosi et al. (eds.), *Mathematical Topics on Representations of Ordered Structures and Utility Theory*, Studies in Systems, Decision and Control 263,
https://doi.org/10.1007/978-3-030-34226-5_15

only with the case where X is unbounded. However, it is not hard to see that the case of bounded X can be handled if we replace local nonsatiation by the weaker condition of local nonsatiation away from preferred vectors (ones that are maximal with respect to the preference-indifference relation $\succcurlyeq$):

LN If $x \in X$ is not a preferred vector, then for each $\varepsilon > 0$ there exists $y \in X$ such that $y \succ x$ and $\|x - y\| < \varepsilon$.

Is there an analogous method of treating the constructive version of the Arrow-Hahn theorem [6, 7] in order to extend its applicability to the case where X is unbounded? In this paper we show not only that this is possible, but also that it can be done in a far more general context, in which X is a locally compact metric space with a new, geometric condition replacing that of convexity in the Euclidean context. Moreover, we carry out this work within the framework of (Bishop's) constructive mathematics, in which each proof is algorithmic in nature. We begin by outlining a few salient features of this variety of constructive mathematics.

2 Constructive Mathematics in (Very) Brief

The fundamental difference between constructive mathematics and the *classical mathematics* practised by the great majority of mathematicians is the interpretation of disjunction and existence. To prove, constructively, a disjunction 'P or Q', we must either produce a proof of P or else produce one of Q; it is not enough to rule out the possibility of both P and Q being false. To prove 'there exists x with the property $P(x)$', we must produce a construction of a certain object x, together with a construction showing that $P(x)$ holds. These interpretations lead one to regard the law of excluded middle (and, incidentally, the axiom of choice, which implies it [13]) as essentially nonconstructive.

In practice, as was shown by Errett Bishop in the ground-breaking monograph [4], you do not need any specific notion of algorithm, or logical metatheorems, to develop mathematics constructively: it suffices to work with intuitionistic, not the usual classical, logic, within a suitable set- or type-theoretic framework such as constructive Zermelo-Fraenkel set theory [1] or Martin-Löf type theory [15].[1] The resulting mathematics is known as **Bishop-style constructive mathematics**, or **BISH** for short. In this paper, however, we shall work in the logically informal manner adopted by Bishop in his development of analysis. As a result, the reader will recognise the theorems as ones of analysis, rather than mathematical logic. We shall need a few facts from Bishop's development, for details of which we refer to [4, 5, 10].

First, a set S is **finitely enumerable** (respectively, **finite**) if for some natural number N there exists a mapping (respectively, one-one mapping) of $\{1, 2, \ldots, N\}$ onto

---

[1] There is also a constructive version [2] of Morse's formalised set theory [18].

S. Finite clearly implies finitely enumerable; the converse is essentially nonconstructive. A phrase like 'choose finitely many objects' is another way of saying 'choose a finitely enumerable set of objects'. A set S is **inhabited** if there exists, constructively, an element of S. Inhabitedness is a stronger property than that of being not empty.

The set $\mathbf{R}$ of real numbers is, contrary to common misconceptions, both Cauchy complete and uncountable. However, the law of trichotomy has to be replaced by **cotransitivity**: if $a > b$, then for all x in $\mathbf{R}$, either $a > x$ or $x > b$. Also, although $(\neg(x > 0) \Rightarrow x \leq 0)$ is derivable, the proposition $(\neg(x \geq 0) \Rightarrow x < 0)$ is constructively equivalent to **Markov's Principle**: for each binary sequence $(a_n)_{n \geq 1}$, if it is impossible that $a_n = 0$ for each n, then there exists (we can find!) n such that $a_n = 1$. Since Markov's Principle embodies an unbounded search, it is not a part of Bishop-style constructive mathematics.[2]

In the theory of metric spaces, we have to be careful in choosing which of several classically equivalent notions are the ones with constructive content. For example, the Heine-Borel theorem requires principles, due to Brouwer (see [12]), that Bishop-style analysis does not adopt; and sequential compactness, even for the pair set $\{0, 1\}$, is essentially nonconstructive. Accordingly, when we refer to a metric space[3] (X, ρ) as **compact**, we mean that it is both totally bounded and complete. For each $\varepsilon > 0$, a compact set can be written as the union of finitely many compact subsets, each of diameter $< \varepsilon$. Also, for us a **locally compact** metric space is one in which each bounded subset is contained in a compact set; in that case, for each $a \in X$ and all but countably many positive r, the closed ball[4] $\overline{B}(a, r)$ with centre a and radius r is compact, but in general we cannot remove the 'for all but countably many' restriction on r. A mapping $f : X \to Y$ between a locally compact metric space X and a metric space Y is defined to be **continuous** if it is uniformly continuous on each compact subset of X, in which case it is uniformly continuous on each bounded set. If f is a continuous, real-valued function on a compact metric space X, then $\inf f$ and $\sup f$ exist, and for all but countably many real numbers $\alpha > \inf f$, the set $\{x \in X : f(x) \leq \alpha\}$ is compact.

One important metric notion that is irrelevant to the classical mathematician is that of locatedness: a subset S of X is **located** (in X) if the distance

$$\rho(x, S) \equiv \inf\{\rho(x, s) : s \in S\}$$

exists for each $x \in X$. By convention, we consider an empty set S to be located, with $\rho(x, S) = \infty$. A locally compact subset of a metric space is located; a closed, located subset of a locally compact metric space is locally compact.

---

[2]Markov's Principle is allowed in Markov's recursive constructive mathematics, but not in Brouwer's intuitionistic mathematics. Since each of these two varieties of mathematics can be regarded as an augmentation of Bishop-style constructive mathematics, Markov's Principle cannot be derived in **BISH**.

[3]We shall use ρ to denote the metric of any metric space other than $\mathbf{R}$.

[4]We denote the open (resp. closed) ball with centre x and radius r in a metric space by $B(x, r)$ (resp. $\overline{B}(x, r)$).

If $\varepsilon > 0$ and S is any, not necessarily located, subset of X, we write '$\rho(x, S) < \varepsilon$' to denote that there exists $s \in S$ such that $\rho(x, s) < \varepsilon$, and '$\rho(x, S) > 0$' to denote that there exists $r > 0$ such that $\rho(x, s) \geq r$ for all $s \in S$.[5] If T is also a subset of X, then '$\rho(x, S) = \rho(x, T)$' is our shorthand for the sentence: if either of the quantities $\rho(x, S)$ and $\rho(x, T)$ exists, then each exists and the two are equal.

We define the **metric complement** of S (in X) to be

$$X - S \equiv \{x \in X : \rho(x, S) > 0\}.$$

If the context is clear, we simply write $-S$ for the metric complement of S in X.

The significance of locatedness arises from the constructive failure of the classical least-upper-bound principle **R**, which implies the law of excluded middle. The **constructive least-upper-bound principle** states that an inhabited, bounded subset S of **R** has a supremum if and only if for all $a, b \in \mathbf{R}$ with $a > b$, either a is an upper bound for S or else there exists $x \in S$ with $x > b$.

Lemma 1 *Let S be a subset of a metric space X, and let $x \in X$. Suppose that for each $\varepsilon > 0$ there exist points $s_1, \ldots, s_n$ of S such that for each $s \in S$, there exists i with $\rho(x, s) > \rho(x, s_i) - \varepsilon$. Then $\rho(x, S)$ exists.* ([6], Lemma 4).

As classically, a subset D of a metric space X is **dense** if $\rho(x, D) < \varepsilon$ for each $x \in X$ and each $\varepsilon > 0$.

Lemma 2 *Let X be a metric space, D a dense subset of X, and S a located subset of X such that S°, the interior of S, is dense in S. Then $S \cap A$ is dense in S.*

Proof Given $x \in S$ and $\varepsilon > 0$, construct $y \in S^\circ$ such that $\rho(x, y) < \varepsilon/2$. There exist $r > 0$ and $a \in D$ such that $B(y, r) \subset S$ and $\rho(y, a) < \min\{r, \varepsilon/2\}$. Then $\rho(x, a) \leq \rho(x, y) + \rho(y, a) < \varepsilon$. $\qquad\square$

We now introduce two definitions that appear to be new to the (constructive) literature. We say that a metric space X has

PBC the **pointwise ball-centre distance property** if for each $r > 0$ and all ξ and x in X with $\rho(\xi, x) \geq r$, there exists y such that $\rho(x, y) \leq r$ and $\rho(\xi, y) < \rho(\xi, x)$;

UBC the **uniform ball-centre distance property on compact sets** if for each compact $K \subset X$ and each $r > 0$, there exists $\alpha \in (0, 1)$ such that if $\xi, x \in K$ and $\rho(\xi, x) \geq r$, then $\rho(\xi, y) < \rho(\xi, x) - \alpha$ for some y with $\rho(x, y) \leq r$.

Clearly, **UBC** implies **PBC**. A straightforward contradiction argument using sequential compactness shows classically that **PBC** implies **UBC**.

Proposition 3 *Every inhabited, convex, locally compact subset X of the Euclidean space $\mathbf{R}^N$ has the uniform ball-center distance property on compact sets.*

---

[5]Note that the statement '$\rho(x, S) < \varepsilon$' does not imply that $\rho(x, S)$ exists.

Proof Without loss of generality, we may assume that $0 \in X$ and consider only a compact ball $K \equiv \overline{B}(0, R)$ in X of radius $R > 0$. We may further assume that $0 < r < R$. Let

$$t = r/3R \text{ and } \alpha = tr/2.$$

Given $\xi, x \in K$ such that $\|\xi - x\| \geq r$, and setting $y = (1 - t)x + t\xi$, we see that $y \in K$, by the convexity of K. Also,

$$\|x - y\| = t \|x - \xi\| \leq 2Rt < r.$$

On the other hand, $\|\xi - y\| = (1 - t) \|\xi - x\|$, so

$$\|\xi - x\| = \|\xi - y\| + t \|\xi - x\| \geq \|\xi - y\| + tr > \|\xi - y\| + \alpha$$

and therefore $\|\xi - y\| < \|\xi - x\| - \alpha$.   □

3   Preference Relations

Let X be a set equipped with a binary relation $\neq_X$ of **inequality** such that

- $x \neq_X y \Rightarrow \neg(x = y)$ and
- $x \neq_X y \Rightarrow y \neq_X x$.

Commonly, this inequality relation will be stronger than the **denial inequality**, defined by $x \neq y$ if and only if $\neg(x = y)$. The **standard inequality relation on R** is given by $x \neq_{\mathbf{R}} y$ if and only if $|x - y| > 0$; more generally, the **standard inequality on a metric space** (X, ρ) is defined by $x \neq_X y$ if and only if $\rho(x, y) > 0$. Note that $\neq_{\mathbf{R}}$ is equivalent to the denial inequality if and only if one is prepared to accept Markov's Principle.

Now let $\succ$ be a binary relation on X. Define an associated binary relation $\succcurlyeq$ as follows:

$$x \succcurlyeq y \text{ if and only if } \forall_{z \in X}(y \succ z \Rightarrow x \succ z).$$

Then $x \succcurlyeq x$ for all $x \in X$, and $\succcurlyeq$ is transitive. For each $x \in X$ define the corresponding **upper contour set**

$$C(x) \equiv \{y \in X : y \succcurlyeq x\}$$

and **strict upper contour set**

$$C^+(x) \equiv \{y \in X : y \succ x\}.$$

We say that $\succ$ is a **preference relation**, or **strict weak order**, if it has the following three properties:

P1 If $x \succ y$, then $x \neq_X y$.
P2 If $x \succ y$, then for each $z \in X$ either $x \succ z$ or $z \succ y$ (cotransitivity).
P3 There exist $x, y \in X$ such that $x \succ y$ (non-triviality).

In that case, $x \succ x$ is contradictory, $\succ$ is transitive, $x \succ y$ entails $x \succcurlyeq y$, and $x \succcurlyeq y$ if and only if $y \succ x$ is contradictory; moreover, if either $x \succ y \succcurlyeq z$ or $x \succcurlyeq y \succ z$, then $x \succ z$. Note that, as is shown by the strict weak order $>$ on $\mathbf{R}$, we cannot expect to prove that $x \succ y$ if $y \succcurlyeq x$ is contradictory. If S is a subset of X, we write '$x \succ S$' to denote that $x \succ s$ for each $s \in S$.

By a **utility function** for, or representing, a preference relation $\succ$ on X we mean a mapping $u : X \to \mathbf{R}$ such that $x \succ y$ if and only if $u(x) > u(y)$.

We say that a point a of X is

- **preferred** if $a \succcurlyeq x$ for each x in X;
- **non-preferred** if $a \neq_X x$ for each preferred point of X.

We denote by $P_\succ$ the set of preferred points for $\succ$.

Lemma 4 *Let $\succ$ be a preference relation on a metric space X. If $C^+(x)$ is inhabited, then x is non-preferred.*

Proof If $z \succ x$, then for each $y \in P_\succ$ we have $y \succcurlyeq z \succ x$, so $y \succ x$ and therefore $x \neq_X y$. ☐

Lemma 5 *Let $\succ$ be a preference relation on a metric space X, such that $P_\succ$ is inhabited, complete, and located. Then $x \in X$ is non-preferred if and only if $x \in -P_\succ$.*

Proof By Bishop's lemma ([10], Proposition 3.1.1), there exists $y \in P_\succ$ such that if $\rho(x, y) > 0$, then $\rho(x, P_\succ) > 0$. It follows immediately that if x is non-preferred, then $x \in -P_\succ$. The converse is trivial. ☐

A preference relation $\succ$ on a metric space X is said to be **continuous** if for each $x \in X$ the sets $C^+(x)$ and

$$L^-(x) \equiv \{y \in X : x \succ y\}$$

are open in X; in which case the sets $C(x)$ and

$$L(x) \equiv \{y \in X : y \succcurlyeq x\}$$

are closed in X. For constructive purposes, we shall need the stronger condition of **uniform continuity of $\succ$ on compact sets**:

UC *If $a, b \in X$, $a \succ b$, and $K \subset X$ is compact, then there exists $r > 0$ such that for all x, y in K with $\rho(x, y) < r$, either $a \succ x$ or $y \succ b$.*

If X is locally compact, then **UC** implies the continuity of $\succ$. To see this, let $a \in X$ and $\xi \in C^+(a)$. Construct a compact set $K \subset X$ such that if $\rho(x, \xi) < 1$, then $x \in K$.

By **UC**, there exists a positive $r < 1$ such that if $x, y \in K$ and $\rho(x, y) < r$, then either $\xi \succ x$ or $x \succ a$. If $\rho(\xi, x) < r$, then $x \in K$, so either $\xi \succ \xi$, which is ruled out by **P1**, or else, as must be the case, $x \succ a$. Hence $B(\xi, r) \subset C^+(a)$. It follows that $C^+(a)$ is open. A similar argument shows that $L^-(a)$ is open.

A classical contradiction argument proves that if X is locally compact and $\succ$ is continuous, then it satisfies **UC**; it is not possible to prove this constructively (see [6], Example 1). However, if u is a continuous utility function on a locally compact space X, then the corresponding preference relation $\succ$ satisfies **UC**. For in that case, if $a, b \in X, a \succ b$, and $K \subset X$ is compact, then $u(a) - u(b) > 0$; so there exists $r > 0$ such that if $x, y \in K$ and $\rho(x, y) < r$, then $u(x) - u(y) < u(a) - u(b)$. Hence $u(a) - u(x) > u(b) - u(y)$, and therefore either $u(a) - u(x) > 0$ or $0 > u(b) - u(y)$. Then either $u(a) > u(x)$ or $u(y) > u(b)$, so either $a \succ x$ or $y \succ b$.

Lemma 6 *Let $\succ$ be a continuous preference relation on a metric space X. Then $P_\succ$ is closed. If also X is complete, then $P_\succ$ is complete.*

Proof If $(x_n)_{n \geq 1}$ is a sequence in $P_\succ$ converging to a limit $x_\infty \in X$, then $P_\succ = C(x_1)$ and is therefore closed, by the continuity condition on $\succ$; so $x_\infty \in P_\succ$. Moreover, if X is complete, then $P_\succ$, being closed in X, is also complete. □

Lemma 7 *Let $\succ$ be a continuous preference relation on a complete metric space X, such that $P_\succ$ is inhabited and located. Then the set of non-preferred points is $-P_\succ$.*

Proof Apply Lemmas 6 and 5. □

We say that $\succ$ is

- **nonsatiated at** $a \in X$ if there exists $x \in C^+(a)$;
- **locally nonsatiated at** $a \in X$ if for each $\varepsilon > 0$ there exists $x \in X$ such that $x \succ a$ and $\rho(a, x) < \varepsilon$;
- **locally nonsatiated** if it is locally nonsatiated at each point of X;
- **locally nonsatiated at non-preferred points**, if it is locally nonsatiated at each non-preferred point of X;
- **locally nonsatiated at the subset** S of X if for each $\varepsilon > 0$ there exists $x \in X$ with $\rho(x, S) < \varepsilon$ and $x \succ S$.

When S is a singleton, this last condition reduces to local nonsatiation at the unique point of S.

If S, T are subsets of a metric space X, such that for some $\varepsilon > 0$,

$$\{x \in X : \rho(x, S) < \varepsilon\} \subset T,$$

then we say that S is **well contained** in T, and we write $S \subset\subset T$. Classically, if a continuous preference relation $\succ$ on X is locally nonsatiated at non-preferred points and has a nonempty set of preferred points, then it is locally nonsatiated at each compact set well contained in $-P_\succ$. To see this, consider such a compact set K. For each $x \in X$, $C(x)$ is closed, as therefore is $K \cap C(x)$. On the other hand, for each finite subset $\{x_1, \ldots, x_n\}$ of K there exists—classically—k such that $x_k \succcurlyeq x_i$ for each i; whence $x_k \in \bigcap_{i=1}^n (K \cap C(x_i))$. It follows from the classical finite-intersection property of the compact set K that there exists $\xi \in \bigcap_{x \in K}(K \cap C(x))$. Since ξ, being a member of K, is non-preferred, for each $\varepsilon > 0$ there exists $x \in X$ with $\rho(x, K) \le \rho(x, \xi) < \varepsilon$ and $x \succ \xi \succcurlyeq y$ for each $y \in K$. Hence $x \succ K$.

Example 2 of [6] shows that in the recursive interpretation of **BISH**, a preference relation represented by continuous utility function on a convex, locally compact subset X of **R** can be locally nonsatiated at each point, but not at each compact subset, of X.

Lemma 8 *Let $\succ$ be a preference relation on a metric space X, such that $P_\succ$ is inhabited, complete, and located. If $C^+(x)$ is inhabited, then $x \in -P_\succ$.*

Proof Apply Lemmas 4 and 5. $\qquad\qquad\qquad\qquad\qquad\qquad\qquad\qquad\qquad\qquad\square$

Lemma 9 *Let X be a metric space, and $\succ$ a preference relation on X that is locally nonsatiated at each compact set well contained in $-P_\succ$. Then for each compact $K \subset\subset -P_\succ$ and each $\varepsilon > 0$, there exist $a, b \in X$ such that $\rho(a, K) < \varepsilon$, $\rho(b, K) < \varepsilon$, and $b \succ a \succ K$.*

Proof There exists $r > 0$ such that $x \in -P_\succ$ whenever $\rho(x, K) < r$. We may assume that $\varepsilon < r$. By local nonsatiation at K, there exists $a \in X$ with $\rho(a, K) < \varepsilon$ and $a \succ K$. Then $a \in P_\succ$, so by local nonsatiation at $\{a\}$, there exists $b \in X$ with $\rho(a, b) < \varepsilon - \rho(a, K)$ and $b \succ a$. Hence $\rho(b, K) \le \rho(a, b) + \rho(a, K) < \varepsilon$ and $b \succ a \succ K$. $\qquad\qquad\qquad\qquad\qquad\qquad\qquad\qquad\square$

Proposition 10 *Let $\succ$ be a preference relation on a metric space X, such that $P_\succ$ is either empty or else inhabited and located. Suppose that $\succ$ is locally nonsatiated at each compact set well contained in $-P_\succ$. Let $a \in X$ be non-preferred, and K a compact subset of X such that $K - P_\succ$ is inhabited. Then for each $\varepsilon > 0$, there exists a finitely enumerable subset F of $C^+(a)$ with the following property:*

$$\text{For each } x \in C(a) \cap K \text{ there exists } z \in F \text{ with } z \succ x \text{ and } \rho(x, z) < \varepsilon. \quad (*)$$

Proof We consider only the case in which $P_\succ$ is inhabited and located, as the proof in the case where $P_\succ$ is empty is similar but simpler. We may assume that there exists $x_0 \in K$ such that $\rho(x_0, P_\succ) > \varepsilon$. Then there exists α such that $\varepsilon/16 < \alpha < \varepsilon/4$ and

$$L \equiv \{x \in K : \rho(x, P_\succ) \ge \alpha\}$$

is (inhabited and) compact. Cover L by finitely many compact subsets $L_1, \ldots, L_n$, each of diameter less than $7\varepsilon/8$. For each $i \le n$, $L_i \subset L \subset\subset -P_\succ$ and therefore there exists y_i such that $\rho(y_i, L_i) < \varepsilon/16$ and $y_i \succ L_i$; choosing $x \in L_i$ such that $\rho(y_i, x) < \varepsilon/16$, we have

$$\rho(y_i, P_\succ) \ge \rho(x, P_\succ) - \rho(x, y_i) > \alpha - \frac{\varepsilon}{16} > 0.$$

Thus $\{y_i\} \subset\subset -P_{\succ}$, and therefore there exists z_i such that $\rho(z_i, y_i) < \varepsilon/16$ and $z_i \succ y_i$; then $\rho(z_i, L_i) \leq \rho(z_i, y_i) + \rho(y_i, L_i) < \varepsilon/8$. Using the cotransitivity of $\succ$, write $\{1, \dots, n\}$ as a union of two sets S, T such that if $i \in S$ then $z_i \succ a$, and if $i \in T$ then $a \succ y_i$.

Next, let $\{k_1, \dots, k_m\}$ be an $\varepsilon/4$-approximation to K; write $\{1, \dots, m\}$ as a union of two sets A, B such that if $j \in A$ then $\rho(k_j, P_{\succ}) > \varepsilon/2$, and if $j \in B$ then $\rho(k_j, P_{\succ}) < 3\varepsilon/4$. For each $j \in B$ choose $\eta_j \in P_{\succ}$ such that $\rho(k_j, \eta_j) < 3\varepsilon/4$. Define a finitely enumerable set

$$F \equiv \{z : \exists_{i \in S}(z = z_i) \text{ or } \exists_{j \in B}(z = \eta_j)\},$$

and note that, by our choice of z_i, η_j and our hypothesis that a is non-preferred, $F \subset C^+(a)$.

Now consider any $x \in C(a) \cap K$. Either $\rho(x, P_{\succ}) < \varepsilon/4$ or else $\rho(x, P_{\succ}) > \alpha$. In the first case, choosing j such that $\rho(x, k_j) < \varepsilon/4$, we have $\rho(k_j, P_{\succ}) \leq \rho(x, k_j) + \rho(x, P_{\succ}) < \varepsilon/2$, so $j \notin A$; hence $j \in B$, $\eta_j \in F \cap P_{\succ}$, and $\rho(k_j, \eta_j) < 3\varepsilon/4$. Then $\rho(x, \eta_j) \leq \rho(x, k_j) + \rho(k_j, \eta_j) < \varepsilon$; moreover, since, by hypothesis, a is non-preferred, local nonsatiation ensures that $\eta_j \succ a$. In the case $\rho(x, P_{\succ}) > \alpha$ we have $x \in L$, so there exists i such that $x \in L_i$. Either $z_i \succ a$ or $a \succ y_i$. If $a \succ y_i$, then $a \succ L_i$, so $a \succ x$, a contradiction; whence $i \notin T$, so $i \in S$, $z_i \succ a$, and $z_i \in F$. Moreover,[6]

$$\rho(x, z_i) \leq \rho(z_i, L_i) + \mathsf{diam}(L_i) < \frac{\varepsilon}{8} + \frac{7\varepsilon}{8} = \varepsilon.$$

In either case there exists $z \in F$ with $z \succ x$ and $\rho(x, z) < \varepsilon$. □

Corollary 11 *Under the hypotheses of Proposition 10, if a is non-preferred, then $C^+(a)$ is dense in $C(a)$.*

Proof Given x in $C(a)$, we have either $\rho(x, P_{\succ}) < \varepsilon$ or $\rho(x, P_{\succ}) > 0$. In the first case, there exists a preferred point y such that $\rho(x, y) < \varepsilon$; since a is non-preferred, there exists $z \succ a$, so $y \succcurlyeq z \succ a$ and therefore $y \in C^+(a)$. In the second case, x is non-preferred, so there exists y such that $\rho(x, y) < \varepsilon$ and $y \succ x \succcurlyeq a$; then $y \in C^+(a)$. □

Theorem 12 *Let $\succ$ be a preference relation on a locally compact metric space X, such that $P_{\succ}$ is either empty of else inhabited and located. Suppose that $\succ$ is locally nonsatiated at each compact set well contained in $-P_{\succ}$. Then for each non-preferred point of X, the upper contour set and strict upper contour set are located.*

Proof Again we consider only the case where $P_{\succ}$ is inhabited and located. Let b be any point of X, and a a non-preferred point. There exists $R > \rho(a, b)$ such that $K \equiv \overline{B}(b, R)$ is compact. Since a is non-preferred, our local nonsatiation hypothesis implies that there exists $\xi \in C^+(a)$ such that $\rho(b, \xi) < R$. Given $\varepsilon > 0$, construct a finitely enumerable subset F of $C^+(a)$ with the property (*) in Proposition 10.

---

[6]We denote the diameter of a set S by $\mathsf{diam}\, S$.

For each $x \in C(a)$, either $\rho(b, x) > \rho(b, \xi) > \rho(b, \xi) - \varepsilon$ or else $\rho(b, x) < R$. In the latter case, $x \in C(a) \cap K$ and therefore there exists $z \in F$ such that $z \succ a$ and $\rho(x, z) < \varepsilon$; then $\rho(b, x) > \rho(b, z) - \varepsilon$. Thus $S \equiv F \cup \{\xi\}$ is a finitely enumerable subset of $C^+(a)$ such that for each $x \in C(a)$ there exists $z \in S$ such that $\rho(b, x) > \rho(b, z) - \varepsilon$. Since $\varepsilon > 0$ and $x \in C(a)$ are arbitrary, we see from Lemma 1, that $\rho(b, C(a))$ exists. Hence, as b is arbitrary, $C(a)$ is located.   □

Corollary 13 *Under the hypotheses of Theorem 12, if X is locally compact and the preference relation $\succ$ is continuous, then the upper contour set of each non-preferred point is locally compact.*

Proof In view of Theorem 12 and the continuity of $\succ$, the upper contour sets of non-preferred points are closed, located subsets of a locally compact space and are therefore themselves locally compact.   □

For our general constructive version of the Arrow-Hahn Theorem, we need the preference relation $\succ$ on X to have the condition of **preferred-point approximation on compact sets**:

UP   *For each compact $K \subset X$ and each $\varepsilon > 0$, there exists $\delta > 0$ such that if $a \in K$ and $\rho(a, P_\succ) < \delta$, then $\rho(x, P_\succ) < \varepsilon$ for all $x \in C(a) \cap K$.*

This condition captures, uniformly on compact sets and formally, the idea that if a is near the set of preferred points, then so is $C(a)$. Note that **UP** holds trivially if $P_\succ$ is empty—in particular, if $\succ$ is locally nonsatiated.

Classically, **UP** holds for every continuous preference relation. For if $\succ$ is continuous and **UP** is false, then there exist a compact set $K \subset X$ and $\alpha > 0$ such that for each positive integer n, there exist $a_n \in K$ and $x_n \in C(a_n) \cap K$ with $\rho(a_n, P_\succ) < 1/n$ and $\rho(x_n, P_\succ) \geq \alpha$. Passing to subsequences, we may assume that $(a_n)_{n \geq 1}$ and $(x_n)_{n \geq 1}$ converge to respective limits a and x in K. Then $\rho(a, P_\succ) = 0$ and $\rho(x, P_\succ) \geq \alpha$. By the continuity of $\succ$, $P_\succ$ is closed, so $a \in P_\succ$; moreover, since $x_n \succeq a_n$ for each n, we have $x \succeq a$. Hence $x \in C(a) = P_\succ$ and therefore $\rho(x, P_\succ) = 0$, a contradiction.

For the reader familiar with recursive analysis, we provide in the Appendix a recursive counterexample to the proposition, 'every preference relation represented by a uniformly continuous function on a compact metric space has the property **UP**'. Thus example shows that the proposition cannot be proved in **BISH**.

Lemma 14 *Let X be a metric space, and $\succ$ a preference relation on X with preferred-point approximation on compact sets, such that $P_\succ$ is either empty or else inhabited and located. Let $a \in X$, and let $b \succeq a$ be such that $\rho(b, P_\succ) > 0$. Then $\rho(a, P_\succ) > 0$.*

Proof The case where $P_\succ$ is empty is trivial. If $P_\succ$ is inhabited and located, then by preferred-point approximation, there exists $\delta > 0$ such that for each x in the compact set $\{a, b\}$, if $\rho(x, P_\succ) < \delta$, then $\rho(y, P_\succ) < \rho(x, P_\succ)$ for all $y \in C(a) \cap \{a, b\}$. In particular, if $\rho(a, P_\succ) < \delta$, then $\rho(b, P_\succ) < \rho(b, P_\succ)$, which is absurd. Hence $\rho(a, P_\succ) \geq \delta > 0$.   □

4 The Arrow-Hahn Theorem Revisited

The classical Arrow-Hahn theorem is one of many providing conditions under which a continuous utility function exists for a continuous preference relation (see [8]). Here is our generalisation of the constructive version of that theorem [6, 7].

Theorem 15 *Let X be a locally compact metric space with the uniform ball-centre distance property on compact sets, such that each open ball is dense in the corresponding closed ball. Let $\succ$ be a preference relation on X that has the following properties:*

  (i) *uniform continuity on compact sets;*
 (ii) *preferred-point approximation on compact sets;*
(iii) *local nonsatiation at each compact set well contained in $-P_\succ$.*

Suppose also that $P_\succ$ is either empty or else inhabited and located. Then for each $x \in X$, the upper contour set $C(x)$ is locally compact. Moreover, if $x_0 \in X$, and $(r_n)_{n \geq 1}$ is an increasing sequence of positive numbers diverging to infinity such that for each n the closed ball $B_n \equiv \overline{B}(x_0, r_n)$ in X is compact, then

$$u_n(x) = \sup\{\rho(\xi, C(x)) : \xi \in B_n\}$$

exists for each n, and

$$u \equiv \sum_{n=1}^{\infty} 2^{-n} \frac{u_n}{1 + u_n}$$

defines a continuous utility function representing $\succ$.

This theorem generalises the original constructive Arrow-Hahn theorem in at least two significant aspects: it replaces the closed convex subset X in the latter by a locally compact metric space with the uniform ball-centre distance property on compact sets, and it admits the possibility of satiation points. Moreover, in view of Proposition 3, we readily see that the original Arrow-Hahn theorem is a special case of Theorem 15 in which $P_\succ$ is empty.

The proof of Theorem 15 requires a number of technical preliminaries.

Lemma 16 *Let $\succ$ be a preference relation on a metric space X, such that $P_\succ$ is either empty or else inhabited and located. Suppose that $\succ$ has properties (i)–(iii), of Theorem 15. Let K be a compact subset of X such that $K - P_\succ$ is inhabited. Then for each $\varepsilon > 0$ there exists $\delta > 0$ such that if $a \in K$, $b \in K - P_\succ$, $\rho(a, b) < \delta$, and $x \in C(a) \cap K$, then $\rho(x, y) < \varepsilon$ for some $y \succ b$.*

Proof We consider only the case where $P_\succ$ is inhabited and located. Fix x_0 in $K - P_\succ$. Given $\varepsilon > 0$, first use preferred-point approximation to construct r such that $0 < r < \rho(x_0, P_\succ)$ and such that if $a \in K$ and $\rho(a, P_\succ) < r$, then $\rho(x, P_\succ) < \varepsilon/2$ for all $x \in C(a) \cap K$. Next, find s such that $0 < s < \min\{r, \rho(x_0, P_\succ)\}$ and

$$L \equiv \{x \in K : \rho(x, P_\succ) \geq s\}$$

is (inhabited, since $x_0 \in L$) and compact. Write L as a union of finitely many compact sets $B_1, \ldots, B_n$, each of diameter at most ε. Then each $B_i \subset\subset -P_\succ$. Using Lemma 9, for each i we can find points a_i and b_i of $-P_\succ$ such that $\rho(a_i, B_i) < \varepsilon, \rho(b_i, B_i) < \varepsilon$, and $b_i \succ a_i \succ B_i$. Since $\succ$ is uniformly continuous on compact sets, there exists $\delta > 0$ such that if $x, y \in K$ and $\rho(x, y) < \delta$, then for each i, either $b_i \succ x$ or $y \succ a_i$. Consider $a \in K$ and $b \in K - P_\succ$ with $\rho(a, b) < \delta$. For each $x \in C(a) \cap K$, either $\rho(x, P_\succ) < r$ or $\rho(x, P_\succ) > s$. In the former case there exists $y \in P_\succ$ with $\rho(x, y) < \varepsilon$; since b is non-preferred and therefore $\succ$ is locally nonsatiated there, $y \succ b$. In the case $\rho(x, P_\succ) > s$, we have $x \in L$, so there exists i such that $x \in B_i$. Then $\rho(x, b_i) \leq \mathrm{diam}\, B_i < \varepsilon$ and $b_i \succ a_i \succ x$. Either $b_i \succ b$ or $a \succ a_i$. But the latter alternative entails $a \succ x$, a contradiction, so the former alternative must hold.   □

Lemma 17 *Let X be a locally compact metric space with the uniform ball-centre distance property on compact sets, and let $\succ$ be a preference relation on X that is uniformly continuous on compact sets. Let $B \subset X$ be compact, and let a, b be points of B with $a \succ b$. Then there exists $\alpha > 0$ such that if $x \succcurlyeq a, \xi \in B$, and $b \succcurlyeq \xi$, then $\rho(\xi, x) > \rho(\xi, y) + \alpha$ for some $y \succ b$.*

Proof Fix $s > \mathrm{diam}\, B$ and $R > 3s$ such that $K \equiv \overline{B}(a, 2R)$ is compact. By the uniform continuity of $\succ$ on compact sets, there exists $r > 0$ such that if $x, y \in K$ and $\rho(x, y) < r$, then either $a \succ x$ or $y \succ b$. On the other hand, by the uniform ball-centre distance property on compact sets, there exists α with $0 < \alpha < s$, such that if $\xi, x \in K$ and $\rho(\xi, x) \geq r$, then $\rho(\xi, y) < \rho(\xi, x) - \alpha$ for some y with $\rho(x, y) \leq r$. Let $x \succcurlyeq a, \xi \in B$, and $b \succcurlyeq \xi$. Either $\rho(a, x) > R$ or $\rho(a, x) < 2R$. In the former case,

$$\rho(\xi, x) \geq \rho(a, x) - \rho(a, \xi) > R - \mathrm{diam}\, B$$
$$> 3s - s = 2s > \mathrm{diam}\, B + s > \rho(\xi, a) + \alpha,$$

so, as $a \succ b$ by hypothesis, we can take $y = a$. In the case $\rho(a, x) < 2R$, we have $\xi, x \in K$. If $\rho(\xi, x) < r$, then either $a \succ x$ or $\xi \succ b$, a disjunction which is false. Hence $\rho(\xi, x) \geq r$ and therefore there exists y such that $\rho(x, y) \leq r$ and $\rho(\xi, y) < \rho(\xi, x) - \alpha$. Then $\rho(\xi, x) > \rho(\xi, y) + \alpha$, and either $a \succ x$, which is absurd, or else, as must be the case, $y \succ b$.   □

The next result is the crux of the proof of Theorem 15.

Proposition 18 *Let X be a locally compact metric space with the uniform ball-centre distance property on compact sets, such that each open ball in X is dense in the corresponding closed ball. Let $\succ$ be a preference relation on X that satisfies hypotheses (i)–(iii) of Theorem 15, and suppose that $P_\succ$ is either empty or else inhabited and located. Then for each $x \in X$, the upper contour set $C(x)$ is locally compact. Moreover, if $K \subset X$ is compact, then*

(a) *the function* $u_K : x \rightsquigarrow \sup\{\rho(\xi, C(x)) : \xi \in K\}$ *is well defined and continuous on* X;
(b) *if* $z \in K$ *and* $a \succ b$, *then* $u_K(a) > u_K(b)$; *and*
(c) *if* $u_K(a) > u_K(b)$, *then* $a \succ b$.

Proof First observe that, since $P_\succ$ is located and either empty or else inhabited and located, $D \equiv P_\succ \cup -P_\succ$ is dense in X. By Lemma 6 and Corollary 13, $C(x)$ is locally compact for each $x \in D$. On the other hand, since $\succ$ is nontrivial, Lemma 7 ensures that $-P_\succ$ is inhabited.

Dividing the remainder of the proof into steps, we deal only with the harder case in which $P_\succ$ is inhabited and located.

STEP 0. *For each* $\xi \in X$ *and each compact ball* $B \equiv \overline{B}(c, t)$ *in* X *with* $t > 0$, *there exist a compact ball* K *and* $r > 0$ *such that*

(i) $x \in K$ *whenever* $\rho(x, B) \leq r$,
(ii) *if* $x \in B$ *and* $z \in -K$, *then* $\rho(\xi, z) \geq \rho(\xi, x) + t$, *and*
(iii) $\rho(\xi, C(x)) = \rho(\xi, C(x) \cap K)$ *for each* $x \in B$.

Let

$$\sigma = \sup\{\rho(\xi, x) : x \in B\}.$$

There exists $r > 2\sigma$ such that $K \equiv \overline{B}(c, r + 2t)$ is compact. If $\rho(x, B) < r$, then there exists $y \in B$ such that $\rho(x, y) < r$, so

$$\rho(x, c) \leq \rho(x, y) + \rho(c, y) < r + t < r + 2t$$

and therefore $x \in K$. This proves (i). Next, if $x \in B$ and $z \in -K$, then

$$\rho(\xi, z) \geq \rho(c, z) - \rho(c, x) - \rho(\xi, x)$$
$$\geq r + 2t - t - \rho(\xi, x) > 2\sigma + t - \rho(\xi, x)$$
$$\geq 2\rho(\xi, x) + t - \rho(\xi, x) = \rho(\xi, x) + t,$$

which proves (ii).

Now let $x \in B$, suppose that $\rho(\xi, C(x))$ exists, and let $0 < \varepsilon < t$. If $\rho(\xi, C(x))$ exists, then for each ε with $0 < \varepsilon < t$, there exists $z \in C(x)$ such that

$$\rho(\xi, z) < \rho(\xi, C(x)) + \varepsilon < \rho(\xi, x) + t.$$

In view of (ii), $z \notin -K$, so $\rho(z, K) = 0$ and therefore $z \in K$. It now follows that $\rho(\xi, C(x)) = \rho(\xi, C(x) \cap K)$. On the other hand, suppose that $d \equiv \rho(\xi, C(x))$ exists. If $y \in C(x) - K$, then $\rho(\xi, y) \geq \rho(\xi, x) + t \geq d + t$. Hence if $y \in C(x)$ and $\rho(\xi, y) < d$, then $y \notin -K$, so $y \in K$ and therefore $\rho(\xi, y) \geq d$. This contradiction ensures that $\rho(\xi, y) \geq d$ for all $y \in C(x)$. It readily follows that $\rho(\xi, C(x))$ exists and equals $\rho(\xi, C(x) \cap K)$, which completes the proof of (iii).

STEP 1. *For each compact ball $B \equiv B(c, t) \subset X$ of positive radius, the mapping $f : (\xi, x) \rightsquigarrow \rho(\xi, C(x))$ is uniformly continuous on $B \times (B \cap D)$.*

Since the open ball corresponding to B is dense in B, it follows from Lemma 2 that $B \cap D$ is dense in B. Construct $r > 0$ and the compact ball K as in Step 0. Given $\varepsilon > 0$, apply Lemma 16 to compute $\delta_1 > 0$ such that if $a \in K$, $b \in K - P_\succ$, $\rho(a, b) < \delta_1$, and $x \in C(a) \cap K$, then $\rho(x, y) < \varepsilon$ for some $y \succ b$. Applying preferred-point approximation, compute $\delta_2 > 0$ such that if $x \in K$ and $\rho(x, P_\succ) < \delta_2$, then $\rho(y, P_\succ) < \varepsilon$ for all $y \in C(x) \cap K$. Let $\delta = \min\{\varepsilon, \delta_1, \delta_2\} > 0$. Consider $a, b \in B \cap D$ such that $\rho(a, b) < \delta$. We first prove that for any $\xi \in B$,

$$|\rho(\xi, C(a)) - \rho(\xi, C(b))| < \varepsilon. \tag{1}$$

If $a \in P_\succ$, then $\rho(b, P_\succ) \leq \rho(a, b) < \delta_2$, so for each $y \in C(b) \cap K$ we have $\rho(y, P_\succ) < \varepsilon$ and therefore

$$\rho(\xi, C(a)) = \rho(\xi, P_\succ) \leq \rho(\xi, y) + \rho(y, P_\succ) < \rho(\xi, y) + \varepsilon.$$

Hence

$$\rho(\xi, C(b)) \leq \rho(\xi, P_\succ)$$
$$= \rho(\xi, C(a))$$
$$\leq \inf\{\rho(\xi, y) : y \in C(b) \cap K\} + \varepsilon$$
$$= \rho(\xi, C(b) \cap K) + \varepsilon = \rho(\xi, C(b)) + \varepsilon,$$

the last step using (iii) of Step 0. Reference (1) follows immediately. Interchanging the roles of a and b, we obtain (1) in the case $b \in P_\succ$.

Now consider the remaining case, where $a, b \in B - P_\succ$. Since $\rho(a, b) < \delta_1$, for each $x \in C(a) \cap K$ there exists $y \succ b$ with $\rho(x, y) < \varepsilon$. Then

$$\rho(\xi, C(b)) \leq \rho(\xi, x) + \rho(x, C(b))$$
$$\leq \rho(\xi, x) + \rho(x, y) < \rho(\xi, x) + \varepsilon.$$

Hence, by (iii) of Step 0,

$$\rho(\xi, C(b)) \leq \inf\{\rho(\xi, x) : x \in C(a) \cap K\} + \varepsilon = \rho(\xi, C(a)) + \varepsilon.$$

Similarly, $\rho(\xi, C(a)) \leq \rho(\xi, C(b)) + \varepsilon$. Hence (1) holds in this case also.

To complete the proof of Step 1, let $a, b \in B \cap D$ and $\xi, \eta \in B$ be such that $\rho((\xi, a), (\eta, b)) < \delta$. Then $\rho(\xi, \eta) < \varepsilon$ and $\rho(a, b) < \delta$, so

$$|\rho(\xi, C(a)) - \rho(\eta, C(b))| \leq |\rho(\xi, C(a)) - \rho(\xi, C(b))| + |\rho(\xi, C(b)) - \rho(\eta, C(b))|$$
$$\leq \varepsilon + \rho(\xi, \eta) < 2\varepsilon.$$

Since $\varepsilon > 0$ is arbitrary, the proof of Step 1 is complete.

STEP 2. *For each* $x \in X$, *$C(x)$ is located in* X. *Moreover, if* B *is a compact ball of positive radius, then the mapping* $f : (\xi, x) \rightsquigarrow \rho(\xi, C(x))$ *is uniformly continuous on* B.

If B is a compact ball of positive radius, then, by Step 1, the mapping f : $(\xi, x) \rightsquigarrow \rho(\xi, C(x))$ is uniformly continuous on the dense subset $B \times (B \cap D)$ of B; it therefore extends to a uniformly continuous mapping, also written f, of B into the set $\mathbf{R}^{0+}$ of nonnegative reals. Construct $r > 0$ and a compact ball K as in Step 0, and let $\xi, a \in B$. We prove that $\rho(\xi, C(a))$ exists and equals $f(\xi, a)$. First suppose that $\rho(\xi, x) < f(\xi, a)$ for some $x \in C(a)$. By the continuity of f on $B \times (B \cap D)$ and the density of $B \cap D$ in B, there exists $b \in B \cap D$ such that $\rho(\xi, x) < \rho(\xi, C(b))$. Since $P_> \subset C(b)$, we see that $\rho(\xi, x) < \rho(\xi, P_>)$; thus $\rho(x, P_>) \geq \rho(\xi, P_>) - \rho(\xi, x) > 0$. It follows from Lemma 14 that $\rho(a, P_>) > 0$, so $a \in -P_> \subset B \cap D$. Thus $f(\xi, a) = \rho(\xi, C(a)) \leq \rho(\xi, x)$, a contradiction. We conclude that $f(\xi, a) \leq \rho(\xi, x)$ for each $x \in C(a)$. It remains to prove that for each $\varepsilon > 0$ there exists $x \in C(a)$ such that $\rho(\xi, x) < f(a, \xi) + \varepsilon$. To that end, apply preferred-point approximation to construct $\delta > 0$ such that if $z \in K$ and $\rho(z, P_>) < \delta$, then $\rho(x, P_>) < \varepsilon/2$ for all $x \in C(z) \cap K$. Either $\rho(a, P_>) > 0$, in which case $a \in B - P_> \subset B \cap D$, $f(\xi, a) = \rho(\xi, C(a))$, and the desired $z \in C(a)$ clearly exists; or else $\rho(a, P_>) < \delta$. In the latter case, by the uniform continuity of f on $B \times (B \cap D)$ and the density of $B \cap D$ in B, there exists $b_1 \in B \cap D$ so close to a that $\rho(\xi, C(b_1)) = f(\xi, b_1) < f(\xi, a) + \varepsilon/2$ and $\rho(b_1, P_>) < \delta$. Since, by Step 0, $\rho(\xi, C(b_1)) = \rho(\xi, C(b_1) \cap K)$, there exists $y \in C(b_1) \cap K$ with $\rho(\xi, y) < f(\xi, a) + \varepsilon/2$. By our choice of δ, there exists $x \in P_>$ with $\rho(x, y) < \varepsilon/2$. Then $x \in C(a)$ and $\rho(\xi, x) \leq \rho(\xi, y) + \rho(x, y) < f(\xi, a) + \varepsilon$. This completes the proof that $\rho(\xi, C(a))$ exists and equals $f(\xi, a)$. Since X is a union of compact balls of positive radius, it follows that for all $\xi, x \in X$, $\rho(\xi, C(x))$ exists and equals $f(\xi, x)$.

STEP 3. *If* $B \subset X$ *is a compact ball, then*

$$u_B(x) \equiv \sup\{\rho(\xi, C(x)) : \xi \in B\}$$

defines a continuous mapping u_B *of* X *into* $\mathbf{R}^{0+}$.

Let B, K be compact balls such that B has positive radius and $B \subset K$. By Step 2, the mapping $(\xi, x) \rightsquigarrow \rho(\xi, C(x))$ is uniformly continuous on $K \times K$; so for each $x \in K$ the mapping $\xi \rightsquigarrow \rho(\xi, C(x))$ is uniformly continuous on B, and therefore $u_B(x)$ exists. Moreover, for each $\varepsilon > 0$ there exists $\delta > 0$ such that if $x, x' \in K$ and $\rho(x, x') < \delta$, then for each $\xi \in B$, $\left| \rho(\xi, C(x)) - \rho(\xi, C(x')) \right| < \varepsilon$ and therefore

$$\rho(\xi, C(x)) < \rho(\xi, C(x')) + \varepsilon \leq u_B(x') + \varepsilon.$$

Hence

$$u_B(x) = \sup\{\rho(\xi, C(x)) : \xi \in B\} \leq u_B(x') + \varepsilon.$$

Likewise, $u_B(x') \leq u_B(x) + \varepsilon$. Since $\varepsilon > 0$ is arbitrary, u_B is uniformly continuous on K. Since X is a union of compact balls that include B, u_B is continuous on X.

STEP 4. *Let $B \subset X$ be compact and a, b points of B such that $a \succ b$. Then $u_B(a) >$*
$u_B(b)$.

By Lemma 17, there exists $\alpha > 0$ such that if $x \succcurlyeq a, \xi \in B$, and $b \succcurlyeq \xi$, then there
exists $y \succ b$ such that

$$\rho(\xi, x) > \rho(\xi, y) + \alpha \geq \rho(\xi, C(b)) + \alpha.$$

Thus if $\xi \in B$ and $b \succcurlyeq \xi$, then

$$\rho(\xi, C(a)) = \inf\{\rho(\xi, x) : x \succcurlyeq a\} \geq \rho(\xi, C(b)) + \alpha$$

and therefore

$$\begin{aligned} u_B(a) &= \sup\{\rho(z, C(a)) : z \in B\} \\ &\geq \rho(\xi, C(a)) \geq \rho(\xi, C(b)) + \alpha \geq \alpha. \end{aligned}$$

Now, either $u_B(a) > u_B(b)$ or $u_B(a) < u_B(b) + \alpha$. In the latter case there exists
$\xi \in B$ such that $u_B(a) < \rho(\xi, C(b)) + \alpha$; then $\rho(\xi, C(b)) > 0$ and so $b \succcurlyeq \xi$. It
follows from the foregoing that $u_B(a) \geq \rho(\xi, C(b)) + \alpha$, a contradiction that ensures
that we must have $u_B(a) > u_B(b)$.

STEP 5. *If $B \subset X$ is compact and $u_B(a) > u_B(b)$, then $a \succ b$.*

There exists $\xi \in B$ such that $\rho(\xi, C(a)) > u_B(b) \geq \rho(\xi, C(b))$, and therefore
there exists $x \in C(b)$ with $\rho(\xi, x) < \rho(\xi, C(a)) \leq \rho(\xi, P_\succ)$. Thus $\rho(x, P_\succ) > 0$;
whence, by Lemma 9, there exists y such that $y \succ x$ and $\rho(x, y) < \rho(\xi, C(a)) -$
$\rho(\xi, x)$. Then $\rho(\xi, y) \leq \rho(\xi, x) + \rho(x, y) < \rho(\xi, C(a))$ and $y \succ b$. Either $y \succ a$
or $a \succ b$. The former alternative is ruled out, since it implies that $\rho(\xi, C(y)) \geq$
$\rho(\xi, C(a))$. This completes the proof of Step 5 and that of the proposition.   □

The passage from Proposition 18 to Theorem 15 is relatively straightforward:

PROOF OF THEOREM 15   By Proposition 18(a), each u_n is continuous on X. Since
the series defining $u(x)$ is uniformly convergent on X, it follows that u is continuous.
Consider $x, y \in X$ with $x \succ y$. By Proposition 18(b), $u_n(x) \geq u_n(y)$ for each n. But,
choosing N such that $y \in B_N$, we see from Proposition 18(b), that $u_N(x) > u_N(y)$.
Hence $u(x) > u(y)$. Conversely, if $u(x) > u(y)$, then $u_n(x) > u_n(y)$ for some n, so
$x \succ y$, by Proposition 18(c).   □

5   Concluding Remarks

The classical Arrow-Hahn existence theorem in [3] has two obvious drawbacks: it
deals only with preferences on closed, convex subsets of $\mathbf{R}^N$, and the local nonsatiation hypothesis forces those subsets to be unbounded. The same holds for the
constructive Arrow-Hahn theorem in [7] (Theorem 3). In our work above we have

shown, in a fully constructive way, how both of the drawbacks can be disposed of, leading to a new, more general Arrow-Hahn theorem for a preference relation $\succ$ on a locally compact metric space X. In order to do this, we placed two geometric conditions on X—namely, that every open ball be dense in the corresponding closed ball, and that X have the uniform ball-distance property on compact sets. We also replaced the local nonsatiation condition used in [7] (Theorem 3) by local nonsatiation at compact sets well contained in $-P_\succ$, and added the requirement of preferred-point approximation on compact sets.

Although our Theorem 15 and its proof are classically valid, with a little extra work we can turn it into a more succinct classical theorem, as follows. First, we note that with classical logic, $P_\succ$ is trivially either empty or else inhabited and located (this is not the case constructively; see our Appendix), and every closed ball in a locally compact space is compact. Next, a metric space that is locally compact in the classical sense (that is, every point has a compact neighbourhood) is locally compact in our sense precisely when it is separable; see (3.18.3) in [11]. From this, the remarks preceding Lemmas 8 and 14, and Theorem 15 we arrive at the **classical extended Arrow-Hahn representation theorem**:

Let X be a separable locally compact metric space with the pointwise ball-center distance property, such that each open ball is dense in the corresponding closed ball. Let $\succ$ be a continuous preference relation on X that is locally nonsatiated at non-preferred points. Let $a \in X$, and for each n let

$$u_n(x) = \sup\{\rho(\xi, C(x)) : \xi \in \overline{B}(a, n)\}.$$

Then

$$u \equiv \sum_{n=1}^{\infty} 2^{-n} \frac{u_n}{1 + u_n}$$

defines a continuous utility function representing $\succ$.

Finally, we observe that the ball-centre distance property, in either pointwise or uniform version, seems worthy of further investigation as a generalisation of convexity. In particular, it may prove useful in optimisation and approximation theory.

Acknowledgements Thanking Ghanshyam Mehta for our stimulating work and conversations, I wish him a long, happy, healthy retirement.

Appendix: Two Proofs of Nonconstructivity

We prove that each of the following classically valid statements is essentially nonconstructive:

(a) If a preference relation on a compact space is represented by a continuous utility function, then it has the property **UP**.
(b) If a preference relation $\succ$ on $\mathbf{R}$ is represented by a continuous utility function and satisfies hypotheses (i)–(iii) of Theorem 15, then $P_\succ$ is either empty or else inhabited and located.

To deal with (a), observe that in the recursive interpretation of **BISH** there exists a uniformly continuous function $f : [0, 1] \to [0, 1]$ such that sup $f = 1$ and $f(x) < 1$ for each $x \in [0, 1]$; this is a simple consequence of Corollary (2.9) in Chap. 5 of [9] (see, alternatively, [14]). Taking $X = [0, 3]$, define a uniformly continuous function $u : X \to [0, 1]$ such that

- $u(0) = 0$, $u(1) = 1$, and $u(3/2) = f(0)$,
- u is linear on each of the intervals $[0, 1]$, $[1, 3/2]$, $[3/2, 2]$, and
- $u(x) = f(3 - x)$ for $2 \le x \le 3$.

Let $\succ$ be the preference relation induced on $[0, 3]$ by u. Then $P_\succ = \{1\}$. If a is any point of $[0, 1)$, then $u(a) < 1$, so there exist points x of $[2, 3]$ such that $u(x) = f(3 - x) > u(a)$. For each such x we have $\rho(x, P_\succ) \ge 1$. Thus we have a recursive counterexample to (a), which is therefore essentially nonconstructive.

Turning now to (b), let $(a_n)_{n \ge 1}$ be an increasing binary sequence with $a_1 = 0$. This time, define a continuous mapping u on the locally compact space $\mathbf{R}$ such that for each positive integer n,

- $u(x) = x$ if $x \le k$ and $a_k = 0$,
- $u(x) = k$ if $x \ge k$, $a_k = 0$, and $a_{k+1} = 1$.

Then the preference relation $\succ$ induced on $\mathbf{R}$ by u satisfies hypotheses (i)–(iii) of Theorem 15. Indeed, condition (i) was dealt with in the paragraph preceding Lemma 6. For condition (ii), let $K \subset \mathbf{R}$ be compact, $a \in K$, and $0 < \varepsilon < 1$. Compute a positive integer N such that $K \subset [0, N)$, and consider first the case $a_{N+1} = 0$. If $y \in P_\succ$, then $y \ge a_{N+1}$ and so $|a - y| \ge 1$. Thus if $\rho(a, P_\succ) < \varepsilon$, then *ex falso* $\rho(x, P_\succ)$ for all $x \in C(a)$. Now consider the case $a_{N+1} = 1$, in which there exists $k \le N$ such that $a_k = 0$ and $a_{k+1} = 1$. Since $x = k$ for all $x \ge k$, we have $P_\succ = [k, \infty)$; so if $\rho(a, P_\succ) < \varepsilon$, then $a > k - \varepsilon$. Thus if $x \in C(a)$ and $x < k$, then

$$x = u(x) \ge u(a) = a > k - \varepsilon.$$

For each $x \in C(a)$, since either $x < k$ or $x > k - \varepsilon$, it follows that $x > k - \varepsilon$ and therefore $\rho(x, P_\succ) < \varepsilon$. Thus $\succ$ satisfies condition (ii) of Theorem 15. It is an easy exercise to show that it satisfies (iii). However, $P_\succ$ is empty if and only if $a_n = 0$ for all n, and $P_\succ$ is inhabited if and only if there exists n with $a_n = 1$. Thus if (b) holds, then we can prove the essentially nonconstructive omniscience principle **LPO**: for each binary sequence, either all the terms are 0 or else there exists a term equal to 1. For more on **LPO**, consult [4, 9].

References

1. Aczel, P., Rathjen, M.J.: Notes on Constructive Set Theory, draft available at http://www1. maths.leeds.ac.uk/~rathjen/book.pdf
2. Alps, R.A., Bridges, D.S.: Morse Set Theory as a Foundation for Constructive Mathematics, monograph in preparation (2019)

3. Arrow, K.J., Hahn, F.H.: General Competitive Analysis. Oliver and Boyd, Edinburgh (1971)
4. Bishop, E.A.: Foundations of Constructive Analysis. McGraw-Hill, New York (1967)
5. Bishop, E.A., Bridges, D.S.: Constructive analysis. Grundlehren der Math. Wiss. **279**. Springer, Heidelberg (1987)
6. Bridges, D.S.: The constructive theory of preference relations on a locally compact space. Indag. Math. **92**(2), 141–165 (1989)
7. Bridges, D.S.: The constructive theory of preference relations on a locally compact space—II. Math. Soc. Sci. **27**, 1–9 (1994)
8. Bridges, D.S., Mehta, G.B.: Representations of Preference Orderings. Lecture Notes in Economics and Mathematical Systems, vol. 422. Springer, Heidelberg-Berlin-New York (1996)
9. Bridges, D.S., Richman, F.: Varieties of Constructive Mathematics. London Mathematical Society Lecture Notes, vol. 97. Cambridge University Press (1987)
10. Bridges, D.S., Vîţă, L.S.: Techniques of Constructive Analysis, Universitext. Springer, New York (2006)
11. Dieudonné, J.: Foundations of Modern Analysis. Academic Press, New York (1960)
12. Dummett, M.A.E.: Elements of Intuitionism, Oxford Logic Guides, vol. 39. Clarendon Press, Oxford (2000)
13. Goodman, N.D., Myhill, J.: Choice implies excluded middle. Zeit. math. Logik und Grundlagen Math. **24**, 461 (1978)
14. Julian, W.H., Richman, F.: A uniformly continuous function on [0, 1] that is everywhere different from its infimum. Pac. J. Math. **111**, 333–340 (1984)
15. Martin-Löf, P.: Intuitionistic Type Theory (Notes by Giovanni Sambin of a series of lectures given in Padua, June 1980). Bibliopolis, Napoli (1984)
16. Mehta, G.B.: A new extension procedure for the Arrow-Hahn theorem. Int. Econ. Rev. **18**, 779–782 (1977)
17. Mehta, G.B.: The Euclidean distance approach to continuous utility functions. Quart. J. Econ. **106**, 975–977 (1991)
18. Morse, A.P.: A Theory of Sets, 2nd edn. Academic Press, Orlando, FL (1986)

Continuous Utility Representation of Fuzzy Preferences

Vicki Knoblauch

Abstract We define continuity for utility functions representing fuzzy preferences and prove that, as in the case of crisp preferences, fuzzy preferences representable by a utility function are representable by a continuous utility function.

Keywords Fuzzy relations · Fuzzy preferences · Continuous utility functions

1 Introduction

Debreu [7] proved that preferences that can be represented by a utility function can be represented by a continuous utility function. We will show that the same holds for fuzzy preferences, for which Agud et al. [1] characterized utility representability in the context of the following definitions of fuzzy preferences and utility functions.

Fuzzy preferences over a set X of alternatives are modeled as a collection $\mathcal{R} = \{\mathcal{R}_\alpha\}_{\alpha \in [0,1]}$ of binary relations on X. The binary relation $\mathcal{R}_\alpha$ is to be interpreted as follows: $(x, y) \in \mathcal{R}_\alpha$ if x is weakly preferred to y with probability greater than or equal to α.

A *utility function* representing fuzzy preferences $\mathcal{R}$ on X is a function $U : X \times [0, 1] \to [0, 1]$ such that for each $\alpha \in [0, 1]$, $U(\cdot, \alpha)$ is a utility function representing $\mathcal{R}_\alpha$. A utility function can provide a conceptual snapshot of a fuzzy preference relation $\mathcal{R}$, possibly revealing structure not apparent when only the raw data $\{\mathcal{R}_\alpha\}_{\alpha \in [0,1]}$ is considered.

A utility function U is *continuous* if, for every α, $U(\cdot, \alpha)$ is a continuous function of x, where the topology on X is that generated by the asymmetric part of $\mathcal{R}_\alpha$, and, for every x, $U(x, \cdot)$ is a continuous function of α. On a first reading of the definition, one might ask, "Why not simply require the function U to be continuous on its domain, $X \times [0, 1]$?" The answer is that there is no appropriate topology for

V. Knoblauch (✉)
Department of Economics, University of Connecticut, 327 Oak Hall, 365 Fairfield Way, Unit 1063, Storrs, CT 06269-1063, USA
e-mail: vicki.knoblauch@uconn.edu

© Springer Nature Switzerland AG 2020
G. Bosi et al. (eds.), *Mathematical Topics on Representations of Ordered Structures and Utility Theory*, Studies in Systems, Decision and Control 263,
https://doi.org/10.1007/978-3-030-34226-5_16

$X \times [0, 1]$; for each α, the topology of interest on $X \times \{\alpha\}$ is that generated by the asymmetric part of $\mathcal{R}_\alpha$, and because these topologies may vary greatly and have little relationship to one another, there is no useful way to amalgamate them into a single topology on $X \times [0, 1]$. This claim will be discussed further in Sect. 4. By Debreu's result mentioned above, given a utility function representing fuzzy preferences $\mathcal{R}$, we can trivially construct a utility function U representing $\mathcal{R}$ such that every $U(\cdot, \alpha)$ is continuous, but we need a very different construction to attain continuity of every $U(\cdot, \alpha)$ *and* every $U(x, \cdot)$.

Crisp preference relations can be associated with a subclass of all fuzzy preference relations in an obvious way; given a crisp preference relation in the form of a binary relation $\mathcal{B}$ on a set X, construct a corresponding fuzzy preference relation $\mathcal{R} = \{\mathcal{R}_\alpha\}_{\alpha \in [0,1]}$ by setting $\mathcal{R}_\alpha = \mathcal{B}$ if $0 < \alpha \leq 1$ and $\mathcal{R}_0 = X \times X$. If $\mathcal{B}$ is representable by a utility function, then so is $\mathcal{R}$; and if $\mathcal{R}$ is continuously representable, then so is $\mathcal{B}$. Therefore our fuzzy preference continuity theorem is a generalization of Debreu's result.

As Agud et al. [1] point out, the α-cut utility function is not the only approach to utility representation of fuzzy preferences. For example, Fono and Salles [9] characterize fuzzy preferences representable by a continuous utility function defined to have domain X and range $[0, 1]$, and to satisfy $v(x) \leq v(y)$ if and only if $\text{Prob}(x\mathcal{R}y) \leq \text{Prob}(y\mathcal{R}x)$. Their result has an advantage over ours in that they sometimes find continuous utility representability for fuzzy preferences that are not even representable under the α-cut definition of utility functions. However, an α-cut utility function provides more information than the more widely applicable version. Using U for the former and v for the latter, if fuzzy preferences $\mathcal{R}$ are representable by U and v, then v and $U(\cdot, 1)$ are order equivalent. Therefore v has missed all the information contained in the $U(\cdot, \alpha)$'s for $0 < \alpha < 1$ or, put another way, every such v represents infinitely many fuzzy preference relations with substantially different α-cut representations.

An important application of fuzzy preferences is the rationalization of fuzzy choice functions. Alternatively, rationality for fuzzy choice functions can be defined in terms of axioms for rational behavior. Alcantud and Díaz [2] explore relationships between fuzzy choice axioms and the rationalization of fuzzy choice correspondences by fuzzy relations. Others who have combined these two approaches to rational choice include Basu [5] and Banerjee [3].

Foundational studies in the field of fuzzy preferences include: Nurmi [10], Dutta [8], Barret et al. [4], Billot [6] and Richardson [11].

The paper is organized as follows. Section 2 contains definitions. Section 3 contains two continuous representation theorems, the first for countable sets of alternatives, the second without this restriction. Section 4 consists of comments.

2 Preliminaries

A *binary relation* $\mathcal{B}$ on a set X is a subset of $X \times X$. As is customary, we will sometimes write $x\mathcal{B}y$ in place of $(x, y) \in \mathcal{B}$. A binary relation $\mathcal{B}$ is *antisymmetric* if $x\mathcal{B}y\mathcal{B}x$ implies $x = y$.

Fuzzy preferences $\mathcal{R}$ over a set of alternatives X, which we will also refer to as a *fuzzy preference relation* on X, consists of a collection $\mathcal{R} = \{\mathcal{R}_\alpha\}_{\alpha \in [0,1]}$ of binary relations on X satisfying

(1) $x\mathcal{R}_\alpha x$ for all $x \in X$, $\alpha \in [0, 1]$,
(2) $\mathcal{R}_0 = X \times X$,
(3) for $x, y \in X$ and $\alpha \in [0, 1]$, $x\mathcal{R}_\alpha y$ if $\alpha \leq \sup\{\beta \in [0, 1]: x\mathcal{R}_\beta y\}$.

These conditions on the $\mathcal{R}_\alpha$'s allow us to interpret $x\mathcal{R}_\alpha y$ as "x is weakly preferred to y with probability greater than or equal to α". We denote by $\mathcal{P}_\alpha$ and $\mathcal{I}_\alpha$ the asymmetric or strict part of $\mathcal{R}_\alpha$ and the symmetric or indifference part of $\mathcal{R}_\alpha$, respectively.

An *α-cut utility function* representing a fuzzy preference relation $\mathcal{R}$ on X is a function $U: X \times [0, 1] \to [0, 1]$ such that for $x, y \in X$, $\alpha \in [0, 1]$, $U(x, \alpha) \leq U(y, \alpha)$ if and only if $y\mathcal{R}_\alpha x$. In what follows, we will omit the term "α-cut", since we will not be dealing with any other form of utility function for fuzzy preferences. Notice that, because every fuzzy preference relation $\mathcal{R}$ we will deal with has a utility function representation, its associated $\mathcal{R}_1$ is complete. Therefore $\mathcal{R}$ tells us not only the probability that x is weakly preferred to y for $x, y \in X$, but also the probability that x is strictly preferred to y.

A utility function U representing fuzzy preferences $\mathcal{R}$ over a set X is said to be *continuous* if $U(\cdot, \alpha)$ is a continuous function of x for every $\alpha \in [0, 1]$ and $U(x, \cdot)$ is a continuous function of α for every $x \in [0, 1]$. The continuity of $U(\cdot, \alpha)$ is relative to the topology on its domain X generated by $\mathcal{P}_\alpha$, that is, the topology with subbasis $\{L_x\}_{x \in X} \cup \{G_x\}_{x \in X}$, where $L_x = \{y \in X: x\mathcal{P}_\alpha y\}$ and $G_x = \{y \in X: y\mathcal{P}_\alpha x\}$.

3 Continuous Utility Representation

We first prove that utility representability implies continuous utility representability for fuzzy preferences on *countable* alternative sets.

Theorem 1 *If a fuzzy preference relation on a countable set can be represented by a utility function, it can be represented by a continuous utility function.*

Proof Suppose $\mathcal{R}$ is a representable fuzzy preference relation on a countable set X. Without loss of generality, we can assume X is infinite and write $X = \{x_i\}_{i=1}^{+\infty}$. By the definition of utility function representability of fuzzy preference relations, $\mathcal{R}_\alpha$ is represented by a utility function for all $\alpha \in [0, 1]$ and therefore $\mathcal{R}_\alpha$ is complete and transitive for all $\alpha \in [0, 1]$.

Debreu [7] showed that for crisp preferences, such as $\mathcal{R}_\alpha$, representability by a utility function implies representability by a continuous utility function. However, this turns out not to be a good start towards establishing continuous representability for $\mathcal{R}$, which requires not only continuity of $U(\cdot, \alpha)$ for each α, but also continuity of $U(x, \cdot)$ for each $x \in X$.

Construction of $U : X \times [0, 1] \to [0, 1]$

Without loss of generality we assume $\mathcal{R}_1$ is antisymmetric and $x_1 P_\alpha x_i P_\alpha x_2$ for $i > 2$ and $\alpha \in (0, 1]$, which may require adding two alternatives, x_1 and x_2, to X. For $i > 2$ let $l(i)$ and $g(i)$ be the positive integers such that

$x_{l(i)}$ is the $\mathcal{P}_1$-maximal element of $\{x_j \in X : j < i \text{ and } x_i \mathcal{P}_1 x_j\}$,

$x_{g(i)}$ is the $\mathcal{P}_1$-minimal element of $\{x_j \in X : j < i \text{ and } x_j \mathcal{P}_1 x_i\}$.

For $1 \le i, j \le n$, let $\alpha_{i,j} = \alpha_{j,i} = \max\{\alpha \in [0, 1] : x_i \mathcal{I}_\alpha x_j\}$.

Using the representability of $\mathcal{R}$, we now construct $U : X \times [0, 1] \to [0, 1]$, a continuous utility function representing $\mathcal{R}$. The construction proceeds by induction on the subscript of the x's. Let $U(x_1, \alpha) = \alpha$ for all α and let $U(x_2, \alpha) = 0$ for all α. Fix $k > 2$ and suppose the construction of $U : \{x_1, x_2, \ldots, x_{k-1}\} \times [0, 1] \to [0, 1]$ has been completed. We construct $U(x_k, \alpha)$ for $0 \le \alpha \le 1$ as follows.

Case 1. $\alpha_{k,l(k)} \ge \alpha_{k,g(k)}$ Let

$$\alpha_k^* = \begin{cases} \inf\{\alpha_{m,k} : m > k \text{ and } x_{g(k)} \mathcal{P}_1 x_m \mathcal{P}_1 x_k\} & \text{if inf is greater than } \alpha_{k,l(k)} \\ \alpha_{k,l(k)} + 1/k & \text{otherwise} \end{cases}$$

with the convention that $\inf(\emptyset) = 0$.

Let $U(x_k, \alpha)$ be a continuous function of α satisfying

$$U(x_k, \alpha) = U(x_{l(k)}, \alpha) \quad \text{if } 0 \le \alpha \le \alpha_{k,l(k)},$$

$$U(x_{l(k)}, \alpha) < U(x_k, \alpha) < \big(U(x_{l(k)}, \alpha) + U(x_{g(k)}, \alpha)\big)/2 \quad \text{if } \alpha_{k,l(k)} < \alpha < \alpha_k^*,$$

$$U(x_k, \alpha) = \big(U(x_{l(k)}, \alpha) + U(x_{g(k)}, \alpha)\big)/2 \quad \text{if } \alpha_k^* \le \alpha \le 1.$$

Case 2. $\alpha_{k,g(k)} \ge \alpha_{k,l(k)}$
Let

$$\beta_k^* = \begin{cases} \inf\{\alpha_{m,k} : m > k \text{ and } x_k \mathcal{P}_1 x_m \mathcal{P}_1 x_{l(k)}\} & \text{if inf is greater than } \alpha_{k,g(k)} \\ \alpha_{k,g(k)} + 1/k & \text{otherwise} \end{cases}$$

Let $U(x_k, \alpha)$ be a continuous function of α satisfying

$$U(x_k, \alpha) = U(x_{g(k)}, \alpha) \quad \text{if } 0 \le \alpha \le \alpha_{k,g(k)},$$

$$\big(U(x_{l(k)}, \alpha) + U(x_{g(k)}, \alpha)\big)/2 < U(x_k, \alpha) < U(x_{g(k)}, \alpha) \quad \text{if } \alpha_{k,g(k)} < \alpha < \beta_k^*,$$

$$U(x_k, \alpha) = \big(U(x_{l(k)}, \alpha) + U(x_{g(k)}, \alpha)\big)/2 \quad \text{if } \beta_k^* \le \alpha \le 1.$$

There are three facts to establish.

U is a utility function representation for $\mathcal{R}$.

Suppose $\alpha_0 \in [0, 1]$, $x_j, x_k \in X$ and $k > j$. By the construction of $U|_{\{x_1, x_2, \ldots, x_k\} \times [0,1]}$,

if $x_j \mathcal{I}_{\alpha_0} x_k$, then $\alpha_0 \le \alpha_{j,k}$ and $U(x_j, \alpha_0) = U(x_k, \alpha_0)$;

if $x_j \mathcal{P}_{\alpha_0} x_k$, then $\alpha_0 > \alpha_{j,k}$ and $U(x_j, \alpha_0) > U(x_k, \alpha_0)$;

if $x_k \mathcal{P}_{\alpha_0} x_j$, then $\alpha_0 > \alpha_{j,k}$ and $U(x_k, \alpha_0) > U(x_j, \alpha_0)$.

For each x_k, $U(x_k, \alpha)$ is a continuous function of α.

Using induction on the subscript of the x's, suppose the construction of $U(x_k, \alpha)$ was governed by Case 1. We can conclude that for $k > 2$,

on $[0, \alpha_{k,l(k)}]$, $U(x_k, \alpha) = U(x_{l(k)}, \alpha)$, a continuous function of α;

on $[\alpha_k^*, 1]$, $U(x_k, \alpha) = \big(U(x_{l(k)}, \alpha) + U(x_{g(k)}, \alpha)\big)/2$, a continuous function of α;

on $[\alpha_{k,l(k)}, \alpha_k^*]$ continuity of $U(x_k, \alpha)$ was attainable as claimed in the construction because the only requirement other than continuity was that on $(\alpha_{k,l(k)}, \alpha_k^*)$, $U(x_k, \alpha)$ must remain strictly between $U(x_{l(k)}, \alpha)$ and $\big(U(x_{l(k)}, \alpha) + U(x_{g(k)}, \alpha)\big)/2$, two continuous functions with disjoint graphs on the interval $(\alpha_{k,l(k)}, 1)$.

A symmetric argument holds if the construction of $U(x_k, \alpha)$ was governed by Case 2.

For each α_0, $U(x, \alpha_0)$ is a continuous function of x.

Suppose $\alpha_0 \in [0, 1]$, $x_{i_0} \in X$ and $(i_j)_{j=1}^{+\infty}$ is a sequence of positive integers such that

$$x_{i_j} \to x_{i_0} \text{ in the topology induced by } \mathcal{P}_\alpha. \tag{1}$$

We will prove that $U(x_{i_j}, \alpha_0) \to U(x_0, \alpha_0)$. Without loss of generality we can assume

$$i_j > i_0 \text{ for all } j \tag{2}$$

and either

$$x_{i_0} \mathcal{P}_{\alpha_0} x_{i_{j+1}} \mathcal{P}_{\alpha_0} x_{i_j} \text{ and } U(x_{g(i_j)}, \alpha_0) = U(x_{i_0}, \alpha_0) \text{ for all } j \tag{3}$$

or

$$x_{i_j} \mathcal{P}_{\alpha_0} x_{i_{j+1}} \mathcal{P}_{\alpha_0} x_{i_0} \text{ and } U(x_{l(i_j)}, \alpha_0) = U(x_{i_0}, \alpha_0) \text{ for all } j \tag{4}$$

and either

$$\alpha_{i_j, g(i_j)} \ge \alpha_{i_j, l(i_j)} \text{ for all } j \tag{5}$$

or

$$\alpha_{i_j, l(i_j)} \ge \alpha_{i_j, g(i_j)} \text{ for all } j. \tag{6}$$

Case A. (2), (3) and (5) hold.

By Case 2 of the construction of $U(x_k, \alpha)$, for each $j > 1$,

$$U(x_{i_j}, \alpha_0) \geq \big(U(x_{l(i_j)}, \alpha_0) + U(x_{g(i_j)}, \alpha_0)\big)/2 \geq \big(U(x_{i_{j-1}}, \alpha_0) + U(x_{i_0}, \alpha_0)\big)/2,$$

where the second inequality holds because $U(x_{l(i_j)}, \alpha_0) \geq U(x_{i_{j-1}}, \alpha_0)$ by the definition of $l(i_j)$ and $U(x_{g(i_j)}, \alpha_0) = U(x_{i_0}, \alpha_0)$ by (3). Therefore $U(x_{i_j}, \alpha_0) \to U(x_{i_0}, \alpha_0)$.

Notice that the proof for Case A did not use the Case 2 definition of β_k^*. That part of the definition will allow us to appeal to the symmetry of Cases 1 and 2 when we address Case C below.

Case B. (2), (3) and (6) hold.
Case B.1. For infinitely many j,

$$\alpha_{i_j}^* = \inf\{\alpha_{m,i_j} : \; m > i_j \text{ and } x_{g(i,j)}\mathcal{P}_1 x_m \mathcal{P}_1 x_{i_j}\} > \alpha_{i_j, l(i_j)},$$

where the first inequality is a reminder of the Case 1 definition of $\alpha_{i_j}^*$. Then for all such j, $\alpha_{i_j}^* < \alpha_{i_j, i_{j+1}} < \alpha_0$, where the second inequality holds by (3), so that by the Case 1 definition of $U(x_{i_j}, \alpha_0)$,

$$U(x_{i_j}, \alpha_0) = \big(U(x_{l(i_j), \alpha_0}) + U(x_{g(i_j), \alpha_0})\big)/2 \geq \big(U(x_{i_{j-1}}, \alpha_0) + U(x_{i_0}, \alpha_0)\big)/2.$$

It follows that $U(x_{i_j}, \alpha_0) \to U(x_{i_0}, \alpha_0)$.
Case B.2. For sufficiently large j, Case B.1 does not hold.
Then there exists $\epsilon > 0$ such that for infinitely many j, $\alpha_{j,j+1} < \alpha_0 - \epsilon$. For each such j with $1/j < \epsilon$, by the Case 1 definition,

$$U(x_j, \alpha_0) = \big(U(x_{l(i_j)}, \alpha_0) + U(x_{g(i_j)}, \alpha_0)\big)/2 \geq \big(U(x_{i_{j-1}}, \alpha_0) + U(x_0, \alpha_0)\big)/2.$$

It follows that $U(x_{i_j}, \alpha_0) \to U(x_{i_0}, \alpha_0)$.
Case C. (2), (4) and either (5) or (6) hold.
Then by the symmetry of the Case 1 and Case 2 definitions of $U(x_k, \alpha)$, the arguments of Cases A and B also establish $U(x_{i_j}, \alpha_0) \to U(x_{i_0}, \alpha_0)$. ∎

We begin the process of eliminating the assumption that X is countable from the statement of Theorem 1 with two definitions and two lemmas. Suppose $\mathcal{R}$ is a representable fuzzy preference relation on X.

Definition 1 For $\alpha \in [0, 1]$ a subset A of X is $\mathcal{R}_\alpha$-*dense* in X if $x, y \in X$ and $x\mathcal{P}_\alpha y$ together imply $x\mathcal{R}_\alpha x_i \mathcal{R}_\alpha y$ for some $x_i \in A$.

Definition 2 For $\alpha \in [0, 1]$ an $\mathcal{R}_\alpha$-*gap* of X is an ordered pair (S, T) of sets from the collection of equivalence classes into which $\mathcal{I}_\alpha$ partitions X satisfying
  (1) $x \in S$ and $y \in T$ together imply $x\mathcal{P}_\alpha y$ and
  (2) $x \in S$ and $y \in T$ together imply for all $z \in X$, not$(x\mathcal{P}_\alpha z\mathcal{P}_\alpha y)$.

Then S and T are *endpoints* of the $\mathcal{R}_\alpha$-gap (S, T) and each $x \in S$ ($y \in T$) is a *representative* of S (of T).

Let Q be the set of rationals in $[0, 1]$. For each $q \in Q$, by the representability of $\mathcal{R}_q$, there are only countably many $\mathcal{R}_q$-gaps. Let A be a countable subset of X such that A is $\mathcal{R}_1$-dense in X and for each $q \in Q$, A contains the $\mathcal{R}_1$-minimal representative of each S endpoint of each $\mathcal{R}_q$-gap in X if that minimum exists, and the $\mathcal{R}_1$-maximal representative of each T endpoint of each $\mathcal{R}_q$-gap in X if that maximum exists. By the representability of $\mathcal{R}$, the following lemma holds.

Lemma 1 *For every $\alpha \in [0, 1]$, A is $\mathcal{R}_\alpha$-dense in X and A contains a representative of each endpoint of each $\mathcal{R}_\alpha$-gap in X.*

Now suppose $\mathcal{R}$ is a representable fuzzy preference relation on a countable set A. Besides showing that a utility function U constructed in the proof of Theorem 1 represents $\mathcal{R}$ and is continuous, the proof of Theorem 1 also establishes the following lemma, which says, roughly, that if two sequences in A approach each other according to $\mathcal{P}_\alpha$ with no element of A between them, then their images under $U(\cdot, \alpha)$ will approach each other.

Lemma 2 *Suppose $(l_j)_{j=1}^{+\infty}$ and $(k_j)_{j=1}^{+\infty}$ are increasing sequences of positive integers such that for all j, $x_{k_j} \mathcal{P}_\alpha x_{k_{j+1}} \mathcal{P}_\alpha x_{l_{j+1}} \mathcal{P}_\alpha x_{l_j}$ and such that if $y \in A$ then there exists $J > 0$ such that $x_{l_j} \mathcal{P}_\alpha y$ or $y \mathcal{P}_\alpha x_{k_j}$. Then $\lim_{j \to +\infty} U(x_{l_j}, \alpha) = \lim_{j \to +\infty} U(x_{k_j}, \alpha)$.*

We can now prove the main result.

Theorem 2 *If a fuzzy preference relation can be represented by a utility function, then it can be represented by a continuous utility function.*

Proof Suppose $\mathcal{R}$ is a representable fuzzy preference relation on X. Let A be a countable subset of X such that for each $\alpha \in [0, 1]$, A is $\mathcal{R}_\alpha$-dense in X and A contains a representative of each endpoint of each $\mathcal{R}_\alpha$-gap in X. The existence of A is guaranteed by Lemma 1.

Next construct a continuous utility function $U: A \times [0, 1] \to [0, 1]$ representing $\mathcal{R}|_A$ as in the proof of Theorem 1. Extend U to $X \times [0, 1]$ by setting

$$U(x, \alpha) = \inf\{U(x_i, \alpha): \ x_i \in A \text{ and } x_i \mathcal{R}_\alpha x\}.$$

We will show that, for $x \in X$ and $\alpha \in [0, 1]$, $x_{i_j} \to x$ in the $\mathcal{P}_\alpha$-topology on X implies $U(x_{i_j}, \alpha) \to U(x, \alpha)$.

It is sufficient to show that, for $\alpha \in [0, 1]$ and $x \in X$ such that for all $x_i \in A$, $\text{not}(x \mathcal{I}_\alpha x_i)$, if $(i_j)_{j=1}^{+\infty}$ is an increasing sequence of positive integers such that $x_{i_j} \to x$ in the $\mathcal{P}_\alpha$-topology on X and $x \mathcal{P}_\alpha x_{i_{j+1}} \mathcal{P}_\alpha x_{i_j}$ for all j, then $U(x_{i_j}, \alpha) \to U(x, \alpha)$. For $\alpha \in [0, 1]$, by Lemma 2 and the definition of $U(x, \alpha)$,

$$U(x_{i_j}, \alpha) \to U(x, \alpha) = \inf\{U(x_i, \alpha): \ x_i \in A \text{ and } x_i \mathcal{R}_\alpha x\}.$$

Summarizing, by the definition of U, U is a utility function representing $\mathcal{R}$ and for $x \in X$, $U(x, \alpha)$ is an infimum of continuous functions of α and is therefore an upper semi-continuous function of α. We have used Lemma 2 to show that for

$\alpha \in [0, 1]$, $U(x, \alpha)$ is a continuous function of x and for $x \in X$, $U(x, \alpha)$ is a supremum of continuous functions of α and is therefore a lower semi-continuous function of α. Since for $x \in X$, $U(x, \alpha)$ is both upper and lower semi-continuous, $U(x, \alpha)$ is a continuous function of α.   ∎

4 Concluding Remarks

We have proven that, as in the case of crisp preferences, fuzzy preferences that are representable by a utility function are representable by a continuous utility function. As was stated in the introduction, we adopted, as our definition of continuity for a utility function $U : X \times [0, 1] \to [0, 1]$ representing fuzzy preferences $\mathcal{R}$, the requirement that for $x \in X$, $U(x, \alpha)$ must be a continuous function of α and for $\alpha \in [0, 1]$, $U(x, \alpha)$ must be a continuous function of x with respect to the topology on X induced by $\mathcal{P}_\alpha$. The most obvious alternative is to define a topology τ on $X \times [0, 1]$ that is the product of a topology τ^1 on X and the Euclidean topology τ^2 on $[0, 1]$ and to deem U continuous if it is continuous as a function from $X \times [0, 1]$ with topology τ to $[0, 1]$ with topology τ^2. As was also mentioned earlier, the problem is to choose an appropriate topology τ^1. The difficulty is that the relevant topology on $X \times \{\alpha\}$ is the topology induced by $\mathcal{P}_\alpha$; for different values of α, these topologies can differ widely. We will see that for $\mathcal{R}$ in Example 1, if the topology τ^1 is too coarse, then *no* utility function representing $\mathcal{R}$ can be continuous with respect to the product topology τ. Example 2 will demonstrate why this difficulty cannot in general be overcome by choosing a finer topology τ^1.

In both examples, for $0 \le \alpha \le 1$, let τ_α be the topology on X induced by $\mathcal{P}_\alpha$.

Example 1 Suppose $X = [0, 1]$,
for $0 \le \alpha \le 1/2$, $x \mathcal{R}_\alpha y$ if $x, y \in [0, 1]$,
for $1/2 < \alpha \le 1$, $x \mathcal{R}_\alpha y$ if $x \ge y$
and
$$\tau^1 = \tau_0 = \{X, \emptyset\}.$$
Then $\tau = \{X \times O : O \in \tau_2\}$.

Suppose U represents $\mathcal{R}$. Let $S = \{(x, \alpha) \in X \times [0, 1] : U(x, \alpha) > U(\frac{2}{3}, \frac{3}{4})\}$.

Since U represents $\mathcal{R}$, $(\frac{5}{6}, \frac{3}{4}) \in S$ but $(\frac{2}{3}, \frac{3}{4}) \notin S$. Therefore, $U^{-1}((U(\frac{2}{3}, \frac{3}{4}), 1]) = S \notin \tau$. In summary, owing to the coarseness of τ^1 compared to $\tau_{\frac{3}{4}}$, no utility function representing $\mathcal{R}$ is continuous with respect to τ.

Next we present an example of a fuzzy preference relation $\mathcal{R}$ on a set X such that $\tau_{\frac{1}{2}}$ is not a refinement of τ_1 and τ_1 is not a refinement of $\tau_{\frac{1}{2}}$. Then we discuss the implications of this example for the possibility of a useful continuity result in the context of a product topology on $X \times [0, 1]$.

Example 2 Suppose $X = [0, 1]$,

 for $0 \leq \alpha \leq 1$, $x\mathcal{R}_\alpha y$ if $0 \leq y \leq 1/2$ or $1/2 < x \leq 1$

and

 for $1/2 < \alpha \leq 1$, $x\mathcal{R}_\alpha y$ if $x \geq y$.

Then $\{y \in X : \frac{2}{3}\mathcal{P}_{\frac{1}{2}}y\} = [0, \frac{1}{2}] \in \tau_{\frac{1}{2}} \setminus \tau_1$ and $\{y \in X : \frac{2}{3}\mathcal{P}_1 y\} = [0, \frac{2}{3}) \in \tau_1 \setminus \tau_{\frac{1}{2}}$.

From Example 1 we know that, in order for there to exist a utility function U that represents a fuzzy preference relation $\mathcal{R}$ on $X = [0, 1]$ and is continuous with respect to the product of τ^1 on X and the Euclidean topology on $[0, 1]$, τ^1 must be at least as fine as τ_α for every α. Following Example 2, we could construct a fuzzy preference relation $\mathcal{R}$ with an infinite set of α's such that for no $\alpha_1 \neq \alpha_2$ in this set is τ_{α_1} a refinement of τ_{α_2}. For such an $\mathcal{R}$, any τ^1 for which there exists a utility function U that represents $\mathcal{R}$ and is continuous with respect to τ must be so fine that continuity is nearly meaningless.

There are obvious opportunities for further study: given a concept, question or result relating to representation or continuous representation of crisp preferences, is there a *meaningful* and *interesting* analog for fuzzy preferences in the α-cut setting? Possibilities include two-function representation of interval orders, discernible difference representation of semi-orders, single-peaked preferences and preference aggregation.

References

1. Agud, L., Catalán, R.G., Díaz, S., Induráin, E., Montes, S.: Numerical representability of fuzzy total preorders. Int. J. Comput. Intell. Syst. **5**(6), 996–1009 (2012)
2. Alcantud, J.C.R., Díaz, S.: Rational fuzzy and sequentially fuzzy choice. Fuzzy Sets Syst. **315**, 76–98 (2017)
3. Banerjee, A.: Fuzzy choice functions, revealed preference and rationality. Fuzzy Sets Syst. **70**, 31–43 (1995)
4. Barret, C.R., Pattanaik, P.K., Salles, M.: On choosing rationally when preferences are fuzzy. Fuzzy Sets Syst. **34**, 197–212 (1990)
5. Basu, K.: Fuzzy revealed preference theory. J. Econ. Theory **32**(2), 212–227 (1984)
6. Billot, A.: An existence theorem for fuzzy utility functions: a new elementary proof. Fuzzy Sets Syst. **74**, 271–276 (1995)
7. Debreu, G.: Continuity properties of Paretian utility. Int. Econ. Rev. **5**, 285–293 (1964)
8. Dutta, B.: Fuzzy preferences and social choice. Math. Soc. Sci. **13**, 215–229 (1987)
9. Fono, L.A., Salles, M.: Continuity of utility functions representing fuzzy preferences. Soc. Choice Welf. **37**, 669–682 (2011)
10. Nurmi, H.: Approaches to collective decision making with fuzzy preference relations. Fuzzy Sets Syst. **6**, 249–259 (1981)
11. Richardson, G.: The structure of fuzzy preferences. Soc. Choice Welf. **15**, 359–369 (1998)

Applications into Economics

Ranking opportunity sets; Money-metric utility functions; Intergenerational justice; Utility stream; Risk aversion; State-dependent preferences

Ranking Opportunity Sets Taking into Account Similarity Relations Induced by Money-Metric Utility Functions

Carmen Vázquez

Abstract We use money-metric utility functions as a cardinal utility to define a binary similarity relation between alternatives. That relation is taken into account in ranking opportunity sets, with information about diversity of alternatives being used to reduce the size of large opportunity sets. We provide characterization results for generalizations of several rules proposed for finite sets of alternatives.

Keywords Similarities between alternatives · Diversity of alternatives · Money-metric utility · Ranking opportunity sets

1 Introduction

The objective of this paper is to use a similarity relation to reduce the size of large opportunity sets so as to establish a ranking in an easier and more suitable manner. The money-metric utility function is used as a cardinal utility function to define the binary similarity relation between commodity bundles.

There have been many papers on the utility functions theory that are involved in the 'distance approach'. We refer the reader to Mehta [11] for further information. The so-called 'money-metric' utility functions provide a structure that enables the measurement of preferences to be taken into account.

Mckenzie [9] introduces the expenditure function as a tool for use in demand theory. Mehta [10] analyzes in detail the structure of a money-metric function, which is a p-section of the Mckenzie expenditure function. The money-metric function is viewed from a new perspective and is constructed in a natural manner by using certain metric operations.

Substantial improvements have been made thanks to two anonymous referees.

C. Vázquez (✉)
Facultade de Ciencias Económicas e Empresariais, Departamento de Matemáticas,
Universidade de Vigo, Campus Lagoas-Marcosende, 36310 Vigo, Spain
e-mail: cvazquez@uvigo.es

© Springer Nature Switzerland AG 2020
G. Bosi et al. (eds.), *Mathematical Topics on Representations of Ordered Structures and Utility Theory*, Studies in Systems, Decision and Control 263,
https://doi.org/10.1007/978-3-030-34226-5_17

313

Various authors have investigated the conditions under which the Mckenzie expenditure function is a utility representation and those under which it is a continuous function. See, for example, Honkapohja [6], Samuelson [15] or Weymark [18]. We use the results of Alcantud and Manrique [2], who establish a framework where the money-metric is a continuous utility defined on a consumption set $X \subset \mathbb{R}^n$.

We observe that metric utility functions can be thought of as cardinal utility functions, and this enables us to define a binary similarity relation between commodity bundles by proceeding as in Vázquez [17]. Such a relation is used in Pattanaik and Xu [13] to provide an axiomatic characterization of a rule for ranking opportunity sets. We implement and extend that rule for the case of not-necessarily-finite subsets of alternatives in a finite-dimensional space. This generalization permits more economic applications.

The paper is organized as follows. Section 2 deals with preliminary matters. Section 3 presents the similarity relation. Section 4 provides a method for ranking opportunity sets and establishes the axioms used in their characterization. Section 5 looks at other methods for ranking opportunity sets by using the similarity relation. Finally, Sect. 6 gives some comments and suggestions for further research.

2 Preliminaries

A binary relation on a set X is denoted by $R \subset X \times X$. The notation xRy is frequently used instead of $(x, y) \in R$ in order to simplify the exposition. The following are some standard properties of a binary relation R.

Reflexivity: xRx, for all $x \in X$.

Transitivity: For all $x, y, z \in X$, xRy, yRz imply xRz.

Completeness: For all $x, y \in X$, xRy or yRx.

Symmetry: For all $x, y \in X$, xRy implies yRx.

Consider an individual, namely the agent, who states his/her individual preferences out of the opportunities available to him/her, which are referred to as alternatives.

Let $X \subset \mathbb{R}^n$ be the set of basic alternatives available to the agent. Let R be the preference relation of the agent, which is assumed to be an ordering defined on the set X, i.e. a complete, transitive binary relation, where xRy means that x is at least as good as y. The indifference relation corresponding to R is denoted by I, while the strict preference relation is denoted by P. They are defined in the standard fashion.

The preference relation R is said to be continuous if the lower-contour set of x, $\{y \in X : xRy\}$, and the upper-contour set of x, $\{y \in X : yRx\}$, are closed in X, for each $x \in X$.

Let $x, y \in \mathbb{R}^n$. Write $x \leq y$ if $x_i \leq y_i$, for all $i = 1, \ldots, n$, $x < y$ if $x \leq y$ and, for some i, $x_i < y_i$, and $x \ll y$ if $x_i < y_i$, for all $i = 1, \ldots, n$. A subset $X \subset \mathbb{R}^n$ is said to be bounded from below if there exists $v \in \mathbb{R}^n$ such that $v \leq x$ for all $x \in X$.

The set-theoretic difference of two sets A and B is denoted by $A \setminus B$.

A real valued function $u: X \longrightarrow \mathbb{R}$ such that xRy if and only if $u(x) \geq u(y)$ is said to be a utility function for the preference relation R. For each $A \subset X$, write $u(A) = \{u(a) : a \in A\} \subset \mathbb{R}$.

In a context where the intensity of preference is available, that is, a context where the notions of being 'slightly worse' and 'much better' are reasonable, an essentially numerical utility function that simultaneously describes both the preferences out of the possible outcomes and their intensities is needed. *Cardinal utility function* here means a utility function where not only does the fact $u(x) < u(y)$ provide information about the preference of y over x, but also the difference $u(y) - u(x)$ reflects the intensity of that preference.

A utility function u is said to be a money-metric utility function if, given a price vector $p \in \mathbb{R}^n$, the utility of x, $u(x)$, is obtained as the result of minimizing the linear functional $p.y$ on the upper-contour set of x, $\{y \in X : yRx\}$. Note that, as $\mathbb{R}^n$ models 'baskets of goods', given an alternative $y \in X$, $y = (y_1, \ldots, y_n)$, if the unit price of the i-commodity is p_i, with $i = 1, \ldots, n$, the value of alternative y becomes the real number $p_1 y_1 + \cdots + p_n y_n = p.y$. Next the minimum of $p.y$ on the upper-contour set is obtained.

Here, $u(x)$ is equal to the p-section, $M(p, .)$ of the Mckenzie expenditure function, $M(p, x) = \min\{p.y : yRx, y \in X\}$. Samuelson [15] coined the term 'money-metric' to refer the p-section of the Mckenzie expenditure function.

3 The Similarity Relation

The consumption set, X, is assumed to be a subset of the Euclidean n-space, which is interpreted as the set that represents the n possible states or peculiarities that determine each alternative.

Let $p \in \mathbb{R}^n$ be a price vector, $p \gg 0$, interpreted here as the list of values assigned to each unit of the peculiarities described by the n-tuples that form the alternatives. Note that this can be a monetary value, i.e. the market price, or a subjective value, i.e. a numerical representation of its emotional importance for the agent. The money-metric or the p-section of the Mckenzie function, M_p, is defined by the expression $M_p(x) = \min\{p.y : yRx, y \in X\}$. $M_p(x)$ specifies the minimum income needed at these prices, p, to obtain a utility level at least as large as x.

X is assumed to be an open set bounded from below set R to be a continuous preference relation. Under these conditions, Alcantud and Manrique [2] establish that M_p is a utility for R. Note that this function M_p is bounded from below: given that X is bounded from below, there exists $v \in \mathbb{R}^n$ such that $v \leq y$ for all $y \in X$. Then, as $p \gg 0$, for each $i = 1, \ldots n$, $p_i v_i \leq p_i y_i$ and $p.v \leq p.y$ is obtained. Thus $M_p(x) \geq p.v$, for all $x \in X$.

This utility function can be considered as a cardinal utility in the sense that changes between outcomes can be ranked in order of desirability. M_p enables concepts of measurement to be used because it describes the intensity of preferences in terms of monetary or emotional value.

Following Vázquez [17] and Lauwers et al. [8], given that preferences can be measured by the agent, alternatives that differ to a certain degree in utility may be seen as similar. That degree can be represented by a number $r > 0$ which can be determined by the ability of the agent to discern. Another way of interpreting this number could be for the agent taking into account his/her circumstances not to worry about a determined variation (r) in utility. The interpretation must fit into the economic interpretation assumed previously for the price vector p.

In any case, for all $x, y \in X$, $x S y$ is to be interpreted as 'x being similar to y' iff $|M_p(x) - M_p(y)| \leq r$. Otherwise, the expression $\neg x S y$ must be used, which is interpreted as 'x being dissimilar to y'. For $x \in X$ and $A \subset X$, write $x S A$ when $x S a$ for all $a \in A$. For all non-empty set $A \subset X$, say that A is homogeneous iff $a S a'$, for all $a, a' \in A$.

Thus, S is a reflexive, symmetric, but not necessarily transitive binary relation defined on X. Thus 'being similar to' is not necessarily an equivalence relation, but a special partition could be defined based on it. For each non-empty set of alternatives $A \subset X$, a similarity-based partition is constructed. If A is homogeneous define $A_1 = A$, otherwise proceed as follows. As $M_p(A)$ is bounded from below, it is possible to denote $a_1 = \inf M_p(A)$ and $A_1 = \{a \in A : |M_p(a) - a_1| \leq r\}$. Trivially A_1 is homogeneous.

Let $A_1^c = A \setminus A_1$. Since A_1^c is bounded from below, it can be considered, in a similar way, that $a_2 = \inf M_p(A_1^c)$ and the homogeneous set $A_2 = \{a \in A : |M_p(a) - a_2| \leq r\}$. Next consider $A_2^c = A \setminus (A_1 \cup A_2)$ to obtain a_3 and A_3. Continue with the same procedure while $A_i^c \neq \emptyset$. For each set $A \subset X$, this procedure generates a set of real numbers $M(A) = \{a_1, a_2, \ldots\}$ and a set of subsets $\phi(A) = \{A_1, A_2, \ldots\}$, that could be either infinite or finite with the same cardinality, i.e. $\#M(A) = \#\phi(A) \in \mathbb{N} \cup \{\infty\}$.

Hence, for each non-empty set of alternatives $A \subset X$, the similarity relation leads to a partition $\phi(A)$ that could be defined as the unique class $\{A_1, A_2, \ldots\}$ that verifies the following statements:

- A_i is a non-empty, homogeneous subset of A, for all i,
- $\bigcup_i A_i = A$,
- A_i, A_j are disjoint for all $i \neq j$, i.e. $A_i \cap A_j = \emptyset$,
- if $i < j$, then $x P y$, for all $x \in A_j, y \in A_i$,
- for each $y \in A_j$, if $i < j$, then there exists $x \in A_i$ such that $\neg x S y$.

Observe that the construction of the partition associated with each set A fits for any bounded-from-bellow utility function on X. We use the money-metric utility function because it is a well-known class of utility functions, which is very important in the literature and has various possibilities as regards economic interpretation from the measurement of the preferences of the agent.

4 Ranking Opportunity Sets by Using a Simple Similarity-Based Ordering

Let $\mathcal{P}(X)$ be the set of all non-empty subsets of X, which are referred to as opportunity sets. The elements of $\mathcal{P}(X)$ are the sets of available alternatives that the agent may be faced with. The purpose of the agent is to establish an ordering (a transitive and complete binary relation), $\succeq$, on $\mathcal{P}(X)$. This ordering is interpreted as the agent's preference ranking on opportunity sets.

We adapt the ordering characterized by Pattanaik and Xu [13] to our context. They consider a finite set of alternatives X and for each $A \in \mathcal{P}(X)$ they refer to each partition satisfying the first four statements described above with the lowest possible cardinal as a smallest similarity-based partition (SSP) of A. Note that any partition $\phi(A)$ falls within that case. Next they establish for all $A, B \in \mathcal{P}(X), A \succeq B$ iff $\#\phi(A) \geq \#\phi(B)$. Finally they characterize this ordering with respect to three independent axioms.

We draw on the purpose described by this ordering. Pattanaik and Xu [13] argue that the extent of diversity between the alternatives in the opportunity set is an important consideration in judging an agent's freedom of choice. The model that they propose incorporates information about the potential degree of diversity between alternatives.

In the same spirit, we seek to extend their work to our context, where $X \subset \mathbb{R}^n$ is not necessarily a finite set. For each $A \in \mathcal{P}(X)$, we define $D(A) = \#\phi(A)$ as a measure of diversity between the alternatives in A. Note that $D(A) = \infty$ may be possible. If so we regard such a measure of diversity in A in the sense that for any level of utility there is an alternative in A with a utility level better than the stated level, or indeed much better than it.

We now introduce a new version of the simple similarity-based ordering. For all $A, B \in \mathcal{P}(X)$, we define $A \succeq B$ iff $D(A) \geq D(B)$.

The next issue is to explore the properties to be used to characterize this rule, which we call the *preference for diversity ordering* or the D-rule for short.

4.1 Mimic Property

An additional definition is required to adapt one of the initial axioms used by Pattanaik and Xu [13]. They define an interesting notion: one set *mimics* another. For all $A, B \in \mathcal{P}(X)$ such that A is homogeneous, they say that A does not mimic B iff, for all SSP of B, $\varphi(B) = \{B_1, \ldots, B_k\}$, there exists $a \in A$ such that for all $i = 1, \ldots, k$, $\neg aSB_i$. They observe that for all $A, B \in \mathcal{P}(X)$ such that A is homogeneous and A does not mimic B, every SSP of $A \cup B$ will have more sets than every SSP of B.

We assert that the converse property is also true. That is, it can be shown that

- A does not mimic B iff $\#\varphi(A \cup B) > \#\varphi(B)$,
- A mimics B iff $\#\varphi(A \cup B) = \#\varphi(B)$.

Observe that the procedure for constructing a new partition of $A \cup B$ in the sense of Pattanaik and Xu [13] consists of distributing the elements of set A among the subsets of the initial partition of B by picking them out as being similar to the subset to which they are added. If the idea is that the new partition of $A \cup B$ be suitable in our sense, a partition of B in that sense must be chosen first and the elements of A must be added to whichever of the similar subsets has the lowest index (if there is more than one subset in these conditions).

The following example illustrates this. Let $X = \{x, y, z\}$ such that $xPyPz, xSy$, ySz and $\neg xSz$. Consider $A = \{y\}$ and $B = \{x, z\}$. The sole SSP of B is $\varphi(B) = \{\{x\}, \{z\}\}$. The requirement that A should mimic B is met and the SSP of $A \cup B$, in the sense of Pattanaik and Xu [13], are $\varphi(A \cup B) = \{\{z\}, \{x, y\}\}$ and $\varphi'(A \cup B) = \{\{y, z\}, \{x\}\}$, but in our sense only the first partition is suitable, because it is the only one that meets our the five requirements.

Obviously, 'mimicry' is not necessarily a symmetric binary relation on $\mathcal{P}(X)$. In the example mentioned, A mimics B but B does not mimic A.

4.2 Axioms

The following is a list of axioms used in our characterization result. We first present their formal definitions and then discuss their contents and some of their implications.

The three independent axioms used by Pattanaik and Xu [13] are INS, SM and SC.

Indifference between no-choice situations (INS): For all $x, y \in X$, $\{x\} \sim \{y\}$.

S-monotonicity (SM): For all homogeneous set $A \in \mathcal{P}(X)$ and for all $x \in X \setminus A$,
If xSA then $A \cup \{x\} \sim A$.
If $\neg xSA$ then $A \cup \{x\} \succ A$.

S-composition (SC): For all $A, B, C, D \in \mathcal{P}(X)$ such that $A \cap C = B \cap D = \emptyset$, C and D are both homogeneous and C does not mimic A, then
$A \succeq B$ and $C \succeq D$ implies $A \cup C \succeq B \cup D$.
$A \succ B$ and $C \succeq D$ implies $A \cup C \succ B \cup D$.

Remark SC is proposed by Pattanaik and Xu [13] as a weaker version of an axiom introduced by Sen [16]. We observe that in a context where X is a finite set and C and D are both homogeneous, INS and SM imply $C \sim D$. In fact, for each $a \in A$ such that A is a finite and homogeneous set, SM implies $A \sim \{a\}$. In any case, observe that condition $C \succeq D$ cannot be dropped in axiom SC when they are considered as independent conditions.

We replace SM by the stronger conditions ISS and SSM and add a new axiom concerning the union of infinite sets, RU. We therefore consider the following axioms.

Indifference between no-choice situations (INS): For all $x, y \in X$, $\{x\} \sim \{y\}$.

Indifference between similar situations (ISS): For all homogeneous set $A \in \mathcal{P}(X)$ and for all $a \in A$, $A \sim \{a\}$.

S-simple monotonicity (SSM): For all $A, B, C \in \mathcal{P}(X)$ such that $A, C \subset B$, $A \cap C = \emptyset$ and C is homogeneous and it does not mimic A, then $B \succ A$.

S-composition (SC): For all $A, B, C, D \in \mathcal{P}(X)$ such that $A \cap C = B \cap D = \emptyset$, C and D are both homogeneous and C does not mimic A, then:

$A \succeq B$ and $C \succeq D$ implies $A \cup C \succeq B \cup D$.

$A \succ B$ and $C \succeq D$ implies $A \cup C \succ B \cup D$.

Robustness with respect to infinite set union (RU): For all $A \in \mathcal{P}(X)$ and all sequence of homogeneous and pairwise disjoint sets $B_i \in \mathcal{P}(X)$, $i \in \mathbb{N}$, if $A \succeq \bigcup_{i=1}^{k} B_i$ for all $k \in \mathbb{N}$ then $A \succeq \bigcup_{i=1}^{\infty} B_i$.

Note that ISS and SSM imply the SM axiom. Let $A \in \mathcal{P}(X)$ be a homogeneous set and $x \in X \setminus A$ such that xSA. Consider $a \in A$, thus $A \cup \{x\}$ is homogeneous and ISS implies $A \cup \{x\} \sim \{a\}$ and $A \sim \{a\}$. Thus $A \cup \{x\} \sim A$. If $A \in \mathcal{P}(X)$ is a homogeneous set and $x \in X \setminus A$ such that $\neg xSA$, consider $B = A \cup \{x\}$ and $C = \{x\}$. Thus, SSM implies $A \cup \{x\} \succ A$.

4.3 Characterization Result

In this section we provide a characterization of the preference for diversity ordering.

Theorem 1 $\succeq$ is the D-rule if and only if $\succeq$ satisfies INS, ISS, SSM, SC and RU.

Proof It can be checked easily that the D-rule satisfies INS, ISS, SSM, SC and RU. Conversely, assume that $\succeq$ satisfies these properties. It suffices to prove that, for all $A, B \in \mathcal{P}(X)$, if $D(A) = D(B)$ then $A \sim B$ and if $D(A) > D(B)$ then $A \succ B$.

We first show that if $D(A) = D(B)$ then $A \sim B$. There are two possible cases:

(a) $D(A) = D(B) = k \in \mathbb{N}$,

(b) $D(A) = D(B) = \infty$.

In the first case, $\phi(A) = \{A_1, \ldots A_k\}$ and $\phi(B) = \{B_1, \ldots B_k\}$. Given $a \in A_1$, $b \in B_1$ whatever, we have that INS and ISS imply $A_1 \sim \{a\} \sim \{b\} \sim B_1$. Since A_2 does not mimic A_1 and B_2 does not mimic B_1, by SC, it follows that $A_1 \cup A_2 \sim B_1 \cup B_2$. The repeated use of the above argument gives $\bigcup_{i=1}^{k} A_i \sim \bigcup_{i=1}^{k} B_i$, that is $A \sim B$.

Now consider the second case, $D(A) = D(B) = \infty$, that is, there exist A_i, B_i for all $i \in \mathbb{N}$ and $\phi(A) = \{A_1, A_2 \ldots\}$, $\phi(B) = \{B_1, B_2 \ldots\}$. By using again the above argument $\bigcup_{i=1}^{k} A_i \sim \bigcup_{i=1}^{k} B_i$, for all $k \in \mathbb{N}$, and by axiom SSM it is clear that $A \succ \bigcup_{i=1}^{k} A_i$, for all $k \in \mathbb{N}$. Hence by RU, $A \succeq \bigcup_{i=1}^{\infty} B_i = B$. By an analogous argument $B \succeq \bigcup_{i=1}^{\infty} A_i = A$ and $A \sim B$ is obtained.

We now show that if $D(A) > D(B)$ then $A \succ B$. There are two possibilities: $D(A) \in \mathbb{N}$ or $D(A) = \infty$. Assume that $D(A) = k \in \mathbb{N}$ holds, so $\phi(A) = \{A_1, \ldots A_k\}$ and $\phi(B) = \{B_1, \ldots B_l\}$, with $l < k$. Using the above argument and axiom SSM it is now clear that $A \succ \bigcup_{i=1}^{l} A_i \sim \bigcup_{i=1}^{l} B_i = B$ and $A \succ B$ is obtained.

Finally assume $D(A) = \infty$ and $D(B) = l \in \mathbb{N}$. Therefore there exists A_i for all $i \in \mathbb{N}$ and $\phi(A) = \{A_1, A_2 \ldots\}$, $\phi(B) = \{B_1, \ldots, B_l\}$. The same reasoning shows that $A \succ \bigcup_{i=1}^{l+1} A_i \succ \bigcup_{i=1}^{l} A_i \sim \bigcup_{i=1}^{l} B_i = B$ and hence $A \succ B$. This completes the proof.

5 Other Rules for Ranking Opportunity Sets by Using the Similarity Relation

The idea of using the similarity relation to reduce the size of large opportunity sets without giving up the diversity of their alternatives can be used to apply other known rules for ranking finite opportunity sets in our context.

Moreover, there are circumstances where a ranking of sets can be considered appropriate if it is an extension of R to $\mathcal{P}(X)$, i.e. for all $x, y \in X$, xRy iff $\{x\} \succeq \{y\}$. In this section we explore several orderings that are extensions of R. Note that the preference for diversity ordering studied in the previous section does not satisfy this condition.

For example, the cardinalities and the preference ranking of the worst elements according to R of two sets can be used as the only information required to rank those sets. In that sense we adapt to our context the cardinality-first lexicographic relation and the preference-first lexicographic relation described by Bossert et al. [5]. We now define new relations, namely the *cardinality-first ordering* and the *preference-first ordering*, which for all $A, B \in \mathcal{P}(X)$ give:

$A \succeq_C B$ iff $D(A) > D(B)$ or $D(A) = D(B)$ and $a_1 \geq b_1$,

$A \succeq_P B$ iff $a_1 > b_1$ or $a_1 = b_1$ and $D(A) \geq D(B)$.

5.1 Axioms

We reformulate some axioms used by Bossert et al. [5] in the way required by our framework.

S-monotonicity (SM): For all $A, B, C \in \mathcal{P}(X)$ such that $A, C \subset B$, $A \cap C = \emptyset$, C is homogeneous and does not mimic A and $a_1 = b_1$, then $B \succ A$.

S-dominance (SD): For all $A, B \in \mathcal{P}(X)$ homogeneous or verifying $D(A) = D(B) = \infty$ such that $a_1 > b_1$, then $A \succ B$.

S-indirect indifference principle (SIIP): For all $A, B \in \mathcal{P}(X)$ homogeneous or verifying $D(A) = D(B) = \infty$ such that $a_1 = b_1$, then $A \sim B$.

S-weak independence (SWIND): For all $A, B, C, D \in \mathcal{P}(X)$ such that $A \cap C = B \cap D = \emptyset$, CPA, DPB, with C and D both being homogeneous and do not mimic A and B respectively, then $A \succeq B$ iff $A \cup C \succeq B \cup D$.

S-indirect preference principle (SIPP): For all $A, B \in \mathcal{P}(X)$ such that $A \subset B$, A is homogeneous and $a_1 > b_1$, then $A \succ B \setminus A$.

S-priority of freedom (SPF): For all $A, B, C \in \mathcal{P}(X)$ homogeneous and pairwise disjoint such that $a_1 > \min\{b_1, c_1\}$ then $B \cup C \succ A$.

In the names of the above axiom definitions we use the prefix S to stand for similarity, because they involve the changes required to transform the axioms used by Bossert et al. [5] to our context. The original axioms concern sets of alternatives and we adapt them considering those sets as a union of homogeneous subsets.

5.2 Lemmas

The following lemmas provide a proof for each of the corresponding characterizations theorems. Additionally, these results are of importance for the possible practical application of the relations described.

Lemma 1 *Let $\succeq$ be an ordering on $\mathcal{P}(X)$ satisfying SM, SIIP and SWIND. Let $A, B \in \mathcal{P}(X)$ be such that $D(A) = D(B)$ and $a_1 = b_1$, then $A \sim B$.*

Proof Suppose $A \neq B$ and $D(A) = D(B) = k \in \mathbb{N}$, that is $\phi(A) = \{A_1, \ldots, A_k\}$ and $\phi(B) = \{B_1, \ldots, B_k\}$. As $a_1 = b_1$, then SIIP implies $A_1 \sim B_1$. Repeated application of SWIND gives $\bigcup_{i=1}^{k} A_i \sim \bigcup_{i=1}^{k} B_i$. That is, $A \sim B$.
    In case $D(A) = D(B) = \infty$, SIIP implies $A \sim B$.

Lemma 2 *Let $\succeq$ be an ordering on $\mathcal{P}(X)$ satisfying SM, SIIP and SWIND. Let $A, B \in \mathcal{P}(X)$ be such that $D(A) > D(B)$ and $a_1 = b_1$, then $A \succ B$.*

Proof From $D(A) > D(B)$, $D(B) \in \mathbb{N}$ is obtained, that is $\phi(B) = \{B_1, \ldots, B_l\}$. SM implies $A \succ \bigcup_{i=1}^{l} A_i$ and SIIP together with SWIND assure $\bigcup_{i=1}^{l} A_i \sim \bigcup_{i=1}^{l} B_i$. Thus $A \succ B$.

Lemma 3 *Let $\succeq$ be an ordering on $\mathcal{P}(X)$ satisfying SD and SWIND. Let $A, B \in \mathcal{P}(X)$ be such that $D(A) = D(B)$ and $a_1 > b_1$, then $A \succ B$.*

Proof Suppose $D(A) = D(B) = k \in \mathbb{N}$. As $a_1 > b_1$, then SD implies $A_1 \succ B_1$. If $k = 1$, the proof is complete. If not, a repeated application of SWIND yields $A \succ B$.
    In case $D(A) = D(B) = \infty$, SD implies $A \succeq B$.

Lemma 4 *Let $\succeq$ be an ordering on $\mathcal{P}(X)$ satisfying SD, SM and SWIND. Let $A, B \in \mathcal{P}(X)$ be such that $D(A) > D(B)$ and $a_1 > b_1$, then $A \succ B$.*

Proof Let $l = D(B) \in \mathbb{N}$. By SM and the first part of the proof of Lemma 3, $A \succ \bigcup_{i=1}^{l} A_i \succ \bigcup_{i=1}^{l} B_i = B$.

Lemma 5 *Let $\succeq$ be an ordering on $\mathcal{P}(X)$ satisfying SIPP and SWIND. Let $A, B \in \mathcal{P}(X)$ be such that $D(A) < D(B)$ and $a_1 > b_1$, then $A \succ B$.*

Proof We have $\phi(A) = \{A_1, \ldots, A_k\}$ and there are two possibilities for $\phi(B)$, it could be $D(B) = \infty$ or $D(B) = l \in \mathbb{N}$ with $k < l$. In any case, we consider $C = A_1 \cup B_1 \cup \left(\bigcup_{i \geq k+1} B_i\right)$, and SIPP implies $A_1 \succ C \setminus A_1$. That is, $A_1 \succ B_1 \cup \left(\bigcup_{i \geq k+1} B_i\right)$. Using SWIND repeatedly for $k - 1$ times, gives $A = A_1 \cup A_2 \cup \ldots \cup A_k \succ B_1 \cup \left(\bigcup_{i \geq k+1} B_i\right) \cup B_2 \cup \ldots \cup B_k = B$.

Lemma 6 *Let $\succeq$ be an ordering on $\mathcal{P}(X)$ satisfying SM, SPF and SWIND. Let $A, B \in \mathcal{P}(X)$ be such that $D(A) > D(B)$ and $a_1 < b_1$, then $A \succ B$.*

Proof Suppose $D(A) \in \mathbb{N}$, then $\phi(A) = \{A_1, \ldots, A_k\}$ and $\phi(B) = \{B_1, \ldots, B_l\}$, with $k > l$. SPF yields $A_1 \cup A_2 \succ B_1$, if $k = 2$, then $l = 1$ and we are done. If $k > 2$, consider $r = k - l + 1$. In case $r = 2$, trivially $\bigcup_{i=1}^{r} A_i \sim A_1 \cup A_2$ and, in case $r > 2$, SM implies $\bigcup_{i=1}^{r} A_i \succ A_1 \cup A_2$. Therefore, we conclude that $\bigcup_{i=1}^{r} A_i \succeq A_1 \cup A_2$ and then $\bigcup_{i=1}^{r} A_i \succ B_1$. By repeated application of SWIND, we obtain $A \succ B$.

In case $D(A) = \infty$ and $D(B) = l \in \mathbb{N}$, SM and the previous case guarantee that $A \succ \bigcup_{i=1}^{l+1} A_i \succ B$.

5.3 Characterization Results

We provide characterization results for the ranking rules defined in this section.

Theorem 2 $\succeq$ *is the preference-first ordering,* $\succeq_P$, *if and only if* $\succeq$ *satisfies SM, SIIP, SWIND and SIPP.*

The proof of the necessary part of this theorem is straightforward. The sufficient part follows from Lemmas 1 to 5. (Note that SIIP implies the SD axiom).

Theorem 3 $\succeq$ *is the cardinality-first ordering,* $\succeq_C$, *if and only if* $\succeq$ *satisfies SM, SD, SIIP, SWIND and SPF.*

The proof of the necessary part of this theorem is straightforward. The sufficient part follows from Lemmas 1 to 4 in addition to Lemma 6.

6 Final Remarks

The similarity relation defined is a semiorder on X, which is different from but compatible with the preference relation, in the same way as the similarity relation defined in Pattanaik and Xu [13]. That notion turns out to be helpful in ranking large opportunity sets: grouping the elements of the alternative sets to be compared by taking into account their similarities, and maybe also their preferences, is useful for ordering sets consisting of a large number of alternatives.

The concept of a similarity relation is not new in the literature. Rubinstein [14] assumes that people use similarity relations when evaluate lotteries. Aizpurua et al. [1] interpret these relations as a way of modeling the imperfect powers of discrimination of the human mind and study the relationship between preferences and similarities. Binder [4] provides axiomatic characterizations of two classes of similarity revelation rules.

Other authors address the notion of similarity in different ways. For example, Mesiar [12] introduces a general approach to the fuzzy utility functions based on dissimilarity functions or Hu [7] introduces traditional similarity measures for multi-criteria collaborative filtering approaches and describes the neighborhood method using the proposed preference-relation-based similarity measures.

Coming back to our framework, as a suggestion for further research, we propose adapting the leximax criterion studied by Ballester and De Miguel [3] to our context as another interesting case of an ordering that is an extension of the preference relation on alternatives. First of all, for each $A \in \mathcal{P}(X)$, $\mathcal{F}(A)$ is denoted as the family of all the subsets C of A such that $D(C) \in \mathbb{N}$, i.e. the subsets C of A such that they are reduced to a finite set after the similarity relation is applied to grouping alternatives. Assuming $C \in \mathcal{F}(A)$, gives $M(C) = \{c_1, c_2, \ldots, c_k\}$, with $k = D(C)$, defined in Sect. 4. Following Ballester and De Miguel [3], for all $A, B \in \mathcal{F}(X)$ we consider $A \succeq_L B$ when

there exists $i \leq D(A)$ such that $a_i > b_i$ and $a_j > b_j$ for all $j < i$,

or $\quad D(A) \geq D(B)$ and $a_i = b_i$ for all $i \leq D(B)$.

Finally, we are in a position to extend the leximax criterion to the family of all sets of $\mathcal{P}(X)$. Consider $A, B \in \mathcal{P}(X)$, we define $A \succeq_L^* B$ iff for all $B_f \in \mathcal{F}(B)$ there exist $A_f \in \mathcal{F}(A)$ such that $A_f \succeq_L B_f$.

It could be useful to have an axiomatic characterization of this relation. However, that is beyond the scope of the present work and calls for further and independent research.

References

1. Aizpurua, J.M., Nieto, J., Uriarte, J.R.: Choice procedure consistent with similarity relations. Theory Decis. **29**, 235–254 (1990)
2. Alcantud, J.C.R., Manrique, A.: Continuous representation by a money-metric function. Math. Soc. Sci. **41**, 365–373 (2001)
3. Ballester, M.A., De Miguel, J.R.: Extending an order to the power set: the leximax criterion. Soc. Choice Welf. **21**, 63–71 (2003)
4. Binder, C.: Preference and similarity between alternatives. RMM-Journal **5**, 120–132 (2014)
5. Bossert, W., Pattanaik, P.K., Xu, Y.: Ranking opportunity sets: an axiomatic approach. J. Econ. Theory **63**, 326–345 (1994)
6. Honkapohja, S.: Representing consumer's tastes in terms Mckenzie expenditure functions: a tribute to Aami Nyberg on his 60th birthday. Acta Academiae Oeconomicae Helsingiensis Series A **52**, 31–44 (1987)
7. Hu, Y.C.: Recommendation using neighborhood methods with preference-relation-based similarity. Inf. Sci. **284**, 18–30 (2014)
8. Lauwers, L., Potoms, T., Vázquez, C.: Addendum to: "Ranking opportunity sets on the basis of similarities of preferences: a proposal". Math. Soc. Sci. **88**, 1–2 (2017)
9. Mckenzie, L.: Demand theory without a utility index. Rev. Econ. Stud. **24**, 185–189 (1957)
10. Mehta, G.B.: Metric utility functions. J. Econ. Behav. Organ. **26**, 289–298 (1995)
11. Mehta, G.B.: Preference and utility. In: Barberá, S., Hammond, P., Seidl, C. (eds.) Handbook of Utility Theory, pp. 1–47. Kluwer Academic Publishers, Dordrecht (1998)
12. Mesiar, R.: Fuzzy set approach to the utility, preference relations, and aggregation operators. Eur. J. Oper. Res. **176**, 414–422 (2007)
13. Pattanaik, P.K., Xu, Y.: On diversity and freedom of choice. Math. Soc. Sci. **40**, 123–130 (2000)

14. Rubinstein, A.: Similarity and decision-making under risk (Is there a utility theory resolution to the Allais paradox?). J. Econ. Theory **46**, 145–153 (1988)
15. Samuelson, P.A.: Complementary: an essay on the 40th anniversary of the Hicks-Allen revolution in demand theory. J. Econ. Lit. **12**(4), 1255–1289 (1974)
16. Sen, A.K.: Welfare, preference and freedom. J. Econ. **50**, 15–29 (1991)
17. Vázquez, C.: Ranking opportunity sets on the basis of similarities of preferences: a proposal. Math. Soc. Sci. **67**, 23–26 (2014)
18. Weymark, J.: Money-metric utility functions. Int. Econ. Rev. **28**, 219–232 (1985)

The Interplay Between Intergenerational Justice and Mathematical Utility Theory

José Carlos R. Alcantud and Alfio Giarlotta

Abstract Intergenerational justice has produced nice arguments in relation to the existence and the non-existence of utility representations of social welfare orderings on infinite chains of social states. Rather surprisingly, nearly all results neglect the extensive literature on mathematical utility theory. This chapter is an attempt to draw a bridge between both fields. In fact, we show that techniques and results from the latter discipline may be quite enlightening in the analysis of the intergenerational aggregation of utilities. Conversely, conclusions about social welfare binary relations that are relevant in the intergenerational debate facilitate the investigation of their Richter–Peleg representability. The interplay between mathematical utility theory and intergenerational justice becomes apparent.

Keywords Intergenerational justice · Utility stream · Mathematical utility theory · Lexicographic preference · Pareto ordering

1 Introduction

Economists consistently use models with an infinite horizon in a variety of scenarios: optimal economic growth, the analysis of environmental issues, the demand of renewable resources, or any other problem without a natural termination date. The periods are often referred to as 'generations', although it may well be that they are years or any other measure of time. Suppose each period is characterized by a real number, which represents welfare, endowments, consumption, or generic utility. Whenever

J. C. R. Alcantud (✉)
BORDA Research Unit and Multidisciplinary Institute of Enterprise (IME),
University of Salamanca, Salamanca, Spain
e-mail: jcr@usal.es
URL: http://diarium.usal.es/jcr

A. Giarlotta
Department of Economics and Business, University of Catania, Catania, Italy
e-mail: giarlott@unict.it

© Springer Nature Switzerland AG 2020   325
G. Bosi et al. (eds.), *Mathematical Topics on Representations of Ordered Structures and Utility Theory*, Studies in Systems, Decision and Control 263,
https://doi.org/10.1007/978-3-030-34226-5_18

a policy-maker has the need to order such programs (or infinite-horizon sequences of numbers), she must be enabled to evaluate the (possibly infinite) trade-offs of the available alternatives. Therefore, any acceptable ordering must satisfy some requirements of consistency. These restrictions may either be of an ethical nature, or require economic efficiency, or impose technical restrictions like continuity.

Preferences and utilities are the traditional tools to achieve a prioritization among social welfare programs. Thus, the problem of intergenerational aggregation of welfare by real-valued functions is naturally related to mathematical utility theory. Put shortly, this type of approach investigates the representation of infinite-horizon sequences of numbers by means of utilities that satisfy certain postulates.

Not surprisingly, modern approaches to intergenerational aggregation of welfare acknowledge the influence of seminal investigations on mathematical utility theory. However, the explicit application of the theoretical knowledge about mathematical utility theory into the debate on intergenerational justice is very limited. At the risk of missing some cases, below we provide a list of the best suited examples of this kind that we are aware of.

In 1960, Koopmans [65] claims that insufficient attention has been given to the idea of a preference for advancing the timing of future satisfaction, and he advocates a definition in terms of an ordinal utility function. Koopmans [65] cites Debreu's and Wold's papers, however making sure that their techniques are *not* applicable to his assumptions in preparation of his main result (see [65, Footnote 5]). In 1965, Diamond [40] gives sufficient conditions for the existence of continuous utility functions representing preferences over an infinite future; in his analysis, he uses the product topology as well as the supremum topology. Diamond [40, Sect. 3] (see also Lauwers [68, Lemma 2]) uses Debreu's Lemma [39] to prove his 'existence theorem'.[1]

Other papers on intergenerational aggregation of utilities use arguments that closely resemble those of mathematical utility theory. Asheim, Mitra, and Tungodden [13] guarantee the existence of a utility function for those complete preorders on a set—included the set $[0, 1]^{\mathbb{N}}$—that satisfy some mild efficiency properties plus a restricted version of continuity (which is weaker than ordinary supnorm continuity). Their line of thought (see Lemma 1 and the proof of Proposition 2 in [13]) seems to be inspired by Wold's [92] approach.[2]

Mitra and Ozbek [78] begin their work on the representation of monotone complete preorders on the space of sequences $[0, 1]^{\mathbb{N}}$, with an account of techniques to characterize the representability of a preference ordering. Their first paragraph acknowledges the early work on this topic by Wold.[3] Mitra and Ozbek derive a number of theorems in the literature—inclusive of Wold's and Diamond's—from their main results (cf. [78, Sect. 5]). They divide their analysis into two parts. In Sect. 3,

---

[1]Beardon and Mehta [23] and Herden and Mehta [60, 61] give important insights about Debreu's lemma.

[2]Section 1 of Beardon and Mehta [23] gives a critical assessment of the role of this approach in the development of mathematical utility theory, as well as a clear presentation of its foundations.

[3]Mitra and Ozbek [78] also recommend Beardon and Mehta [23] for a comprehensive discussion of Wold's contribution.

Mitra and Ozbek are concerned with weak representations (i.e., with the property "if x is weakly preferred to y, then the utility of x is at least as high as the utility of y", like, for example, Jaffray [63]).[4] Then, in Sect. 4, they give a sufficient condition for the existence of utility ("if and only if") representations.

Finally, let us mention Lauwers [69, Proposition 1] and Sakai [83, Theorem 1], who also study utility representations of complete preorders on l_∞, the set of all bounded sequences of real numbers.

The references given above are just a short sample of remarkable papers in the technical analysis of intergenerational justice. There are many additional contributions that use a utility approach to intergenerational justice, such as [2, 3, 7, 19, 38, 42, 83, 84]. In particular, some specific forms of the utility representations have been characterized. For example, Asheim, Mitra and Tungodden [13], Koopmans [66] and Lauwers [68] characterize the discounted utilitarian rule, and Lauwers [68] axiomatizes the Rawlsian infimum rule.

The above discussion may give the overall impression that aggregation of infinite sequences by utilities is the prevailing trend. This is by no means true. A larger literature has been produced to analyze social welfare relations, that is, binary relations that respect a collection of postulates and can be used to compare infinite programs. Svensson [87] gives a positive result on the existence of continuous and complete preorders that satisfy equity and a strong efficiency postulate; however, as Lauwers [71] and Zame [93] emphasize, his conclusion cannot be used for policy-making, because it is of a non-constructive nature. Asheim and Tungodden [14] characterize two versions of the Rawlsian leximin principle, and two versions of the utilitarian overtaking criterion (see also Atsumi [16] and von Weizsäcker [91]).

Furthermore, several contributions are concerned with ethical properties other than the standard 'equal treatment of all generations' axiom. 'Hammond equity for the future' is a property that is peculiar to this setting: this is a variation of the classical 'Hammond equity axiom', introduced by Asheim and Tungodden [15] (see also Asheim, Mitra and Tungodden [12] and Banerjee [17]). Non-interference properties are the subject of many studies, too [4, 73–75]. Principles of distributive justice include the 'Pigou–Dalton transfer principle' [3, 41, 59, 83, 84], 'Altruistic Equity' [3, 84], and the 'Strong Equity principle' [28] among many others.

This chapter aims at providing a fresh perspective of intergenerational aggregation of utilities, insofar as it revisits some known results in the light of the terminology and achievements of mathematical utility theory. We hope that our conclusions may be useful to both researchers in utility theory and intergenerational justice. The formers will find new applications of their expertise to an important field of research in economics, in particular because we reveal the importance of 'Richter–Peleg representations' in intergenerational justice. The latters will find a concise background for the driving arguments used to represent infinite programs by numerical utilities.

---

[4]Weak utilities for strict preferences are the subject of Alcantud [1] and Peleg [80], among many others. These weak representations have the property "if x is strictly preferred to y, then the utility of x is strictly higher than the utility of y".

This chapter is organized as follows. Section 2 recalls basic terminology and definitions. Section 3 contains a revision of some known possibility and impossibility results in intergenerational justice, which are respectively analyzed in Sects. 3.1 and 3.2. Section 4 investigates the existence of Richter–Peleg representations for suitable social welfare relations, which are relevant in intergenerational justice. Section 5 concludes the paper by suggesting future directions of research.

2 Preliminaries

In this section, first we introduce basic terminology for preferences, utilities, and intergenerational aggregation of utilities (cf. Bridges and Mehta [27]). Then, we dwell on lexicographic preferences and their relationship with non-representability. Finally, we recall the main notions that are needed to analyze the problem of aggregating infinite utility streams.

2.1 Preference Relations and Utility Representations

Hereafter, X is a nonempty (possibly infinite) set of alternatives (courses of actions, etc.). Further, R is a reflexive relation on X, where $x R y$ means that x is 'at least as good as' y (or x is 'weakly preferred to' y). The asymmetric part of R is denoted by $R^>$, and is defined by $x R^> y$ if $x R y$ and not $y R x$, to be read as x is 'strictly preferred to' y. The symmetric part of R is denoted by $R^=$, and is defined by $x R^= y$ if $x R y$ and $y R x$, to be read as x is 'indifferent to' y. Finally, the symmetric part of the complement of R is denoted by $R^\perp$, and is defined by $x R^\perp y$ if $\neg(x R y)$ and $\neg(y R x)$, to be read as x is 'incomparable to' y. Accordingly, we call R *weak preference*, $R^>$ *strict preference*, $R^=$ *indifference*, and $R^\perp$ *incomparability*. Notice that R is the (disjoint) union of $R^>$ and $R^=$; furthermore, R and $R^\perp$ are disjoint.

For each element $x \in X$, the *lower section* of x is the nonempty set $x^\downarrow = \{y \in X : x R y\}$, and the *upper section* of x is the set $x^\uparrow = \{y \in X : y R x\}$. A subset D of X is *downward closed* when $x^\downarrow \subseteq D$ for each $x \in D$.[5]

A *preorder* is a reflexive and transitive binary relation R on X; in this case the pair (X, R) is a *preordered set*. A *total (or complete) preorder* is a connected preorder R, that is, $x R y$ or $y R x$ holds for all distinct $x, y \in X$ (hence $R^\perp$ is empty). A *linear order* is an antisymmetric total preorder R; in this case the pair (X, R) is a *chain*. A *subchain* of a chain (X, R) is a pair $(Y, R \restriction_Y)$, where Y is a subset of X, and $R \restriction_Y$ is the restriction of the linear order R to Y.

Morphisms (general and special) in the category of preordered sets are defined as follows:

---

[5] A downward closed set is sometimes called an *ideal*.

Definition 1 Let (X, R) and (Y, S) be two preordered sets. An *order-homomorphism* (for short, *homomorphism*) is a mapping $f: (X, R) \to (Y, S)$ such that, for all $x, x' \in X$,

$$x R x' \implies f(x) S f(x') . \tag{1}$$

In particular, an *order-embedding* (for short, *embedding* or *monomorphism*) of (X, R) into (Y, S) is an injective homomorphism $f: (X, R) \to (Y, S)$ such that, for all $x, x' \in X$,

$$x R x' \iff f(x) S f(x') . \tag{2}$$

Finally, an *order-isomorphism* (for short, *isomorphism*) is an onto embedding $f: (X, R) \to (Y, S)$: in this case, we say that (X, R) and (Y, S) are *isomorphic*, and denote this fact by $(X, R) \cong (Y, S)$.

The next two definitions recall the main notions of representability for preordered sets.

Definition 2 Let (X, R) be a preordered set. A *Richter–Peleg representation* of (X, R) is a mapping $w: (X, R) \to (\mathbb{R}, \geqslant)$ such that, for all $x, y \in X$,

$$x R y \implies w(x) \geqslant w(y) \quad \text{and} \quad x R^> y \implies w(x) > w(y) . \tag{3}$$

In this case, we also say that the preorder R is *Richter–Peleg representable*.

Thus, a Richter–Peleg representation of a preordered set (X, R) is a special homomorphism from (X, R) to the chain $(\mathbb{R}, \geqslant)$, which preserves not only the weak preference but also the strict one (see Minguzzi [77] and Alcantud et al. [5]). However, a Richter–Peleg representation does not convey full information about the preordered set, insofar as the ordering cannot be retrieved from the mapping. To perfectly reflect the primitive ordering by means of the chain of real numbers, we need the following stronger form of representation:

Definition 3 Let (X, R) be a preordered set. A *utility representation* of (X, R) is a mapping $u: (X, R) \to (\mathbb{R}, \geqslant)$ such that, for all $x, y \in X$,

$$x R y \iff u(x) \geqslant u(y) . \tag{4}$$

We call u a *utility representation* of (X, R). If such a representation exists, then we say that (X, R) is *representable* (equivalently, R is *representable*); otherwise, it is *non-representable*.

Notice that a utility representation u of a preordered set (X, R) need not be injective; however, if (X, R) is a chain, then u is an embedding. Observe also that if a preorder has a utility representation, then it also has a Richer-Peleg representation, but the converse does not hold, in general. It is well-known that, however, the two representations are equivalent for *total* preorders:

Lemma 1 *A total preorder is Richter–Peleg representable if and only if it is representable.*

In order to deal with the non-representability of many preference relations that arise in practice, weaker notions of representability have been introduced over time. A *multi-utility representation* of a preordered set (X, R) is a family $\mathbf{V}$ of real-valued homomorphisms, such that for all $x, y \in X$, $x R y$ is equivalent to the joint satisfaction of $v(x) \geqslant v(y)$ for each $v \in \mathbf{V}$. When $\mathbf{V}$ is a countable (resp. finite) family of homomorphisms, we say that $\mathbf{V}$ is a *countable* (resp. *finite*) multi-utility representation of (X, R). If we fix a topology τ on X and $\mathbf{V}$ is a family of upper semicontinuous (resp. lower semicontinuous) homomorphisms, then $\mathbf{V}$ is an *upper semicontinuous* (resp. *lower semicontinuous*) multi-utility representation of (X, R). The early development of multi-utilities owes to Levin [72], Evren and Ok [45] and Bosi and Herden [25] among others. Subsequent achievements appeared in, e.g., Bosi and Herden [26] and Pivato [81].[6]

A *Richter–Peleg multi-utility representation* of a preordered set (X, R) is a multi-utility representation $\mathbf{V}$ of (X, R), such that every $v \in \mathbf{V}$ is a Richter–Peleg representation of (X, R). It also characterizes the strict part $R^{>}$ of R, in the sense that for all $x, y \in X$, $x R^{>} y$ if and only if $v(x) > v(y)$ for each $v \in \mathbf{V}$. Alcantud, Bosi and Zuanon [5, Theorem 3.1] prove that for any preordered set (X, R), the existence of Richter–Peleg multi-utility representations is equivalent to the existence of Richter–Peleg representations of the preordered set. In addition, they show that every preorder on a countable set has a countable Richter–Peleg multi-utility representation, and the existence of a countable multi-utility representation of a preordered set implies the existence of a Richter–Peleg representation of it.

2.2 Lexicographic Preferences and Non-representability

In this section we deal with special types of preference orderings, called 'lexicographic'. As we shall see, this is one of the possible causes for which a preference relation fails to be representable by a utility function. The reader may consult the (dated but always valuable) survey by Fishburn [46] for a general overview on lexicographic preferences. See also Giarlotta [47] for a broad survey on some recent trends in preference, utility, and choice (which deals with lexicographic preferences in Sect. 3.1).

---

[6]A special type of multi-utility representation is called 'modal', and enables one to describe two interconnected preference relations with a different 'modality'. Formally, a *modal representation* of a pair (R, S) of binary relation on the same set X is a family $\mathbf{M} = \left\{ u_h^k : h \in H \text{ and } k \in K_h \right\}$ of mappings $h_h^k : X \to \mathbb{R}$ indexed over the set $\bigcup_{h \in H} \{(h, k) : k \in K_h\}$ such that, for all $x, y \in X$, $x R y$ if and only if $u_h^k(x) \geq u_h^k(y)$ for all *modes* $h \in H$ and *extents* $k \in K_h$, whereas $x S y$ if and only if there is a mode $h \in H$ such that $u_h^k(x) \geq u_h^k(y)$ for all extents $k \in K_h$. Pairs of preferences that admit a modal representation are characterized by suitable properties: see Theorem 4.3 in Giarlotta and Greco [51], and the successive contributions on *bi-preferences* and *NaP-preferences* [8, 33, 49, 50, 55–58].

To start, we describe the lexicographically ordered set $\mathbb{R}^2_{\text{lex}} = (\mathbb{R}^2, \succcurlyeq_{\text{lex}})$, as well as lexicographic orderings, in general. The chain $\mathbb{R}^2_{\text{lex}}$ has a special place in the history of mathematical economics: in fact, it is the preference relation that was used in 1954 by Debreu [39] to disprove the inveterate belief of economists that preferences and utilities may be treated as synonyms.

Definition 4 The *lexicographic plane* is the chain $(\mathbb{R}^2, \succcurlyeq_{\text{lex}})$, where $\succcurlyeq_{\text{lex}}$ is the linear order defined as follows for each $(x, y), (x', y') \in \mathbb{R}^2$: $(x, y) \succcurlyeq_{\text{lex}} (x', y')$ if either $x > x'$, or $x = x'$ and $y \geqslant y'$. For simplicity, we denote the lexicographic plane by $\mathbb{R}^2_{\text{lex}}$. More generally, if $(X, \succcurlyeq_X)$ and $(Y, \succcurlyeq_Y)$ are two chains, then their *lexicographic product* $X \times_{\text{lex}} Y = (X \times Y, \succcurlyeq_{\text{lex}})$ is defined by $(x, y) \succcurlyeq_{\text{lex}} (x', y')$ if either $x \succ_X x'$, or $x = x'$ and $y \succcurlyeq_Y y'$. The definition of $\mathbb{R}^n_{\text{lex}}$ is similar for any $n \geq 3$. Even more generally, if $\mathcal{C} = \{(X_i, \succcurlyeq_i) : i \in I\}$ is any family of chains indexed over a well-ordered set $(I, \trianglelefteq)$, the *lexicographic product of* $\mathcal{C}$ is a chain $\prod^{\text{lex}}_{i \in I} X_i = (\prod_{i \in I} X_i, \succcurlyeq_{\text{lex}})$, whose strict linear order $\succ_{\text{lex}}$ is defined by $(x_i)_{i \in I} \succ_{\text{lex}} (y_i)_{i \in I}$ if there is a least $j \in (I, \trianglelefteq)$ such that $x_j \succ_j y_j$ and $x_i = y_i$ for all $i \trianglelefteq j$.

Due to an imperfect communication in the scientific community, until the early 1950s economists were considering all preference relations as representable in $\mathbb{R}$. In other words, the concepts of 'preference' and 'utility' were (wrongly) considered equivalent. Even Hicks [62] (p. 19) was still claiming in 1956 that

> If a set of items is strongly ordered, it is such that each item has a place of its own in the order; it could, in principle, be given a number.

For an overview of the topic, the reader is referred to Bridges and Mehta [27] and Mehta [76].

Beardon et al. [21, 22] systematically analyze the structure of complete transitive preferences that fail to be representable. In [21], they obtain a striking subordering classification of these preferences, which can be suggestively rephrased as follows:

Theorem 1 ([21, 22]) *A chain is non-representable in $\mathbb{R}$ if and only if it is* (i) *long, or* (ii) *planar, or* (iii) *wild.*

The next definition explains the terminology used in Theorem 1.

Definition 5 A totally preordered set is said to be:

– *long* if it contains a copy of the first uncountable ordinal[7] ω_1 or its reverse ordering $\omega_1{}^*$ (otherwise, it is called *short*);

---

[7]An *ordinal* is a well-ordered set $(X, <)$ such that each $x \in X$ is equal to its initial segment $\{y \in X : y < x\}$. The finite ordinals are the natural numbers. The first infinite ordinal is the set $\omega = \omega_0$ of all natural numbers, endowed with the usual order. The first uncountable ordinal is the set ω_1 of all countable ordinals, endowed with the natural order. The famous *continuum hypothesis*, formulated by George Cantor in 1878, says that the cardinality of $\mathbb{R}$ is equal to $\aleph_1$ (the cardinality of ω_1). In 1963, Paul Cohen [36, 37] proved that the continuum hypothesis is independent from the axioms of ZFC (Zermelo–Fraenkel axiomatic set theory, plus the Axiom of Choice), in sense that there are models in which it is true, and models in which it is false (because $|\mathbb{R}| = 2^{\aleph_0} > \aleph_1$ holds). See the classical textbook by Kunen [67] for ZFC axiomatic set theory.

– *planar* if it contains a copy of a non-representable subchain of $\mathbb{R}^2_{\text{lex}}$;
– *wild* if it contains a copy of an *Aronszajn line*, which is an uncountable chain such that neither ω_1 nor $\omega_1{}^*$ nor an uncountable subchain of $\mathbb{R}$ embeds into it.[8]

Several types of preference relations that are typically considered in the economic literature cannot be represented using the set of real numbers (with the usual ordering) as codomain of utility functions. A first example of this kind is the infinite-dimensional commodity space $L^\infty(\mu)$ of essentially bounded measurable functions on a measure space; in fact, in most models used in general equilibrium theory [24], this linear ordering is too large to be embedded in $\mathbb{R}$. Another example of a linear preference that is not embeddable in $\mathbb{R}$ for cardinal reasons is pertinent to the topic of this paper, that is, the space $(\mathbb{R}^n)^{\mathbb{R}}$ of all functions from $\mathbb{R}$ to the commodity space $\mathbb{R}^n$, which is used in the evaluation of infinite utility streams [40].

The non-representability of many classical preference relations has motivated the search for other types of chains, different from the real numbers, which can be used as a mirror for representing preference (cf. Mehta [76]). In 1961, Chipman [35] was already arguing that utility is essentially a vector and not a real number, and advocated the use of lexicographic products as codomain of utility functions (in particular the chain $\{0, 1\}^\alpha_{\text{lex}}$, where α is any ordinal number). Many other contributions in the same direction followed: for instance, Wakker [89] considers the lexicographic product $\mathbb{R} \times_{\text{lex}} \{0, 1\}$, Knoblauch [64] examines preference representations using the lexicographic power $\mathbb{R}^n_{\text{lex}}$, Campión et al. [29] use the 'long line' (that is, the lexicographic product $\omega_1 \times_{\text{lex}} [0, 1)$ deprived of its initial point, where ω_1 is the first uncountable ordinal).[9]

Additional contributions on the topic introduce suitable notions related to the representability of preferences in a topological setting, often using lexicographic orderings as a codomain. For instance, Herden and Mehta [61], in the process of analyzing the famous Debreu's *Open Gap Lemma* [39],[10] introduce the notion of a *Debreu chain*: this is a linear ordering $(X, \succcurlyeq)$ such that each subchain Y of X *continuously* order-embeds into X, where X and Y are endowed with the 'order topology' $\tau_\succcurlyeq$ (i.e., the topology having the open intervals as a basis) and its restriction to Y, respectively. In other words, a Debreu chain $(X, \succcurlyeq)$ has the property that if there is a utility function with values in it, then there is another utility function that is also continuous. The ordered metric space $(\mathbb{R}, \geqslant, d)$ is the prototype of a Debreu chain. The long line—as well as some generalized forms of it, see Definition 31 in [32]—, endowed with the order topology, is a Debreu chain, too.

---

[8]The term 'wild' is not the one originally used by the authors. Furthermore, Beardon et al. [21] separately consider the non-representability of *Souslin lines*, which are nonempty linear orderings with no minimum or maximum, densely ordered, complete, satisfying the countable chain condition (that is, every pairwise disjoint family of open intervals is countable), and yet not isomorphic to $\mathbb{R}$ (equivalently, non-separable). It is well-known that the existence of Souslin lines is independent of the axioms of ZFC, whereas the existence of Aronszajn lines is guaranteed under no additional assumptions: see Kunen [67].

[9]The importance of the long line in economic theory is widely acknowledged [44, 79].

[10]For a recent generalization of the Debreu Gap Lemma, see Estevan [43].

Caserta et al. [32] consider two weaker versions of the Debreu property. Specifically, they define: (i) a *locally Debreu chain*, that is, a linear ordering $(X, \succcurlyeq)$ such that for each $Y \subseteq X$ and $y \in Y$, there is an order-embedding $f_y \colon (Y, \tau_{\succcurlyeq}) \to (X, \tau_{\succcurlyeq})$, which is continuous at y; and (ii) a *pointwise Debreu chain*, that is, a linear ordering $(X, \succcurlyeq)$ such that for each $Y \subseteq X$ and $y \in Y$, there is an order-homomorphism $f_y \colon (Y, \tau_{\succcurlyeq}) \to (X, \tau_{\succcurlyeq})$, which is continuous at y and 'injective at y' (i.e., the preimage $f^{-1}(y)$ of y is a singleton). Obviously, any Debreu chain is locally Debreu, and any locally Debreu chain is pointwise Debreu; however, the converse is false in both cases. For instance, the Cantor set (which can be identified with the lexicographic power $\{0, 1\}_{\mathrm{lex}}^{\omega}$, endowed with the order topology) and the set of irrational numbers (which can be identified with the lexicographic power $\mathbb{Z}_{\mathrm{lex}}^{\omega}$, endowed with the order topology) are locally Debreu but not Debreu: see Examples 23 and 36 in [32].[11] Furthermore, the lexicographic plane $\mathbb{R}_{\mathrm{lex}}^2$, endowed with the order topology, is pointwise Debreu but not locally Debreu: see Example 26 in [32].[12]

To extend the range of 'suitably representable' preferences (in a wider sense), Giarlotta [48] introduces the notion of the *representability number* $\mathrm{repr}_{\mathbb{R}}(X)$ of an arbitrary chain $(X, \succcurlyeq)$: this is the least ordinal α such that $(X, \succcurlyeq)$ order-embeds into the lexicographic power $\mathbb{R}_{\mathrm{lex}}^{\alpha}$. He shows, for instance, that $\mathrm{repr}_{\mathbb{R}}(\mathbb{R}_{\mathrm{lex}}^{\alpha}) = \alpha$ for any nonzero ordinal α, $\mathrm{repr}_{\mathbb{R}}(\kappa) = \kappa$ for any regular[13] uncountable cardinal κ, and $\mathrm{repr}_{\mathbb{R}}(A) = \omega_1$ for any Aronszajn line or Souslin line A. On a similar topic, Giarlotta and Watson [53, 54] analyze forms of representability of preferences that use suitable lexicographic structures.

2.3 Axioms for Intergenerational Justice

In our analysis of the intergenerational aggregation of utilities, $X \subseteq \mathbb{R}^{\mathbb{N}}$ represents a domain of utility sequences, that is, infinitely long utility streams. As usual, we denote an arbitrary element of X by $x = (x_n)_{n \in \mathbb{N}}$. By $\langle y \rangle = (y)_{n \in \mathbb{N}}$ we mean the constant sequence $(y, y, \dots)$. Further, by a slight abuse of notation, we denote by $(x, \langle y \rangle)$ and $(x_0, \dots, x_k, \langle y \rangle)$ the two eventually constant sequences $(x, y, y, \dots)$ and $(x_0, \dots, x_k, y, y, \dots)$, respectively. For the sake of clarity, the classical notions of 'coordinate-wise dominance' (or 'Pareto dominance') will be emphasized with a subscript: thus, we define, for any $x, y \in X$,

---

[11] Here ω denotes the first infinite ordinal, that is, $\mathbb{N}$ endowed with the usual linear order.

[12] More generally, the following fact holds: *For any chain X with at least two points, and any limit ordinal α, (i) if X is short and α is countable, then $X_{\mathrm{lex}}^{\alpha}$ is pointwise Debreu, and (ii) if α is uncountable, then $X_{\mathrm{lex}}^{\alpha}$ is not pointwise Debreu* [32, 52].

[13] A cardinal κ is *regular* if $\mathrm{cf}(\kappa) = \kappa$, where $\mathrm{cf}(\kappa)$ is the *cofinality* of κ, defined as the least ordinal (in fact, cardinal) that 'maps cofinally' into κ (i.e., there is an order-homomorphism $f \colon \mathrm{cf}(\kappa) \to \kappa$ with the property that, for each $\xi < \kappa$, there exists $\gamma < \mathrm{cf}(\kappa)$ such that $\xi \leqslant f(\gamma)$).

$$x \geqslant_{\text{Par}} y \quad \overset{\text{def}}{\Longleftrightarrow} \quad x_n \geqslant y_n \text{ for all } n \in \mathbb{N} \qquad (\text{weak Pareto preference}),$$

$$x >_{\text{Par}} y \quad \overset{\text{def}}{\Longleftrightarrow} \quad x \geqslant_{\text{Par}} y \text{ and } x \neq y \qquad (\text{strict Pareto preference}),$$

$$x \gg_{\text{Par}} y \quad \overset{\text{def}}{\Longleftrightarrow} \quad x_n > y_n \text{ for all } n \in \mathbb{N} \qquad (\text{strong Pareto preference}).$$

Whenever clear from context, we shall however drop the subscript, and simply write $x \geqslant y, x > y$, or $x \gg y$.

A *social welfare function on* X is a mapping $W: X \to \mathbb{R}$, and a *social welfare relation on* X is a binary relation on X. The main aim of the literature on the topic is to determine whether there exist social welfare functions (or, more frequently, relations) on X that satisfy some 'natural' properties. Typically, the axioms that may hold for a social welfare function are collected into two main categories: 'equity' properties and 'efficiency' properties.

Equity axioms are, in turn, of two kinds: 'procedural' and 'consequentialist'. Anonymity is the most used property of procedural equity, requiring social indifference to hold whenever two streams of utilities are obtained from each other by a finite number of local switches. On the other hand, consequentialist equity axioms are designed to implement preference for egalitarian allocations of utilities among generations. In what follows, we describe a few of the plethora of properties examined in the literature.

Let us emphasize from the outset that all axioms will be stated for social welfare *functions* W on X. The conclusion of all these properties is always either "$W(x) = W(y)$" or "$W(x) \geqslant W(y)$" or "$W(x) > W(y)$", that is, how the social evaluation of two infinite streams x, y of utilities compare to each other. These axioms can also be formulated for social welfare *relations* R on X: simply replace "$W(x) \geqslant W(y)$" by "$x R y$", "$W(x) = W(y)$" by "$x R^= y$", and "$W(x) > W(y)$" by "$x R^> y$".

The first equity axiom encodes a principle of 'equal treatment of all generations': it states that any change in the order of assignment of utilities to generations does not matter, as long as it only affects a finite number of generations.[14]

◊ **Axiom An** (Anonymity): If x is a finite permutation of y, then $W(x) = W(y)$.

The second equity principle compares two streams of utilities that *only* display a conflict between two generations (every other generation receiving the same utility in both streams): in this case, the stream such that the least favoured generation is better off must be weakly preferred to the other one.

◊ **Axiom HE** (Hammond Equity): If $y_i > x_i > x_j > y_j$ for some $i, j \in \mathbb{N}$, and $x_n = y_n$ for all $n \neq i, j$, then $W(x) \geqslant W(y)$.

The next axiom is an inequality aversion principle having a cardinal spirit[15]: it states that a non-costly transfer of utility from a generation to a poorer one must strictly increase intergenerational welfare, as long as it is not so large as to reverse their relative ranking.

---

[14] An axiom of **Strong Anonymity** reinforces **Anonymity** by dropping the finiteness condition, that is, any (finite or infinite) permutation of utilities does not alter the global evaluation of a stream. However, we shall not analyze this stronger property here.

[15] See Bossert et al. [28] (where it is called *strict transfer principle*), Hara et al. [59], and Sakai [82].

◊ **Axiom PDT** (Pigou–Dalton Transfer Principle): If there is $\varepsilon > 0$ such that $y_i - \varepsilon = x_i \geqslant x_j = y_j + \varepsilon$ for some $i, j \in \mathbb{N}$, and $x_n = y_n$ for $n \neq i, j$, then $W(x) > W(y)$.

Another equity principle with a cardinal spirit is inspired by a philosophy underlying many tax systems (cf. Adam Smith's [86] 'four maxims of taxation'): it requests that reductions in welfare for the rich that are paired up with increases in welfare for the poor must be socially strictly preferred whenever the gain of the poor is greater than the loss of the rich.[16]

◊ **Axiom AE** (Altruistic Equity): If there are $\varepsilon > \delta > 0$ such that $y_i - \delta = x_i \geqslant x_j = y_j + \varepsilon$ for some $i, j \in \mathbb{N}$, and $x_n = y_n$ for $n \neq i, j$, then $W(x) > W(y)$.

Now we turn to efficiency properties. These are axioms encoding the broad principle that suitable modifications of an infinite utility stream should raise social welfare when these changes make every generation at least as well off. The least controversial formulation of this principle is the following property:

◊ **Axiom Mo** (Monotonicity): If $x \geqslant_{\text{Par}} y$, then $W(x) \geqslant W(y)$.

Three additional axioms, which are however less mandatory, are often imposed, too. The first one captures the idea that improving the welfare of *exactly one* generation raises social welfare. The second one requires that an increase of the utility of *all* generations yields an increase of social welfare. The last one imposes that if *at least one* generation increases its utility, then social welfare raises. Notice that the conclusion of these three axioms is always the same, namely a *strict* increase of social welfare.

◊ **Axiom WD** (Weak Dominance): If $x >_{\text{Par}} y$, with $x_i > y_i$ for some $i \in \mathbb{N}$, and $x_n = y_n$ for $n \neq i$, then $W(x) > W(y)$.

◊ **Axiom WP** (Weak Pareto): If $x \gg_{\text{Par}} y$, then $W(x) > W(y)$.

◊ **Axiom SP** (Strong Pareto): If $x >_{\text{Par}} y$, then $W(x) > W(y)$.

Observe that Mo relaxes SP, but is related to neither WD nor WP. Further, SP implies both WD and WP, but the latter two axioms are independent of each other. Notice also that SP is equivalent to the joint satisfaction of Mo and WD, but the conjunction of Mo and WP is weaker than SP. Finally, observe that all social welfare functions that satisfy PDT together with either Mo or WD must also satisfy AE (cf. Alcantud [3]).

In the next section we shall prove some well-known possibility and impossibility results for social welfare function by using social welfare relations. Conversely, one has the following trivial implications:

Lemma 2 *Let R be a preorder on $X \subseteq \mathbb{R}^{\mathbb{N}}$, and $W : (X, R) \to (\mathbb{R}, \geqslant)$ a social welfare function.*

(i) *If W is a homomorphism, and R satisfies An (resp., Mo, HE), then W satisfies An (resp., Mo, HE).*

(ii) *If W is a Richter–Peleg representation of (X, R), and R satisfies WD (resp., SP, HE, WP, PDT, AE), then W satisfies the same properties.*

---

[16]See Alcantud [3] and Sakamoto [84].

3   Intergenerational Utility via Mathematical Utility Theory

This section dwells on two categories of problems related to social welfare functions satisfying suitable axioms: (1) their existence, and (2) their non-existence. We have selected some relevant results on social welfare theory, which we shall prove using results from mathematical utility theory. In the case of impossibility theorems, we shall first identify the source of non-representability (using the classification given in Beardon et al. [21]), thus casting light on these negative results.

3.1   Possibility Results

In this section we give an alternative proof of a result by Basu and Mitra [20] (Theorem 5 p. 81), which states the existence of social welfare functions on $A^{\mathbb{N}}$, with $A \subseteq \mathbb{R}$, satisfying **Weak Dominance** and **Anonymity**. Our proof relies on the following characterization of the Richter–Peleg representability of preorders:

Lemma 3 (Alcantud and Rodríguez-Palmero[11]) *The following statements are equivalent for any preordered set (X, R):*

 (i) *(X, R) has a Richter–Peleg representation;*
 (ii) *there is a countable family $\mathbb{D} = \{D_n\}_{n \in \mathbb{N}}$ of downward closed subsets of X such that, for all $x, y \in X$, $x R^> y$ implies $y \in D_i$ and $x \notin D_i$ for some $i \in \mathbb{N}$.*

The argument we use to derive a social welfare function from Lemma 3 consists of two parts. In the first step, we prove that any Richter–Peleg representation of a suitable preorder is a social welfare function satisfying **WD** and **An**. In the second step, we use an argument similar to that employed by Basu and Mitra [20] to prove that such a preorder satisfies condition (ii) in Lemma 3.

Theorem 2 (Basu and Mitra [20]) *For each $A \subseteq \mathbb{R}$, there is a social welfare function on $A^{\mathbb{N}}$ satisfying **Anonymity** and **Weak Dominance** (but **Monotonicity** fails for such a function).*

Proof Fix $A \subseteq \mathbb{R}$, and let $X := A^{\mathbb{N}}$. Define a binary relation ME on X as follows for all $x, y \in X$:

$$x \, ME \, y \quad \overset{\text{def}}{\iff} \quad x \text{ and } y \text{ are eventually equal and } \sum_{n=0}^{\infty}(x_n - y_n) \geqslant 0. \quad (5)$$

(The reason why this binary relation is denoted by ME—which stands for 'Mean' and 'Equality'—will become clear in Sect. 4.)

CLAIM 1: *ME is a preorder on X.*

Reflexivity is obvious, so we prove transitivity. Let $x, y, z \in X$ be such that $x\,ME\,y$ and $y\,ME\,z$. Thus, x and y are eventually equal, and y and z are eventually equal. It follows that x and z are eventually equal. Further, by hypothesis we have $\sum_{n=0}^{\infty}(x_n - y_n) \geq 0$ and $\sum_{n=0}^{\infty}(y_n - z_n) \geq 0$, where the addends in both summations are eventually 0. It follows that $\sum_{n=0}^{\infty}(x_n - z_n) = \sum_{n=0}^{\infty}(x_n - y_n + y_n - z_n) = \sum_{n=0}^{\infty}(x_n - y_n) + \sum_{n=0}^{\infty}(y_n - z_n) \geq 0$. This proves that $x\,ME\,z$, as claimed.

CLAIM 2: Any Richter–Peleg representation of (X, ME) satisfies An and WD.

Let $W : (X, ME) \to (\mathbb{R}, \geq)$ be an arbitrary Richter–Peleg representation of (X, ME).

To prove An, let $x, y \in X$ be such that $x = \sigma(y)$ for some finite permutation σ of $\mathbb{N}$. It follows that x and y are eventually equal, and $\sum_{n=0}^{\infty}(x_n - y_n) = 0$. This implies that both $x\,ME\,y$ and $y\,ME\,x$ hold, whence (3) in the definition of a Richter–Peleg representation yields $W(x) = W(y)$. Thus W satisfies An.

To prove WD, let $x, y \in X$ be such that $x_i > y_i$ for some $i \in \mathbb{N}$, and $x_n = y_n$ for all $n \neq i$. It follows that x and y are eventually equal and $\sum_{n=0}^{\infty}(x_n - y_n) > 0$, hence $x\,ME^{>}\,y$. Now the second implication in (3) (see Definition 2) yields $W(x) > W(y)$, and so WD holds for W, too.

CLAIM 3: (X, ME) has a Richter–Peleg representation.

Since ME is a preorder on X by Claim 1, it suffices to check that (X, ME) satisfies condition (ii) in Lemma 3. To that end, let $\sim$ be the binary relation on X defined as follows for all $x, y \in X$:

$$x \sim y \overset{\text{def}}{\iff} x \text{ and } y \text{ are eventually equal.}$$

Clearly, $\sim$ is an equivalence relation on X that extends ME (i.e., $x\,MD\,y$ implies $x \sim y$). Using the Axiom of Choice, select a representative $\overline{x} = (\overline{x}_n)_{n \in \mathbb{N}}$ in each equivalence class $[x]_{\sim}$. Observe that $x\,ME\,y$ implies $\overline{x} = \overline{y}$. For each $q \in \mathbb{R}$, let

$$D_q = \left\{ x \in X : \sum_{n=0}^{\infty}(x_n - \overline{x}_n) < q \right\}.$$

We claim that each set D_q is downward closed. To prove that, fix $q \in \mathbb{R}$, and let $x \in D_q$. We shall show that, for each $y \in x^{\downarrow}$, we have $\sum_{n=0}^{\infty}(y_n - \overline{y}_n) < q$, hence $y \in D_q$. Let $y \in x^{\downarrow}$, i.e., $x\,ME\,y$, and so $x \sim y$. This means that x and y are eventually equal, $0 \leq \sum_{n=0}^{\infty}(x_n - y_n) < \infty$, $\sum_{n=0}^{\infty}(x_n - \overline{x}_n) < q$, and $\overline{x} = \overline{y}$. Since all sums are actually finite, we get

$$\sum_{n=0}^{\infty}(x_n - \overline{x}_n) - \sum_{n=0}^{\infty}(y_n - \overline{y}_n) = \sum_{n=0}^{\infty}\left((x_n - \overline{x}_n) - (y_n - \overline{y}_n)\right) = \sum_{n=0}^{\infty}(x_n - y_n) \geq 0.$$

It follows that $\sum_{n=0}^{\infty}(y_n - \overline{y}_n) \leq \sum_{n=0}^{\infty}(x_n - \overline{x}_n) < q$, as claimed. Now set

$$\mathbb{D} = \{D_q\}_{q \in \mathbb{Q}}.$$

Since $\mathbb{D}$ is a countable family of downward closed subsets of X, to complete the proof it suffices to show that for all $x, y \in X$ such that $x \, ME^> y$ there exists $D_q \in \mathbb{D}$ containing y but not x. Indeed, if $x, y \in X$ are such that $x \, ME^> y$, then $\bar{x} = \bar{y}$ and $\sum_{n=0}^{\infty}(x_n - y_n) > 0$, where the latter summation has a finite number of nonzero addends. This implies that there exists $q \in \mathbb{Q}$ such that

$$\sum_{n=0}^{\infty}(x_n - \bar{x}_n) > q > \sum_{n=0}^{\infty}(y_n - \bar{y}_n).$$

It follows that $y \in D_q$ and $x \notin D_q$, and the proof of Claim 3 is complete.

By Lemma 2, any Richter–Peleg representation $W: (X, ME) \to (\mathbb{R}, \geqslant)$ of (X, ME) is a social welfare function on X, which satisfies **Anonymity** and **Weak Dominance** if so does ME. Therefore, the main statement of Theorem 2 readily follows from Claims 1–3.

Finally, we show that **Monotonicity** fails for such a social welfare function W. Indeed, if **Monotonicity** were to hold, then **Weak Dominance** would imply **Strong Pareto**. However, a famous impossibility result by Basu and Mitra [19] (see Theorem 4 below) says that there is no social welfare function on $\{0, 1\}^{\mathbb{N}}$ satisfying **Anonymity** and **Strong Pareto**. $\square$

Note that this proof suffers from the same drawback mentioned by Basu and Mitra [20]. Indeed, since we use the **Axiom of Choice** to simultaneously select a representative in each equivalence class of $\sim$, the proof is non-constructive, and it does not show how to build the associated social welfare function. It still remains an open question whether the **Axiom of Choice** is really necessary to prove Theorem 2.

It is known that, *under suitable domain restrictions*, there are social welfare functions satisfying additional properties, along with **Anonymity** and **Weak Dominance**. Proposition 1 in Basu and Mitra [20] (p. 75) is an instance of such a result. Below we give an alternative proof of this proposition by taking advantage of their construction.

Theorem 3 (*Basu and Mitra [20]*) *For any nonempty $Y \subseteq \mathbb{N}$, there is a social welfare function on $Y^{\mathbb{N}}$ satisfying* **Anonymity**, **Weak Dominance**, *and* **Weak Pareto** (*but* **Monotonicity** *fails for such a function*).

Proof Let Y be a nonempty subset of $\mathbb{N}$; without loss of generality, $|Y| \geqslant 2$. Set $X := Y^{\mathbb{N}}$.

CLAIM 1: *There is $V: X \to \left(-\frac{1}{2}, \frac{1}{2}\right)$ satisfying* **Anonymity** *and* **Weak Dominance**.

Apply Theorem 2 for $A := Y$ to get a function $V': X \to \mathbb{R}$ that satisfies **An** and **WD**. If $\phi: \mathbb{R} \to \left(-\frac{1}{2}, \frac{1}{2}\right)$ is any strictly increasing mapping, then $V := \phi \circ V'$ satisfies the claim.

CLAIM 2: *Let $f: X \to \mathbb{N}$ be the function defined by $f(x) = \min\{x_n : n \in \mathbb{N}\}$ for all $x \in X$. Then $W := f + V$ is a social welfare function on X, which satisfies* **Anonymity**, **Weak Dominance**, *and* **Weak Pareto**.

Anonymity is obvious. To prove WD, let $x, y \in X$ be such that $x_i > y_i$ for some $i \in \mathbb{N}$, and $x_n = y_n$ for all $n \neq i$. It follows that $f(x) \geqslant f(y)$. Since $V(x) > V(y)$ because V satisfies WD, we obtain $W(x) > W(y)$, as required. Finally, we prove WP. Let $x, y \in X$ be such that $x \gg_{\text{Par}} y$. Thus, $x_n > y_n$ for each $n \in \mathbb{N}$, and so $f(x) \geqslant f(y) + 1$. Since $V(x), V(y) \in \left(-\frac{1}{2}, \frac{1}{2}\right)$, we get $W(x) = f(x) + V(x) > f(y) + V(y) = W(y)$, as claimed. $\qquad\square$

Other possibility results in this field are strictly related to advanced results from mathematical utility theory. For instance, Alcantud and Dubey [6, Proposition 1] prove the existence of social welfare relations satisfying SP and a weak no-impatience condition, which have both a countable multi-utility representation (that is continuous with respect to the product topology) and a Richter–Peleg representation. Furthermore, Alcantud, Bosi and Zuanon [5, Theorem 3.1] assure that such relations must also have a countable Richter–Peleg multi-utility representation, which is continuous in the product topology. Finally, Alcantud and Dubey [6, Corollary 3] prove the existence of social welfare orders satisfying An and SP, which have a multi-utility representation that is lower semicontinuous with respect to the Svensson topology.

3.2 Impossibility Results

To start, let us mention two contributions that make use of sophisticated tools from mathematical utility theory in order to produce negative results, namely, Banerjee and Dubey [18] and Alcantud and Dubey [6]. Banerjee and Dubey [18, Propositon 1] prove that there is no An and SP preorder on $[0, 1]^{\mathbb{N}}$ with a Richter–Peleg representation. Relatedly, their Theorem 2 states that there is no An and SP preorder on $[0, 1]^{\mathbb{N}}$ that has a multi-utility representation with the set of utilities being countably infinite.[17] Alcantud and Dubey [6] prove negative results in their Proposition 2 and Corollary 2, too. The first of these two results assures that when $X = \{0, 1\}^{\mathbb{N}}$, social welfare relations that satisfy SP and a no-impatience condition are incompatible with Richter–Peleg representability. Their second result states the impossibility of multi-utility representations that are lower semicontinuous in standard topologies, for any social welfare relation that satisfies WP and this no-impatience condition.

The main goal of this section is, however, different. For the first time we disclose exact reasons for the non-representability of certain social welfare relations in the sense discussed in Sect. 2.2. Beardon et al. [21] discuss lexicographic decomposition of chains, which may display what they call "the simplest germ of non-representability". Specifically, when planar chains order-embed into a complete preorder, the latter cannot be represented by a utility function. In this section we prove that any *complete* preorder on $\{0, 1\}^{\mathbb{N}}$ satisfying An and SP contains a planar chain: a well-known impossibility result by Basu and Mitra [19] will readily follow.

---

[17]Banerjee and Dubey give a direct proof of this fact, however it is a simple corollary of their aforementioned Proposition 1 (see also Alcantud, Bosi and Zuanon [5, Sect. 4]).

The first step of our proof is to identify the planar chain that is needed to prove the claim: this will be the lexicographic product $(0, 1) \times_{\text{lex}} \mathbf{2}$, where $\mathbf{2}$ denotes the unique, up to isomorphisms, chain with two points (that is, $\mathbf{2} = (\{0, 1\}, \geqslant)$, with $1 > 0$).

Lemma 4 *The chain* $(0, 1) \times_{\text{lex}} \mathbf{2}$ *is planar, hence non-representable.*

Proof The argument is similar to Debreu's proof of the non-representability of $\mathbb{R}^2_{\text{lex}}$. □

In set-theoretic topology, $(0, 1) \times_{\text{lex}} \mathbf{2}$ is a very well-known linear ordering, which goes under the name of *double arrow space* when endowed with the order topology. Since no topology is involved here, we call it the *double arrow chain*.[18] Our main result is the following:

Lemma 5 *Any complete preordered set* $(\mathbf{2}^{\mathbb{N}}, \succsim)$ *satisfying* **Anonymity** *and* **Strong Pareto** *embeds the double arrow chain.*

Proof Let $\succsim$ be a complete preorder on $\mathbf{2}^{\mathbb{N}}$ satisfying **An** and **SP**. We show that there is a subchain of $(\mathbf{2}^{\mathbb{N}}, \succsim)$ that is order-isomorphic to $[0, 1] \times_{\text{lex}} \mathbf{2}$.

Sierpiński [85] (p. 82) shows that there is an uncountable family $\mathcal{E} = \{E(z) : z \in (0, 1)\}$ of distinct nested subsets of $\mathbb{N}$ with the property that for any $z, z' \in (0, 1)$ such that $z < z'$, $E(z)$ is a proper subset of $E(z')$, and $E(z') \setminus E(z)$ is infinite. This collection $\mathcal{E}$ with the cardinality of the continuum is one of the main tools in the proof of Theorem 1 in Basu and Mitra [19]. Here we mimic their argument.

For each $z \in (0, 1)$, we define two infinite sequences $a(z) = (a(z)_n)_{n \in \mathbb{N}}$ and $b(z) = (a(z)_n)_{n \in \mathbb{N}}$ in $\mathbf{2}^{\mathbb{N}}$ as follows for each $n \in \mathbb{N}$:

1. $a(z)_n = 1$ if $n \in E(z)$, and $a(z)_n = 0$ otherwise;
2. $b(z)_n = 1$ if either $n \in E(z)$ or $n = \min\{k \in \mathbb{N} : k \notin E(z)\}$, and $b(z)_n = 0$ otherwise.

Observe that each sequence $a(z)$ has an infinite number of 0's and an infinite number of 1's, and the corresponding sequence $b(z)$ is obtained from $a(z)$ by simply replacing the first occurrence of 0 by 1. This implies that $b(z) > a(z)$ holds in the Pareto ordering, and so **SP** (actually **WD** suffices for this) yields $b(z) \succ a(z)$ for all $z \in (0, 1)$.

CLAIM 1: *The following chain of equivalences holds for each* $z, z' \in (0, 1)$:

$$z < z' \iff E(z) \subsetneqq E(z') \iff a(z') \succ b(z).$$

Indeed, $z < z'$ implies $E(z) \subsetneqq E(z')$ by the very definition of the family $\mathcal{E}$. Further, by the argument given in the proof of Theorem 1 of Basu and Mitra [19], axioms **An** and **SP** ensure that $E(z) \subsetneqq E(z')$ implies $a(z') \succ b(z)$. Finally, $a(z') \succ b(z)$

---

[18]The name comes from the fact that each real number in the open unit interval 'is resolved into' ('points at') two consecutive elements. See the classical survey on *resolution of topological spaces* by Watson [90], as well as applications of resolutions to linearly ordered topological spaces [31].

implies $z < z'$, because $z = z'$ is impossible, and $z > z'$ produces the contradiction $a(z) \succ b(z') \succ a(z') \succ b(z) \succ a(z)$. This proves the claim.

Now set $Y = \{a(z) : z \in (0, 1)\} \cup \{b(z) : z \in (0, 1)\}$, and endow it with the induced preorder $\succsim \upharpoonright_Y$.

CLAIM 2: $(Y, \succsim \upharpoonright_Y)$ *is order-isomorphic to the double arrow chain.*

Define a function

$$
\begin{aligned}
f : (Y, \succsim \upharpoonright_Y) &\longrightarrow (0, 1) \times_{\text{lex}} \mathbf{2} \\
a(z) &\longmapsto (z, 0) \\
b(z) &\longmapsto (z, 1)
\end{aligned}
$$

We shall prove that f is an order-isomorphism. It suffices to show that for each $x, y \in Y$, $x \succ y$ implies $f(x) \succ_{\text{lex}} f(y)$. Fix $x, y \in Y$ such that $x \succ y$. We deal separately with the two possible cases.

Case 1: $x = a(z')$ *for some* $z' \in (0, 1)$. The definition of f yields $f(x) = (z', 0)$. Since $x \succ y$, we have $y \neq b(z')$, since otherwise $a(z') = x \succ y = b(z')$, a contradiction. Thus, there is $z \neq z'$ such that either (i) $y = a(z)$ (thus $f(y) = (z, 0)$), or (ii) $y = b(z)$ (thus $f(y) = (z, 1)$). If (i) holds, then we have $a(z') \succ a(z)$, which implies $z' > z$. (In fact, $z > z'$ is impossible, because Claim 1 would yield $y = a(z) \succ b(z') \succ a(z') = x$, a contradiction.) If (ii) holds, then we have $a(z') \succ b(z)$, which implies $z' > z$ by Claim 1. It follows that we always have $z' > z$, hence $f(x) \succ_{\text{lex}} f(y)$, as claimed.

Case 2: $x = b(z')$ *for some* $z' \in (0, 1)$. The definition of f yields $f(x) = (z', 1)$. We deal separately with the three possible subcases: (i) $y = a(z')$; (ii) $y = a(z)$ for some $z \neq z'$; (iii) $y = b(z)$ for some $z \neq z'$. In subcase (i), we have $f(y) = (z', 0)$, hence $f(x) = (z', 1) \succ_{\text{lex}} (z', 0) = f(y)$. In subcase (ii), we have $f(y) = (z, 0)$, hence, since $z' > z$ (because $z > z'$ implies $y = a(z) \succ b(z') \succ a(z') = x$ by Claim 1, a contradiction), we get $f(x) = (z', 1) \succ_{\text{lex}} (z, 0) = f(y)$. Finally, in subcase (iii), a similar argument yields $f(x) = (z', 1) \succ_{\text{lex}} (z, 1) = f(y)$.

Since $(Y, \succsim \upharpoonright_Y)$ is a subchain of $(2^{\mathbb{N}}, \succsim)$, the result readily follows from Claim 2. $\square$

Lemmas 4 and 5 imply Theorem 1 of Basu and Mitra [19]:

Theorem 4 (*Basu and Mitra [19]*) *There is no social welfare function on* $2^{\mathbb{N}}$ *satisfying* **Anonymity** *and* **Strong Pareto**.

Proof Toward a contradiction, suppose there is a social welfare function W on $\{0, 1\}^{\mathbb{N}}$ satisfying **An** and **SP**. Such a function W would produce a complete preorder on $\{0, 1\}^{\mathbb{N}}$ satisfying **An** and **SP**, for which W is a utility representation. However, this is impossible by Lemmas 4 and 5. $\square$

Other results from intergenerational justice refer to the impossibility of producing utility representations for complete preorders that satisfy certain properties. It may be worth revealing the reasons for such impossibilities, too. Below we list a short sample of results, which one may wish to explore from the same perspective as in Theorem 4:

Proposition 1 ([2]) *There is no social welfare function on* $[0, 1]^{\mathbb{N}}$ *satisfying* **Pigou–Dalton Transfer** *and* **Weak Pareto**. [19]

Proposition 2 ([2]) *There is no social welfare function on* $[0, 1]^{\mathbb{N}}$ *satisfying* **Hammond Equity** *and* **Weak Pareto**. [20]

Proposition 3 (Basu and Mitra [20], Theorem 4) *There is no social welfare function on* $Y^{\mathbb{N}}$ *satisfying* **Anonymity** *and* **Weak Pareto** *when* $[0, 1] \subseteq Y$.

4 Richter–Peleg Representations in Intergenerational Justice

In Sect. 3.1 we have exhibited some examples to witness the surprising relevance of Richter–Peleg representations in the field of intergenerational justice. In particular, we have so far shown that this is the key concept linking special preorders with the existence of social welfare functions with suitable properties.

There is no doubt that preorders are important in intergenerational justice. In this field, it is often impossible to combine a list of postulates that includes the completeness axiom and most of the existing criteria.[21] Under such circumstances, social welfare functions are of course ruled out. Nevertheless, it is still possible to preserve a part of the information embodied in social welfare relations that are reflexive and transitive, and Richter–Peleg representations are the right tool to do that.

How widespread are Richter–Peleg representations for the (non-representable) preference orderings that play a leading role in the discussions about distributive justice? In Sect. 3.2 we have recalled two previous uses of Richter–Peleg representations in this discipline, namely, Banerjee and Dubey [18] and Alcantud and Dubey [6]. In this section we reverse the approach. In fact, we shall prove that impossibility results in the field of intergenerational justice can be used to prove the failure of the Richter–Peleg representability of some—rather natural—preference orderings on infinite streams of utilities. These orderings are obtained by 'concatenating' different criteria related to different periods of time. The next definition is an example of what we mean by 'concatenation' in this context:

---

[19]The full force of **PDT** and **WP** is not used along the proof. The argument in [2] ensures incompatibility of a weaker version of **PDT** with a weaker version of **WP**.

[20]The argument in [2] ensures incompatibility of **HE** with a weaker version of **WP**.

[21]It is simple to check that no preorder is able to combine **Strong Anonymity**—see Footnote 14—and **SP** (Lauwers [68, Lemma 1]). Vallentyne [88] argues that incompleteness is far less problematic for utilitarianism than **An** and **SP**, and proposes an incomplete criterion 'in the spirit of utilitarianism' that satisfies both axioms. Lauwers [70, Section IV] defines a principle that satisfies them, too: however, his principle is closer to being complete than Vallentyne's criterion. Explicit incompatibilities of the completeness axiom with other sets of properties include Asheim, Mitra, and Tungodden [12, Corollaries 1–3, Propositions 4 and 5], and Sakai [83, Corollary 1].

Definition 6 Define a binary relation MD on X by setting, for all $x, y \in X$:

$$x \, MD \, y \stackrel{\text{def}}{\Longleftrightarrow} \text{ there is } k \in \mathbb{N} \text{ such that } \sum_{i=0}^{k}(x_i - y_i) \geq 0 \text{ and } x_n \geq y_n \text{ for all } n > k. \quad (6)$$

We call MD the *mean-dominance rule* on X.

Remark 1 The mean-dominance rule MD on X is a proper extension of the weak Pareto preference. As a matter of fact, MD strictly extends both the symmetric and the asymmetric part of the weak Pareto preference. To prove this fact for the symmetric part, start by observing that the indifference of the weak Pareto preference is equality (since it is a partial order), and clearly identical sequences are MD-indifferent. Furthermore, MD is invariant under finite permutations: for instance, the two eventually equal streams $x = (0, 1, \langle 0 \rangle)$ and $y = (1, 0, \langle 0 \rangle)$ are such that $x \, MD^= y$. It follows that $MD^=$ is a proper extension of Pareto indifference. Concerning the asymmetric part, it is easy to show that $x >_{\text{Par}} y$ implies $x \, MD^> y$, for all $x, y \in X$. On the other hand, the two eventually equal streams $x = (1, 1, 1, \langle 0 \rangle)$ and $y = (0, 2, 0, \langle 0 \rangle)$ are Pareto-incomparable, and yet $x \, MD^> y$.

Definition 6 has a natural economic interpretation. The preference relation MD is defined by concatenating two distinct evaluation criteria. The first criterion—finite mean—is compensatory in nature: what is lost by a finite number r of generations can be compensated by what is gained by a finite number s of other generations (where $r \neq s$, in general). The second criterion—domination—is totally non-compensatory: there is no possibility to make up for any loss of utility, even by a single generation. These two criteria are applied to distinct but temporally consecutive generations. Specifically, Definition 6 declares an infinite stream x of utilities 'at least as good as' another infinite stream y of utilities if

- there is a common finite horizon of generations for which the average evaluation of x is greater or equal than the average evaluation of y, and
- x dominates y on the residual infinite horizon, since after the initial segment the utility of x for each future generation is greater or equal than the respective utility of y.

We have already described another instance of concatenated preference in Sect. 3.1, namely the relation ME (*mean-equality*): see (5) in the proof of Theorem 2. Upon comparing the definitions (5) and (6), it becomes apparent that the two relations MD and ME are very similar: in fact, we have $ME \subsetneq MD$. The following equivalent reformulation of (5) makes this similarity even more striking:

$$x \, ME \, y \stackrel{\text{def}}{\Longleftrightarrow} \text{ there is } k \in \mathbb{N} \text{ such that } \sum_{i=0}^{k}(x_i - y_i) \geq 0 \text{ and } x_n = y_n \text{ for all } n > k.$$

This justifies the acronym used for the social welfare relation ME.[22]

A possibly surprising fact is that ME is Richter–Peleg representable (by the proof of Theorem 2) whereas MD is not:

Proposition 4 *For each $Y \subseteq \mathbb{R}$ such that $|Y| \geqslant 2$, the mean-dominance preference on $Y^{\mathbb{N}}$ is a preorder, which is not Richter–Peleg representable.*

Proof Fix $Y \subseteq \mathbb{R}$ such that $|Y| \geqslant 2$, and let MD be the mean-dominance preference on $Y^{\mathbb{N}}$. First, we prove that MD is a preorder. Since MD is clearly reflexive, we prove transitivity. Let $x, y, z \in X$ be such that $x\,MD\,y$ and $y\,MD\,z$. Thus,

– there is $k_1 \in \mathbb{N}$ such that $\sum_{i=0}^{k_1}(x_i - y_i) \geqslant 0$ and $x_n \geqslant y_n$ for all $n > k_1$, and
– there is $k_2 \in \mathbb{N}$ such that $\sum_{i=0}^{k_2}(y_i - z_i) \geqslant 0$ and $y_n \geqslant z_n$ for all $n > k_2$.

Set $k := \max\{k_1, k_2\}$. Since $\sum_{i=0}^{k}(x_i - y_i) \geqslant 0$ and $\sum_{i=0}^{k}(y_i - z_i) \geqslant 0$, it follows that $\sum_{i=0}^{k}(x_i - z_i) \geqslant 0$, and $x_n \geqslant z_n$ whenever $n > k$. We conclude that $x\,MD\,z$, as claimed.

To complete the proof, suppose by way of contradiction that W is a Richter–Peleg representation of $(Y^{\mathbb{N}}, MD)$. Since An and SP hold for MD, Lemma 2 implies that W is a social welfare function on $Y^{\mathbb{N}}$ satisfying the same properties. Now a contradiction readily follows from the aforementioned impossibility result by Basu and Mitra [19] (see Theorem 4 below), which says that there is no social welfare function on $\{0, 1\}^{\mathbb{N}}$ (and, in general, on any $Y^{\mathbb{N}}$, where Y has at least 2 elements) satisfying **Anonymity** and **Strong Pareto**. □

Remark 2 Proposition 4 shows that the joint satisfaction of two properties—namely An and SP—deems MD incompatible with Richter–Peleg representability. There are other arguments which enforce such an incompatibility whenever the domain has mild structural properties. For instance, suppose $Y \subseteq \mathbb{R}$ contains a, b, c, d such that $a > b > c > d > 0$ and $c - d > b - c > a - b > 0$. Then the distributional implications of AE can be put into action for settings such us, for instance, $Y = \mathbb{N}$, or $\{0.1, 0.4, 0.6, 0.7\} \subseteq Y$, or $[0, 1] \subseteq Y$, or $|Y| \geqslant 10$. In fact, any Richter–Peleg representation of $(Y^{\mathbb{N}}, MD)$ would be a social welfare function satisfying Mo and AE because so does MD (cf. Lemma 2). However, this is impossible by Theorem 1 in Alcantud [3], which assures that there is no social welfare function on $Y^{\mathbb{N}}$ that satisfies Mo and AE.

The mean-dominance rule considers the mean average as one of the two criteria of evaluation. Actually, the following relation, which is based on an extended form of average, has already been considered in the literature (see Asheim and Tungodden [14, Definition 3]):

---

[22]Both MD and ME use the finite mean as the first (finite) criterion. Thus, it is worth examining the interconnections between these types of social welfare relations and an abstract notion of finite mean. (On the latter topic, see the recent contribution by Campión et al. [30].).

Definition 7 Let $X = A^{\mathbb{N}}$, where A is any nonempty subset of $\mathbb{R}$. Define a binary relation C on X by setting, for all $x, y \in X$:

$$x C y \quad \overset{\text{def}}{\Longleftrightarrow} \quad \text{there is } h \in \mathbb{N} \text{ such that } \sum_{i=0}^{k}(x_i - y_i) \geqslant 0 \ \text{ for all } k > h. \quad (7)$$

The relation C is the *catching-up rule* on X.

The catching-up rule C and the mean-dominance MD rule look alike, but they are not. The appearance of similarity is due to the fact that both of them use a form of average as one of the two evaluation criteria. However, the resemblance ends there. First of all, the order of application is different, since the mean-dominance rule MD uses the average as initial criterion, whereas the catching up rule C uses it as final criterion. Furthermore, the use of the average criterion is not 'pure' in the catching-up rule, because it concerns infinitely many instances of averaging computations. Finally, the first criterion in the catching-up rule is a non-criterion, since everything may happen in the initial segment of generations.

Asheim and Tungodden [14, Sect. 5] examine additional forms of 'overtaking rules', already studied in [16, 91]. They also characterize the relations that include C, and, in particular, they check that C satisfies **An** and **SP** [14, Proposition 4]. The latter fact allows us to show that the catching-up rule fails to be Richter–Peleg representable by essentially replicating the argument used to prove Proposition 4.

Proposition 5 *For each $Y \subseteq \mathbb{R}$ such that $|Y| \geqslant 2$, the catching-up preference C on $Y^{\mathbb{N}}$ is a preorder, which is not Richter–Peleg representable.*

Proof Fix $Y \subseteq \mathbb{R}$ such that $|Y| \geqslant 2$. First, we prove that C is a preorder. Since C is clearly reflexive, we prove transitivity. Let $x, y, z \in X$ be such that $x C y$ and $y C z$. Thus,

- there is $h_1 \in \mathbb{N}$ such that $\sum_{i=0}^{k}(x_i - y_i) \geqslant 0$ for all $k > h_1$, and
- there is $h_2 \in \mathbb{N}$ such that $\sum_{i=0}^{k}(y_i - z_i) \geqslant 0$ for all $k > h_2$.

Set $h := \max\{h_1, h_2\}$. Then $\sum_{i=0}^{k}(x_i - y_i) \geqslant 0$ and $\sum_{i=0}^{k}(y_i - z_i) \geqslant 0$ whenever $k > h$, which implies that $\sum_{i=0}^{k}(x_i - z_i) \geqslant 0$ for all $k > h$. We conclude that $x C z$, as claimed.

To complete the proof, suppose by way of contradiction that W is a Richter–Peleg representation of $(Y^{\mathbb{N}}, C)$. Since **An** and **SP** hold for C, Lemma 2 implies that W is a social welfare function on $Y^{\mathbb{N}}$ satisfying the same properties. Now a contradiction readily follows from Theorem 4. $\qquad\square$

The conclusion of Proposition 5 also holds for the 'overtaking rule', which is defined by Asheim and Tungodden [14, Definition 4] as follows:

Definition 8 Let $X = A^{\mathbb{N}}$, where A is any nonempty subset of $\mathbb{R}$. Define a binary relation O on X by setting, for all $x, y \in X$:

$$x\,O^>y \quad \overset{\text{def}}{\Longleftrightarrow} \quad \text{there is } h \in \mathbb{N} \text{ such that } \sum_{i=0}^{k}(x_i - y_i) > 0 \text{ for all } k > h, \quad (8)$$

$$x\,O^=y \quad \overset{\text{def}}{\Longleftrightarrow} \quad \text{there is } h \in \mathbb{N} \text{ such that } \sum_{i=0}^{k}(x_i - y_i) = 0 \text{ for all } k > h. \quad (9)$$

The relation O is the *overtaking rule* on X.

As announced, we have:

Proposition 6 *For each $Y \subseteq \mathbb{R}$ such that $|Y| \geqslant 2$, the overtaking preference O on $Y^{\mathbb{N}}$ is a preorder, which is not Richter–Peleg representable.*

Proof We can proceed as in Propositions 4 and 5, since Asheim and Tungodden [14, Proposition 5] characterize the relations that include O, and, in particular, they check that O satisfies **An** and **SP**.   □

5   Conclusions and Future Research

In this chapter, we have established the existence of a fruitful interaction between two disciplines, which have mostly developed independently of each other. Mathematical utility theory provides a unified framework for the understanding of some common principles in the analysis of social welfare functions. These functions—which include the classical discounted utilities as a cornerstone—often fail to ensure the right collection of properties that researchers need in the process of equitably aggregating infinite utility streams. This chapter has shown that we can incorporate the central results from mathematical utility theory into their analysis, irrespective of whether there are or there are not social welfare functions with the required features. Despite its simplicity, Lemma 2 provides the key to place Richter–Peleg representability at the very core of some existence results on intergenerational justice. Section 4 has shown that minor changes in the specification of the social welfare relations affect the existence of these representations. In this respect, we have also proven that there are not Richter–Peleg representations of the catching-up and overtaking criteria.

Future research on the topic may develop in several directions; here we mention two of them. First of all, we are actively working on the axiomatization of sequential social rules, of which the mean-equality and the mean-dominance rules are two instances (Alcantud and Giarlotta [9]). Informally, a *sequential social rule* is a binary relation S on $X = \mathbb{R}^{\mathbb{N}}$ (or a suitable subset of it), which can be written as a 'temporal concatenation' of a sequence $(S_1, \ldots, S_p)$ of binary relations on X. The idea underlying a sequential social rule is that of partitioning the infinite horizon into $p + 1 \geq 2$ periods of time, and evaluate any stream of utilities over it by applying a criterion per period. Sequential social rules can be associated to new types of equitability axioms. Furthermore, the weak Pareto preference is obtained as the limit of suitable sequences of sequential social rules.

Another possible direction of research consists of an alternative approach to intergenerational justice, which still uses binary relations to compare streams of endowments. In fact, instead of considering infinite sequences of real numbers, one can use infinite sequences of *sets*, intended as accomplishments to be achieved by each generation. Among these accomplishments, some are more important than others: these are the so-called 'benchmarks'. According to the 'theory of benchmarking' recently developed by Chambers and Miller [34], a *benchmarking rule* is a binary relation $\succsim$ on a suitable family $\mathfrak{R}$ of finite sets of accomplishments such that there exists a distinguished subset Z of accomplishments—the benchmarks—with the following property: $A \succsim B$ if and only if $A \cap Z$ contains $B \cap Z$, for all A, B in $\mathfrak{R}$. Chambers and Miller [34] characterize benchmarking rules as those preorders that extend set-inclusion and satisfy two economically meaningful properties.[23] Upon considering infinite sequences of finite sets of accomplishments, it is possible to use a similar approach to define *benchmarking social rules*, that is, binary relations on infinite sequences of finite sets that are based on the existence a suitable set of accomplishments—the benchmarks. We are currently working on the topic (Alcantud and Giarlotta [10]).

Acknowledgements The authors wish to thank two anonymous referees for their detailed comments, which determined a definite improvement in the overall presentation of the topic. The work of the co-editors, which led to the publication of this well-deserved tribute to Ghanshyam, is enormously appreciated.

References

1. Alcantud, J.C.R.: Weak utilities from acyclicity. Theory Decis. **47**(2), 185–196 (1999)
2. Alcantud, J.C.R.: Inequality averse criteria for evaluating infinite utility streams: the impossibility of Weak Pareto. J. Econ. Theory **147**, 353–363 (2012)
3. Alcantud, J.C.R.: The impossibility of social evaluations of infinite streams with strict inequality aversion. Econ. Theory Bull. **1**, 123–130 (2013)
4. Alcantud, J.C.R.: Liberal approaches to ranking infinite utility streams: when can we avoid interference? Soc. Choice Welf. **41**, 381–396 (2013)
5. Alcantud, J.C.R., Bosi, G., Zuanon, M.: Richter-Peleg multi-utility representations of preorders. Theory Decis. **80**, 443–450 (2016)
6. Alcantud, J.C.R., Dubey, R.S.: Ordering infinite utility streams: efficiency, continuity, and no impatience. Math. Soc. Sci. **72**, 33–40 (2014)
7. Alcantud, J.C.R., García-Sanz, M.D.: Evaluations of infinite utility streams: Pareto-efficient and egalitarian axiomatics. Metroeconomica **64**, 432–447 (2013)
8. Alcantud, J.C.R., Giarlotta, A.: Necessary and possible hesitant fuzzy sets: a novel model for group decision making. Inf. Fus. **46**, 63–76 (2019)
9. Alcantud, J.C.R., Giarlotta, A.: Sequential Social Rules. University of Catania, Mimeo (2019)

---

[23]Very recently, Giarlotta and Watson [58] have revisited, extended, and generalized Chambers and Miller's approach.

10. Alcantud, J.C.R., Giarlotta, A.: Benchmarking Social Rules. University of Catania, Mimeo (2019)
11. Alcantud, J.C.R., Rodríguez-Palmero, C.: Characterization of the existence of semicontinuous weak utilities. J. Math. Econ. **32**, 503–509 (1999)
12. Asheim, G.B., Mitra, T., Tungodden, B.: A new equity condition for infinite utility streams and the possibility of being Paretian. In: Roemer, J., Suzumura, K. (eds.), Intergenerational Equity and Sustainability. Conference Proceedings of the IWEA Roundtable Meeting on Intergenerational Equity, Palgrave, pp. 69–84 (2007)
13. Asheim, G.B., Mitra, T., Tungodden, B.: Sustainable recursive social welfare functions. Econ. Theory **49**, 267–292 (2012)
14. Asheim, G.B., Tungodden, B.: Resolving distributional conflicts between generations. Econ. Theory **24**, 221–230 (2004)
15. Asheim, G.B., and Tungodden, B.: Do Koopmans' postulates lead to discounted utilitarianism? Discussion paper 32/04, Norwegian School of Economics and Business Administration
16. Atsumi, H.: Neoclassical growth and the efficient program of capital accumulation. Rev. Econ. Stud. **32**, 127–136 (1965)
17. Banerjee, K.: On the equity-efficiency trade off in aggregating infinite utility streams. Econ. Lett. **93**, 63–67 (2006)
18. Banerjee, K., Dubey, R.S.: On multi-utility representation of equitable intergenerational preferences. In: Basu, B., Chakravarty, S.R., Chakrabarti, B.K., Gangopadhyay, K. (eds.) Econophysics and Economics of Games, Social Choices and Quantitative Techniques, pp. 175–180. Springer (2010)
19. Basu, K., Mitra, T.: Aggregating infinite utility streams with intergenerational equity: the impossibility of being Paretian. Econometrica **71**, 1557–1563 (2003)
20. Basu, K., and Mitra, T.: Possibility theorems for equitably aggregating infinite utility streams. In: Roemer, J., Suzumura, K. (eds.) Intergenerational Equity and Sustainability. Conference Proceedings of the IWEA Roundtable Meeting on Intergenerational Equity, Palgrave, pp. 69–84 (2007)
21. Beardon, A.F., Candeal, J.C., Herden, G., Induráin, E., Mehta, G.B.: The non-existence of a utility function and the structure of non-representable preference relations. J. Math. Econ. **37**, 17–38 (2002)
22. Beardon, A.F., Candeal, J.C., Herden, G., Induráin, E., Mehta, G.B.: Lexicographic decomposition of chains and the concept of a planar chain. J. Math. Econ. **37**(2), 95–104 (2002)
23. Beardon, A.F., Mehta, G.B.: The utility theorems of Wold, Debreu, and Arrow-Hahn. Econometrica **62**(1), 181–186 (1994)
24. Bewley, T.F.: Existence of equilibria in economies with infinitely many commodities. J. Econ. Theory **4**, 514–540 (1972)
25. Bosi, G., Herden, G.: Continuous multi-utility representations of preorders. J. Math. Econ. **48**, 212–218 (2012)
26. Bosi, G., Herden, G.: On continuous multi-utility representations of semi-closed and closed preorders. Math. Soc. Sci. **79**, 20–29 (2016)
27. Bridges, D.S., Mehta, G.B.: Representations of Preference Orderings. Springer, Berlin (1995)
28. Bossert, W., Sprumont, Y., Suzumura, K.: Ordering infinite utility streams. J. Econ. Theory **135**, 579–589 (2007)
29. Campión, M.J., Candeal, J.C., Induráin, E.: The existence of utility functions for weakly continuous preferences on a Banach space. Math. Soc. Sci. **51**, 227–237 (2006)
30. Campión, M.J., Candeal, J.C., Catalán, R.G., Giarlotta, A., Greco, S., Induráin, E., Montero, J.: An axiomatic approach to finite means. Inf. Sci. **457–458**, 12–28 (2018)
31. Caserta, A., Giarlotta, A., Watson, S.: On resolutions of linearly ordered spaces. Appl. Gen. Topol. **7**(2), 211–231 (2006)
32. Caserta, A., Giarlotta, A., Watson, S.: Debreu-like properties of utility representations. J. Math. Econ. **44**, 1161–1179 (2008)
33. Cerreia-Vioglio, S., Giarlotta, A., Greco, S., Maccheroni, F., Marinacci, M.: Rational preference and rationalizable choice. Econ. Th. forthcoming (2018)

34. Chambers, C.P., Miller, A.D.: Benchmarking. Theor. Econ. **11**, 485–504 (2018)
35. Chipman, J.S.: On the lexicographic representations of preference orderings. In: Chipman, J.S., Hurwicz, L., Richter, M., Sonnenschein, H.F. (eds.) Preference, pp. 276–288. Utility and Demand, Harcourt Brace and Jovanovich, New York (1971)
36. Cohen, P.J.: The Independence of the Continuum Hypothesis. Proc. Natl. Acad. Sci. U.S.A. **50**(6), 1143–1148 (1963)
37. Cohen, P.J.: The Independence of the Continuum Hypothesis II. Proc. Natl. Acad. Sci. U.S.A. **51**(1), 105–110 (1963)
38. Crespo, J., Núñez, C., Rincón-Zapatero, J.P.: On the impossibility of representing infinite utility streams. Econ. Theory **40**, 47–56 (2009)
39. Debreu, G.: Representation of a Preference Ordering by a Numerical Function. In: Thrall, R.M., Coombs, C.H., Davis, R.L. (eds.) Decision Processes, pp. 159–166. Chapter XI. Wiley, N.Y. (1954)
40. Diamond, P.A.: The evaluation of infinite utility streams. Econometrica **33**(1), 170–177 (1965)
41. Dubey, R.S.: A note on social welfare orders satisfying Pigou-Dalton transfer principle. Hitotsubashi J. Econ. **57**, 243–262 (2016)
42. Dubey, R.S., Mitra, T.: On equitable social welfare functions satisfying the weak Pareto axiom: a complete characterization. Int. J. Econ. Theory **7**, 231–250 (2011)
43. Estevan, A.: Generalized Debreu's open gap lemma and continuous representability of biorders. Order **33**(2), 213–229 (2016)
44. Estévez, M., Hervés, C.: On the existence of continuous preference orderings without utility representations. J. Math. Econ. **24**, 305–309 (1995)
45. Evren, O., Ok, E.A.: On the multi-utility representation of preference relations. J. Math. Econ. **47**, 554–563 (2011)
46. Fishburn, P.C.: Lexicographic orders, utilities and decision rules. Manag. Sci. **20**, 1442–1471 (1974)
47. Giarlotta, A.: New trends in preference, utility, and choice: From a mono-approach to a multi-approach. In: Doumpos, M., Figueira, J.R., Greco, S., Zopounidis, C. (eds.) New Perspectives in Multiple Criteria Decision Making. Multiple Criteria Decision Making Series, pp. 3–80. Springer International Publishing, Cham (2019)
48. Giarlotta, A.: The representability number of a chain. Topol. Appl. **150**, 157–177 (2005)
49. Giarlotta, A.: A genesis of interval orders and semiorders: transitive NaP-preferences. Order **31**, 239–258 (2014)
50. Giarlotta, A.: Normalized and strict NaP-preferences. J. Math. Psychol. **66**, 34–40 (2015)
51. Giarlotta, A., Greco, S.: Necessary and possible preference structures. J. Math. Econ. **42**(1), 163–172 (2013)
52. Giarlotta, A., Watson, S.: Pointwise Debreu lexicographic powers. Order **26**(4), 377–409 (2009)
53. Giarlotta, A., Watson, S.: A hierarchy of chains embeddable into the lexicographic power $\mathbb{R}^{\omega}_{\text{lex}}$. Order **30**, 463–485 (2013)
54. Giarlotta, A., Watson, S.: Lexicographic preferences representable by real-branching trees with countable height: a dichotomy result. Ind. Math. **25**, 78–92 (2014)
55. Giarlotta, A., Watson, S.: Well-graded families of NaP-preferences. J. Math. Psychol. **77**, 21–28 (2017)
56. Giarlotta, A., Watson, S.: Necessary and possible indifferences. J. Math. Psychol. **81**, 98–109 (2017)
57. Giarlotta, A., Watson, S.: The interplay between two rationality tenets: extending Schmeidler's theorem to bi-preferences. University of Catania, Mimeo (2019)
58. Giarlotta, A., Watson, S.: Benchmarking: revisited, extended, and generalized. University of Catania, Mimeo (2019)
59. Hara, C., Shinotsuka, T., Suzumura, K., Xu, Y.: Continuity and egalitarianism in the evaluation of infinite utility streams. Soc. Choice Welf. **31**, 179–191 (2008)
60. Herden, G., Mehta, G.B.: Open gaps, metrization and utility. Econ. Theory **7**, 541–546 (1996)
61. Herden, G., Mehta, G.B.: The Debreu Gap Lemma and some generalizations. J. Math. Econ. **40**, 747–769 (2004)

62. Hicks, J.: A Revision of Demand Theory. Clarendon Press, Oxford (1956)
63. Jaffray, J.-Y.: Existence of a continuous utility function: an elementary proof. Econometrica **43**, 981–983 (1975)
64. Knoblauch, V.: Lexicographic orders and preference representation. *J. Math. Econ.* 34, 255–267 (200)
65. Koopmans, T.C.: Stationary ordinal utility and impatience. Econometrica **28**, 287–309 (1960)
66. Koopmans, T.C.: Representation of preference orderings over time. In: McGuire, C.B., Radner, R. (eds.) Decision and Organization. North-Holland, Amsterdam, 79–100 (1972). Econometrica **28**, 287–309 (1960)
67. Kunen, K.: Set Theory. An Introduction to Independence Proofs. North-Holland, Amsterdam (1980)
68. Lauwers, L.: Rawlsian equity and generalised utilitarianism with an infinite population. Econ. Theory **9**, 143–150 (1997)
69. Lauwers, L.: Continuity and equity with infinite horizons. Soc. Choice Welf. **14**, 345–356 (1997)
70. Lauwers, L.: Infinite utility: Insisting on strong monotonicity. Australas. J. Philos. **75**, 222–233 (1997)
71. Lauwers, L.: Ordering infinite utility streams comes at the cost of a non-Ramsey set. J. Math. Econ. **46**, 32–37 (2010)
72. Levin, V.L.: The Monge-Kantorovich problems and stochastic preference relation. Adv. Math. Econ. **3**, 97–124 (2001)
73. Lombardi, M., Miyagishima, K., Veneziani, R.: Liberal egalitarianism and the Harm Principle. Econ. J. **126**, 2173–2196 (2016)
74. Mariotti, M., Veneziani, R.: 'Non-interference' implies equality. Soc. Choice Welf. **32**, 123–128 (2009)
75. Mariotti, M., Veneziani, R.: The liberal ethics of non-interference. Br. J. Pol. Sci. 1–18 (2017)
76. Mehta, G.B.: Preference and utility. In: Barberà, S., Hammond, P., Seidl, C. (eds.) Handbook of Utility Theory, pp. 1–47. Kluwer Academic Publisher, Dordrecht (1998)
77. Minguzzi, E.: Normally preordered spaces and utilities. Order **30**, 137–150 (2013)
78. Mitra, T., Ozbek, M.K.: On representation of monotone preference orders in a sequence space. Soc. Choice Welf. **41**(3), 473–487 (2013)
79. Monteiro, P.K.: Some results on the existence of utility functions on path connected spaces. J. Math. Econ. **16**, 147–156 (1987)
80. Peleg, B.: Utility functions for partially ordered topological spaces. Econometrica **38**(1), 93–96 (1970)
81. Pivato, M.: Multiutility representations for incomplete difference preorders. Math. Soc. Sci. **66**, 196–220 (2013)
82. Sakai, T.: Equitable intergenerational preferences on restricted domains. Soc. Choice Welf. **27**, 41–54 (2006)
83. Sakai, T.: Limit representations of intergenerational equity. Soc. Choice Welf. **47**(2), 481–500 (2016)
84. Sakamoto, N.: Impossibilities of Paretian social welfare functions for infinite utility streams with distributive equity. Hitotsubashi J. Econ. **53**, 121–130 (2012)
85. Sierpiński, W.: Cardinal and Ordinal Numbers. Polish Scientific, Warsaw (1965)
86. Smith, A.: The Wealth of Nations: A Translation into Modern English. Industrial Systems Research (2015)
87. Svensson, L.-G.: Equity among generations. Econometrica **48**, 1251–1256 (1980)
88. Vallentyne, P.: Utilitarianism and infinite utility. Australas. J. Philos. **71**, 212–215 (1993)
89. Wakker, P.P.: Continuity of preference relations for separable topologies. Int. Econ. Rev. **29**, 105–110 (1988)
90. Watson, S.: The Construction of Topological Spaces: Planks and Resolutions. In: Hušek, M., van Mill, J. (eds.) Recent Progress in General Topology, pp. 673–757 North-Holland, Amsterdam (1992)

91. von Weizsäcker, C.C.: Existence of optimal program of accumulation for an infinite time horizon. Rev. Econ. Stud. **32**, 85–104 (1965)
92. Wold, H.: A synthesis of pure demand analysis Part II. Skandinavisk Aktuaritidskrift **26**, 220–263 (1943)
93. Zame, W.R.: Can intergenerational equity be operationalized? Theor. Econ. **2**, 187–202 (2007)

Comparative Risk Aversion for State-Dependent Preferences

John Quiggin and Robert G. Chambers

Abstract This paper shows how dual concepts of risk-aversion, developed by Chambers and Quiggin may be extended to the case of state-dependent preferences, whether or not these preferences are autocomparable in the sense of Karni. We characterize autocomparability as a special case. We show how standard comparative static results, originally derived for the state-independent expected utility model, may be extended to general state-dependent preferences, without the condition of additive separability.

1 Introduction

The idea that preferences over uncertain outcomes may depend on the state of nature in which a given outcome is realised seems intuitively appealing. Common examples include states of nature associated with death, injury, illness or accident. Karni [10] formalized this idea, presenting a state-dependent version of the expected utility model. Karni [11] extended the analysis to generalized expected utility theory.

A crucial issue in any analysis of preferences concerning uncertainty is the concept of risk-aversion. Important aspects of this issue include the definition of risk-aversion, comparisons of the riskiness of alternative state-contingent consumption bundles, comparisons of risk-aversion between individuals and comparisons of risk aversion for a given individual at different wealth levels. In particular, the hypothesis of decreasing absolute risk aversion (DARA) has played a central role in comparative static analysis.

Karni [9–11] analyzes a number of these issues. His central idea is that the set of riskless outcomes may be replaced by a more general reference set, representing the most desirable allocation of possible wealth levels at actuarially fair prices. Karni

J. Quiggin (✉)
University of Queensland, Brisbane, Australia
e-mail: j.quiggin@uq.edu.au

R. G. Chambers
University of Maryland, College Park, USA

© Springer Nature Switzerland AG 2020
G. Bosi et al. (eds.), *Mathematical Topics on Representations of Ordered Structures and Utility Theory*, Studies in Systems, Decision and Control 263,
https://doi.org/10.1007/978-3-030-34226-5_19

develops a notion of DARA for state-contingent preferences. However, this notion is relevant to autocomparable preferences (those characterized by affine reference sets), and, except for the case of additively separable (state-dependent expected utility) preferences, it does not yield sharp comparative static results.

The axiomatic basis of state-dependent utility is further developed by Karni and Schmeidler [14]. Recent applications include Baccelli [2], Karni and Safra [13] and Riedener [20].

The idea that preferences may be state-dependent fits naturally with an analysis of uncertainty based on a representation of random variables as state-contingent consumption or production bundles analogous to consumption and production bundles in standard consumer and producer theory. Chambers and Quiggin [3] show that the state-contingent approach may be used to characterize risk aversion using standard tools of modern consumer theory such as expenditure functions. Chambers and Quiggin [4] develop a range of dual measures of risk aversion.

This paper shows how these concepts of risk-aversion may be extended to the case of state-dependent preferences, whether or not these preferences are autocomparable. We characterize autocomparability as a special case. We show how standard comparative static results, originally derived for the state-independent expected utility model, may be extended to general state-dependent preferences, without the requirement for additive separability.

2   Notation

We consider preferences over random variables represented as mappings from a state space Ω to a convex outcome space $Y \subseteq \Re$. Our focus is on the case where Ω is a finite set $\{1, \ldots, S\}$, and the space of random variables is $Y^S \subseteq \Re^S$. The *unit vector* is denoted $\mathbf{1} = (1, 1, \ldots 1)$, and we define $\mathbf{e}_i$ as the i-th row of the $S \times S$ identity matrix

$$\mathbf{e}_i = (0, \ldots, 1, 0, \ldots, 0).$$

The *certainty ray* is the set $\{c\mathbf{1} : c \in \Re\}$ containing random variables that yield some outcome c with certainty.

Probabilistic beliefs are defined by a vector $\hat{\pi} \in \Re_+^S$ such that $\sum_s \hat{\pi}_s = 1$. The expected value of a random variable $\mathbf{y}$ is

$$E[\mathbf{y}] = \sum_s \hat{\pi}_s y_s$$

If preferences are state-independent, interchanging the outcomes of two states s, s' such that $\hat{\pi}_s = \hat{\pi}_{s'}$ will not affect preferences. More generally, preferences over Y^S will depend only on the *cumulative distribution function*

$$F(c; \mathbf{y}) = \sum_{y_s \leq c} \hat{\pi}_s$$

State-independent preferences are (strictly) risk averse if W is (strictly) quasiconcave and each element of $c\mathbf{1}$ is preferred to any other $\mathbf{y}$ with the same expected value that is if

$$c\mathbf{1} \in \arg\max \left\{ W(y) : \sum_s \hat{\pi}_s y_s = c \right\}$$

The way in which probabilistic beliefs may be elicited from individuals with state-dependent preferences, is discussed by Karni [12] and Grant and Karni [8].

Preferences over state-contingent incomes are given by an ordinal mapping $W : \mathfrak{R}^S \to \mathfrak{R}$. W is continuous, nondecreasing, and quasi-concave in $\mathbf{y}$. The least-as-good sets associated with this preference ordering are given by

$$V(w) = \{\mathbf{y} : W(\mathbf{y}) \geq w\}.$$

Important examples are (state-independent) expected utility preferences

$$W(\mathbf{y}) = \sum_s \hat{\pi}_s u(y_s) \tag{1}$$

(where $u : \mathfrak{R} \to \mathfrak{R}$ is assumed concave), and state-dependent expected utility preferences

$$W(\mathbf{y}) = \sum_s \hat{\pi}_s u^s(y_s) \tag{2}$$

(where each $u^s : \mathfrak{R} \to \mathfrak{R}$ is assumed concave).

We will make extensive use of concepts from duality theory, developed by Gorman [7] and subsequent writers. Cornes [5] is a useful guide to this literature. Chambers and Quiggin [3, 4] extend duality theory to the analysis of choice and production under uncertainty.

For any vector of state-contingent prices, $\mathbf{p} \in \mathfrak{R}^S_{++}$, and given income, m, we can represent W by the indirect utility function

$$I(\mathbf{p}, m) = \max_{\mathbf{y}} \{W(\mathbf{y}) : \mathbf{p}\mathbf{y} \leq m\}.$$

The associated expenditure function is defined:

$$E(\mathbf{p}, w) = \inf \{\mathbf{p}\mathbf{y} : \mathbf{y} \in V(w)\}$$
$$= \inf \{m : I(\mathbf{p}, m) \geq w\}.$$

The expenditure function is linear (more precisely, affine) in income if

$$E\left(\mathbf{p},w\right) = f\left(\mathbf{p}\right) + wg\left(\mathbf{p}\right)$$

for functions f, g homogenous of degree one in $\mathbf{p}$. This gives rise to the Gorman polar form for the indirect utility function.

Definition 1 The Gorman polar form of the indirect utility function is

$$I\left(\mathbf{p},m\right) = \frac{m - f\left(\mathbf{p}\right)}{g\left(\mathbf{p}\right)}$$

for functions f, g homogenous of degree one in $\mathbf{p}$.

We do not assume that the utility and expenditure functions are globally differentiable. This allows the analysis to encompass choice models such as rank-dependent utility [17], which are not Frechet differentiable [15, 23]. To extend the analysis to this case, we apply the notion of the superdifferential [21].

Definition 2 Let $f : Y^S \to \Re$ be a real-valued convex function and let $x_0 \in Y^S$. A vector $v \in Y^S$ is called a supergradient at x_0 if, for all $x \in Y^S$

$$f(x) - f(x_0) \leq v \bullet (x - x_0)$$

The set of all supergradients at x_0 is called the superdifferential at x_0 and is denoted $\partial f(x_0)$. The superdifferential is always a nonempty convex compact set.

Using this concept, we may define $\hat{\mathbf{y}}\left(\mathbf{p}, w\right)$ as the set of state-contingent income vectors that would minimize the cost of achieving welfare level w, given state-contingent prices $\mathbf{p}$.

$$\hat{\mathbf{y}}\left(\mathbf{p},w\right) = \arg\min \left\{\mathbf{p}'\mathbf{y} : W\left(\mathbf{y}\right) \geq w\right\}$$
$$= \partial E\left(\mathbf{p},w\right).$$

For strictly quasiconcave W, $\hat{\mathbf{y}}\left(\mathbf{p},w\right)$ is a singleton.

3 Reference Sets and Risk Premiums

Karni [10] defines the *reference set* as " ...the optimal distribution of wealth across states of nature that is chosen by a risk-averse decision maker facing fair insurance" at the objective probabilities $\hat{\pi}$. In the notation used here, we have

Definition 3 The reference set $\mathbf{y}$ for income m is given by

$$\hat{\mathbf{y}}\left(\hat{\pi}, I\left(\hat{\pi}, m\right)\right) = \arg\min_{\mathbf{y}} \left\{\hat{\pi}'\mathbf{y} : W\left(\mathbf{y}\right) \geq I\left(\hat{\pi}, m\right)\right\}$$

Provided preferences are strictly risk averse, $\hat{\mathbf{y}}\left(\hat{\pi}, m\right)$ is unique.

In the absence of strict risk aversion $\hat{\mathbf{y}}\left(\hat{\pi}, m\right)$ need not be unique. In particular, for risk-neutral preferences, any $\mathbf{y}$ will be an element of the reference set for $m = \hat{\pi}'\mathbf{y}$. However, we will focus attention on the case of strictly risk-averse preferences.

Definition 4 For given $\mathbf{y}$, the element of the reference set yielding welfare $W(\mathbf{y})$ is the reference equivalent $\hat{\mathbf{y}}\left(\hat{\pi}, W(\mathbf{y})\right)$.

Thus the expenditure on the reference equivalent satisfies

$$\hat{\pi}'\hat{\mathbf{y}}\left(\hat{\pi}, W(\mathbf{y})\right) = E\left(\hat{\pi}, W(\mathbf{y})\right)$$

may be interpreted as the reference-equivalent income (or minimal-equivalent income).

The reference set corresponds in a consumer context to the consumer's income-expansion path given $\hat{\pi}$. For given $\hat{\pi}$, the reference set, $\hat{Y}\left(\hat{\pi}\right) \subset \Re^S$, is thus

$$\hat{Y}\left(\hat{\pi}\right) = \cup_w \left\{\hat{\mathbf{y}}\left(\hat{\pi}, w\right)\right\}.$$

For strictly risk-averse, probabilistically sophisticated, state-independent preferences with subjective probabilities $\hat{\pi}$, $\hat{Y}\left(\hat{\pi}\right)$, the reference set for the price vector $\mathbf{p} = \hat{\pi}$, is simply the certainty ray $\{c\mathbf{1} : c \in \Re\}$. This can easily be checked for the case of expected-utility preferences (1). By contrast, for state-dependent expected-utility preferences (2), the reference set will not coincide with the certainty ray when the price vector $\mathbf{p}$ coincides with the decision maker's beliefs $\hat{\pi}$, unless all the utility functions u^s are identical (up to an additive shift).

3.1 Risk Premiums

Following the standard analysis of the state-independent [1, 16] and state-dependent [10] cases, we define the absolute and relative risk premiums.

Definition 5 Let $\mathbf{y} \in \Re^S$ and let $W : \Re^S \to \Re$. The absolute risk premium is

$$a\left(\mathbf{y}; \hat{\pi}\right) = \hat{\pi}'\mathbf{y} - E\left(\hat{\pi}, W(\mathbf{y})\right)$$

For the case of strictly positive outcomes, $W : \Re^S_{++} \to \Re$. The relative risk premium is

$$r\left(\mathbf{y}; \hat{\pi}\right) = \frac{\hat{\pi}'\mathbf{y}}{E\left(\hat{\pi}, W(\mathbf{y})\right)}.$$

The absolute risk premium is the difference between the value of $\mathbf{y}$ using $\hat{\pi}$ and the reference-equivalent income, that is, the minimal expenditure required to reach the same level of preference as $\mathbf{y}$ at prices $\hat{\pi}$. The literature on consumer surplus and other approximations to the compensating and equivalent variations can be applied to yield close approximations for $a\left(\mathbf{y}, \hat{\pi}\right)$ and related measures [6].

We can express the risk premium as the difference between two expenditure functions. Let $y^* \in Y(\hat{\pi})$ and $\pi' y^* = \pi' y$, then

$$a(y; \hat{\pi}) = E(\hat{\pi}, W(y^*)) - E(\hat{\pi}, W(y))$$
$$= E(\hat{\pi}, W(y + (y^* - y))) - E(\hat{\pi}, W(y)),$$

so that we can think of the risk premium as the willingness to pay to avoid the actuarially fair risk $(y^* - y)$.

Observe that by the definition of the expenditure function

$$\hat{\pi}' y \geq E(\hat{\pi}, W(y)),$$

so that $a(y; \hat{\pi}) \geq 0$ with equality if and only if $y \in Y(\hat{\pi})$.

Similar interpretations are available for the relative risk premium, and we may derive $r(y; \hat{\pi}) \geq 1$ with equality only for $y \in Y(\hat{\pi})$.

4 Comparisons of Risk and Risk Aversion

4.1 Risk Orderings for Equal Mean Sets

In state-independent utility, a variety of risk orderings, $\preceq$, are used, where $y \preceq y'$ corresponds to various interpretations of the statement 'y is less risky than y''. In all such orderings, the least risky state-contingent vectors are non-stochastic vectors of the form $c\mathbf{1}$. Risk-averse preferences are then characterized by the requirement that $W(y) \leq W(\mu\mathbf{1})$ where $\mu = \hat{\pi}' y$. Most risk orderings relate only variables with the same mean, and are translation-invariant in the sense that, for all δ, y, y':

$$y \preceq y' \Leftrightarrow y + \delta \mathbf{1} \preceq y' + \delta \mathbf{1}.$$

Thus, riskiness may be seen as a property of deviations from certainty, of the general form

$$\varepsilon = y - (\hat{\pi}' y) \mathbf{1}.$$

Hence, given a risk-ordering $\preceq$ we derive the induced risk ordering $\preceq^*$ on $M = \{\varepsilon : \hat{\pi}' \varepsilon = 0\}$ such that

$$\varepsilon \preceq^* \varepsilon' \Leftrightarrow \varepsilon + \mu\mathbf{1} \preceq \varepsilon' + \mu\mathbf{1}, \quad \forall \mu.$$

It is useful to apply this interpretation to specific risk orderings used in the literature. We follow the notation of Quiggin and Chambers [19]. Consider the following examples.

Example 6 The minimal risk ordering consistent with risk aversion, namely, requiring that receipt of mean income with certainty is preferred to the corresponding risky state-contingent income vector is denoted $\preceq_0$. The only risk-ordering relationships implied by $\preceq_0$ are of the form $\mu(\mathbf{y})\mathbf{1} \preceq_0 \mathbf{y}$. This ordering is translation invariant, and induces the ordering $\preceq_0^*$ on M

$$\mathbf{0} \preceq_0^* \varepsilon, \quad \varepsilon \in M.$$

Example 7 Consider the multiplicative-spread risk ordering, $\preceq_1$, described by

$$\lambda\mathbf{y} + (1 - \lambda) \left(\hat{\pi}'\mathbf{y}\right)\mathbf{1} \preceq_1 \mathbf{y}, \quad 0 \le \lambda \le 1.$$

This ordering is translation invariant, and corresponds to the requirement

$$\lambda\varepsilon \preceq_1^* \varepsilon, \quad \varepsilon \in M, \quad 0 \le \lambda \le 1.$$

Example 8 The monotone spread ordering is given by

$$\mathbf{y} \preceq_m \mathbf{y} + \varepsilon$$

where $\hat{\pi}'\varepsilon = 0$ and ε is comonotonic with $\mathbf{y}$, that is, for any s, t

$$(y_s - y_t)(\varepsilon_s - \varepsilon_t) \ge 0.$$

This ordering is translation invariant and induces the ordering $\preceq_m^*$ on M

$$\varepsilon \preceq_m \varepsilon + \varepsilon'$$

for $\varepsilon, \varepsilon'$ comonotonic.

4.2 General Risk Orderings

To extend this analysis to the case of state-dependent preferences, we replace the certainty ray with $Y\left(\hat{\pi}\right)$ and define

$$M\left(\mathbf{y}, \hat{\pi}\right) = Y\left(\hat{\pi}\right) \cap \left\{\tilde{\mathbf{y}} : \hat{\pi}'\left(\tilde{\mathbf{y}} - \mathbf{y}\right) = 0\right\}$$

as the set of points on the reference set that has the same mean as $\mathbf{y}$. When the reference set is a one-dimensional manifold, $M\left(\mathbf{y}, \hat{\pi}\right)$ is a singleton.

Now we can apply risk orderings on the basis of deviations of the form

$$\varepsilon = \mathbf{y} - M\left(\mathbf{y}, \hat{\pi}\right), \tag{3}$$

yielding

$$M\left(\mathbf{y}, \hat{\pi}\right) \preceq_0 \mathbf{y},$$
$$\lambda \mathbf{y} + (1 - \lambda) M\left(\mathbf{y}, \hat{\pi}\right) \preceq_1 \mathbf{y}, \quad 0 \leq \lambda \leq 1,$$
$$\mathbf{y} \preceq_m \mathbf{y} + \tilde{\varepsilon}$$

where $\hat{\pi}'\tilde{\varepsilon} = 0$ and $\tilde{\varepsilon}$ is comonotonic with some element of $\mathbf{y} - M\left(\mathbf{y}, \hat{\pi}\right)$. Intuitively, the definition of $\preceq_0$ says that points on the reference set are less risky than points off it with the same mean. The definition of $\preceq_1$ says that riskiness increases as we move along any line segment from $M\left(\mathbf{y}, \hat{\pi}\right)$ towards $\mathbf{y}$. The comonotone order has the same properties as in the standard case.

With this construction, it is straightforward to relax the requirement that risk orderings should relate only variables with the same mean. Observe that if $w = I\left(\hat{\pi}, \hat{\pi}'\mathbf{y}\right)$, we have

$$\mathbf{y} = \hat{\mathbf{y}}\left(\hat{\pi}, w\right) + \varepsilon$$

where ε is as in (3). Similarly denote

$$\mathbf{y}' = \hat{\mathbf{y}}\left(\hat{\pi}, w'\right) + \varepsilon'$$

and any of the orderings $\preceq$ discussed above can be extended to a partial order on Y^S defined by

$$\mathbf{y} \preceq \mathbf{y}' \Leftrightarrow \varepsilon \preceq \varepsilon'.$$

In the remainder of this paper, we will use $\preceq_0$, $\preceq_1$ and $\preceq_m$ to denote these more general orderings.

4.3 Risk Aversion

An important reason for defining a generalized risk premium is to permit the comparison of risk aversion across individuals. As Karni [10] observes, comparisons of risk aversion are feasible for individuals who share common beliefs $\hat{\pi}$ and a reference set $Y\left(\hat{\pi}\right)$. Comparing two individuals i and j who have common beliefs and a common reference set, we say that i is (absolutely) more risk-averse than j if for all $\mathbf{y}$

$$a^i\left(\mathbf{y}; \hat{\pi}\right) \geq a^j\left(\mathbf{y}; \hat{\pi}\right). \tag{4}$$

Departures from the objective probabilities incur greater costs in maintaining the reference utility for the more risk-averse person. Equivalently i is more risk-averse than j if for all $\mathbf{y}$

$$\hat{\pi}'\mathbf{y} - E^i\left(\hat{\pi}, W^i\left(\mathbf{y}\right)\right) \geq \hat{\pi}'\mathbf{y} - E^j\left(\hat{\pi}, W^j\left(\mathbf{y}\right)\right),$$

or

$$E^j\left(\hat{\pi}, W^j\left(\mathbf{y}\right)\right) \geq E^i\left(\hat{\pi}, W^i\left(\mathbf{y}\right)\right).$$

This also implies a notion of 'relatively more risk-averse' by a parallel definition. More significantly, it implies that more risk-averse people are characterized by the requirement that given state-claim prices corresponding to their probabilities, they would always spend less on attaining the level of utility offered by $\mathbf{y}$ than less risk-averse individuals.

It may seem counterintuitive that more risk-averse individuals should spend less on their reference equivalent than less risk-averse individual. However, this has an exact parallel in state-independent expected utility theory in more risk-averse individuals having a lower certainty equivalent, for any state-contingent income vector, than less risk-averse individuals. Given that the expansion path is monotonically increasing in w, more risk-averse individuals always find their reference-equivalent consumption bundle 'closer to the origin' in the reference set than less risk-averse individuals.

More general comparisons of risk aversion are also useful. Suppose $\preceq$ is a risk ordering for both i and j. Then, we say that $i \preceq^* j$, stated as i is more risk-averse than j for $\preceq$, for given $\hat{\pi}$, if for all $\mathbf{y}, \mathbf{y}'$

$$\mathbf{y} \preceq \mathbf{y}' \Rightarrow a^i\left(\mathbf{y}'; \hat{\pi}\right) - a^i\left(\mathbf{y}; \hat{\pi}\right) \geq a^j\left(\mathbf{y}'; \hat{\pi}\right) - a^j\left(\mathbf{y}; \hat{\pi}\right) \tag{5}$$

or equivalently

$$E^i\left(\hat{\pi}, W^i\left(\mathbf{y}\right)\right) - E^i\left(\hat{\pi}, W^i\left(\mathbf{y}'\right)\right) \geq E^j\left(\hat{\pi}, W^j\left(\mathbf{y}\right)\right) - E^j\left(\hat{\pi}, W^j\left(\mathbf{y}'\right)\right).$$

This is a standard dual relationship. As usual, for any dual ordering of risk aversion $\preceq^*$ on a family I of individuals with common beliefs $\hat{\pi}$, we can define the double dual $\preceq^{**}$ such that $\mathbf{y} \preceq^{**} \mathbf{y}'$ if and only if for all $i, j \in I$,

$$i \preceq^* j \Rightarrow a^i\left(\mathbf{y}'; \hat{\pi}\right) - a^i\left(\mathbf{y}; \hat{\pi}\right) \geq a^j\left(\mathbf{y}'; \hat{\pi}\right) - a^j\left(\mathbf{y}; \hat{\pi}\right).$$

It is easy to show that $\mathbf{y} \preceq \mathbf{y}' \Rightarrow \mathbf{y} \preceq^{**} \mathbf{y}'$. Quiggin [18] gives conditions under which the reverse implication is true. Hence, there is no substantial loss of generality in denoting generic orderings of risk aversion by $\preceq^*$.

This definition may be restated in the terminology of supermodularity theory [22]. Consider a family I of individuals with common beliefs $\hat{\pi}$ and a common expansion path $Y\left(\hat{\pi}\right)$. If (5) holds whenever $j \preceq^* i$, then the risk premium $a^i\left(\mathbf{y}; \hat{\pi}\right)$ and the expenditure function $E^i\left(\hat{\pi}, W^i\left(\mathbf{y}\right)\right)$ display increasing differences in $i, \mathbf{y}$ relative to the orderings $\preceq$ and $\preceq^*$. Note that the difference between $a^i\left(\mathbf{y}; \hat{\pi}\right)$ and $E^i\left(\hat{\pi}, W^i\left(\mathbf{y}\right)\right)$ is given by $\hat{\pi}'\mathbf{y}$ which is a valuation on $\mathbf{y}$, so that supermodularity of $a^i\left(\mathbf{y}; \hat{\pi}\right)$ is equivalent to supermodularity of $E^i\left(\hat{\pi}, W^i\left(\mathbf{y}\right)\right)$.

The general definition in terms of supermodularity theory reduces to the basic definition (4) if we consider the ordering $\preceq_0$ defined by the sole requirement $\mathbf{y}\left(\hat{\pi}, I\left(\hat{\pi}, m\right)\right) \preceq_0 \mathbf{y}, \forall \mathbf{y}$.

5 Constant, Decreasing and Increasing Risk Aversion

Concepts of constant, decreasing and increasing risk aversion play a crucial role in the literature on problems of economic choice under uncertainty that has been developed for state-independent models of preferences. If models of state-dependent preferences are to be applied to such problems, it is important to consider the extent to which such concepts can be generalized. We will focus on absolute risk aversion. The extension to concepts of constant, decreasing and increasing relative risk aversion is straightforward if the reference set is a ray from the origin, but more complex in other cases.

5.1 Constant Risk Aversion and Linear Risk Tolerance

With state-independent preferences, risk-averse for probabilities $\hat{\pi}$, the reference set for $\hat{\pi}$ is the certainty ray $c\mathbf{1}$. Under constant absolute risk aversion (CARA), for any $\mathbf{p}$, the expansion path is parallel to $c\mathbf{1}$. It follows that, in any standard choice problem, an increase in wealth of δ simply produces a shift of $\delta\mathbf{1}$ in the optimal consumption vector. (A standard choice problem is one involving allocation of wealth across a range of assets or other activities, such that changes in wealth shift the choice set parallel to $c\mathbf{1}$.) Similarly, under constant relative risk aversion (CRRA), the expansion path is a ray from the origin. This leads us to the following definition of these concepts in terms of reference sets.

Definition 9 Preferences satisfy constant absolute risk aversion if, for all $\hat{\pi}$, $Y(\hat{\pi}) + \beta\mathbf{1} \subseteq Y(\hat{\pi})$, $\beta \in \Re$. Preferences satisfy constant relative risk aversion if, for all $\hat{\pi}$, $\mu Y(\hat{\pi}) \subseteq Y(\hat{\pi})$, $\mu > 0$.

Ordinal preferences, therefore, satisfy CARA if there exists a utility normalization such that, for all π,

$$E(\pi, w) = E(\pi, 0) + w\pi'\mathbf{1}$$
$$= E(\pi, 0) + w$$

under the normalization that $\pi'\mathbf{1} = 1$ (which is harmless for probabilities). Hence,

$$Y(\hat{\pi}) = \partial E(\hat{\pi}, 0) + \cup_w \{w\mathbf{1}\}.$$

The primal implication is that

$$V(w) = V(0) + w\mathbf{1},$$

which is equivalent to the definition in terms of reference sets.

Similarly, ordinal preferences satisfy CRRA if there exists a utility normalization ($w > 0$) such that for all π,

$$E\left(\pi,w\right) = wE\left(\pi,1\right),$$

so that

$$Y\left(\hat{\pi}\right) = \cup_w \left\{w\partial E\left(\hat{\pi}, 1\right)\right\}$$

The primal implication is that

$$V\left(w\right) = wV\left(1\right).$$

We, therefore, generalize the notion of CARA as follows:

Definition 10 Preferences display nonlinear CARA if

$$V\left(w\right) = V\left(0\right) + g\left(w\right),$$

with $g : \Re \to \Re^S$ and $g\left(0\right) = \mathbf{0}$.

Thus, preferences display nonlinear CARA if and only if

$$E\left(\pi,w\right) = E\left(\pi,0\right) + \pi'g\left(w\right),$$

so that preferences are risk averse with respect to $\hat{\pi}$ if and only if

$$\hat{Y}\left(\hat{\pi}\right) = \partial E\left(\hat{\pi}, 0\right) + \cup_w \left\{g\left(w\right)\right\}.$$

Under this definition, the expansion paths are 'parallel' to the manifold $\cup_w \left\{g\left(w\right)\right\}$, which allows for nonlinear responsiveness to real wealth (w) changes.

That is, all income effects on state-contingent incomes are independent (in a direct sense) of state-contingent prices. The income effects are measured by $g\left(w\right)$ and by the rate at which w changes with income. Let the vector of partial derivatives of $g\left(w\right)$ with respect to w be denoted $g'\left(w\right) \in \Re^S$. Then, presuming differentiability, the income effects are measured by

$$\frac{\partial}{\partial m}g\left(I\left(\pi,m\right)\right) = g'\left(I\left(\pi,m\right)\right)I_m\left(\pi,m\right)$$

$$= \frac{g'\left(I\left(\pi,m\right)\right)}{E_w\left(\pi, I\left(\pi,m\right)\right)}.$$

This leads to an absolute risk premium of the form

$$a\left(\mathbf{y}; \hat{\pi}\right) = \hat{\pi}'\mathbf{y} - E\left(\hat{\pi}, W\left(\mathbf{y}\right)\right)$$

$$= \hat{\pi}'\mathbf{y} - E\left(\hat{\pi}, 0\right) - \hat{\pi}'g\left(W\left(\mathbf{y}\right)\right).$$

which implies:

Proposition 11 *Preferences display nonlinear CARA if and only if the absolute risk premium* $a\left(\mathbf{y};\hat{\pi}\right)$ *is constant for shifts parallel to the reference set. That is, for any* w *and* $\mathbf{y} \in V\left(0\right)$,

$$a\left(\mathbf{y}+g\left(w\right);\hat{\pi}\right) = a\left(\mathbf{y};\hat{\pi}\right).$$

Proof For $\mathbf{y} \in V\left(0\right)$

$$a\left(\mathbf{y};\hat{\pi}\right) = \hat{\pi}'\mathbf{y} - E\left(\hat{\pi},0\right).$$

Under nonlinear CARA for $\mathbf{y} \in V\left(0\right)$, $\mathbf{y} + g\left(w\right) \in V\left(w\right)$, whence

$$\begin{aligned} a\left(\mathbf{y}+g\left(w\right);\hat{\pi}\right) &= \hat{\pi}'\mathbf{y} + \hat{\pi}'g\left(w\right) - E\left(\hat{\pi},0\right) - \hat{\pi}'g\left(w\right) \\ &= \hat{\pi}'\mathbf{y} - E\left(\hat{\pi},0\right) \\ &= a\left(\mathbf{y};\hat{\pi}\right). \end{aligned}$$ ∎

Another especially convenient class of preferences are the linear risk tolerant preferences, which correspond in the standard consumer case to the class of quasi-homothetic preferences, whose demands assume the Gorman polar form. We have:

Definition 12 Preferences satisfy linear risk tolerance if

$$V\left(w\right) = V^0 + wV^1,$$

where $V^0, V^1 \subset Y^S$.

The importance of the linear risk tolerant (LRT) class for state-dependent preferences emerges in attempts to generalize the Pratt-Arrow notions of decreasing and increasing absolute risk aversion for state-dependent preferences. In his generalizations of these concepts, Karni [10] restricts attention to *autocomparable* preferences. Preferences are *autocomparable* if the reference set is affine.[1]

To admit the possibility of multiple solutions to the expected-value problem, we have:

Definition 13 An individual's preferences are autocomparable for $\hat{\pi}$ if $Y\left(\hat{\pi}\right) = Y^0\left(\hat{\pi}\right) + Y^1\left(\hat{\pi}\right)$ where $Y^0\left(\hat{\pi}\right) = \partial E\left(\hat{\pi},0\right) \subset Y^S$ and $Y^1\left(\hat{\pi}\right)$ is a cone, that is, $\mu Y^1\left(\hat{\pi}\right) \subseteq Y^1\left(\hat{\pi}\right), \mu > 0$.

Preferences can be autocomparable for some $\hat{\pi}$ but not for others. For example, state-independent expected utility preferences are autocomparable for the subjective probabilities that parametrize the expected-utility function, but not necessarily for other probabilities. It is trivial from what has gone before that CARA, CRRA, and LRT preferences are all autocomparable for all possible probability distributions. Generalized CARA preferences are only autocomparable if the manifold $\cup_w \left\{g\left(w\right)\right\}$ is a cone, but in this case, such preferences can always be renormalized to be CARA preferences.

---

[1] Karni [10] uses the term linear, but his definition is equivalent to requiring the elements of a particular point in the reference set to be affine translates of one another.

5.2   Decreasing and Increasing Absolute Risk Aversion

Quiggin and Chambers [19] show that, like other notions of comparative risk aversion, notions of decreasing and increasing absolute and relative risk aversion are most compactly and usefully expressed in the language of supermodularity theory. With the notions of risk ordering and the reference set, developed above, it is straightforward to extend the definition of decreasing absolute risk aversion (DARA) presented by Quiggin and Chambers [19] for risk-averse state-independent preferences to the case of general reference sets.

Definition 14 Preferences display DARA with respect to a risk ordering $\preceq^*$ on E, for a given π, if $a\left(\hat{\mathbf{y}}\left(\hat{\pi}, w\right) + \varepsilon; \hat{\pi}\right)$ is submodular in w and ε.

The discussion of autocomparability presented above suggests an alternative approach, closer to that adopted by Karni [10]. Suppose that the reference set for $\hat{\pi}$ takes the affine form

$$Y\left(\hat{\pi}\right) = Y^0\left(\hat{\pi}\right) + Y^1\left(\hat{\pi}\right). \tag{6}$$

Under CARA, this will be true for all π, so Y^1 is independent of π. Hence, for $\mathbf{y}^1 \in Y^1\left(\hat{\pi}\right)$,

$$V\left(W\left(\mathbf{y}+c\mathbf{y}^1\right)\right) = V\left(W\left(\mathbf{y}\right)\right) + c\mathbf{y}^1$$

for all $\mathbf{y}, c$ and, in particular, all $\hat{\mathbf{y}} \in \hat{Y}\left(\hat{\pi}\right)$.

With decreasing risk-aversion, the addition of consumption $c\mathbf{y}^1\left(\hat{\pi}\right)$ should make the individual more willing to accept movement away from the reference set, implying that, for all $\hat{\mathbf{y}} \in \hat{Y}\left(\hat{\pi}\right)$

$$V\left(W\left(\hat{\mathbf{y}}\right)\right) + c\mathbf{y}^1\left(\hat{\pi}\right) \subseteq V\left(W\left(\hat{\mathbf{y}} + c\mathbf{y}^1\left(\hat{\pi}\right)\right)\right). \tag{7}$$

This kind of comparison is feasible only for autocomparable preferences.

These two approaches can be related using the following proposition, which shows that the definition based on autocomparable preferences is a special case of the supermodularity definition.

Proposition 15 *Suppose that condition (6) is satisfied. Then preferences satisfy (7) if and only if $a\left(\mathbf{y} + c\mathbf{y}^1\left(\hat{\pi}\right); \hat{\pi}\right)$ is submodular in c and $\mathbf{y}$ with respect to the ordering $\preceq_0$.*

Proof Since the ordering $\preceq_0$ requires only $\hat{\mathbf{y}}\left(\hat{\pi}, \hat{\pi}\mathbf{y}\right) \preceq_0 \mathbf{y}$, and $a\left(\hat{\mathbf{y}}\left(\hat{\pi}, \hat{\pi}\mathbf{y}\right); \hat{\pi}\right) = 0$, $a\left(\mathbf{y} + m\mathbf{y}^1\left(\hat{\pi}\right); \hat{\pi}\right)$ is submodular in c and $\mathbf{y}$ with respect to the ordering $\preceq_0$ if and only if $a\left(\mathbf{y} + c\mathbf{y}^1\left(\hat{\pi}\right); \hat{\pi}\right)$ is decreasing in c which will be true if and only if (7) holds. ∎

6   Implications for Asset Demand

We can now extend the results for the standard two-asset portfolio demand problem
to the case of preferences with an affine expansion set for π. Consider an individual
with reference set

$$\hat{y}\left(\hat{\pi}, w\right) = \left\{y\left(\hat{\pi}, 0\right) + w y^1\left(\hat{\pi}\right)\right\}, \tag{8}$$

stochastic endowment $\mathbf{e}$ and income or wealth m, that can be allocated between two
assets, 1 and 2, with payoffs $\mathbf{y}^1$ and $\mathbf{y}^2$ where $\mathbf{y}^1$ is the direction of the reference set
and

$$\hat{\pi}\mathbf{y}^2 = \hat{\pi}\mathbf{y}^1 = 1.$$

We will treat asset 1 as numeraire with price equal to 1, and denote the price of asset
2 by $p_2 < 1$. Thus, $\mathbf{y}^1$ and $\mathbf{y}^2$ correspond respectively to the safe asset and the risky
asset in the standard analysis. Let α denote purchases of the risky asset.

Thus the optimisation problem is

$$\max_a W\left(\mathbf{e} + \frac{(m - \alpha)}{p_2}\mathbf{y}^1 + \alpha\mathbf{y}^2\right).$$

We will assume that the base allocation $\mathbf{y}^0$ is in the span of the market and more
precisely that there exist b^1, b^2 such that

$$\mathbf{y}^0 = \mathbf{e} + b_1\mathbf{y}^1 + b_2\mathbf{y}^2.$$

In particular, this encompasses the special case $\mathbf{e} = \mathbf{y}^0$, $b_1 = b_2 = 0$.

We then obtain

Proposition 16 *Assume preferences display DARA with respect to $\preceq_1$. Then the
following are sufficient conditions for an increase in optimal purchases of the asset:*

(i)   an increase in wealth w;
(ii) a reduction in the asset price p_2;
(iii) the replacement of the return vector $\mathbf{y}^2$ by $\mathbf{y}^{2\prime}$ where $\mathbf{y}^{2\prime} \preceq_1 \mathbf{y}^2$.

Proof Consider the initially optimal α and some $\alpha' < \alpha$. Let

$$\mathbf{y} = \mathbf{e} + \frac{(m - \alpha)}{p_2}\mathbf{y}^1 + \alpha\mathbf{y}^2$$

$$\mathbf{y}' = \mathbf{e} + \frac{(m - \alpha')}{p_2}\mathbf{y}^1 + \alpha'\mathbf{y}^2.$$

Observe that $\mathbf{y}' \preceq_1 \mathbf{y} - d\mathbf{y}^1$, where

$$d = \left(\alpha - \alpha'\right)\left(1 - p_2\right).$$

The optimality of α implies

$$a\left(\mathbf{y}; \hat{\pi}\right) - a\left(\mathbf{y}'; \hat{\pi}\right) < \delta.$$

∎

Now consider a shift from initial wealth m to $m + \delta$. If preferences display DARA with respect to $\preceq_1$, then

$$a\left(\mathbf{y}+\delta\mathbf{y}^1; \hat{\pi}\right) - a\left(\mathbf{y}'+\delta\mathbf{y}^1; \hat{\pi}\right) < \delta$$

and so the optimal allocation for wealth $m + \delta$ cannot be $\alpha' < \alpha$. The other cases are proved similarly.

7 Concluding Comments

Models of state-dependent preferences have a number of attractive properties. Applicability of such models has, however, been limited by the lack of analytical tools comparable to those available for state-independent preferences.

Using the state-contingent approach and the tools of modern consumer and producer theory, much of the analysis of risk aversion developed for state-independent preferences may be extended to the case of state-dependent preferences. In particular, it is possible to make *comparisons* of risk aversion with respect to a range of risk ordering.

Because state-contingent representations of problems involving uncertainty exhibit the symmetry between production and consumption familiar from analysis under certainty, a natural extension of the results derived here is to consider their applicability to problems of production under uncertainty, with the reference set being reinterpreted as an expansion path, and risk-aversion being interpreted as the existence of a cost premium for deviations from the expansion path. This issue will be addressed in future work.

References

1. Arrow, K.: Aspects of the Theory of Risk-Bearing. Yrjo Jahnsson Lecture, Helsinki (1965)
2. Baccelli, J.: Do bets reveal beliefs. Synthese **194**(9), 3393–3419 (2017)
3. Chambers, R.G., Quiggin, J.: Uncertainty, Production, Choice and Agency: The State-Contingent Approach. Cambridge University Press, New York (2000)
4. Chambers, R.G., Quiggin, J.: Dual approaches to the analysis of risk aversion. Economica **74**(294), 189–213 (2007)
5. Cornes, R.: Duality and Modern Economics. Cambridge University Press, Cambridge, UK (1992)
6. Diewert, W.E.: Exact and superlative welfare indicators. Econ. Inq. **30**, 565–582 (1992)
7. Gorman, W.M.: Community preference fields. Econometrica **21**, 63–80 (1953)

8. Grant, S., Karni, E.: A theory of quantifiable beliefs. J. Math. Econ. **40**(5), 515–546 (2004)
9. Karni, E.: Risk aversion for state-dependent preferences: measurement and applications. Int. Econ. Rev. **24**(3), 637–647 (1983)
10. Karni, E.: Decision Making Under Uncertainty: The Case of State-Dependent Preferences. Harvard University Press, Cambridge, Mass (1985)
11. Karni, E.: Generalized expected utility analysis of risk aversion with state-dependent preferences. Int. Econ. Rev. **28**(1), 229–240 (1987)
12. Karni, E.: Elicitation of subjective probabilities when preferences are state-dependent. Int. Econ. Rev. **40**(2), 479–486 (1999)
13. Karni, E., Safra, Z.: A theory of stochastic choice under uncertainty. J. Math. Econ. **63**, 164–173 (2016)
14. Karni, E., Schmeidler, D.: An expected utility theory for state-dependent preferences. Theory Decis. **81**, 467–478 (2016)
15. Machina, M.: Choice under uncertainty: problems solved and unsolved. J. Econ. Perspect. **1**(1), 121–154 (1987)
16. Pratt, J.: Risk aversion in the small and in the large. Econometrica **32**(1), 122–136 (1964)
17. Quiggin, J.: A theory of anticipated utility. J. Econ. Behav. Organ. **3**, 323–343 (1982)
18. Quiggin, J.: Efficient sets with and without the expected utility hypothesis: a generalization. J. Math. Econ. **21**, 395–399 (1992)
19. Quiggin, J., Chambers, R.: Supermodularity and the comparative statics of risk. Theory Decis. **62**, 91–117 (2007)
20. Riedener, S.: An axiomatic approach to axiological uncertainty. Philos. Stud. 1–22 (2018)
21. Rockafellar, R.T.: Convex Analysis. Princeton University Press, Princeton, NJ (1970)
22. Topkis, D.M.: Supermodularity and Complementarity. Princeton University Press, Princeton, NJ (1998)
23. Wang, T.: Lp-Fréchet differentiable preference and "Local Utility" analysis. J. Econ. Theory **61**(1), 139–159 (1993)

Printed in the United States
By Bookmasters